Natural Computing Series

Gabriel Ciobanu · Gheorghe Păun · Mario J. Pérez-Jiménez (Eds.)

Applications of Membrane Computing

With 99 Figures and 24 Tables

Editors
Gabriel Ciobanu
Institute of Computer Science of the
Romanian Academy
Blvd. Copou 8
700506 Iaşi, Romania
gabriel@iit.tuiasi.ro

Gheorghe Păun
Research Group on Natural Computing
Department of Computer Science and
Artificial Intelligence
University of Seville
Avda. Reina Mercedes s/n
41012 Seville, Spain
gpaun@us.es

Mario J. Pérez-Jiménez
Research Group on Natural Computing
Department of Computer Science and
Artificial Intelligence
University of Seville
Avda. Reina Mercedes s/n
41012 Seville, Spain
marper@us.es

Series Editors
G. Rozenberg (Managing Editor)
rozenber@liacs.nl
Th. Bäck, J.N. Kok, H.P. Spaink
Leiden Institute of Advanced
Computer Science
Leiden University
Niels Bohrweg 1
2333 CA Leiden, The Netherlands

A.E. Eiben
Vrije Universiteit Amsterdam
The Netherlands

ACM Computing Classification (1998): D.1, F.1, F.4, I.1, I.6, J.3–4.

ISBN-13 978-3-642-06401-2 e-ISBN-13 978-3-540-29937-0

Springer is a part of Springer Science+Business Media

springeronline.com

Softcover reprint of the hardcover 1st edition 2006

Cover Design: KünkelLopka, Werbeagentur, Heidelberg

Preface

Membrane computing is a branch of natural computing which investigates computing models abstracted from the structure and functioning of living cells and from their interactions in tissues or higher order biological structures.

Briefly, a membrane system is a distributed computing model processing multisets of objects either in the compartments of a cell-like hierarchical arrangement of membranes (hence a structure of compartments which corresponds to a rooted tree), or in a tissue-like structure consisting of "cells" placed in the nodes of an arbitrary graph. Both the objects of the membranes, the membranes, and the links among them evolve according to some rules. For instance, the multisets of objects evolve mainly by means of rewriting rules, which have the form of usual chemical equations (several objects "react" and get transformed into some product objects). A crucial aspect of this processing is the resulting communication of objects through membranes, between regions of the same cell, between cells, or between cells and their environment.

A detailed introduction to membrane computing is provided in the first chapter of the book. This research area was initiated recently, at the end of 1998, with the aim of learning ideas, tools, techniques, and models from the biology of the cell that could turn out to be useful (or at least interesting) for the purpose of computing. The new field has flourished during the last five years; details can be found in the monograph *Membrane Computing. An Introduction*, published in 2002 in the same series as the present volume, and at the membrane systems (also called P systems) Website `http://psystems.disco.unimib.it`). Many classes of P systems, inspired by either biological or mathematical considerations, were introduced, and their power (in comparison with various classes of Turing machines) and efficiency (the possibility to address/solve computationally hard problems) were extensively investigated.

Moreover, especially in recent years, it has turned out that membrane computing has significant potential to be applied to various problems of biology as well as to linguistics, theoretical computer science (sorting and ranking, 2D languages) and applied computer science (computer graphics, cryptography, approximate algorithms for optimization problems).

This book presents some applications of membrane computing, organized in three main categories (biology, computer science, and linguistics), beginning with the introductory chapter mentioned above, and ending with a chapter about membrane computing software (useful in applications).

Before looking at these applications, let us discuss the *attractiveness of membrane computing as a modeling framework*, with an implicit reference to applications in biology.

First, there are several essential features genuinely relevant to membrane computing that are of interest for many applications:

- *distribution* (with important issues related to system-part interaction and emergent behavior nonlinearly resulting from the composition of local behaviors),
- *discrete mathematics* (continuous mathematics, especially systems of differential equations, has a glorious history of applications in many disciplines, such as astronomy, physics, and meteorology, but has failed to prove adequate for linguistics, and cannot cover more than local processes in biology because of the complexity of the processes and, in many cases, because of the imprecise character of the processes; a basic question is whether the biological reality is of continuous or discrete nature, as languages proved to be, with the latter ruling out the usefulness of many tools from continuous mathematics),
- *algorithmicity* (by definition, P systems are computability models of the same type as Turing machines or other classic representations of algorithms, and, as a consequence, they can be easily simulated on computers),
- *scalability/extensibility* (this is one of the main difficulties of using differential equations in biology),
- *transparency* (multiset rewriting rules are nothing other than reaction equations as customarily used in chemistry and biochemistry, without any "mysterious" notation or "mysterious" behavior),
- *massive parallelism* (a dream of computer science, a commonplace in biology),
- *nondeterminism* (let us compare the "program" of a P system, i.e., a set of rules localized in certain regions and without any imposed ordering, with the rigid sequences of instructions of programs written in typical programming languages),
- *communication* (with the astonishing and still not completely understood way in which life is coordinating the multitude of processes taking place in cells, tissues, organs, and organisms, in contrast with the costly way of coordinating/synchronizing computations in parallel electronic computing architectures, where the communication time becomes prohibitive with the increase in the number of processors).

The majority of papers on applications of membrane computing are concerned with applications to biology. This is quite natural in view of the fact that the theory of membrane systems as a model of computation originated in biology as an abstraction of the structure and functioning of biological membranes. This supports our optimism about the (potential) success of these applications in modeling various aspects of biological reality, and should be contrasted with various models "transferred by force" from one area of science to another, where the original model was meant for totally different and incompatible phenomena (and consequently the modeling effort was unsuccessful).

Concerning applications reported till now, and hence applications reported in this volume, they are carried out at different levels. In many cases, what is actually used is the *language* of membrane computing, which involves (at least) three different aspects: (i) the long list of newly formulated concepts, (ii) the mathematical formalism of membrane computing, and (iii) the graphical way to represent membranes, cell-like structures, tissue-like structures, and so on.

Concerning graphical representation, we want to point out that not only the standard features (such as the hierarchical or tissue-like arrangements of membranes, the objects inhabiting the compartments, and the flow of information through communication channels) but also the evolution rules of the systems are part of the graphical representation. This makes visualization of the "evolution engine" transparent.

Another level of application is to use tools, techniques, and results obtained through research on membrane computing. These applications may aim at either

- solving problems already formulated by biologists, albeit informally or within a different model, or
- suggesting entirely new problems (problem areas) that become (more) transparent and interesting through the insights provided by the model of membrane computing.

Applications of all these types have been reported in the literature on membrane computing, and several of them are presented in this book. As already indicated, most applications concern biology. Moreover, applications to a number of other domains are also presented here. In particular, included are applications to computer graphics (where compartmentalization seems to add a significant efficiency to well known techniques based on L systems; see Chapter 9), linguistics (both as a representation language for various concepts related to language evolution, dialogue, semantics, Chapter 13, and making use of the parallelism for solving parsing problems in an efficient way, Chapter 14), management (again, mainly at the level of the formalism and the graphical language; such applications are not presented in the book, but references can be found on the Website mentioned above), sorting and ranking algorithms (Chapter 8), approximate algorithms for solving optimization problems

(Chapter 11), cryptography (attacking a well known public-key cryptographic system; see Chapter 10), and **NP**-complete problems (the standard time-space trade-off; see Chapter 12).

The applications to biology in most cases follow a standard scenario. One examines a piece of reality, e.g., from the biochemistry of the cell, one writes a P system modeling the respective processes, one writes a program simulating this P system (or uses one of the existing programs for this purpose), and one performs a large number of experiments with the program (this is much cheaper than conducting laboratory experiments), tuning certain parameters and observing the evolution of the system (usually, following the population of certain objects). Illustrations of this strategy are given in Chapter 2 (modeling mechanosensitive channels), Chapter 4 (respiration in bacteria), Chapter 5 (cell-mediated immunity), Chapter 6 (photosynthesis), and Chapter 3 (gene expression regulation). Chapter 3 deserves special attention: it discusses a part of a more general research plan to develop a theory of discrete dynamical systems intimately related to both P systems and applications to biology.

Chapter 15 provides an overview of the available membrane computing software. Actually, these computer programs are not real "implementations" of P systems. This is due to the difficulties related to the nondeterminism and the parallelism; however, there are attempts to implement P systems on dedicated, reconfigurable hardware, as done by Petreska & Teuscher, or on a cluster of computers, as done by Ciobanu & Guo, or in a distributed fashion, as reported by Syropoulos et al. (references can be found in the bibliography of Chapter 15). At this moment, there is no bio-implementation of P systems.

Actually, as illustrated by the contributions to this volume, a lot remains to be done. It is the hope of the editors that this volume will contribute to and motivate future research in membrane computing. It is also our belief that membrane computing can offer a useful variety of tools, techniques, and models for a wide spectrum of applications.

*

We are much indebted to all contributors to this volume, admiring their patience and thanking them for their work during the many adjustments of the chapters. Special thanks are due to Springer, in particular to Mrs. Ingeborg Mayer and Mr. Ronan Nugent, for very pleasant and efficient cooperation.

January 2005

Gabriel Ciobanu
Gheorghe Păun
Mario J. Pérez-Jiménez

Contents

Chapter 1
Introduction to Membrane Computing

Gheorghe Păun

Institute of Mathematics of the Romanian Academy
PO Box 1-764, 014700 Bucureşti, Romania
george.paun@imar.ro

Research Group on Natural Computing
Department of Computer Science and Artificial Intelligence
University of Sevilla
Avda. Reina Mercedes s/n, 41012 Sevilla, Spain
gpaun@us.es

Summary. This is a comprehensive (and friendly) introduction to membrane computing (MC), meant to offer both computer scientists and non-computer scientists an up-to-date overview of the field. That is why the set of notions introduced here is rather large, but the presentation is informal, without proofs and with rigorous definitions given only for the basic types of P systems – symbol object P systems with multiset rewriting rules, systems with symport/antiport rules, systems with string objects, tissue-like P systems, and neural-like P systems. Besides a list of (biologically inspired or mathematically motivated) ingredients/features which can be used in systems of these types, we also mention a series of results, as well as a series of research trends and topics.

1 (The Impossibility of) A Definition of Membrane Computing

Membrane computing (MC) is an area of computer science aiming to abstract computing ideas and models from the structure and the functioning of living cells, as well as from the way the cells are organized in tissues or higher order structures.

In short, MC deals with distributed and parallel computing models, processing multisets of symbol objects in a localized manner (evolution rules and evolving objects are encapsulated into compartments delimited by membranes), with an essential role played by the communication between compartments (and with the environment). Of course, this is just a rough description of a membrane system – hereafter called P system – of the very basic type, as many different classes of such devices exist.

The essential ingredient of a P system is its *membrane structure*, which can be a hierarchical arrangement of membranes, as in a cell (hence described by a tree), or a net of membranes (placed in the nodes of a graph), as in a tissue or a neural net. The intuition behind the notion of a membrane is a three-dimensional vesicle from biology, but the concept itself is generalized/idealized to interpreting a membrane as a *separator* of two regions (of Euclidean space), a finite "inside" and an infinite "outside," providing the possibility of *selective communication* between the two regions.

The variety of suggestions from biology and the range of possibilities to define the architecture and the functioning of a membrane-based multiset processing device are practically endless, and already the literature of MC contains a very large number of models. Thus, MC is not merely a theory related to a specific model, it is a *framework* for devising compartmentalized models. Because the domain is rather young (the trigger paper is [64], circulated first on the Web, though related ideas were considered before, in various contexts), and as a genuine feature, based on both the biological background and the mathematical formalism used, not only are there already many types of proposed P systems, but also the flexibility and the versatility of P systems seem, in principle, to be unlimited.

This last observation, as well as the rapid development and enlargement of the research in this area, make impossible a short and faithful presentation of membrane computing.

However, there are series of notions, notations, and models which are already "standard," which have stabilized and can be considered as basic elements of MC. This chapter is devoted to presenting mainly such notions and models, together with their notations.

The presentation will be both historically and didactically organized, introducing mainly notions first investigated in this area, or simple notions able to quickly offer an idea of membrane computing to the reader not familiar with the domain.

The reader has surely noticed that the discussion refers mainly to computer science (goals), and much less to biology. MC was not initiated as an area aiming to provide models to biology, models of the cell in particular. At this moment, after considerable development at the theoretical level, the domain is not yet fully prepared to offer such models to biology, though this has been an important direction of the recent research, and considerable advances toward such achievements have been reported. The present volume is a proof of this assertion.

2 Membrane Computing as Part of Natural Computing

Before entering into more specific elements of MC, let us spend some time with the relationship of this area with, let us say, the "local" terminology, the "outside." We have said above that MC is part of computer science. However, the *genus proximus* is natural computing, the general attempt to learn ideas, models, and paradigms useful to computer science from the way nature – life, especially – "computes" in various circumstances where substance and information processing can be interpreted as computation. Classic bio-inspired branches of natural computing are genetic algorithms (more generally, evolutionary computing) and neural computing. Both have long histories, which can be traced to the unpublished works of Turing, many applications, and a huge bibliography. Both are proof that "it is worth learning from biology," supporting the optimistic observation that during many billions of years nature/life has adjusted certain tools and processes which, correctly abstracted and implemented in computer science terms, can prove to be surprisingly useful in many applications.

A more recent branch of natural computing, with an enthusiastic beginning and as yet unconfirmed computational applicability (we do not discuss here the by-products, such as the nanotechnology related developments), is DNA computing, whose birth is related to the Adleman experiment [1] of solving a (small) instance of the Hamiltonian path problem by handling DNA molecules in a laboratory. According to Hartmanis [39, 40], this was a *demo* that we can compute with biomolecules, a big event for computability. However, after one decade of research, the domain is still preparing its tools for a possible future practical application and looking for a new breakthrough idea, similar to Adleman's one from 1994.

Both evolutionary computing and DNA computing are inspired from and related to DNA molecules. Neural computing considers the neurons as simple finite automata linked in specific types of networks. Thus, these "neurons" are not interpreted as cells, with an internal structure and life, but as "dots on a grid", with a simple input-output function. (The same observation holds true for cellular automata, where again the "cells" are "dots on a grid," interacting only among themselves, in a rigid structure.) None of these domains considers the cell itself as its main object of research; in particular, none of these domains pays any attention to membranes and compartmentalization – and this is the point where membrane computing enters the stage. Thus, MC can be seen as an extension of DNA (or, more generally, molecular) computing, from the "one processor" level to a distributed computing model.

3 Laudation to the Cell (and Its Membranes)

Life (as we know it on earth in the traditional meaning of the term, that investigated by biology) is directly related to cells; everything alive consists of cells or has to do in a direct way with cells. The cell is the smallest "thing" unanimously considered *alive*. It is very small and very intricate in its structure and functioning, has elaborate internal activity and complex interaction with the neighboring cells and with the environment. It is fragile and robust at the same time, with a way to organize (control) the biochemical (and informational) processes developed during billions of years of evolution.

Cell means membranes. The cell itself is defined – separated from its environment – by a membrane, the external one. Inside the cell, several membranes enclose "protected reactors," compartments where specific biochemical processes take place. In particular, a membrane encloses the nucleus (of eukaryotic cells), where the genetic material is placed. Through vesicles enclosed by membranes one can transport packages of molecules from a part of the cell (e.g., from the Golgi apparatus) to other parts of the cell in such a way that the transported molecules are not "available" during their journey to neighboring chemicals.

The membranes allow a selective passage of substances between the compartments delimited by them. This can be a simple selection by size in the case of small molecules, or a much more intricate selection, through protein channels which do not only select but can also move molecules from a low concentration to a higher concentration, perhaps coupling molecules, through so-called symport and antiport processes.

Moreover, the membranes of a cell do not delimit only compartments where specific reactions take place in solution, *inside* the compartments, but many reactions in a cell develop *on the membranes*, catalyzed by the many proteins bound to them. It is said that when a compartment is too large for the local biochemistry to be efficient, life creates membranes, both in order to create smaller "reactors" (small enough that, through the Brownian motion, any two of the enclosed molecules can collide – hence, react – frequently enough) and in order to create further "reaction surfaces." Anyway, biology contains many fascinating facts from a computer science point of view, and the reader is encouraged to check the validity of this assertion, e.g., through [2, 53, 7].

Life means surfaces inside surfaces, as can be learned from the title of [41], while S. Marcus puts it in an equational form [56]: *Life = DNA software + membrane hardware*.

There are cells living alone (unicellular organisms, such as ciliates, bacteria, etc.), but in general the cells are organized as tissues, organs, organisms, and communities of organisms. All these suppose a specific organization, starting with the direct communication/cooperation among neighboring cells and ending with the interaction with the environment at various levels. Together with the internal structure and organization of the cell, these suggest a lot of ideas, exciting from a mathematical point of view, and potentially useful from a computability point of view. Some of them have already been explored in MC, but many more still await research efforts (for example, the brain, the best "computer" ever invented, is still a major challenge for mathematical modeling).

4 Some General Features of Membrane Computing Models

It is worth mentioning from the beginning, besides the essential use of membranes/compartmentalization, some of the basic features of models investigated in this field.

We have mentioned above the notion of a multiset. The compartments of a cell contain substances (ions, small molecules, macromolecules) swimming in an aqueous solution. There is no ordering there; everything is close to everything; the concentration matters, i.e., the population, *the number of copies of each molecule* (of course, we are abstracting/idealizing here, departing from the biological reality). Thus, the suggestion is immediate: to work with sets of objects whose multiplicities matters; hence, with *multisets*. This is a data structure with peculiar characteristics, not new but not systematically investigated in computer science.

A multiset can be represented in many ways, but the most compact one is in the form of a string. For instance, if the objects a, b, and c are present, respectively, in 5, 2, and 6 copies each, we can represent this multiset by the string $a^5b^2c^6$; of course, all permutations of this string represent the same multiset.

The string representation of multisets and the biochemical background, where standard chemical reactions are common, suggest processing the multisets from the compartments of our computing device by means of rewriting-like rules; this means rules of the form $u \to v$, where u and v are multisets of objects (represented by strings). Continuing the previous example, we can consider a rule $aab \to abcc$. It indicates that two copies of object a and a copy of object b react, and, as a result of this reaction, we get back a copy of a as well as the copy of b (hence b behaves here as a catalyst), and we produce two new copies of c. If this rule is applied once to the multiset $a^5b^2c^6$, then, because aab are "consumed" and then $abcc$ are "produced," we obtain the multiset $a^4b^2c^8$. Similarly, by using the rule $bb \to aac$, we get the multiset a^7c^7, which contains no occurrence of object b.

Two important problems arise here. The first one is related to the *nondeterminism*. Which rules should be applied and to which objects? The copies of an object are considered identical, so we do not distinguish among them; whether to use the first rule or the second one is a significant issue, especially because they cannot be both used at the same time (for the multiset mentioned), as they compete for the "reactant" b. The standard solution to this problem in membrane computing is that *the rules and the objects are chosen in a nondeterministic manner* (at random, with no preference; more rigorously, we can say that any possible evolution is allowed).

This is also related to the idea of *parallelism*. Biochemistry is not only (to a certain degree) nondeterministic, but it is also (to a certain degree) parallel. If two chemicals can react, then the reaction does not take place for only two molecules of the two chemicals, but, in principle, for all molecules. This is the suggestion supporting the maximal parallelism used in many classes of P systems: at each step, all rules which can be applied have to be applied to all possible objects. We will come back to this important notion later, but now we illustrate it only with the previous multiset and pair of rules. Using these rules in the maximally parallel manner means either using the first rule twice (thus involving four copies of a and both copies of b) or using the second rule once (it consumes both copies of b, hence the first rule cannot be used at the same time). In the first case, one copy of a remains unused (and the same for all copies of c), and the resulting multiset is $a^3b^2c^{10}$; in the second case, all copies of a and c remain unused, and the resulting multiset is a^7c^7. Note that in the latter case the maximally parallel application of rules corresponds to the *sequential* (one at a time) application of the second rule.

There are also other types of rules used in MC (e.g., symport and antiport rules), but we will discuss them later. Here we conclude with the observation that MC deals with models which are intrinsically *discrete* (basically, working with multisets of objects, with the multiplicities being natural numbers) and evolve through *rewriting-like* (we can also say reaction-like) rules.

5 Computer Science Related Areas

Rewriting rules are standard rules for handling strings in formal language theory (although other types of rules, such as insertion, deletion, context-adjoining, are also used both in formal language theory and in P systems). Similarly, working with strings modulo the ordering of symbols is another old idea: commutative languages (investigated, e.g., in [28]) are nothing other than the permutation closure of languages. In turn, the multiplicity of symbol occurrences in a string corresponds to the Parikh image of the string, which directly leads to vector addition systems, Petri nets, register machines, and formal power series.

Parallelism is also considered in many areas of formal languages, and it is the main feature of Lindenmayer systems. These systems deserve a special discussion here, since they are a well developed branch of formal language theory inspired by biology, specifically, by the development of multi-cellular organisms (which can be described by strings of symbols). However, for L systems the cells are considered as symbols; their organization in (mainly linear) patterns, not their structure, is investigated. P systems can be seen as dual to L systems, as they zoom into the cell, distinguishing the internal structure and the objects evolving inside it, maybe also distinguishing (when "zooming enough") the structure of the objects, which leads to the category of P systems with string objects.

However, a difference exists between the kind of parallelism in L systems and that in P systems: in L systems the parallelism is *total* – all symbols of a string are processed at the same time; in P systems we work with a maximal parallelism – we process as many objects as possible, but not necessarily all of them.

Still closer to MC are the multiset processing languages, the most known of them being Gamma [8, 9]. The standard rules of Gamma are of the form $u \rightarrow v(\pi)$, where u and v are multisets and π is a predicate which should be satisfied by the multiset to which the rule $u \rightarrow v$ is applied. The generality of the form of rules ensures great expressivity and, in a direct manner, computational universality. What Gamma does not have (at least in the initial versions) is distributivity. Then, MC restricts the form of rules, on the one hand as imposed by the biological roots and on the other hand in search of mathematically simple and elegant models.

Membranes appear even in Gamma-related models, and this is the case with CHAM, the Chemical Abstract Machine of Berry and Boudol, [12], the direct ancestor of membrane systems; however, the membranes of CHAM are not membranes as in cell biology, but correspond to the contents of membranes, i.e., multisets, and lower level membranes together, while the goals and the approach are completely different, directed to the algebraic treatment of the processes these membranes can undergo. From this point of view, of goals and tools, CHAM has a recent counterpart in the so-called *brane calculus* (of course, "brane" comes from "membrane") from [17] (see also [74] for a related approach), where process algebra is used for investigating the processes taking place *on* membranes and *with* membranes of a cell.

The idea of designing a computing device based on compartmentalization through membranes was also suggested in [55].

Many related areas and many roots, with many common ideas and many differences! To some extent, MC is a synthesis of some of these ideas, integrated in a framework directly inspired by cell biology, paying deserved attention to membranes (and hence to distribution, hierarchization, communication, localization, and other related concepts), aiming – in the basic types of devices – to find computing models, as elegant (minimalistic) as possible, as powerful as possible (in comparison with Turing machines and their subclasses), and as efficient as possible (able to solve computationally hard problems in feasible time).

6 The Cell-like Membrane Structure

We move now toward presenting in a more precise manner the computing models investigated in our area, and we start by introducing one of the fundamental ingredients of a P system, namely, the *membrane structure.*

The meaning of this notion is illustrated in Figure 1, and this is what we can see when looking (through mathematical glasses, hence abstracting as much as necessary in order to obtain a formal model) at a standard cell.

Thus, as suggested by Figure 1, a membrane structure is a hierarchically arranged set of membranes, contained in a distinguished external membrane (corresponding to the plasma membrane and usually called the *skin* membrane). Several membranes can be placed inside the skin membrane (they correspond to the membranes present in a cell, around the nucleus, in Golgi apparatus, vesicles, mitochondria, etc.); a membrane without any other membrane inside it is said to be *elementary*. Each membrane determines a compartment, called a *region*, the space delimited by it from above and from below by the membranes placed directly inside, if any exist. Clearly, the membrane-region correspondence is one-to-one; that is why we sometimes use the terms interchangeably.

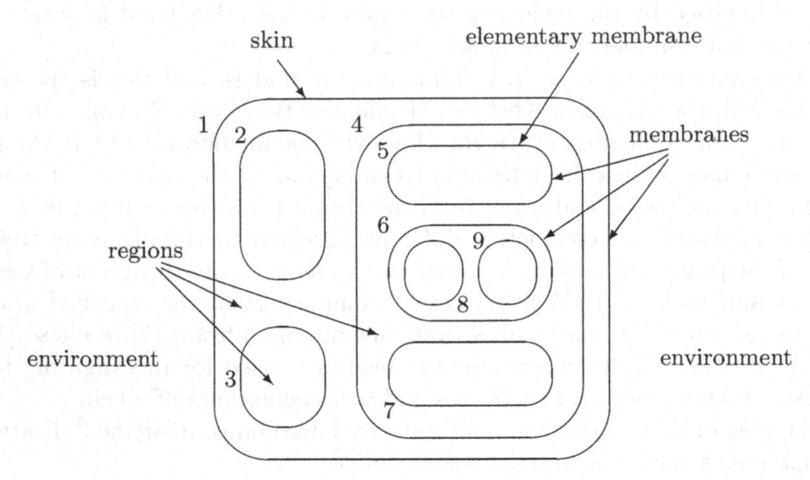

Fig. 1. A membrane structure.

Usually, the membranes are identified by *labels* from a given set of labels. In Figure 1, we use numbers, starting with number 1 assigned to the skin membrane (this is the standard labeling, but the labels can be more informative "names" associated with the membranes). Also, in the figure the labels are assigned in a one-to-one manner to membranes, but this is possible only in the case of membrane structures which cannot grow (indefinitely), otherwise several membranes would have the same label (we will later see such cases). Due to the membrane-region correspondence, we identify by the same label a membrane and its associated region.

Clearly, the hierarchical structure of membranes can be represented by a rooted tree; Figure 2 gives the tree which describes the membrane structure in Figure 1. The root of the tree is associated with the skin membrane and the leaves are associated with the elementary membranes. In this way, various graph-theoretic notions are brought onto the stage, such as the distance in the tree, the level of a membrane, the height/depth of the membrane structure, as well as terminology such as parent/child membrane, ancestor, etc.

Directly suggested by the tree representation is the symbolic representation of a membrane structure, by strings of labeled matching parentheses. For instance, a string corresponding to the structure from Figure 1 is the following:

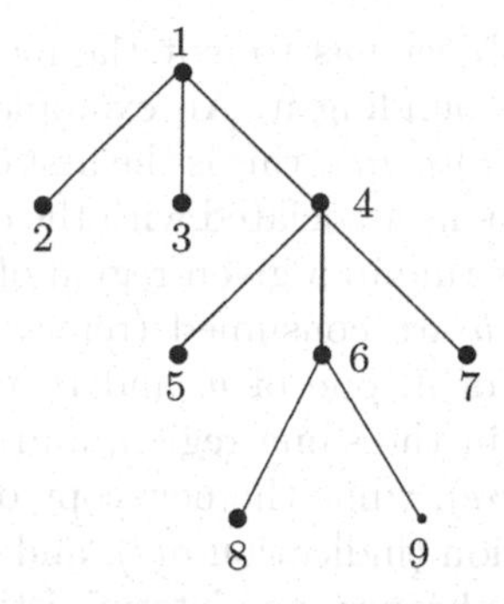

Fig. 2. The tree describing the membrane structure from Figure 1.

$$[_1\ [_2\ ]_2\ [_3\ ]_3\ [_4\ [_5\ ]_5\ [_6\ [_8\ ]_8\ [_9\ ]_9\ ]_6\ [_7\ ]_7\ ]_4\ ]_1. \qquad (*)$$

An important aspect should now be noted: the membranes of the same level can float around, that is, the tree representing the membrane structure is not oriented; in terms of parentheses expressions, two subexpressions placed at the same level represent the same membrane structure. For instance, in the previous case, the expression

$$[_1\ [_3\ ]_3\ [_4\ [_6\ [_8\ ]_8\ [_9\ ]_9\ ]_6\ [_7\ ]_7\ [_5\ ]_5\ ]_4\ [_2\ ]_2\ ]_1$$

is a representation of the same membrane structure, equivalent to $(*)$.

7 Evolution Rules and the Way to Use Them

In the basic variant of P systems, each region contains a multiset of symbol objects, which correspond to the chemicals swimming in a solution in a cell compartment. These chemicals are considered here as unstructured; that is why we describe them with symbols from a given alphabet.

The objects evolve by means of *evolution rules*, which are also localized, associated with the regions of the membrane structure. Actually, there are three main types of rules: (1) multiset-rewriting rules (one calls them, simply, evolution rules), (2) communication rules, and (3) rules for handling membranes.

In this section we present the first type of rules. They correspond to the chemical reactions possible in the compartments of a cell; hence they are of the form $u \to v$, where u and v are multisets of objects. However, in order to make the compartments cooperate, we have to move objects across membranes, and for this we add *target indications* to the objects produced by a rule as above (to the objects from multiset v). These indications are *here*, *in*, and *out*, with the meanings that an object associated with the indication *here* remains in the same region, one associated with the indication *in* goes immediately into an adjacent lower membrane, nondeterministically chosen, and *out* indicates that the object has to exit the membrane, thus becoming an element of the region surrounding it. An example of an evolution rule is $aab \rightarrow (a, here)(b, out)(c, here)(c, in)$ (this is the first of the rules considered in Section 4, with target indications associated with the objects produced by rule application). After using this rule in a given region of a membrane structure, two copies of a and one of b are consumed (removed from the multiset of that region), and one copy of a, one of b, and two of c are produced; the resulting copy of a remains in the same region, and the same happens with one copy of c (indications *here*), while the new copy of b exits the membrane, going to the surrounding region (indication *out*), and one of the new copies of c enters one of the child membranes, nondeterministically chosen. If no such child membrane exists, that is, the membrane with which the rule is associated is elementary, then the indication *in* cannot be followed, and the rule cannot be applied. In turn, if the rule is applied in the skin region, then b will exit into the environment of the system (and it is "lost" there, since it can never come back). In general, the indication *here* is not specified (an object without an explicit target indication is supposed to remain in the same region where the rule is applied).

It is important to note that in this initial type of system we do not provide similar rules for the environment, since we do not care about the objects present there; later we will consider types of P systems where the environment also takes part in system evolution.

A rule such as the one above, with at least two objects in its left hand side, is said to be *cooperative*; a particular case is that of *catalytic* rules, of the form $ca \rightarrow cv$, where c is an object (called catalyst) which assists the object a to evolve into the multiset v; rules of the form $a \rightarrow v$, where a is an object, are called *non-cooperative.*

The rules can also have the form $u \rightarrow v\delta$, where δ denotes the action of *membrane dissolving*: if the rule is applied, then the corresponding membrane disappears and its contents, object and membranes alike, are left free in the surrounding membrane; the rules of the dissolved membrane disappear with the membrane. The skin membrane is never dissolved.

The communication of objects through membranes evokes the fact that biological membranes contain various (protein) channels through which the molecules can pass (in a passive way, due to concentration difference, or in an active way, with consumption of energy), in a rather selective manner. However, the fact that the communication of objects from a compartment to a neighboring compartment is controlled by the "reaction rules" is mathematically attractive, but is not quite realistic from a biological point of view; that is why variants were also considered where the two processes are separated: the evolution is controlled by rules as above, without target indications, and the communication is controlled by specific rules (e.g., by symport/antiport rules).

It is also worth noting that evolution rules are stated in terms of *names of objects*, while their application/execution is done using *copies of objects* – remember the example from Section 4, where the multiset $a^5b^2c^6$ was processed by a rule of the form $aab \rightarrow a(b, out)c(c, in)$, which, in the maximally parallel manner, is used twice, for the two possible sub-multisets aab.

We have arrived in this way at the important feature of P systems, concerning *the way of using the rules*. The key phrase in this respect is: *in the maximally parallel manner, nondeterministically choosing the rules and the objects.*

Specifically, this means that we assign objects to rules, nondeterministically choosing the objects and the rules until no further assignment is possible. Mathematically stated, we look to the *set* of rules, and try to find a *multisct* of rules, by assigning multiplicities to rules, with two properties: (i) the multiset of rules is *applicable* to the multiset of objects available in the respective region; that is, there are enough objects to apply the rules a number of times as indicated by their multiplicities; and (ii) the multiset is *maximal*, i.e., no further rule can be added to it (no multiplicity of a rule can be increased), because of the lack of available objects.

Thus, an evolution step in a given region consists of finding a maximal applicable multiset of rules, removing from the region all objects specified in the left hand sides of the chosen rules (with multiplicities as indicated by the rules and by the number of times each rule is used), producing the objects from the right hand sides of the rules, and then distributing these objects as indicated by the targets associated with them. If at least one of the rules introduces the dissolving action δ, then the membrane is dissolved, and its contents become part of the parent membrane, provided that this membrane was not dissolved at the same time; otherwise we stop at the first upper membrane which was not dissolved (the skin membrane at least remains intact).

8 A Formal Definition of a Transition P System

Systems based on multiset-rewriting rules as above are usually called *transition P systems*, and we preserve here this terminology (although "transitions" are present in all types of systems).

Of course, when presenting a P system we have to specify the alphabet of objects (a usual finite nonempty alphabet of abstract symbols identifying the objects), the membrane structure (it can be represented in many ways, but the one most used is by a string of labeled matching parentheses), the multisets of objects present in each region of the system (represented in the most compact way by strings of symbol objects), the sets of evolution rules associated with each region, and the indication about the way the output is defined (see below).

Formally, a *transition P system* (of degree $m \geq 1$) is a construct of the form

$$\Pi = (O, C, \mu, w_1, w_2, \dots, w_m, R_1, R_2, \dots, R_m, i_o),$$

where:

1. O is the (finite and nonempty) alphabet of *objects*,
2. $C \subset O$ is the set of *catalysts*,
3. μ is a membrane structure, consisting of m membranes, labeled $1, 2, \ldots, m$; we say that the membrane structure, and hence the system, is *of degree* m,
4. $w_1, w_2, \ldots, w_m$ are strings over O representing the *multisets of objects* present in regions $1, 2, \ldots, m$ of the membrane structure,
5. $R_1, R_2, \ldots, R_m$ are finite *sets of evolution rules* associated with regions $1, 2, \ldots, m$ of the membrane structure,
6. i_o is either one of the labels $1, 2, \ldots, m$, and the respective region is the *output region* of the system, or it is 0, and the result of a computation is collected in the environment of the system.

The rules are of the form $u \rightarrow v$ or $u \rightarrow v\delta$, with $u \in O^+$ and $v \in (O \times Tar)^*$, where[1] $Tar = \{here, in, out\}$. The rules can be cooperative (with u arbitrary), non-cooperative (with $u \in O - C$), or catalytic (of the form $ca \rightarrow cv$ or $ca \rightarrow cv\delta$, with $a \in O - C, c \in C$, and $v \in ((O - C) \times Tar)^*$); note that the catalysts never evolve and never change the region, they only help the other objects to evolve.

A possible restriction about the region i_o in the case when it is an internal one is to consider only regions enclosed by elementary membranes for output (that is, i_o should be the label of an elementary membrane of μ).

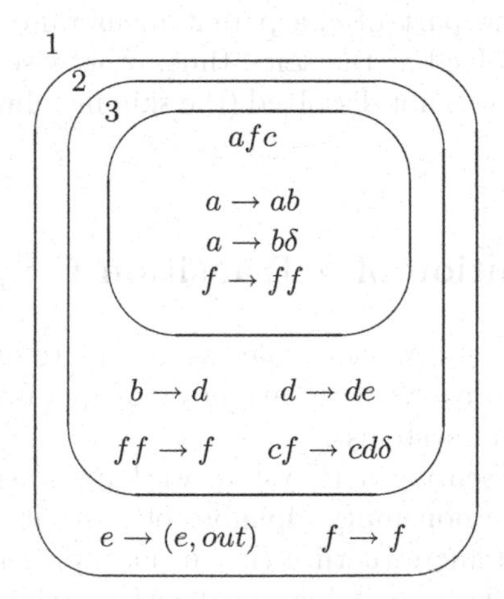

Fig. 3. The initial configuration of a P system, rules included.

In general, the membrane structure and the multisets of objects from its compartments identify a *configuration* of a P system. The *initial configuration*

[1] By V^* we denote the set of all strings over an alphabet V, the empty string λ included, and by V^+ we denote the set $V^* - \{\lambda\}$ of all nonempty strings over V.

is given by specifying the membrane structure and the multisets of objects available in its compartments at the beginning of a computation, that is, by $(\mu, w_1, \ldots, w_m)$. During the evolution of the system, by applying the rules, both the multisets of objects and the membrane structure can change. We will see how this is done in the next section; here we conclude with an example of a P system, represented in Figure 3. It is important to note that adding the set of rules to the initial configuration, placed in the corresponding regions, we have a complete and concise presentation of the system (the indication of the output region can also be added in a suitable manner, for instance, writing "output" inside it).

9 Defining Computations and Results of Computations

In their basic variant, membrane systems are synchronous devices, in the sense that a global clock is assumed, which marks the time for all regions of the system. In each time unit a transformation of a configuration of the system – we call it *transition* – takes place by applying the rules in each region in a *nondeterministic* and *maximally parallel manner*. As explained in the previous sections, this means that the objects to evolve and the rules governing this evolution are chosen in a nondeterministic way, and this choice is "exhaustive" in the sense that, after the choice is made, no rule can be applied in the same evolution step to the remaining objects.

A sequence of transitions constitutes a *computation*. A computation is successful if it halts, reaches a configuration where no rule can be applied to the existing objects, and the output region i_o still exists in the halting configuration (in the case where i_o is the label of a membrane, it can be dissolved during the computation, but the computation is then no longer successful). With a successful computation we can associate a *result* in various ways. If we have an output region specified, and this is an internal region, then we have an *internal output*: we count the objects present in the output region in the halting configuration and this number is the result of the computation. When we have $i_o = 0$, we count the objects which leave the system during the computation, and this is called *external output*. In both cases the result is a number. If we distinguish among different objects, then we can have as the result a vector of natural numbers. The objects which leave the system can also be arranged in a sequence according to the time when they exit the skin membrane, and in this case the result is a string (if several objects exit at the same time, then all their permutations are accepted as a substring of the result). Note that non-halting computations provide no output (we cannot know when a number is "completely computed" before halting); if the output membrane is dissolved during the computation, then the computation aborts, and no result is obtained (of course, this makes sense only in the case of internal output).

A possible extension of the definition is to consider a *terminal* set of objects, $T \subseteq O$, and to count only the copies of objects from T, discarding the objects from $O - T$ present in the output region. This allows some additional leeway in constructing and "programming" a P system, because we can ignore some auxiliary objects (e.g., the catalysts).

Because of the nondeterminism of the application of rules, starting from an initial configuration we can get several successful computations, and hence several results. Thus, a P system *computes* (one also says *generates*) a set of numbers (or a set of vectors of numbers, or a language, depending on the way the output is defined). The case when we get a language is important in view of the qualitative difference between the "loose" data structure we use inside the system (vectors of numbers) and the data structure of the result, strings, where we also have a "syntax," a positional information.

For a given system Π we denote by $N(\Pi)$ the set of numbers computed by Π in the above way. When we consider the vector of multiplicities of objects from the output region, we write $Ps(\Pi)$. In turn, in the case where we take as (external) output the strings of objects leaving the system, we denote the language of these strings by $L(\Pi)$.

Let us illustrate the previous definitions by examining the computations of the system from Figure 3, with the output region being the environment.

We have objects only in the central membrane, that with label 3; hence only here can we apply rules. Specifically, we can repeatedly apply the rule $a \rightarrow ab$ in parallel with $f \rightarrow ff$, and in this way the number of copies of b grows each step by one, while the number of copies of f is doubled in each step. If we do not apply the rule $a \rightarrow b\delta$ (again in parallel with $f \rightarrow ff$), which dissolves the membrane, then we can continue in this way forever. Thus, in order to ever halt, we have to dissolve membrane 3. Assume that this happens after $n \geq 0$ steps of using the rules $a \rightarrow ab$ and $f \rightarrow ff$. When membrane 3 is dissolved, its contents ($n+1$ copies of b, 2^{n+1} copies of f, and one copy of the catalyst c) are left free in membrane 2, which can now start using its rules. In the next step, all objects b become d. Let us examine the rules $ff \rightarrow f$ and $cf \rightarrow cd\delta$. The second rule dissolves membrane 2, and hence passes its contents to membrane 1. If among the objects which arrive in membrane 1 there is at least one copy of f, then the rule $f \rightarrow f$ from region 1 can be used forever and the computation never stops; moreover, if the rule $ff \rightarrow f$ is used at least once, in parallel with the rule $cf \rightarrow cd\delta$, then at least one copy of f is present. Therefore, the rule $cf \rightarrow cd\delta$ should be used only if region 2 contains only one copy of f (note that, because of the catalyst, the rule $cf \rightarrow cd\delta$ can be used only for one copy of f). This means that the rule $ff \rightarrow f$ was used always for all available pairs of f, that is, at each step the number of copies of f is divided by 2. This is already done once in the step where all copies of b become d, and will be done from now on as long as at least two copies of f are present. Simultaneously, at each step, each d produces one copy of e. This process can continue until we get a configuration with only one copy of f present; at that step we have to use the rule $cf \rightarrow cd\delta$, hence also membrane 2 is dissolved. Because we have applied the rule $d \to de$, in parallel for all copies of d (there are $n+1$ such copies) during $n+1$ steps, we have $(n+1)(n+1)$ copies of e, $n+2$ copies of d (one of them produced by the rule $cf \to cd\delta$), and one copy of c in the skin membrane of the system (the unique membrane still present). The objects e are sent out, and the computation halts. Therefore, we compute in this way the number $(n+1)^2$ for $n \geq 0$, that is, $N(\Pi) = \{n^2 \mid n \geq 1\}$.

10 Using Symport and Antiport Rules

The multiset rewriting rules correspond to reactions taking place in the cell, *inside* the compartments. However, an important part of the cell activity is related to the passage of substances through membranes, and one of the most interesting ways to handle this trans-membrane communication is by coupling molecules. The process by which two molecules pass together across a membrane (through a specific protein channel) is called *symport*; when the two molecules pass simultaneously through a protein channel, but in opposite directions, the process is called *antiport*.

We can formalize these operations in an obvious way: (ab, in) or (ab, out) are symport rules, stating that a and b pass together through a membrane, entering in the former case and exiting in the latter case; similarly, $(a, out; b, in)$ is an antiport rule, stating that a exits and, at the same time, b enters the membrane. Separately, neither a nor b can cross a membrane unless we have a rule of the form (a, in) or (a, out), called, for uniformity, the *uniport* rule.

Of course, we can generalize these types of rules, by considering symport rules of the form (x, in) and (x, out), and antiport rules of the form $(z, out; w, in)$, where x, z, and w are multisets of arbitrary size; one says that $|x|$ is the *weight* of the symport rule, and $\max(|z|, |w|)$ is the *weight* of the antiport rule[2].

Now, such rules can be used in a P system instead of the target indications *here*, *in*, and *out*: we consider multiset rewriting rules of the form $u \to v$ (or $u \to v\delta$) without target indications associated with the objects from v, as well as symport/antiport rules for communication of the objects between compartments. Such systems, called *evolution-communication* P systems, were considered in [18] (for various restricted types of rules of the two forms).

Here, we do not go down that direction, but stay closer both to the chronological evolution of the domain and to the mathematical minimalism, and we check whether we can compute using only communication, that is, only symport and antiport rules. This leads to considering one of the most interesting classes of P systems, which we formally introduce here.

A *P system with symport/antiport rules* is a construct of the form

[2] By $|u|$ we denote the length of the string $u \in V^*$ for any alphabet V.

$$\Pi = (O, \mu, w_1, \ldots, w_m, E, R_1, \ldots, R_m, i_o),$$

where:

1. O is the alphabet of objects,
2. μ is the membrane structure (of degree $m \geq 1$, with the membranes labeled $1, 2, \ldots, m$ in a one-to-one manner),
3. $w_1, \ldots, w_m$ are strings over O representing the multisets of objects present in the m compartments of μ in the initial configuration of the system,
4. $E \subseteq O$ is the set of objects supposed to appear in the environment in arbitrarily many copies,
5. $R_1, \ldots, R_m$ are the (finite) sets of rules associated with the m membranes of μ,
6. $i_o \in H$ is the label of a membrane of μ, which indicates the *output* region of the system.

The rules from R can be of two types, symport rules and antiport rules, of the forms specified above.

The rules are used in the nondeterministic maximally parallel manner. We define transitions, computations, and halting computations in the usual way. The number (or the vector of multiplicities) of objects present in region i_o in the halting configuration is said to be computed by the system by means of that computation; the set of all numbers (or vectors of numbers) computed in this way by Π is denoted by $N(\Pi)$ (by $Ps(\Pi)$, respectively).

We note here a new component of the system, the set E of objects which are present in the environment in arbitrarily many copies; because we move objects only across membranes and because we start with finite multisets of objects present in the system, we cannot increase the number of objects necessary for the computation if we do not provide a supply of objects, and this can be done by considering the set E. Because the environment is supposedly inexhaustible, the objects from E are inexhaustible; regardless of how many of them are brought into the system, arbitrarily many remain outside.

Another new feature is that this time the rules are associated with membranes, and not with regions, and this is related to the fact that each rule governs communication through a specific membrane.

The P systems with symport/antiport rules have a series of attractive characteristics: they are fully based on biological types of multiset processing rules; the environment plays a direct role in the evolution of the system; the computation is done only by communication, no object is changed, and the objects move only across membranes; no object is created or destroyed, and hence the conservation law is observed (as given in the previous sections, this is not valid for multiset rewriting rules because, for instance, rules of the form $a \to aa$ or $ff \to f$ are allowed).

11 An Example (Like a Proof...)

Because P systems with symport/antiport rules constitute an important class of P systems, it is worth considering an example; however, instead of a simple example, we directly give a general construction for simulating a *register machine*. In this way, we also introduce one of the widely used proof techniques for the universality results in this area. (Of course, the biologist can safely skip this section.)

Informally speaking, a register machine consists of a specified number of counters (also called registers) which can hold any natural number, and which are handled according to a program consisting of labeled instructions; the counters can be increased or decreased by 1 – the decreasing possible only if a counter holds a number greater than or equal to 1 (we say that it is nonempty) – and checked whether they are nonempty.

Formally, a (nondeterministic) *register machine* is a device $M = (m, B, l_0, l_h, R)$, where $m \geq 1$ is the number of counters, B is the (finite) set of instruction labels, l_0 is the initial label, l_h is the halting label, and R is the finite set of instructions labeled (hence uniquely identified) by elements from B. The labeled instructions are of the following forms:

- $l_1 : (\mathtt{add}(r), l_2, l_3)$, $1 \leq r \leq m$ (add 1 to counter r and go nondeterministically to one of the instructions with labels l_2, l_3),
- $l_1 : (\mathtt{sub}(r), l_2, l_3)$, $1 \leq r \leq m$ (if counter r is not empty, then subtract 1 from it and go to the instruction with label l_2, otherwise go to the instruction with label l_3).

A counter machine generates a k-dimensional vector of natural numbers in the following manner: we distinguish k counters as output counters (without loss of generality, they can be the first k counters), and we start computing with all m counters empty, with the instruction labeled l_0; if the label l_h is reached, then the computation *halts* and the values of counters $1, 2, \ldots, k$ are the vector generated by the computation. The set of all vectors from $\mathbf{N}^k$ generated in this way by M is denoted by $Ps(M)$. If we want to generate only numbers (1-dimensional vectors), then we have the result of a computation in counter 1, and the set of numbers computed by M is denoted by $N(M)$. It is known (see [60]) that nondeterministic counter machines with $k + 2$ counters can compute any set of Turing computable k-dimensional vectors of natural numbers (hence machines with three counters generate exactly the family of Turing computable sets of numbers).

Now, a register machine can be easily simulated by a P system with symport/antiport rules. The idea is illustrated in Figure 4, where we have represented the initial configuration of the system, the rules associated with the unique membrane, and the set E of objects present in the environment.

The value of each register r is represented by the multiplicity of object $a_r, 1 \leq r \leq m$, in the unique membrane of the system. The labels from B,

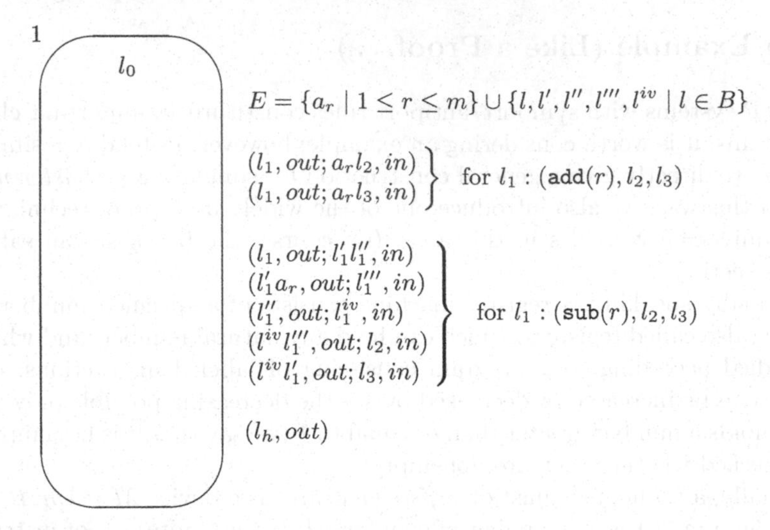

Fig. 4. An example of a symport/antiport P system.

as well as their primed versions, are also objects of our system. We start with the unique object l_0 present in the system. In the presence of a label object l_1 we can simulate the corresponding instruction $l_1 : (\mathtt{add}(r), l_2, l_3)$ or $l_1 : (\mathtt{sub}(r), l_2, l_3)$.

The simulation of an add instruction is clear, so we discuss only a sub instruction. The object l_1 exits the system in exchange of the two objects $l_1' l_1''$ (rule $(l_1, out; l_1' l_1'', in)$). In the next step, if any copy of a_r is present in the system, then l_1' has to exit (rule $(l_1' a_r, out; l_1''', in)$), thus diminishing the number of copies of a_r by one, and bringing inside the object l_1'''; if no copy of a_r is present, which corresponds to the case when the register r is empty, then the object l_1' remains inside. Simultaneously, rule $(l_1'', out; l_1^{iv}, in)$ is used, bringing inside the "checker" l_1^{iv}. Depending on what this object finds in the system, either l_1''' or l_1', it introduces the label l_2 or l_3, respectively, which corresponds to the correct continuation of the computation of the register machine.

When the object l_h is introduced, it is expelled into the environment and the computation stops.

Clearly, the (halting) computations in Π directly correspond to (halting) computations in M; hence $Ps(M) = Ps(\Pi)$.

12 A Large Panoply of Possible Extensions

We have mentioned the flexibility and the versatility of the formalism of MC, and we have already mentioned several types of systems, making use of several types of rules, with the output of a computation defined in various ways.

We continue here in this direction, by presenting a series of possibilities for changing the form of rules and/or the way of using them. The motivation for such extensions comes both from biology, i.e., from the desire to capture more and more biological facts, and from mathematics and computer science, i.e., from the desire to have more powerful or more elegant models.

First, let us return to the basic target indications, *here, in,* and *out*, associated with the objects produced by rules of the form $u \rightarrow v$; *here* and *out* indicate precisely the region where the object is to be placed, but *in* introduces a degree of nondeterminism in the case where there are several inner membranes. This nondeterminism can be avoided by indicating also the label of the target membrane, that is, using target indications of the form in_j, where j is a label. An intermediate possibility, more specific than *in* but not completely unambiguous like in_j, is to assign the *electrical polarizations*, $+, -$, and 0, to both objects and membranes. The polarizations of membranes are given from the beginning (or can be changed during the computation), the polarization of objects is introduced by rules, using rules of the form $ab \rightarrow c^+c^-(d^0, tar)$. The charged objects have to go to any lower level membrane of opposite polarization, while objects with neutral polarization either remain in the same region or get out, depending on the target indication $tar \in \{here, out\}$ (this is the case with d in the previous rule).

A spectacular generalization, considered recently in [26], is to use indications in_j, for any membrane j from the system; hence the object is "teleported" immediately at any distance in the membrane structure. Also, commands of the form in^* and out^* were used, with the meaning that the object should be sent to (one of) the elementary membranes from the current membrane or to the skin region, respectively, no matter how far the target is.

We have considered the membrane dissolution action, represented by the symbol δ; we may imagine that such an action decreases the thickness of the membrane from the normal thickness, 1, to 0. A dual action can be also used, of increasing the thickness of a membrane, from 1 to 2. We indicate this action by τ. Assume that δ also decreases the thickness from 2 to 1, that the thickness cannot have values other than 0 (membrane dissolved), 1 (normal thickness), and 2 (membrane impermeable), and that when both δ and τ are introduced simultaneously in the same region, by different rules, their actions cancel, and the thickness of the membrane does not change. In this way, we can nicely control the work of the system: if a rule introduces a target indication *in* or *out* and the membrane which has to be crossed by the respective object has thickness 2, and hence is non-permeable, then the rule cannot be applied.

Let us look now to the catalysts. In the basic definition they never change their state or their place like ordinary objects do. A "democratic" decision is to also let the catalysts evolve within certain limits. Thus, *mobile catalysts* were proposed, moving across membranes like any object (but not themselves changing). The catalysts were then allowed to change their state, for instance, oscillating between c and $\bar{c}$. Such a catalyst is called *bistable*, and the natural generalization is to consider k-stable catalysts, allowed to change along k

given forms. Note that the number of catalysts is not changed; we do not produce or remove catalysts (provided they do not leave the system), and this is important in view of the fact that the catalysts are in general used for inhibiting the parallelism (a rule $ca \rightarrow cv$ can be used simultaneously at most as many times as copies of c are present).

There are several possibilities for controlling the use of rules, leading in general to a decrease in the degree of nondeterminism of a system. For instance, a mathematically and biologically motivated possibility is to consider a *priority* relation on the set of rules from a given region, in the form of a partial order relation on the set of rules from that region. This corresponds to the fact that certain reactions/reactants are more active than others, and can be interpreted in two ways: as a competition for reactants/objects, or in a strong sense. In the latter sense, if a rule r_1 has priority over a rule r_2 and r_1 can be applied, then r_2 cannot be applied, regardless of whether rule r_1 leaves objects which it cannot use. For instance, if $r_1 : ff \rightarrow f$ and $r_2 : cf \rightarrow cd\delta$, as in the example from Section 8, and the current multiset is $fffc$, because rule r_1 can be used, consuming two copies of f, we do not also use the second rule for the remaining fc. In the weak interpretation of the priority, the use of the second rule is allowed: the rule with the maximal priority takes as many objects as possible, and, if there are objects still remaining, the next rule in the decreasing order of priority is used for as many objects as possible, and we continue in this way until no further rule can be added to the multiset of applicable rules.

Also coming directly from bio-chemistry are the rules with *promoters* and *inhibitors*, written in the form $u \rightarrow v|_z$ and $u \rightarrow v|_{\neg z}$, respectively, where u, v, and z are multisets of objects; in the case of promoters, the rule $u \rightarrow v$ can be used in a given region only if all objects from z are present in the same region, and they are different from the (copies of) objects from u; in the inhibitors case, no object from z should be present in the region and be different from the objects from u. The promoting objects can evolve at the same time by other rules, or by the same rule $u \rightarrow v$ but by another instance of it (e.g., $a \rightarrow b|_a$ can be used twice in a region containing two copies of a, with each instance of $a \rightarrow b|_a$ acting on one copy of a and promoted by the other copy, but it cannot be used in a region where a appears only once).

Interesting combinations of rewriting-communication rules are those considered in [77], where rules of the following three forms are proposed: $a \rightarrow (a, tar)$, $ab \rightarrow (a, tar_1)(b, tar_2)$, and $ab \rightarrow (a, tar_1)(b, tar_2)(c, come)$, where a, b, and c are objects, and tar, tar_1, and tar_2 are target indications of the forms *here, in,* and *out*, or in_j, where j is the label of a membrane. Such a rule just moves objects from one region to another, with rules of the third type usable only in the skin region; $(c, come)$ means that a copy of c is brought into the system from the environment. Clearly, these rules are different from the symport/antiport rules; for instance, the two objects ab from a rule $ab \rightarrow (a, tar_1)(b, tar_2)$ start from the same region, and can go in different directions, one up and the other down, in the membrane structure.

We are left with one of the most general types of rule, introduced in [11] under the name *boundary rules*, directly capturing the idea that many reactions take place on the inner membranes of a cell, depending maybe on the contents of both the inner and the outer regions adjacent to that membrane. These rules are of the form $xu[_i vy \rightarrow xu'[_i v'y$, where x, u, u', v, v', and y are multisets of objects and i is the label of a membrane. Their meaning is that in the presence of the objects from x outside and from y inside the membrane i, the multiset u from outside changes to u', and, simultaneously, the multiset v from inside changes to v'. The generality of this kind of rule is apparent, and it can be decreased by imposing various restrictions on the multisets involved.

There also are other variants considered in the literature, especially in the way of controlling the use of the rules, but we do not continue here in that direction.

13 P Systems with Active Membranes

We pass now to presenting a class of P systems, which, together with the basic transition systems and the symport/antiport systems, is one of the three central types of cell-like P systems considered in membrane computing. As in the above case of boundary rules, they start from the observation that membranes play an important role in the reactions which take place in a cell, and, moreover, they can evolve themselves, either by changing their characteristics or by dividing.

This last idea especially has motivated the class of *P systems with active membranes*, which are constructs of the form

$$\Pi = (O, H, \mu, w_1, \ldots, w_m, R),$$

where:

1. $m \geq 1$ (the initial degree of the system);
2. O is the alphabet of *objects*;
3. H is a finite set of *labels* for membranes;
4. μ is a *membrane structure*, consisting of m membranes initially having neutral polarizations labeled (not necessarily in a one-to-one manner) with elements of H;
5. $w_1, \ldots, w_m$ are strings over O, describing the *multisets of objects* placed in the m regions of μ;
6. R is a finite set of *developmental rules*, of the following forms:
   (a) $[_h a \rightarrow v]_h^e$, for $h \in H, e \in \{+, -, 0\}, a \in O, v \in O^*$
   (object evolution rules, associated with membranes and depending on the label and the charge of the membranes but not directly involving the membranes, in the sense that the membranes are neither taking part in the application of these rules nor are they modified by them);

(*b*) $a[_h\ ]_h^{e_1} \rightarrow [_h b]_h^{e_2}$, for $h \in H, e_1, e_2 \in \{+, -, 0\}, a, b \in O$
(*in* communication rules; an object is introduced in the membrane, and possibly modified during this process; also the polarization of the membrane can be modified, but not its label);

(*c*) $[_h a\ ]_h^{e_1} \rightarrow [_h\ ]_h^{e_2} b$, for $h \in H, e_1, e_2 \in \{+, -, 0\}, a, b \in O$
(*out* communication rules; an object is sent out of the membrane, and possibly modified during this process; also the polarization of the membrane can be modified, but not its label);

(*d*) $[_h a\ ]_h^{e} \rightarrow b$, for $h \in H, e \in \{+, -, 0\}, a, b \in O$
(dissolving rules; in reaction with an object, a membrane can be dissolved, while the object specified in the rule can be modified);

(*e*) $[_h a\ ]_h^{e_1} \rightarrow [_h b\ ]_h^{e_2} [_h c\ ]_h^{e_3}$, for $h \in H, e_1, e_2, e_3 \in \{+, -, 0\}, a, b, c \in O$
(division rules for elementary membranes; in reaction with an object, the membrane is divided into two membranes with the same label, and possibly of different polarizations; the object specified in the rule is replaced in the two new membranes possibly by new objects; the remaining objects are duplicated and may evolve in the same step by rules of type (*a*)).

The objects evolve in the maximally parallel manner, used by rules of type (*a*) or by rules of the other types, and the same is true at the level of membranes, which evolve by rules of types (*b*)–(*e*). Inside each membrane, the rules of type (*a*) are applied in parallel, with each copy of an object used by only one rule of any type from (*a*) to (*e*). Each membrane can be involved in only one rule of types (*b*)–(*e*) (the rules of type (*a*) are not considered to involve the membrane where they are applied). Thus, in total, the rules are used in the usual nondeterministic maximally parallel manner, in a bottom-up way (we use first the rules of type (a), and then the rules of other types; in this way, in the case of dividing membranes, the result of using first the rules of type (a) is duplicated in the newly obtained membranes). Also, as usual, only halting computations give a result, in the form of the number (or the vector) of objects expelled into the environment during the computation.

The set H of labels has been specified because it is possible to allow the change of membrane labels. For instance, a division rule can be of the more general form

(e') $[_{h_1} a\ ]_{h_1}^{e_1} \rightarrow [_{h_2} b\ ]_{h_2}^{e_2} [_{h_3} c\ ]_{h_3}^{e_3}$,
for $h_1, h_2, h_3 \in H, e_1, e_2, e_3 \in \{+, -, 0\}, a, b, c \in O$.

The change of labels can also be considered for rules of types (b) and (c). Also, we can consider the possibility of dividing membranes into more than two copies, or even of dividing non-elementary membranes (in such a case, all inner membranes are duplicated in the new copies of the membrane).

It is important to note that in the case of P systems with active membranes, the membrane structure evolves during the computation, not only by decreasing the number of membranes, due to dissolution operations (rules of type (d)), but also by increasing the number of membranes by division. This increase can be exponential in a linear number of steps: using a division rule successively n steps, due to the maximal parallelism, we get 2^n copies of the same membrane. This is one of the most investigated ways of obtaining an exponential working space in order to trade time for space and solve computationally hard problems (typically **NP**-complete problems) in feasible time (typically polynomial or even linear time).

Some details can be found in Section 20, but we illustrate here the way of using membrane division in such a framework with an example dealing with the generation of all 2^n truth assignments possible for n propositional variables.

Assume that we have the variables $x_1, x_2, \ldots, x_n$; we construct the following system (of degree 2):

$$\begin{aligned}
\Pi &= (O, H, \mu, w_1, w_2, R),\\
O &= \{a_i, c_i, t_i, f_i \mid 1 \le i \le n\} \cup \{\texttt{check}\},\\
H &= \{1, 2\},\\
\mu &= [_1[_2\ ]_2]_1,\\
w_1 &= \lambda,\\
w_2 &= a_1 a_2 \ldots a_n c_1,\\
R &= \{[_2 a_i]_2^0 \to [_2 t_i]_2^0 [_2 f_i]_2^0 \mid 1 \le i \le n\}\\
&\quad \cup \{[_2 c_i \to c_{i+1}]_2^0 \mid 1 \le i \le n-1\}\\
&\quad \cup \{[_2 c_n \to \texttt{check}]_2^0,\ [_2\texttt{check}]_2^0 \to \texttt{check}[_2\ ]_2^+\}.
\end{aligned}$$

We start with the objects $a_1, \ldots, a_n$ in the inner membrane and we divide this membrane repeatedly by means of the rules $[_2 a_i]_2^0 \to [_2 t_i]_2^0 [_2 f_i]_2^0$; note that the object a_i used in each step is nondeterministically chosen, but each division replaces that object by t_i (for *true*) in one membrane and with f_i (for *false*) in the other membrane; hence after n steps the configuration obtained is the same regardless of the order of expanding the objects. Specifically, we get 2^n membranes with label 2, each one containing a truth assignment for the n variables. Simultaneously with the division, we have to use the rules of type (a) which update the "counter" c; hence at each step we increase by one the subscript of c. Therefore, when all variables have been expanded, we get the object check in all membranes (the rule of type (a) is used first, and after that the result is duplicated in the newly obtained membranes). In step $n+1$, this object exits each copy of membrane 2, changing its polarization to positive; this is meant to signal the fact that the generation of all truth assignments is completed, and we can start checking the truth values of (the clauses of) the propositional formula.

The previous example was chosen also to show that the polarizations of membranes are not used while generating the truth assignments, though they might be useful after that; till now, this is the case in all polynomial time solutions to **NP**-complete problems obtained in this framework, in particular for solving SAT (satisfiability of propositional formulas in the conjunctive normal form). An important *open problem* in this area is whether or not the polarizations can be avoided. This can be done if other ingredients are considered, such as label changing or division of non-elementary membranes, but without adding such features the best result obtained so far is that from [3] where it is proved that the number of polarizations can be reduced to two.

14 A Panoply of Possibilities for Having a Dynamical Membrane Structure

Membrane dissolving and dividing are only two of the many possibilities of handling the membrane structures. One additional possibility investigated early is membrane *creation*, based on rules of the form $a \to [_h v]_h$, where a is an object, v is a multiset of objects, and h is a label from a given set of labels. Using such a rule in a membrane j, we create a new membrane, with label h, having inside the objects specified by v. Because we know the label of the new membrane, we know the rules which can be used in its region (a "dictionary" of possible membranes is given, specifying the rules to be used in any membrane with labels in a given set). Because rules for handling membranes are of a more general interest (e.g., for applications), we illustrate them in Figure 5, where the reversibility of certain pairs of operations is also made visible.

For instance, converse to membrane division, the operation of *merging* the contents of two membranes can be considered; formally, we can write such a rule in the form $[_{h_1} a]_{h_1} [_{h_2} b]_{h_2} \to [_{h_3} c]_{h_3}$, where a, b, and c are objects and h_1, h_2, and h_3 are labels (we have considered the general case, where the labels can be changed).

Actually, the merging operation can also be considered as the reverse of the *separation* operation, formalized as follows: let $K \subseteq O$ be a set of objects; a separation with respect to K is done by a rule of the form $[_{h_1}\ ]_{h_1} \to [_{h_2} K]_{h_2} [_{h_3} \neg K]_{h_3}$, with the meaning that the content of membrane h_1 is split into two membranes, with labels h_2 and h_3, the first one containing all objects in K and the second one containing all objects not in K.

The operations of *endocytosis* and *exocytosis* (we use these general names, although in biology there are distinctions depending on the size of the objects and the number of objects moved; phagocytosis, pinocytosis, etc.) are also simple to formalize. For instance, $[_{h_1} a]_{h_1} [_{h_2}\ ]_{h_2} \to [_{h_2} [_{h_1} b]_{h_1}]_{h_2}$, for $h_1, h_2 \in H, a, b \in V$, is an endocytosis rule, stating that an elementary membrane labeled h_1 enters the adjacent membrane labeled h_2 under the control of object a; the labels h_1 and h_2 remain unchanged during this process; however, the object a may be modified to b. Similarly, the rule $[_{h_2} [_{h_1} a]_{h_1}]_{h_2} \to [_{h_1} b]_{h_1} [_{h_2}\ ]_{h_2}$, for $h_1, h_2 \in H, a, b \in V$, indicates an exocytosis operation: an elementary membrane labeled h_1 is sent out of a membrane labeled h_2 under the control of object a; the labels of the two membranes remain unchanged, but the object a from membrane h_1 may be modified during this operation.

Finally, let us mention the operation of *gemmation*, by which a membrane is created inside a membrane h_1 and sent to a membrane h_2; the moving membrane is dissolved inside the target membrane h_2, thus releasing its contents there. In this way, multisets of objects can be transported from a membrane to another one in a protected way: the enclosed objects cannot be processed by the rules of the regions through which the travelling membrane passes. The travelling membrane is created with a label of the form $@_{h_2}$, which indicates that it is a temporary membrane, having to get dissolved inside the membrane with label h_2. Corresponding to the situation from biology, in [13, 14] one considers only the case where the membranes h_1, h_2 are adjacent and placed directly in the skin membrane, but the operation can be generalized.

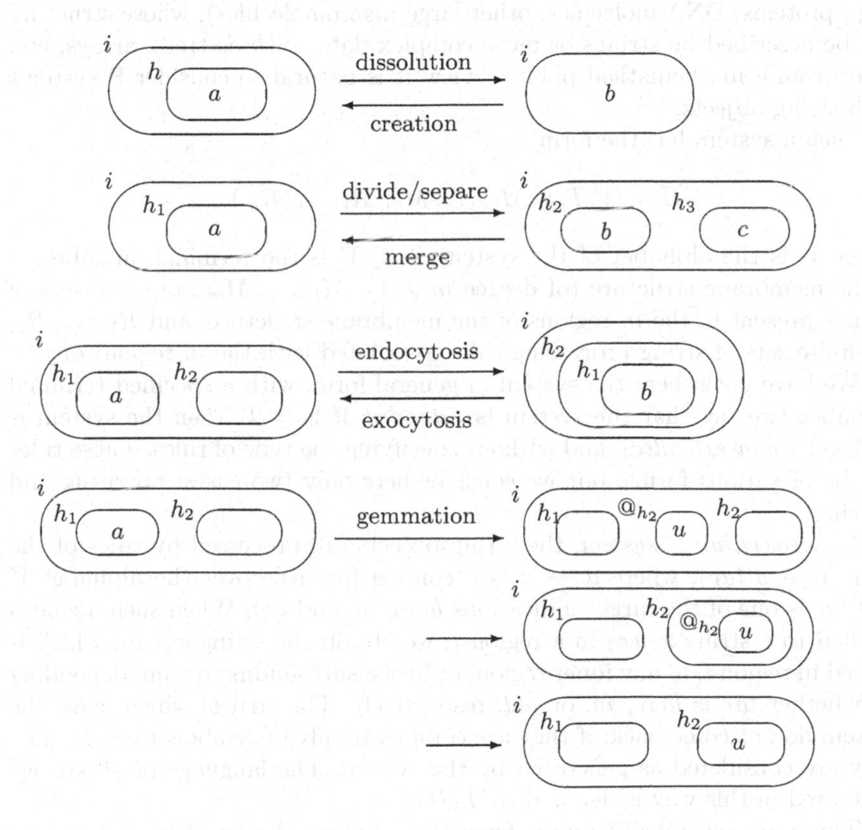

Fig. 5. Membrane handling operations.

A gemmation rule is of the form $a \to [_{@_{h_2}} u]_{@_{h_2}}$, where a is an object and u a multiset of objects (but it can be generalized by creating several travelling membranes at the same time, with different destinations); the result of applying such a rule is as illustrated in the bottom of Figure 5. Note that the crossing of one membrane takes one time unit (it is supposed that the travelling membrane finds the shortest path from the region where it is created to the target region).

Several other operations with membranes were considered, e.g., in the context of applications to linguistics, as well as in [47] and in other papers, but we do not enter into further details here.

15 Structuring the Objects

In the previous classes of P systems, the objects were considered atomic, identified only by their name, but in a cell many chemicals are complex molecules (e.g., proteins, DNA molecules, other large macromolecules), whose structure can be described by strings or more complex data, such as trees, arrays, etc. Also, from a mathematical point of view it is natural to consider P systems with string objects.

Such a system has the form

$$\Pi = (V, T, \mu, M_1, \ldots, M_m, R_1, \ldots, R_m),$$

where V is the alphabet of the system, $T \subseteq V$ is the terminal alphabet, μ is the membrane structure (of degree $m \geq 1$), $M_1, \ldots, M_m$ are finite sets of strings present in the m regions of the membrane structure, and $R_1, \ldots, R_m$ are finite sets of string-processing rules associated with the m regions of μ.

We have given here the system in general form, with a specified terminal alphabet (we say that the system is *extended*; if $V = T$, then the system is said to be *non-extended*), and without specifying the type of rules. These rules can be of various forms, but we consider here only two cases: rewriting and splicing.

In a *rewriting P system*, the string objects are processed by rules of the form $a \to u(tar)$, where $a \to u$ is a context-free rule over the alphabet V and tar is one of the target indications *here, in,* and *out.* When such a rule is applied to a string $x_1 a x_2$ in a region i, we obtain the string $x_1 u x_2$, which is placed in region i, in any inner region, or in the surrounding region, depending on whether tar is *here, in,* or *out,* respectively. The strings which leave the system do not come back; if they are composed only of symbols from T, then they are considered as generated by the system. The language of all strings generated in this way is denoted by $L(\Pi)$.

There are several differences from the previous classes of P systems: we work with *sets* of string objects, not with multisets; in order to introduce a string in the language $L(\Pi)$ we do not need to have a halting computation, because the strings do not change after leaving the system; each string is processed by only one rule (the rewriting is sequential at the level of strings), but in each step all strings from all regions which can be rewritten by local rules are rewritten by one rule.

In a *splicing P system*, we use splicing rules such as those in DNA computing [38, 70], that is, of the form $u_1\#u_2\$u_3\#u_4$, where u_1, u_2, u_3, and u_4 are strings over V. For four strings $x, y, z, w \in V^*$ and a rule $r : u_1\#u_2\$u_3\#u_4$, we write

$$(x, y) \vdash_r (z, w) \text{ if and only if } x = x_1u_1u_2x_2, \; y = y_1u_3u_4y_2,$$
$$z = x_1u_1u_4y_2, \; w = y_1u_3u_2x_2,$$
$$\text{for some } x_1, x_2, y_1, y_2 \in V^*.$$

We say that we splice x and y at the sites u_1u_2 and u_3u_4, respectively, and the result of the splicing (obtained by recombining the fragments obtained by cutting the strings as indicated by the sites) are the strings z and w.

In our case we add target indications to the two resulting strings, that is, we consider rules of the form $r : u_1\#u_2\$u_3\#u_4(tar_1, tar_2)$, with tar_1 and tar_2 one of *here, in,* and *out.* The meaning is as standard: after splicing the strings x, y from a given region, the resulting strings z, w are moved to the regions indicated by tar_1, tar_2, respectively. The language generated by such a system consists again of all strings over T sent into the environment during the computation, without considering only halting computations.

We do not give here an example of a rewriting or a splicing P system, but we move on to introducing an important extension of rewriting rules, namely, *rewriting with replication* [49]. In such systems, the rules are of the form $a \to (u_1, tar_1)||(u_2, tar_2)|| \dots ||(u_n, tar_n)$, with the meaning that by rewriting a string x_1ax_2 we get n strings, $x_1u_1x_2, x_1u_2x_2, \dots, x_1u_nx_2$, which have to be moved in the regions indicated by targets $tar_1, tar_2, \dots, tar_n$, respectively. In this case we work again with halting computations, and the motivation is that if we do not impose the halting condition, then the strings $x_1u_ix_2$ evolve completely independently; hence we can replace the rule $a \to (u_1, tar_1)||(u_2, tar_2)|| \dots ||(u_n, tar_n)$ with n rules $a \to (u_i, tar_i), 1 \leq i \leq n$, without changing the language; that is, replication makes a difference only in the halting case.

The replicated rewriting is important because it provides the possibility to replicate strings, thus enlarging the workspace, and indeed this is one of the frequently used ways to generate an exponential workspace in linear time, used then for solving computationally hard problems in polynomial time.

Besides these types of rules for string processing, other kinds of rules were also used, such as insertion and deletion, context adjoining in the sense of Marcus contextual grammars [63], splitting, conditional concatenation, and so on, sometimes with motivations from biology, where several similar operations can be found, e.g., at the genome level.

16 Tissue-like P Systems

We pass now to consider a very important generalization of the membrane structure, passing from the cell-like structure, described by a tree, to a tissue-like structure, with the membranes placed in the nodes of an arbitrary graph (which corresponds to the complex communication networks established among adjacent cells by making their protein channels cooperate, moving molecules directly from one cell to another, [53]). Actually, in the basic variant of tissue-like P systems, this graph is virtually a total one; what matters is the communication graph, dynamically defined during computations. In short, several (elementary) membranes – also called cells – are freely placed in a common environment; they can communicate either with each other or with the environment by symport/antiport rules. Specifically, we consider antiport rules of the form $(i, x/y, j)$, where i, j are labels of cells (or at most one is zero, identifying the environment), and x, y are multisets of objects. This means that the multiset x is moved from i to j at the same time as the multiset y is moved from j to i. If one of the multisets x, y is empty, then we have, in fact, a symport rule. Therefore, the communication among cells is done either directly, in one step, or indirectly, through the environment: one cell throws some objects out and other cells can take these objects in the next step or later. As in symport/antiport P systems, the environment contains a specified set of objects in arbitrarily many copies. A computation develops as standard, starting from the initial configuration and using the rules in the nondeterministic maximally parallel manner. When halting, we count the objects from a specified cell, and this is the result of the computation.

The graph plays a more important role in so-called *tissue-like P systems with channel-states*, [33], which are constructs of the form

$$\Pi = (O, T, K, w_1, \ldots, w_m, E, syn, (s_{(i,j)})_{(i,j)\in syn}, (R_{(i,j)})_{(i,j)\in syn}, i_o),$$

where O is the alphabet of *objects*, $T \subseteq O$ is the alphabet of *terminal* objects, K is the alphabet of *states* (not necessarily disjoint of O), $w_1, \ldots, w_m$ are strings over O representing the initial multisets of objects present in the cells of the system (it is assumed that we have m cells, labeled with $1, 2, \ldots, m$), $E \subseteq O$ is the set of objects present in arbitrarily many copies in the environment, $syn \subseteq \{(i, j) \mid i, j \in \{0, 1, 2, \ldots, m\}, i \neq j\}$ is the set of links among cells (we call them *synapses*; 0 indicates the environment) such that for $i, j \in \{0, 1, \ldots, m\}$ at most one of $(i, j), (j, i)$ is present in syn, $s_{(i,j)}$ is the *initial state* of the synapse $(i, j) \in syn$, $R_{(i,j)}$ is a finite set of rules of the form $(s, x/y, s')$, for some $s, s' \in K$ and $x, y \in O^*$, associated with the synapse $(i, j) \in syn$, and, finally, $i_o \in \{1, 2, \ldots, m\}$ is the *output* cell.

We note the restriction that there is at most one synapse among two given cells, and the synapse is given as an ordered pair (i, j) with which a state from K is associated. The fact that the pair is ordered does not restrict the communication between the two cells (or between a cell and the environment), because we work here in the general case of antiport rules, specifying simultaneous movements of objects in the two directions of a synapse.

A rule of the form $(s, x/y, s') \in R_{(i,j)}$ is interpreted as an antiport rule $(i, x/y, j)$ as above, acting only if the synapse (i, j) has the state s; the application of the rule means (1) moving the objects specified by x from cell i (from the environment, if $i = 0$) to cell j, at the same time with the move of the objects specified by y in the opposite direction, and (2) changing the state of the synapse from s to s'.

The computation starts with the multisets specified by $w_1, \ldots, w_m$ in the m cells; in each time unit, a rule is used on each synapse for which a rule can be used (if no rule is applicable for a synapse, then no object passes over it and its state remains unchanged). Therefore, the use of rules is sequential at the level of each synapse, but it is parallel at the level of the system: all synapses which can use a rule must do so (the system evolves synchronously). The computation is successful if and only if it halts and the result of a halting computation is the number of objects from T present in cell i_o in the halting configuration (the objects from $O-T$ are ignored when considering the result). The set of all numbers computed in this way by the system Π is denoted by $N(\Pi)$. Of course, we can compute vectors, by considering the multiplicity of objects from T present in cell i_o in the halting configuration.

A still more elaborated class of systems, called *population P systems*, were investigated in a series of papers by F. Bernardini and M. Gheorghe (see, e.g., [10]) with motivations related to the dynamics of cells in skin-like tissues, populations of bacteria, and colonies of ants. These systems are highly dynamical; not only the links between cells, corresponding to the channels from the previous model with states assigned to the channels, can change during the evolution of the system, but also the cells can change their names, can disappear (get dissolved), and can divide, thus producing new cells; these new cells inherit, in a well specified sense, links with the neighboring cells of the parent cell. The generality of this model makes it rather attractive for applications in areas such as those mentioned above, related to tissues, populations of bacteria, etc.

17 Neural-like P Systems

The next step in enlarging the model of tissue-like P systems is to consider more complex cells, for instance, moving the states from the channels between cells to the cells themselves – while still preserving the network of synapses. This suggests the neural motivation of these attempts, aiming to capture something from the intricate structure of neural networks of the way the neurons are linked and cooperate in the human brain.

We do not recall the formal definition of a neural-like P system, but we refer to [67] for details, and here we present only the general idea behind these systems.

We again use a population of cells (each one identified by its label) linked by a specified set of synapses. This time, each cell has at every moment a state from a given finite set of states, contents in the form of a multiset of objects from a given alphabet of objects, and a set of rules for processing these objects.

The rules are of the form $sw \rightarrow s'(x, here)(y, go)(z, out)$, where s, s' are states and w, x, y, z are multisets of objects; in state s, the cell consumes the multiset w and produces the multisets x, y, z; the objects from multiset x remain in the cell, those of multiset y have to be communicated to the cells toward which there are synapses starting in the current cell; a multiset z, with the indication *out*, is allowed to appear only in a special cell, designated as the output cell, and for this cell the use of the previous rule entails sending the objects of z to the environment.

The computation starts with all cells in specified initial states, with initially given contents, and proceeds by processing the multisets from all cells, simultaneously, according to the local rules, redistributing the obtained objects along synapses and sending a result into the environment through the output cell; a result is accepted only when the computation halts.

Because of the use of states, there are several possibilities for processing the multisets of objects from each cell. In the *minimal* mode, a rule is chosen and applied once to the current pair (state, multiset). In the *parallel* mode, a rule is chosen, e.g., $sw \rightarrow s'w'$, and used in the maximally parallel manner: the multiset w is identified in the cell contents in the maximal manner, and the rule is used for processing all these instances of w. Finally, in the *maximal* mode, we apply in the maximally parallel manner all rules of the form $sw \rightarrow s'w'$, that is, with the same states s and s' (note the difference with the parallel mode, where in each step we choose a rule and we use only that rule as many times as possible).

There are also three ways to move the objects between cells (of course, we only move objects produced by rules in multisets with the indication *go*). Assume that we have applied a rule $sw \rightarrow s'(x, here)(y, go)$ in a given cell i. In the *spread* mode, the objects from y are nondeteterministically distributed to all cells j such that (i, j) is a synapse of the system. In the *one* mode, all the objects from y are sent to one cell j, provided that the synapse (i, j) exists. Finally, we can also *replicate* the objects of y, and each object from y is sent to all cells j such that (i, j) is an available synapse.

Note that the states ensure a powerful way to control the work of the system, that the parallel and maximal modes are efficient ways to process the multisets, and that the replicative mode of distributing the objects provides the possibility of increasing exponentially the number of objects in linear time. Altogether, these features make the neural-like P systems both very powerful and very efficient computing devices. However, this class of P systems still waits for a systematic investigation – maybe starting with questioning their very definition, and changing this definition in such a way as to capture more realistic brain-like features.

18 Other Ways of Using a P System; P Automata

In all previous sections we have considered the various types of P systems as *generative devices*: starting from an initial configuration, because of the nondeterminism of using the rules, we can proceed along various computations, at the end of which we get a result; in total, all successful computations provide a set of numbers, of vectors or numbers, or a language (set of strings), depending on the way the result of a computation is defined. This grammar oriented approach is only one possibility, mathematically attractive and theoretically important, but not useful from a practical point of view when dealing with specific problems to be solved and specific functions to be computed. However, a P system can also be used for computing functions and for solving problems (in a standard algorithmic manner).

Actually, besides the generative approach, there are two other general (related) ways of using a P system: in the *accepting* mode and in the *transducer* mode. In both cases, an input is provided to the system in a way depending on the type of systems at hand. For instance, in a symbol object P system, besides the initial multisets present in the regions of the membrane structure, we can introduce a multiset w_0 in a specified region, adding the objects of w_0 to the objects present in that region. The computation proceeds, and if it halts, then we say that the input is accepted (or recognized). In the transducer mode, we have not only to halt, but also to collect an output from a specified output region, internal to the system or the environment.

Now, an important distinction appears between systems which behave deterministically (at each moment at most one transition is possible, that is, either the computation stops, or it continues in a unique mode) and those which work in a nondeterministic way. Such a distinction does not make much sense in the generative mode, especially if only halting computations provide a result at the end: such a system can generate only a single result. In the case of computing functions or solving problems (e.g., decidability problems), the determinism is obligatory.

Again a distinction is in order: actually, we are not interested in the way the system behaves, deterministically or nondeterministically, but in the uniqueness and the reliability of the result. If, for instance, we ask whether or not a propositional formula in conjunctive normal form is satisfiable or not, we do not care how the result is obtained, but we want to make sure that it is the right one. Whether or not the truth assignments were created as in the example from Section 13, expanding the variables in a random order, is not relevant; what is important is that after n steps we get the same configuration. This brings to the stage the important notion of *confluence*. A system is *strongly confluent* if, starting from the initial configuration and behaving in a way which we do not care, it after a while reaches a configuration from where the computation continues in a deterministic way. Because we are only interested in the result of computations (e.g., in the answer, yes or no, to a decidability problem), we can relax the previous condition, to a *weak con-*

fluence property: regardless of how the system works, it always halts and all halting computations provide the same result. These notions will be invoked when discussing the efficiency of P systems, as in Section 20.

Let us consider here in some detail the accepting mode of using a P system. Given, for instance, a transition P system Π, let us denote by $N_a(\Pi)$ the set of all numbers accepted by Π in the following sense: we introduce a^n, for a specified object a, into a specified region of Π, and we say that n is accepted if and only if there is a computation of Π, starting from this augmented initial configuration, which halts. In the case of systems taking objects from the environment, such as the symport/antiport or the communicative ones [77], we can consider that the system accepts/recognizes the sequence of objects taken from the environment during a halting computation (if several objects are brought into the system at the same time, then all their permutations are accepted as substrings of the accepted string). Similar strategies can be followed for all types of systems, tissue-like and neural-like included (but P automata were first introduced in the symport/antiport case in [30]; see also [32]).

The above set $N_a(\Pi)$ was defined in general, for nondeterministic systems, but, clearly, in the accepting mode the determinism can be imposed (the nondeterminism is moved to the environment, to the "user," which provides an input, unique, but nondeterministically chosen, from which the computation starts). Note that the example of a P system with symport/antiport rules from Section 11 works in the same manner for an accepting register machine (a number is introduced in the first register and is accepted if and only if the computation halts); in such a case, the add instructions can be deterministic, that is, with labels l_2, l_3 identical (one simply writes $l_1 : (\texttt{add}(r), l_2)$, with the continuation unique), and for this case the P system itself is deterministic.

19 Universality

The initial goal of membrane computing was to define computability models inspired from the cell biology, and indeed a large part of the investigations in this area was devoted to producing computing devices and examining their computing power, in comparison with the standard models in computability theory, Turing machines and their restricted variants. As it turns out, most of the classes of P systems considered are equal in power to Turing machines. In a rigorous manner, we have to say that they are Turing complete (or computationally complete), but because the proofs are always constructive, starting the constructions used in these proofs from universal Turing machines or from equivalent devices, we obtain universal P systems (able to simulate any other P system of the given type after introducing a "code" of the particular system as an input in the universal one). That is why we speak about *universality* results and not about computational completeness.

All classes of systems considered above, whether cell-like, tissue-like, or neural-like, with symbol objects or string objects, working in the generative or the accepting modes, with certain combinations of features, are known to be universal. The cell turns out to be a very powerful "computer," both when standing alone and in tissues.

In general, for P systems working with symbol objects, these universality results are proved by simulating computing devices known to be universal, and which either work with numbers or do not essentially use the positional information from strings. This is true/possible for register machines, matrix grammars (in the binary normal form), programmed grammars, regularly controlled grammars, and graph-controlled grammars (but not for arbitrary Chomsky grammars and for Turing machines, which can be used only in the case of string objects). The example from Section 11 illustrates a universality proof for the case of P systems with symport/antiport rules (with rules of sufficiently large weight; see below stronger results from this point of view).

We do not enter here into details other than specifying some notations which are already standard in membrane computing and, after that, mentioning some universality results of particular interest.

As for notations, the family of sets $N(\Pi)$ of numbers (we hence use the symbol N) generated by P systems of a specified type (P), working with symbol objects (O), having at most m membranes, and using features/ingredients from a given *list* is denoted by NOP_m(*list-of-features*). If we compute sets of vectors, we write $PsOP_m(\ldots)$, with Ps coming from "Parikh set." When the systems work in the accepting mode, one writes $N_aOP_m(\ldots)$, and when string objects are used, one replaces N with L (from "languages") and O with S (from "strings"), thus obtaining families $LSP_m(\ldots)$. The case of tissue-like systems is indicated by adding the letter t before P, thus obtaining $NOtP_m(\ldots)$, while for neural-like systems one uses instead the letter n. When the number of membranes is not bounded, the subscript m is replaced by $*$, and this is a general convention, used also for other parameters.

Now, the list of features can be taken from an endless pool: cooperative rules are indicated by *coo*; catalytic rules are indicated by *cat*, noting that the number of catalysts matters, and hence we use cat_r in order to indicate that we use systems with at most r catalysts; bistable catalysts are indicated by $2cat$ ($2cat_r$, if at most r catalysts are used); mobile catalysts are indicated by $Mcat$. When using a priority relation, we write *pri*. For the actions δ, τ we write simply δ, τ. Membrane creation is represented by *mcre*; endocytosis and exocytosis operations are indicated by *endo*, *exo*, respectively. In the case of P systems with active membranes, one directly lists the types of rules used, from (a) to (e), as defined and denoted in Section 13.

For systems with string objects, one write *rew*, $repl_d$, and *spl* for indicating that one uses rewriting rules, replicated rewriting rules (with at most d copies of each string produced by replication), and splicing rules, respectively.

In the case of (cell-like or tissue-like) systems using symport/antiport rules, we have to specify the maximal weight of the used rules, and this is done by writing $sym_p, anti_q$, meaning that symport rules of weight at most p and antiport rules of weight at most q are allowed.

There are many other features, with notations of the same type (as mnemonic as possible), which we do not recall here. Sometimes, when it is important to show in the name of the discussed family that a specific feature fe is *not* allowed, one writes nFe – for instance, one writes $nPri$ for not using priorities (note the capitalization of the initial name of the feature), $n\delta$, etc.

Specific examples of families of numbers (we do not consider here sets of vectors or languages, although, as we have said above, a lot of universality results are known for all cases) appear in the few universality results which we recall below. In these results, NRE denotes the family of Turing computable sets of numbers (the notation comes from the fact that these numbers are the length sets of recursively enumerable languages, those generated by Chomsky type-0 grammars or by many types of regulated rewriting grammars and recognized by Turing machines). The family NRE is also the family of sets of numbers generated/recognized by register machines. When dealing with vectors of numbers, hence with the Parikh images of languages (or with the sets of vectors generated/recognized by register machines), we write $PsRE$.

Here are some universality results (for the proofs, see the papers mentioned):

1. $NRE = NOP_1(cat_2)$ [31].
2. $NRE = NOP_3(sym_1, anti_1) = NOP_3(sym_2, anti_0)$ [4].
3. $NRE = NOP_3((a), (b), (c))$ [54].
4. $NRE = NSP_3(repl_2)$ [50].

In all these results, the number of membranes sufficient for obtaining the universality is pretty small. Actually, in all cases when the universality holds (and the code of a particular system is introduced in a universal system in such a way that the membrane structure is not modified), the hierarchy on the number of membranes collapses, because a number of membranes as large as the degree of the universal system suffices.

Still, "the number of membranes matters," as we read already in the title of [43]: there are (sub-universal) classes of P systems for which the number of membranes induces an infinite hierarchy of families of sets of numbers (see also [44]).

20 Solving Computationally Hard Problems in Polynomial Time

The computational power (the "competence") is only one of the important questions to be dealt with when defining a new computing model. The other fundamental question concerns the computing *efficiency*, the resources used for solving problems. In general, the research in natural computing is especially concerned with this issue. Because P systems are parallel computing devices, it is expected that they can solve hard problems in an efficient manner, and this expectation is confirmed for systems provided with ways for producing an exponential workspace in linear time.

We have discussed above three basic ways to construct such an exponential space in cell-like P systems, namely, membrane division (the separation operation has the same effect, as do other operations which replicate partially or totally the contents of a membrane), membrane creation (combined with the creation of exponentially many objects), and string replication. Similar possibilities are offered by cell division in tissue-like systems and by object replication in neural-like systems. Also the possibility to use a pre-computed exponential workspace, unstructured and non-active (e.g., with the regions containing no object) was considered.

In all these cases polynomial or pseudo-polynomial solutions to **NP**-complete problems were obtained. The first problem addressed in this context was SAT [66] (the solution was improved in several respects in other subsequent papers), but similar solutions are reported in the literature for the Hamiltonian Path and the Node Covering problems, the problem of inverting one-way functions, the Subset-sum problem and the Knapsack problem (note that the last two are numerical problems, where the answer is not of the yes/no type as in decidability problems), and for several other problems. Details can be found in [67, 72], as well as in the Web page of the domain [82].

Roughly speaking, the framework for dealing with complexity matters is that of *accepting P systems with input*: a family of P systems of a given type is constructed starting from a given problem, and an instance of the problem is introduced as an input in such systems; working in a deterministic mode (or a *confluent* mode: some nondeterminism is allowed, provided that the branching converges after a while to a unique configuration, or, in the case of weak confluence, all computations stop in a determined time and give the same result) in a given time one of the answers yes/no is obtained in the form of specific objects sent to the environment. The family of systems should be constructed in a uniform mode (starting from the size of problem instances) by a Turing machine working in polynomial time. A more relaxed framework is that where a *semi-uniform* construction is allowed, carried out in polynomial time by a Turing machine, but starting from the instance to be solved (the condition to have a polynomial time construction ensures the "honesty" of the construction: the solution to the problem cannot be found during the construction phase).

This direction of research is very active at the present moment. More and more problems are being considered, the membrane computing complexity classes are being refined, characterizations of the **P**$\neq$**NP** conjecture have been obtained in this framework, and improvements are being looked for. An important recent result concerns the fact that **PSPACE** was shown to be included in $\mathbf{PMC}_D$, the family of problems which can be solved in polynomial time by P systems with the possibility of dividing both elementary and non-elementary membranes. The **PSPACE**-complete problem used in this proof was QSAT (see [77, 5] for details).

There also are many *open problems* in this area. We have mentioned already the intriguing question about whether polynomial solutions to **NP**-complete problems can be obtained through P systems with active membranes without polarizations (and without label changing possibilities of other additional features). In general, the borderline between *efficiency* (the possibility to solve **NP**-complete problems in polynomial time) and *non-efficiency* is a challenging topic. Anyway, we know that membrane division cannot be avoided ("Milano theorem": a P system without membrane division can be simulated by a Turing machine with a polynomial slowdown; see [80, 81]).

21 Focusing on the Evolution

Computational power is of interest to theoretical computer science, and computational efficiency is of interest to practical computer science, but neither is of direct interest to biology. Actually, this last statement is not correct: if a biologist is interested in simulating a cell – and this seems to be a major concern of biology today (see [48, 42] and other sources) – then the generality of the model (its comparison with the Turing machine and its restrictions) is directly linked to the possibility of algorithmically solving questions about the model. An example: is a given configuration reachable from the initial configuration? Imagine that the initial configuration represents a healthy cell and we are interested in knowing whether a sickness state is ever reached. Then, if both healthy and non-healthy configurations can be reached, the question arises whether we can find the "bifurcation configurations," and this is again a reachability issue. The relevance of such a "purely theoretical" problem is clear, and its answer depends directly on the generality (hence the power) of the model. Then, of course, the time needed for answering the question is a matter of computational complexity. So, both the power and the efficiency are, indirectly, of interest also to biologists, so we (the biologists, too) should be more careful when asserting that a given type of "theoretical" investigation is not of interest to biology.

Still, the immediate concern of biological research is the evolution of biological systems, their life, whatever this means, and not the result of a specific evolution. Alternatively stated, halting computations are of interest to computer science, whereas of direct interest to biology is the computation/evolution itself. Although membrane computing was not intended initially to deal with such issues, a series of recent investigations indicate a strong tendency toward considering P systems as dynamical systems. This does not concern only the fact that, besides the rules for object evolution, a more complete panoply of possibilities was imagined for making the membrane structure also evolve, with specific developments in the case of tissue-like and population P systems, where also the links between cells are evolving; but this concerns especially the formulation of questions which are typical for dynamical systems study. Trajectories, periodicity and pseudo-periodicity, stability, attractors, basins, oscillations, and many other concepts were brought in the framework of membrane computing – and the enterprise is not trivial, as these concepts were initially introduced in areas handled by means of continuous mathematics tools (mainly differential equations). A real program of defining discrete dynamical systems, with direct application to the dynamics of P systems, was started by V. Manca and his collaborators; we refer to Chapter 3 for details.

22 Recent Developments

Of course, the specification "recent" is risky, as it can soon become obsolete, but still we want to mention here some directions of research and some results which were not presented before – after repeating the fact that topics such as complexity classes and polynomial solutions to hard problems, dynamical systems approaches, and population P systems (in general, systems dealing with populations of cells, as in tissue-like or neural-like systems) are of strong current interest which will probably lead to significant theoretical and practical results. To these trends we can add another general and yet not very structured topic: using non-crisp mathematics, handling uncertainty by means of probabilistic, fuzzy set, and rough set theories.

However, we want here to also point out a few more precise topics.

One of them concerns the role of time in P systems. The synchronization and the existence of a global clock are too strong assumptions (from a biological point of view). What about P systems where there exist no internal clocks and all rules have different times to be applied? This can mean both that the duration needed by a rule to be applied can differ from the duration of another rule and the extreme possibility that the duration is not known. In the first case, we can have a timing function assigning durations to rules; in the second case even such information is missing. How does the power of a system depend on the timing function? Are there time-free systems, which generate the same set of numbers regardless of what time function associates durations with its rules? Such questions are addressed in a series of papers by M. Cavaliere and D. Sburlan; see, e.g., [22, 23].

Another powerful idea explored by M. Cavaliere and his collaborators is that of coupling a simple bio-inspired *system*, *Sys*, such as a P system without large computing power, with an *observer Obs*, a finite state machine which analyzes the configurations of the system *Sys* through the evolutions; from each configuration either a symbol or nothing (that is, the "result" of that configuration is the empty string λ) is produced; in a stronger variant, the observer can also reject the configuration and hence the system evolution, trashing it. The couple (Sys, Obs), for various simple systems and multiset processing finite automata, proved to be a very powerful computing device, universal even for very weak systems Sys. Details can be found in [19, 20].

An idea recently explored is that of trying to bound the number of objects used in a P system, and still computing all Turing computable numbers. The question can be seen as "orthogonal" to the usual questions concerning the number of membranes and the size of rules, since, intuitively, one of these parameters should be left free in order to codify and handle an arbitrary amount of information by using a limited number of objects. The first results of this type were given in [69] and they are surprising: in formal terms, we have $NRE = NOP_4(obj_3, sym_*, anti_*)$ (P systems with four membranes and symport and antiport rules of arbitrary weight are universal even when using only three objects). In turn, two objects (but without a bound on the number of membranes) are sufficient in order to generate all sets of vectors computed by so-called (see [36]) partially blind counter machines (for sets of numbers the result is not so interesting, because partially blind counter machines accept only semilinear sets of numbers, while the sets of vectors they accept can be non-semilinear).

Other interesting topics recently investigated which we only list here concern the reversibility of computations in P systems [52], energy accounting (associating quanta of energy to objects or to rules handled during the computation) [35, 34, 51], relations with grammar systems and with colonies [68], descriptional complexity, and non-discrete multisets [61, 27].

We close this section by mentioning the notion of *Sevilla carpet* introduced in [25], which proposes a way to describe the time-and-space complexity of a computation in a P system by considering the two-dimensional table of all rules used in each time unit of a computation. This corresponds to the Szilard language from language theory, with the complication now that we use several rules at the same step, and each rule is used several times. Considering all the information concerning the rules, we can get a global evaluation of the complexity of a computation, as illustrated, for instance, in [75] and [37].

23 Closing Remarks

The present chapter should be seen as a general overview of membrane computing, with the choice of topics intended to be as pertinent as possible, but, of course, not completely free of a subjective bias. The reader interested in further technical details, formal definitions, proofs, research topics and open problems, or details concerning the applications (and the software behind them) is advised to consult the relevant chapters of the book, as well as the comprehensive web page from `http://psystems.disco.unimib.it`. A complete bibliography of membrane computing can be found there, with many papers available for downloading (in particular, one can find there the proceedings volumes of the yearly Workshops on Membrane Computing, as well as of the yearly Brainstorming Weeks on Membrane Computing).

References

1. L.M. Adleman: Molecular Computation of Solutions to Combinatorial Problems. *Science*, 226 (November 1994), 1021–1024.
2. B. Alberts, A. Johnson, J. Lewis, M. Raff, K. Roberts, P. Walter: *Molecular Biology of the Cell*, 4th ed. Garland Science, New York, 2002.
3. A. Alhazov, R. Freund: On the Efficiency of P Systems with Active Membranes and Two Polarizations. In [59], 147-161.
4. A. Alhazov, M. Margenstern, V. Rogozhin, Y. Rogozhin, S. Verlan: Communicative P Systems with Minimal Cooperation. In [59], 162–178.
5. A. Alhazov, C. Martín-Vide, L. Pan: Solving a PSPACE-Complete Problem by P Systems with Restricted Active Membranes. *Fundamenta Informaticae*, 58, 2 (2003), 67–77.
6. A. Alhazov, C. Martín-Vide, Gh. Păun, eds.: *Pre-proceedings of Workshop on Membrane Computing*, WMC 2003, Tarragona, Spain, July 2003. Technical Report 28/03, Rovira i Virgili University, Tarragona, 2003.
7. I.I. Ardelean: The Relevance of Biomembranes for P Systems. *Fundamenta Informaticae*, 49, 1–3 (2002), 35–43.
8. J.-P. Banâtre, A. Coutant, D. Le Métayer: A Parallel Machine for Multiset Transformation and Its Programming Style. *Future Generation Computer Systems*, 4 (1988), 133–144.
9. J.-P. Banâtre, P. Fradet, D. Le Métayer: Gamma and the Chemical Reaction Model: Fifteen Years After. In [16], 17–44.
10. F. Bernardini, M. Gheorghe: Population P Systems. *Journal of Universal Computer Science*, 10, 5 (2004), 509–539.
11. F. Bernardini, V. Manca: Dynamical Aspects of P Systems. *BioSystems*, 70, 2 (2003), 85–93.
12. G. Berry, G. Boudol: The Chemical Abstract Machine. *Theoretical Computer Science*, 96 (1992), 217–248.
13. D. Besozzi: *Computational and Modeling Power of P Systems.* PhD Thesis, Univ. degli Studi di Milano, 2004.
14. D. Besozzi, C. Zandron, G. Mauri, N. Sabadini: P Systems with Gemmation of Mobile Membranes. In *Proc. ICTCS 2001*, Torino 2001 (A. Restivo, S.R. Della Rocca, L. Roversi, eds.), LNCS 2202, Springer, Berlin, 2001, 136–153.
15. C. Bonanno, V. Manca: Discrete Dynamics in Biological Models. *Romanian Journal of Information Science and Technology*, 5, 1-2 (2002), 45–67.
16. C.S. Calude, Gh. Păun, G. Rozenberg, A. Salomaa, eds.: *Multiset Processing. Mathematical, Computer Science, and Molecular Computing Points of View. Lecture Notes in Computer Science*, 2235, Springer, Berlin, 2001.
17. L. Cardelli: Brane Calculus. In *Computational Methods in Systems Biology. International Conference CMSB 2004, Paris, France, May 2004, Revised Selected Papers*, LNCS 3082, Springer-Verlag, Berlin, 2005, 257–280.
18. M. Cavaliere: Evolution-Communication P Systems. In [71], 134–145.
19. M. Cavaliere, P. Leupold: Evolution and Observation – A New Way to Look at Membrane Systems. In [57], 70–87.
20. M. Cavaliere, P. Leupold: Evolution and Observation. A Non-standard Way to Generate Formal Languages. *Theoretical Computer Science*, 321, 2-3 (2004), 233–248.

21. M. Cavaliere, C. Martín-Vide, Gh. Păun, eds.: *Proceedings of the Brainstorming Week on Membrane Computing, Tarragona, February 2003*. Technical Report 26/03, Rovira i Virgili University, Tarragona, 2003.
22. M. Cavaliere, D. Sburlan: Time-Independent P Systems. In [59], 239–258.
23. M. Cavaliere, D. Sburlan: Time and Synchronization in Membrane Systems. *Fundamenta Informaticae*, 64 (2005), 65–77.
24. R. Ceterchi, R. Gramatovici, N. Jonoska, K.G. Subramanian: Generating Picture Languages with P Systems. In [21], 85–100.
25. G. Ciobanu, Gh. Păun, Gh. Ştefănescu: Sevilla Carpets Associated with P Systems. In [21], 135–140.
26. L. Colson, N. Jonoska, M. Margenstern: λP Systems and Typed λ-Calculus. In [59], 1–18.
27. A. Cordón-Franco, F. Sancho-Caparrini: Approximating Non-discrete P Systems. In [59], 288–296.
28. S. Crespi-Reghizzi, D. Mandrioli: Commutative Grammars. *Calcolo*, 13, 2 (1976), 173–189.
29. E. Csuhaj-Varjú, J. Kelemen, A. Kelemenová, Gh. Păun, G. Vaszil: Cells in Environment: P Colonies. *Multiple Valued Logic and Soft Computing Journal*, to appear.
30. E. Csuhaj-Varju, G. Vaszil: P Automata or Purely Communicating Accepting P Systems. In [71], 219–233.
31. R. Freund, L. Kari, M. Oswald, P. Sosik: Computationally Universal P Systems Without Priorities: Two Catalysts Are Sufficient. *Theoretical Computer Science*, 330, 2 (2005), 251–266.
32. R. Freund, M. Oswald: A Short Note on Analysing P Systems. *Bulletin of the EATCS*, 78 (2003), 231–236.
33. R. Freund, Gh. Păun, M.J. Pérez-Jiménez: Tissue-Like P Systems with Channel-States. *Brainstorming Week on Membrane Computing*, Sevilla, February 2004, TR 01/04 of Research Group on Natural Computing, Sevilla University, 2004, 206–223, and *Theoretical Computer Science*, 330, 1 (2005), 101–116.
34. P. Frisco: *Theory of Molecular Computing. Splicing and Membrane Systems*. PhD Thesis, Leiden University, The Netherlands, 2004.
35. P. Frisco, S. Ji: Towards a Hierarchy of Info-Energy P Systems. In [71], 302–318.
36. S.A. Greibach: Remarks on Blind and Partially Blind One-Way Multicounter Machines. *Theoretical Computer Science*, 7 (1978), 311–324.
37. M.A. Gutiérrez-Naranjo, M.J. Pérez-Jiménez, A. Riscos-Núñez: On Descriptive Complexity of P Systems. In [59], 321–331.
38. T. Head: Formal Language Theory and DNA: An Analysis of the Generative Capacity of Specific Recombinant Behaviors. *Bulletin of Mathematical Biology*, 49 (1987), 737–759.
39. J. Hartmanis: About the Nature of Computer Science. *Bulletin of the EATCS*, 53 (1994), 170–190.
40. J. Hartmanis: On the Weight of Computation. *Bulletin of the EATCS*, 55 (1995), 136–138.
41. J. Hoffmeyer: Surfaces Inside Surfaces. On the Origin of Agency and Life. *Cybernetics and Human Knowing*, 5, 1 (1998), 33–42.
42. M. Holcombe: Computational Models of Cells and Tissues: Machines, Agents and Fungal Infection. *Briefings in Bioinformatics*, 2, 3 (2001), 271–278.
43. O.H. Ibarra: The Number of Membranes Matters. In [57], 218–231.

44. O.H. Ibarra: On Membrane Hierarchy in P Systems. *Theoretical Computer Science*, 334, 1-3 (2005), 115–129.
45. O.H. Ibarra: On Determinism Versus Nondeterminism in P Systems. *Theoretical Computer Science*, to appear. Available at `http://psystems.disco.unimib.it`.
46. O.H. Ibarra, H.-C.Yen, Z. Dang: The Power of Maximal Parallelism in P Systems. *Proceedings of the Eighth Conference on Developments in Language Theory*, Auckland, New Zealand, 2004 (C.S. Calude, E. Calude, M.J. Dinneed, eds.), LNCS 3340, Springer, Berlin, 2004, 212–224.
47. M. Ionescu, T.-O. Ishdorj: Replicative-Distribution Rules in P Systems with Active Membranes. *Proc. of ICTAC2004, First Intern. Colloq. on Theoretical Aspects of Computing*, Guiyang, China, 2004.
48. H. Kitano: Computational Systems Biology. *Nature*, 420, 14 (2002), 206–210.
49. S.N. Krishna, R. Rama: P Systems with Replicated Rewriting. *Journal of Automata, Languages and Combinatorics*, 6, 3 (2001), 345–350.
50. S.N. Krishna, R. Rama, H. Ramesh: Further Results on Contextual and Rewriting P Systems. *Fundamenta Informaticae*, 64 (2005), 235–246.
51. A. Leporati, C. Zandron, G. Mauri. Simulating the Fredkin Gate with Energy-Based P systems. *Journal of Universal Computer Science*, 10, 5 (2004), 600–619.
52. A. Leporati, C. Zandron, G. Mauri: Universal Families of Reversible P Systems. *Proc. Conf. Universal Machines and Computations 2004*, St.Petersburg, 2004 (M. Margenstern, ed.), LNCS 3354, Springer, Berlin, 2005, 257–268.
53. W.R. Loewenstein: *The Touchstone of Life. Molecular Information, Cell Communication, and the Foundations of Life.* Oxford University Press, New York, Oxford, 1999.
54. M. Madhu, K. Krithivasan: Improved Results About the Universality of P Systems. *Bulletin of the EATCS*, 76 (2002), 162–168.
55. V. Manca: String Rewriting and Metabolism. A Logical Perspective. In *Computing with Bio-molecules. Theory and Experiments* (Gh. Păun, ed.), Springer, Singapore, 1998, 36–60.
56. S. Marcus: Bridging P Systems and Genomics: A Preliminary Approach. In [71], 371–376.
57. C. Martín-Vide, G. Mauri, Gh. Păun, G. Rozenberg, A. Salomaa, eds.: *Membrane Computing. International Workshop, WMC2003, Tarragona, Spain, Revised Papers. Lecture Notes in Computer Science*, 2933, Springer, Berlin, 2004.
58. C. Martín-Vide, Gh. Păun, J. Pazos, A. Rodríguez-Patón: Tissue P Systems. *Theoretical Computer Science*, 296, 2 (2003), 295–326.
59. G. Mauri, Gh. Păun, M.J. Pérez-Jiménez, G. Rozenberg, A. Salomaa, eds.: *Membrane Computing. International Workshop WMC5, Milan, Italy, 2004. Revised Papers, Lecture Notes in Computer Science*, 3365, Springer, Berlin, 2005.
60. M. Minsky: *Computation – Finite and Infinite Machines.* Prentice Hall, Englewood Cliffs, NJ, 1967.
61. T.Y. Nishida: Simulations of Photosynthesis by a K-subset Transforming System with Membranes. *Fundamenta Informaticae*, 49, 1-3 (2002), 249–259.
62. A. Păun, Gh. Păun: The Power of Communication: P Systems with Symport/Antiport. *New Generation Computing*, 20, 3 (2002), 295–306.
63. Gh. Păun: *Marcus Contextual Grammars.* Kluwer, Dordrecht, 1997.
64. Gh. Păun: Computing with Membranes. *Journal of Computer and System Sciences*, 61, 1 (2000), 108–143 (and Turku Center for Computer Science–TUCS Report 208, November 1998, `www.tucs.fi`).

65. Gh. Păun: From Cells to Computers: Computing with Membranes (P Systems). *BioSystems*, 59, 3 (2001), 139–158.
66. Gh. Păun: P Systems with Active Membranes: Attacking NP-Complete Problems. *Journal of Automata, Languages and Combinatorics*, 6, 1 (2001), 75–90.
67. Gh. Păun: *Membrane Computing. An Introduction.* Springer, Berlin, 2002.
68. Gh. Păun: Grammar Systems vs. Membrane Computing: A Preliminary Approach. *Workshop on Grammar Systems*, MTA SZTAKI, Budapest, 2004, 225–245.
69. Gh. Păun, J. Pazos, M.J. Pérez-Jiménez, A. Rodríguez-Patón: Symport/Antiport P Systems with Three Objects Are Universal. *Fundamenta Informaticae*, 64 (2005), 345–358.
70. Gh. Păun, G. Rozenberg, A. Salomaa: *DNA Computing. New Computing Paradigms.* Springer, Berlin, 1998.
71. Gh. Păun, G. Rozenberg, A. Salomaa, C. Zandron, eds.: *Membrane Computing. International Workshop, WMC-CdeA 2002, Curtea de Argeş, Romania, Revised Papers. Lecture Notes in Computer Science*, 2597, Springer, Berlin, 2003.
72. M. Pérez-Jiménez, A. Romero-Jiménez, F. Sancho-Caparrini: *Teoría de la Complejidad en Modelos de Computatión Celular con Membranas.* Editorial Kronos, Sevilla, 2002.
73. B. Petreska, C. Teuscher: A Hardware Membrane System. In [57], 269–285.
74. A. Regev, E.M. Panina, W. Silverman, L. Cardelli, E. Shapiro: BioAmbients – An Abstraction for Biological Compartments. *Theoretical Computer Science*, 325 (2004), 141–167.
75. A. Riscos-Núñez: *Programacion celular. Resolucion eficiente de problemas numericos NP-complete.* PhD Thesis, Univ. Sevilla, 2004.
76. P. Sosik: The Computational Power of Cell Division in P Systems: Beating Down Parallel Computers? *Natural Computing*, 2, 3 (2003), 287–298.
77. P. Sosik, J. Matysek: Membrane Computing: When Communication Is Enough. In *Unconventional Models of Computation 2002* (C.S. Calude, M.J. Dinneen, F. Peper, eds.), LNCS 2509, Springer, Berlin, 2002, 264–275.
78. M. Tomita: Whole-Cell Simulation: A Grand Challenge of the 21st Century. *Trends in Biotechnology*, 19 (2001), 205–210.
79. G. Vaszil: On the Size of P Systems with Minimal Symport/Antiport. In *Pre-Proceedings of Workshop on Membrane Computing, WMC5, Milano, Italy*, June 2004, 422–431.
80. C. Zandron: *A Model for Molecular Computing: Membrane Systems.* PhD Thesis, Univ. degli Studi di Milano, 2001.
81. C. Zandron, C. Ferretti, G. Mauri: Solving NP-Complete Problems Using P Systems with Active Membranes. In *Unconventional Models of Computation* (I. Antoniou, C.S. Calude, M.J. Dinneen, eds.), Springer, London, 2000, 289–301.
82. The Web Page of Membrane Computing: `http://psystems.disco.unimib.it`.

Chapter 2
P System Models for Mechanosensitive Channels

Ioan I. Ardelean[1], Daniela Besozzi*[2], Max H. Garzon[3], Giancarlo Mauri[4], Sujoy Roy[3]

[1] Institute of Biology of the Romanian Academy
Centre of Microbiology, Splaiul Independenţei 296
PO Box 56–53, Bucharest 060031, Romania
ioan.ardelean@ibiol.ro

[2] Università degli Studi di Milano
Dipartimento di Informatica e Comunicazione
Via Comelico 39, 20135 Milano, Italy
besozzi@dico.unimi.it

[3] The University of Memphis, Computer Science
Memphis, TN 38152, U.S.A.
mgarzon/sujoyroy@memphis.edu

[4] Università degli Studi di Milano Bicocca
Dipartimento di Informatica, Sistemistica e Comunicazione
Via Bicocca degli Arcimboldi 8, 20126 Milano, Italy
mauri@disco.unimib.it

Summary. The activity of mechanosensitive channels of large conductance (MscL for short) in cellular membranes is modeled within the framework of P systems. These channels are gated by changes in the pressure exerted against the membrane, which may be due to natural environmental conditions (e.g., rain falling) or to a suction applied by artificial patch clamping. In correspondence to these distinct situations, we present two models for the description of MscL activation and functioning. We present simulations *in silico* of one model and show the emergent behavior of fundamental quantitites. Finally, we discuss several topics for further extensions and development of these models.

1 Introduction

The aim of this work is to define possible models for simulating the activity of a particular kind of transmembrane protein channel present in bacterial cells by using the framework of P systems.

* Corresponding author

It should be emphasized that membrane systems were not initially intended to be a model of the cell; instead, their purpose was to investigate some computational features which can be abstracted from the cellular biology. We stress the fact that the interest is now focused only on the modeling power of P systems; hence we will not address the analysis of theoretical computations. For this aspect, the interested reader is referred to [37].

Several other scientists from different disciplines have recently researched biological simulators and theoretical models. For instance, the E-CELL project [24, 45, 46] was launched in 1996 with the aim of modeling and simulating intracellular processes, such as metabolic pathways, protein synthesis, and membrane transport, and to predict the dynamic behavior of living cells. Another promising software environment is Virtual Cell [29, 39], which enables the construction of compartmental models (with a topological arrangement of compartments and membranes, the molecules inside each compartment, and the reactions between these molecules). For the generation of simulations, both softwares are based on the numerical integration of sets of differential equations.

The approach we propose with this work is totally different, since in our opinion discrete mathematics may be more appropriate than continuous mathematics for describing noncontinuous molecular events (such as channels opening and closure), as already suggested in [1]. This is linked to the intrinsic power and suitability of P systems to elaborate models of specific biological processes, since in any P system one finds a compartmentalized membrane structure, similar to the architecture of the cell, the transformation of elements into different ones and their communication through membranes,highly parallelized processing, and the possibility of adding useful features such as probabilities, promoters, inhibitors, and many others.

It is natural to ask whether the introduction of (discrete) mathematics into biology, as exemplified by the emergence of membrane computing and its cross breeding with biology, could be as important and fruitful for biology as the introduction of instruments and concepts from physics and chemistry since the 19th century. This permanent import flourished into molecular biology. As for the use of mathematics, the situation started to significantly change in the second half of the 20th century with the seminal work of mathematicians who beneficially explored different domains of biology. This work is also an invitation for comments and feedback, as well as for criticism, from both mathematicians and biologists (especially those working with mechanosensitive channels in bacteria), with the aim of benefiting both domains.

Our attempt in using the framework of P systems for modeling cellular structures consists of the definition of two minimal models for the activity of mechanosensitive channels (see also [2, 6]); hence we will not describe any chemical process occurring inside the bacterium. Mechanosensitive channels are protein-based channels gated by mechanical forces; in this chapter, we focus our analysis only on the mechanosensitive channel of large conductance (MscL for short). In Gram-positive and Gram-negative bacteria, MscL is located in the cell membrane. This location in bacteria can be correlated to its physiological function, the protection against severe osmotic downshifts. The major role of these channels under osmotic downshift is to allow the rapid exit of different chemicals, and hence the sudden decrease of the osmotic pressure inside the cell. This event is fundamental for the bacterial cell because, when the difference between osmotic pressure inside and outside the cell is too large, the integrity of the cell can be damaged by disruption of the cell wall and the plasma membrane, followed by cell death [9, 30, 41].

With respect to osmotic pressure, we will focus only on the *hypotonic shock* (osmotic downshift) which is a decrease in the concentration (osmolarity, more precisely) of the extracellular environment [48]. This is an event that happens both in laboratory cultures (when the scientist is adding pure water to the growing medium of a given cell culture) and in natural media when, e.g., a rainfall dilutes a rather small pond.

The large amount of knowledge accumulated so far about MscL opens the possibility of defining P models of their biological dynamics. The availability of data ranging from molecular structure of the protein channel to its functioning under different conditions (cells of wild type or mutant type, addition of substances which modify the response of channel opening) enables the design of mathematical models of a *realistic* type (that is to say models that fit the biological data obtained by laboratory experiments, and vice versa).

We define two distinct models, corresponding to (*in vitro*) patch clamping experiments and (*in vivo*) hypotonic shocks. These systems consist of some basic components: an environment, a region and a membrane tension, which naturally correspond to essential aspects of MscL activity. We introduce probabilities associated with evolution rules in order to achieve a closer resemblance to biological reality. Moreover, we define evolution rules according to different environmental events (a suction applied to a patch membrane or water addition). In both models, objects are never modified by evolution rules, instead they are only exchanged between the internal region and the environment. In addition, by using different target indications, we distinguish between the direct passage of water through the cellular membrane (which happens by osmosis) and the different passage of chemicals which, in contrast, need the opening of mechanically gated channels.

After the definition of the models, we present the simulation environment which has been used to produce simulations *in silico* of an *in vitro* P model and obtain predictions about observable quantities that could be used to test its soundness. We show the emergent behavior of bio-physical quantities (not explicitly programmed in the simulations) which appear to be in line with the observed biological phenomena.

The models we present could be enlarged to also include other processes intervening during osmotic shocks (see Section 9.4), but for the moment we present only the formal simulation of the activity of MscL, and we base our investigation on biological data and results obtained from *E. coli*. However, the models we introduce are general enough to cover distinct conditions for the functioning of MscL in other prokaryotes.

This work is a largely modified and extended version of [6]. In Section 2 we report the biological notions concerning MscL in bacteria and the way they are gated; this data will be used in the subsequent sections as the basis for the definition of P models. In Section 3 we give a few basic notions and notations about P systems, and we define the membrane system elements which will be used in the sequel. In Section 4 and 5 we present the two models mentioned above. In Section 6 we share some notes about *in silico* simulations of the *in vitro* model; we describe the software environment used and show some results obtained, such as the emergent behavior over time of membrane tension, conductance, and the current of channels. In Section 7 we summarize some general properties of the models and we describe the living biological conditions of other prokaryotes, which could be easily modeled by P systems. In Section 8 we suggest some possible behaviors of mechanosensitive channels during subsequent activation cycles. Finally, in Section 9 we focus on some final remarks and future extensions, in particular the analysis of the relations between *in vivo* and *in vitro* models, the localization of multiple occurrences of MscL upon a patch membrane, and the possibility of including in the models the effect of inhibitor or activator substances.

2 Mechanosensitive Channels in *E. Coli*

Mechanosensitive channels (Msc for short), discovered in 1984 in animal cells, are protein-based channels gated by mechanical forces. In the last decade the study of Msc in Gram-negative and Gram-positive bacteria has significantly increased [43, 41, 21]. The first cloned and reconstituted Msc [9] is the one of *large conductance* (MscL) from the bacterium *Escherichia coli*. Mechanosensitive channels with *small* (MscS) and *mini* (MscM) conductances also exist in bacteria [32, 41]; the three types of Msc differ with respect to changes in the conductivity of the membrane when they are opening [41], as well as in the pressure gradient across the patch membrane they need to activate.

In Gram-positive and Gram-negative bacteria, MscL is located in the cell membrane. This location in bacteria can be correlated to its physiological function, the protection against severe osmotic downshifts. The major role of Msc under osmotic downshift is to allow the rapid exit of different chemicals (ions and very small molecules), and hence the sudden decrease of the osmotic pressure inside the cell. Thus, by the opening of the Msc the osmotic pressure inside the cell approaches the osmotic values of the extracellular medium. This event is fundamental for a bacterial cell because, when the difference between osmotic pressure inside the cell and osmotic pressure outside the cell (the so-called *turgor pressure*) is too large, the integrity of the cell can be damaged by the disruption of the cell wall and the plasma membrane, followed by cell death [9, 30, 41]. It is important to realize that the turgor pressure provides the force that expands the cell wall, and it is necessary for the growth of the cell wall and for cell division; that is why cells have developed mechanisms by which the turgor pressure is maintained relatively constant (see reviews [48, 21]).

Under steady state conditions, the retaining force of the peptidoglycan wall in bacteria balances the cell turgor, which for *E. coli* is about 4 atm. During hypotonic shock, water will rapidly enter the cell to balance the upset equilibrium, thus increasing the turgor pressure. In the absence of the mechanisms to reduce turgor, an osmotic shift equivalent, e.g., to 0.3M salt would cause the turgor pressure of *E. coli* to rise by about 11 atm. Cell integrity will be retained as long as the elastic limit of the cell wall and cell membrane is not reached. The proposed hypothesis is that damage to the cell wall and cell membrane is avoided by rapid activation of the mechanosensitive channels. The rapid release of solutes decreases the osmotic potential of the cytoplasm, thus decreasing the driving force for water entry.

In patch clamping experiments, MscL can be activated by pressure gradients across the patches of bacterial spheroplasts (obtained from bacterial cells by the chemical elimination of the cell wall, and also of the external membrane in case of Gram-negative bacteria); purified MscL reconstituted into phospholipid liposomes have produced similar currents. This indicates that MscL can be gated directly by tension transmitted via the lipid bilayer alone [44]. Although these results implicate Msc in bacterial osmoregulation, what was lacking was a bacterial phenotype that would support such a role *in vivo* [21]. Indeed, Blount et al. [8] identified such a phenotype by its inability to grow in media with normal concentration of nutrients. They demonstrated the crucial role of MscL in *E. coli* in the adaptation to large osmotic downshift, and suggested that if the normally tight regulation of MscL gating is disrupted, then cell growth can be severely inhibited. Patch clamp studies in native bacterial membrane substantiate the hypothesis that the mutant channel they obtained is more sensitive to mechanical stress than the native MscL channel. This is why during an hypotonic shock more cytoplasm potassium ions are released from cells expressing the mutant than from the cells expressing the wild type of MscL. This was the first time *in vivo* measurements could be correlated with *in vitro* measurements. For a discussion about the relations between *in vivo* and *in vitro* aspects, see Section 9.1.

2.1 Structure of MscL in *E. Coli*

In this section we report data about MscL in *E. coli*, obtained by patch clamping experiments. The functioning of MscL deduced by means of these data will later be used to define two models which we claim to be valid also for describing the activity of MscL in different prokaryotes.

The MscL protein comprises 136 amino acid residues and its molecular mass is 15 Kda. The secondary structure includes two alpha helical transmembrane domains (`M1` and `M2`) connected by a periplasmic loop that can be divided in two segments S1 and S2. Both the N terminus and the C terminus (by convention they represent the beginning and the end of the amino acid chain, respectively) of MscL are located within the cytoplasm (see [21]). In lipid bilayers MscL protein forms homopentameric or homohexameric channels (for more discussions about the dispute concerning the number of molecules of MscL protein self assembled in artificial membranes see [21, 42]). In this work, we assume a homopentameric structure for the channels, though the models given in next sections can be easily transformed to describe any other channel structure.

According to a model proposed by Sukharev et al. [42, 44], when the MscL is subject to a mechanical stretch, it experiences an increase in the membrane tension, which causes its progression from the initial closed conformation (**C**) to an expanded (but still closed) conformation (**CE**). Then, the first subconducting open state (**SO1**) occurs when one section of the homopentameric structure breaks away. Then, the second subconducting open state (**SO2**) occurs when another section of the homopentameric structure breaks away. The remaining subunits disrupt quickly one by one, thus leading to the fully open state (**O**), which shows short and intermittent subconductances when individual sections of the gate partly occlude the pore. In these conformations, the channel is open but the diameter of the pore can be different, in a range $(0, D]$, where D corresponds to the diameter when the channel is fully open. The channel outer diameter increases primarily during the transition from the closed to the first subconducting conformation (in the full open state the pore outer diameter is 6 nm and inner diameter is 3.6 nm). The transitions among the low subconducting states and the fully open state are relatively independent of tension, which means that the major increase in conductance does not involve substantial changes in the outer dimension of the channel.

The occurrence of multiple conducting states having different pore inner diameters and different conductivities are intensively studied by biologists, and this is the reason why we will consider these multiple functional substates and their flickering among each other.

We list the conformations and the relative notations which will be used (see Figure 1, which is based on a model proposed in [42]):

- the closed conformation, denoted by **C**;
- the expanded closed conformation, denoted by **CE**;
- the subconducting open conformations, denoted by **SOk**, where k subunits (out of five) are open, for $k = 1, 2, 3, 4$;
- the fully open conformation, denoted by **O** (where all five subunits are open).

By using the values reported in [42] we have derived the following conductivity values of the subconducting and open states:

- the conductivity of the state **SO1** is $0.25 \cdot 3.5$ nS, that is, 0.875 nS;
- the conductivity of the state **SO2** is $0.56 \cdot 3.5$ nS, that is, 1.96 nS;

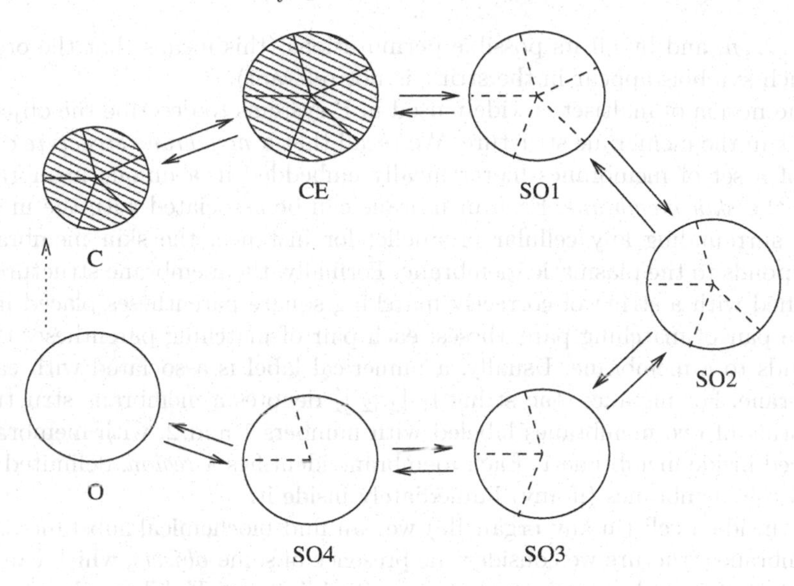

Fig. 1. Activity of mechanosensitive channels: from the closed conformation to the fully open conformation.

- the conductivity of the state **SO3** is $0.74 \cdot 3.5$ nS, that is, 2.59 nS;
- the conductivity of the state **SO4** is $0.89 \cdot 3.5$ nS, that is, 3.115 nS;
- the conductivity of the state **O** is 3.5 nS.

3 P Systems Prerequisites

In this section we recall a few notions and notations from Membrane Computing; for further details and results we refer you to [37].

3.1 Basic Notions and Notations

An alphabet is a finite (nonempty) set of symbols. Given an alphabet V, we denote by V^* the set of all possible strings over V, including the empty string, which is denoted by λ. By $V^+ = V^* \setminus \{\lambda\}$ we denote the set of nonempty strings over V. The length of a string $x \in V^*$, that is, the number of symbols appearing in x, is denoted by $|x|$. For each symbol $a \in V$, $|x|_a$ denotes the number of occurrences of a in x. The set of symbols from V occurring in a string x is denoted by $alph(x)$.

A *multiset* over a set V is a map $M : V \to \mathbf{N}$, where $M(a)$ denotes the multiplicity of the symbol $a \in V$ and $\mathbf{N}$ is the set of natural numbers. If the set V is finite, e.g., $V = \{a_1, \ldots, a_n\}$, then the multiset M can be explicitly represented by the string $w = a_1^{M(a_1)} a_2^{M(a_2)} \ldots a_n^{M(a_n)}$, with $M(a_i) \neq 0$ for all

$i = 1, \ldots, n$, and by all its possible permutations (this means that the order in which symbols appear in the string is not relevant).

The notion of multiset is widely used in P systems to describe the objects present in the membrane structure. We recall that a *membrane structure* consists of a set of membranes hierarchically embedded in a unique membrane called the *skin membrane*. Each membrane can be associated with the membrane surrounding any cellular organelle; for instance, the skin membrane corresponds to the plasmatic membrane. Formally, the membrane structure is identified with a string of correctly matching square parentheses placed in a unique pair of matching parentheses; each pair of matching parentheses corresponds to a membrane. Usually, a numerical label is associated with each membrane. For instance, the string $[_1\ [_2\ ]_2\ ]_1$ denotes a membrane structure consisting of two membranes labeled with numbers 1 and 2, with membrane 2 placed inside membrane 1. Each membrane identifies a *region*, delimited by it and the membranes (if any) immediately inside it.

As inside a cell (in any organelle) we can find biochemical substances, in a membrane structure we consider the presence of some *objects*, which can be multisets of symbols or strings over a specified finite set V. These objects are modified by means of *evolution rules*, which are multiset rewriting rules with target indications associated with the newly introduced objects, of the form $tar \in \{here, out, in\}$. For instance, the rule $a \rightarrow (b, out)$ means that the object a appearing on the left hand side of the rule is rewritten (transformed) into a different symbol b appearing on the right hand side of the rule, and b exits the membrane where it was initially placed. Hence, the rule specifies both the transformation and the transportation processes. Specifically, the target indication determines the region where the object will be communicated after the application of the rule: if $tar = here$, then the object remains in the same region; if $tar = out$, then the object exits from the region where it was placed; if $tar = in$, then the object nondeterministically enters one of the membranes immediately inside the region where the rule is applied, if any inner region exists (otherwise the rule cannot be applied).

At any given time, the membrane structure together with all multisets of objects associated with the regions represent the *configuration* of the system at that time. A transition from one configuration to another is obtained by letting all objects in all regions evolve in a nondeterministically and maximally parallel way; this means that all objects which can be transformed and communicated must evolve, with a simultaneous process involving all membranes and all objects inside the membranes at the same time.

3.2 P Systems and Mechanosensitive Channels

In the following discussion, we will call the model corresponding to patch clamping *in vitro*, and the model where the bacterium is suddenly put in a more diluted environment (in both events of natural or artificial addition of water) *in vivo*.

For the simulation of an activation cycle of MscL we will need to define some basic components of P systems (the membrane structure, multisets of objects, evolution rules) and *ad hoc* additional components, parameters, and particular evolution rules (depending on patch pressure in the *in vitro* model and on water addition in the *in vivo* model).

As said before, MscL act as transmembrane mechanoelectrical switches, opening in response to lipid bilayer stretches and deformations [30] and converting mechanical stresses of the membrane into gating transitions. The channel open probability, as well as the dynamic of closed-to-open transition, are functions of the *membrane tension* (see [32, 41]), an essential parameter that will be considered in both models. In particular, in the following P systems the tension will be described by means of multiple (variable) labels attached to the membrane. Hence, the evolution rules will not only intervene in the transformation and communication of objects, but also in the modification of the label, which is to be interpreted as a key descriptor of the channel status. This is a new interpretation of the membrane label, which is now a fundamental component of the system used to describe a biological significant counterpart (the status of the channel, in this case), and not just an identifier of a couple of square parentheses.

As for the membrane structure to be used, the difference between Gram-positive and Gram-negative bacteria should be pointed out: they are both enclosed by a cell wall and a cell membrane (where MscL are embedded), but in Gram-negative bacteria there is also an external membrane, delimiting a space between itself and the cell membrane (see [1] and the references herein). Since MscL channels are placed in the cell membrane appearing in both kinds of bacteria, we will always consider a membrane structure consisting of a single membrane.

For the sake of completeness, we will also give a description of the solutions inside and outside the membrane, that is we will consider an external *environment* (*Env*) and an inner *region* (*Reg*), which can be briefly defined as follows:

1. the environment is made of solutes (symbols from a given alphabet V_{chem}) and water molecules (each denoted by a special symbol w not appearing in V_{chem});
2. the internal region consists of objects over the same alphabet of the environment and, for simplicity, we assume that no other processes take place inside it.

In what follows, the semi-bracket notation introduced in [5, 4] (with the new concept that rules are locally extensible to the regions around a specified boundary, that is, they are "able to see" around the membrane boundary) will be used to denote the membrane, here labeled with the tension parameter t, which separates the environment and the internal region:

$$Env\ [_t\ Reg.$$

We stress the fact that in the *in vitro* experiments the activation of MscL is determined by the negative pressure artificially applied to the membrane; hence the external and internal solutions do not play any role in the opening mechanism. As we will see later, these two components will play a major role in the *in vivo* model, so they will be defined with more details.

Finally, we give the definition of an activation cycle of MscL, which will be later described in terms of P system models.

Definition 1. A *cycle* is the transition of the MscL from the initial closed state to the open state, passing through the (closed) expanded and subconducting (open) states, and returning finally to the closed state.

4 A P Model for *In Vitro* Activation of MscL

To understand the definition of the P model for *in vitro* activation of MscL, some preliminary observations about membrane tension and patch pressure should be made. The reader is referred to [21] for further details.

The activation of MscL by the negative pressure (i.e., suction) applied to the patch clamp pipette can be described by a Boltzmann distribution function for the channel open probability, which is shown to increase as the suction increases. The applied suction can be also related to the membrane tension (in Section 9.2 we show how the linear dependence of membrane tension to patch pressure can be derived by making plausible discretizations). These notions are also related to the possibility of using the following model to predict the localization of the channels upon the small piece of membrane analyzed during patch clamping experiments (for a detailed explanation see Section 9.2).

In the *in vitro* P model, one fundamental component is a variable membrane parameter called *tension*, which is formally represented as the labeling of the membrane structure. The parameter tension may assume real positive values in the finite set $Tension = \{t_C, t_{CE}, t_{SO1}, t_{SO2}, t_{SO3}, t_{SO4}, t_O, t_L\}$. Each element in this set is used to denote, as follows, the particular value (or the interval of values) of membrane tension and of the corresponding channel status:

- t_C corresponds to the steady state value of the membrane tension (the setting where no patch pressure is applied), as well as to the value at the end of the cycle;
- t_{CE} corresponds to the value that the membrane reaches during the phase of MscL expansion, while the channel remains still closed;
- $t_{SO1}, t_{SO2}, t_{SO3}, t_{SO4}$ correspond to the values of the membrane tension when the channel is partly open and still partly occluded;
- t_O corresponds to the value of the membrane tension when the MscL is fully open;

- t_L corresponds to the value of the membrane tension near or equal to the lytic threshold; whenever reached, the membrane will break and the cell will die.

Transitions among tension values are due to the changes in the pressure applied to the patch membrane (see Figure 2), according to the linear dependence of membrane tension to patch pressure.

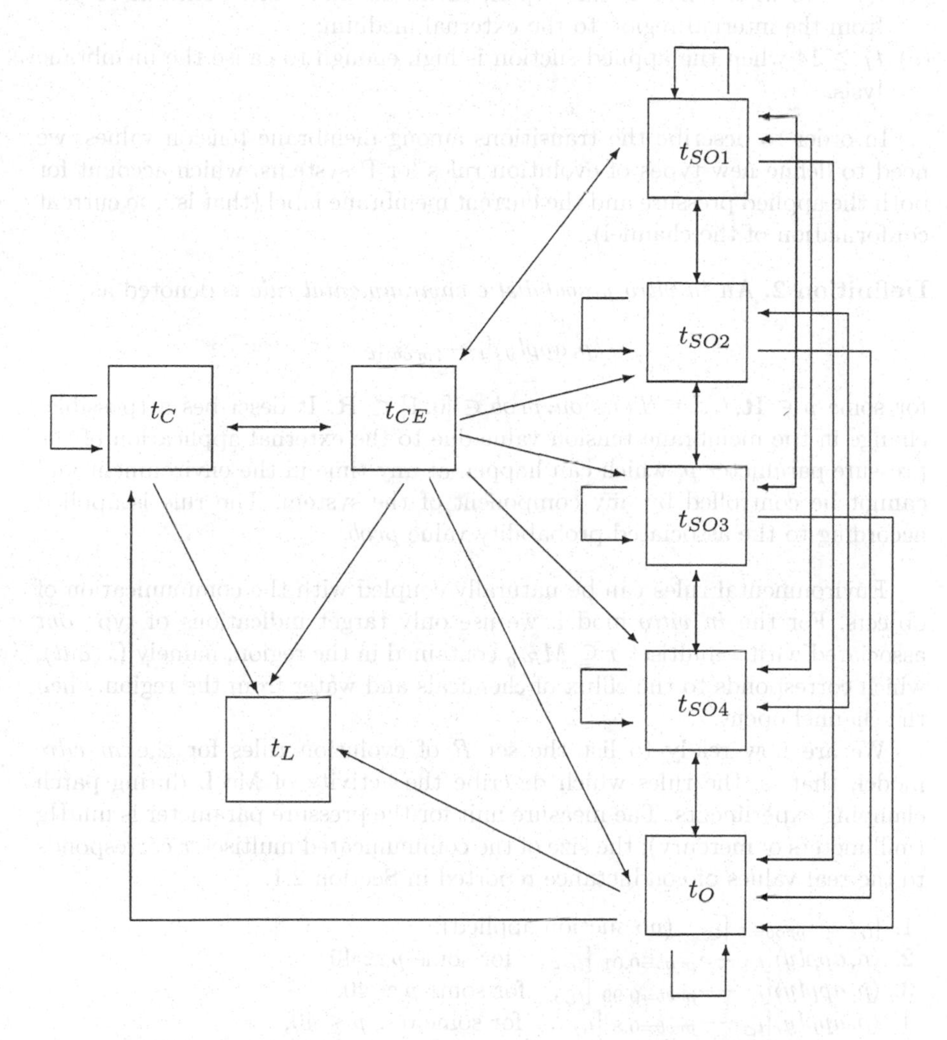

Fig. 2. Membrane tension transitions for *in vitro* model.

According to data collected from *in vitro* experiments on *E. coli* [44], we might consider the following real values (or intervals of values) for the membrane tension (measured in dyne/cm):

(*a*) $t_C \in [0, 10)$ when no suction is applied to the patch membrane;
(*b*) $t_{CE} = 10$ when a suction is applied to the patch membrane; the membrane tension increases and MscL is in the closed expanded substate;
(*c*) $t_{SO1}, t_{SO2}, t_{SO3}, t_{SO4} \in (10, 13)$ when the channel is partly open (solutes and water pass from the internal region to the external medium) and shows a flickering through subconducting states;
(*d*) $t_O = 13$ when MscL is fully open; chemicals and water continue to pass from the internal region to the external medium;
(*e*) $t_L \geq 14$ when the applied suction is high enough to cause the membrane lysis.

In order to describe the transitions among membrane tension values, we need to define new types of evolution rules for P systems, which account for both the applied pressure and the current membrane label (that is, the current conformation of the channel).

Definition 2. An *in vitro probabilistic environmental rule* is denoted as

$$\langle p, apply \rangle [_t \longrightarrow_{prob} [_{t'}$$

for some $p \in \mathbf{R}, t, t' \in Tension, prob \in [0, 1] \subset \mathbf{R}$. It describes a (possible) change in the membrane tension value due to the external application of the pressure parameter p, which can happen at any time in the environment and cannot be controlled by any component of the system. The rule is applied according to the associated probability value *prob*.

Environmental rules can be naturally coupled with the communication of objects. For the *in vitro* model, we use only target indications of type *out* associated with a multiset $x \subseteq M_{Reg}$ contained in the region, namely (x, out), which corresponds to the efflux of chemicals and water from the region when the channel opens.

We are now ready to list the set R of evolution rules for the *in vitro* model, that is, the rules which describe the activity of MscL during patch clamping experiments. The measure unit for the pressure parameter is mmHg (millimeters of mercury); the size of the communicated multiset x corresponds to the real values of conductance reported in Section 2.1.

1. $[_{t_C} \longrightarrow_{prob=1} [_{t_C}$ (no suction applied),
2. $\langle p, apply \rangle [_{t_C} \longrightarrow_{prob=0.01} [_{t_{CE}},$ for some $p \simeq 40$,
3. $\langle p, apply \rangle [_{t_C} \longrightarrow_{prob=0.99} [_{t_C},$ for some $p \simeq 40$,
4. $\langle p, apply \rangle [_{t_C} \longrightarrow_{prob=0.8} [_{t_{CE}},$ for some $0 < p \leq 40$,
5. $\langle p, apply \rangle [_{t_C} \longrightarrow_{prob=0.2} [_{t_C},$ for some $0 < p \leq 40$,
6. $\langle p, apply \rangle [_{t_{CE}} \longrightarrow_{prob=0.05} [_{t_C},$ for some $p \cong 40$,
7. $\langle p, apply \rangle [_{t_{CE}} \longrightarrow_{prob=0.5} [_{t_{SO1}},$ for some $p \cong 40$,
8. $\langle p, apply \rangle [_{t_{CE}} \longrightarrow_{prob=0.20} [_{t_{SO2}},$ for some $p \cong 40$,
9. $\langle p, apply \rangle [_{t_{CE}} \longrightarrow_{prob=0.15} [_{t_{SO3}},$ for some $p \cong 40$,
10. $\langle p, apply \rangle [_{t_{CE}} \longrightarrow_{prob=0.07} [_{t_{SO4}},$ for some $p \cong 40$,

11. $\langle p, apply\rangle [_{t_{CE}} \longrightarrow_{prob=0.03} [_{t_O}$, for some $p \cong 40$,
12. $\langle p, apply\rangle [_{t_{SO1}} (x, out) \longrightarrow_{prob=0.05} [_{t_{CE}}$, for some $p \cong 40, x \subseteq M_{Reg}$,
13. $\langle p, apply\rangle [_{t_{SO1}} (x, out) \longrightarrow_{prob=0.2} [_{t_{SO1}}$, for some $p \cong 40, x \subseteq M_{Reg}$,
14. $\langle p, apply\rangle [_{t_{SO1}} (x, out) \longrightarrow_{prob=0.5} [_{t_{SO2}}$, for some $p \cong 40, x \subseteq M_{Reg}$,
15. $\langle p, apply\rangle [_{t_{SO1}} (x, out) \longrightarrow_{prob=0.15} [_{t_{SO3}}$, for some $p \cong 40, x \subseteq M_{Reg}$,
16. $\langle p, apply\rangle [_{t_{SO1}} (x, out) \longrightarrow_{prob=0.07} [_{t_{SO4}}$, for some $p \cong 40, x \subseteq M_{Reg}$,
17. $\langle p, apply\rangle [_{t_{SO1}} (x, out) \longrightarrow_{prob=0.03} [_{t_O}$, for some $p \cong 40, x \subseteq M_{Reg}$,
18. $\langle p, apply\rangle [_{t_{SO2}} (x, out) \longrightarrow_{prob=0.2} [_{t_{SO1}}$, for some $p \cong 40, x \subseteq M_{Reg}$,
19. $\langle p, apply\rangle [_{t_{SO2}} (x, out) \longrightarrow_{prob=0.15} [_{t_{SO2}}$, for some $p \cong 40, x \subseteq M_{Reg}$,
20. $\langle p, apply\rangle [_{t_{SO2}} (x, out) \longrightarrow_{prob=0.5} [_{t_{SO3}}$, for some $p \cong 40, x \subseteq M_{Reg}$,
21. $\langle p, apply\rangle [_{t_{SO2}} (x, out) \longrightarrow_{prob=0.1} [_{t_{SO4}}$, for some $p \cong 40, x \subseteq M_{Reg}$,
22. $\langle p, apply\rangle [_{t_{SO2}} (x, out) \longrightarrow_{prob=0.05} [_{t_O}$, for some $p \cong 40, x \subseteq M_{Reg}$,
23. $\langle p, apply\rangle [_{t_{SO3}} (x, out) \longrightarrow_{prob=0.02} [_{t_{SO1}}$, for some $p \cong 40, x \subseteq M_{Reg}$,
24. $\langle p, apply\rangle [_{t_{SO3}} (x, out) \longrightarrow_{prob=0.3} [_{t_{SO2}}$, for some $p \cong 40, x \subseteq M_{Reg}$,
25. $\langle p, apply\rangle [_{t_{SO3}} (x, out) \longrightarrow_{prob=0.08} [_{t_{SO3}}$, for some $p \cong 40, x \subseteq M_{Reg}$,
26. $\langle p, apply\rangle [_{t_{SO3}} (x, out) \longrightarrow_{prob=0.5} [_{t_{SO4}}$, for some $p \cong 40, x \subseteq M_{Reg}$,
27. $\langle p, apply\rangle [_{t_{SO3}} (x, out) \longrightarrow_{prob=0.1} [_{t_O}$, for some $p \cong 40, x \subseteq M_{Reg}$,
28. $\langle p, apply\rangle [_{t_{SO4}} (x, out) \longrightarrow_{prob=0.02} [_{t_{SO1}}$, for some $p \cong 40, x \subseteq M_{Reg}$,
29. $\langle p, apply\rangle [_{t_{SO4}} (x, out) \longrightarrow_{prob=0.08} [_{t_{SO2}}$, for some $p \cong 40, x \subseteq M_{Reg}$,
30. $\langle p, apply\rangle [_{t_{SO4}} (x, out) \longrightarrow_{prob=0.3} [_{t_{SO3}}$, for some $p \cong 40, x \subseteq M_{Reg}$,
31. $\langle p, apply\rangle [_{t_{SO4}} (x, out) \longrightarrow_{prob=0.1} [_{t_{SO4}}$, for some $p \cong 40, x \subseteq M_{Reg}$,
32. $\langle p, apply\rangle [_{t_{SO4}} (x, out) \longrightarrow_{prob=0.5} [_{t_O}$, for some $p \cong 40, x \subseteq M_{Reg}$,
33. $\langle p, apply\rangle [_{t_O} (x, out) \longrightarrow_{prob=0.55} [_{t_C}$, for some $p \cong 40, x \subseteq M_{Reg}$,
34. $\langle p, apply\rangle [_{t_O} (x, out) \longrightarrow_{prob=0.15} [_{t_O}$, for some $p \cong 40, x \subseteq M_{Reg}$,
35. $\langle p, apply\rangle [_{t_O} (x, out) \longrightarrow_{prob=0.3} [_{t_{SO4}}$, for some $p \cong 40, x \subseteq M_{Reg}$,
36. $[_{t_O} (x, out) \longrightarrow_{prob=1} [_{t_C}$ (no more suction), for some $x \subseteq M_{Reg}$,
37. $\langle p, apply\rangle [_{t_O} (x, out) \longrightarrow_{prob=P_{(O \to L)}} [_{t_L}$, for some $p > 40, x \subseteq M_{Reg}$,
38. $\langle p, apply\rangle [_{t_O} (x, out) \longrightarrow_{prob=1-P_{(O \to L)}} [_{t_O}$, for some $p > 40, x \subseteq M_{Reg}$,
39. $\langle p, apply\rangle [_{t_C} \longrightarrow_{prob=0.99} [_{t_L}$, for some $p \gg 40$,
40. $\langle p, apply\rangle [_{t_C} \longrightarrow_{prob=0.01} [_{t_{CE}}$, for some $p \gg 40$,
41. $\langle p, apply\rangle [_{t_{CE}} \longrightarrow_{prob=0.9} [_{t_L}$, for some $p \gg 40$,
42. $\langle p, apply\rangle [_{t_{CE}} \longrightarrow_{prob=0.1} [_{t_{SO1}}$, for some $p \gg 40$,
43. $\langle p, apply\rangle [_{t_L} \longrightarrow_{prob=1} \dagger$, for all $p \geq 0$.

Note that in the set R we can distinguish fourteen subsets of rules, each corresponding to the transitions from a class of values of the membrane tension, for a fixed value of the applied suction. The rules belonging to the same subset have associated probability values summing to 1. For instance, the subset of rules 6→11 describes the transitions from the closed expanded conformation (value t_{CE} of the membrane tension) of the channel to other conformations (values $t_C, t_{SO1}, t_{SO2}, t_{SO3}, t_{SO4}, t_O$ of the membrane tension), in response to an applied suction of 40 mmHg. The sum of all associated probabilities is $0.05 + 0.5 + 0.20 + 0.15 + 0.07 + 0.03 = 1$, where each single value denotes which of the other conformations is more likely to be reached, according to the applied suction. Here we stress the fact that we have chosen the values for the associated probabilities so that they have a close resemblance to the biological phenomenon of MscL activity. Moreover, probability values have been adjusted after a first analysis of *in silico* simulation results (see Section 6) in order to obtain a flickering behavior among subconducting states similar to the one recorded during patch clamping experiments.

The simulation of a cycle proceeds in the following way. Initially, the system is in the configuration $M_{Env} \left[_{t_C} M_{Reg}\right.$, which corresponds to the situation where the MscL is closed, the membrane tension is equal to t_C, and no suction is applied to the patch membrane (rule 1). The initial multisets in *Env* and in *Reg* are identical; they are not subject to any modification or communication (so, it is not necessary to indicate them in the rule). At any time, a suction can be artificially applied to the membrane: if the pressure is too high, i.e., $p \gg 40$ mmHg, then with the highest associated probability the membrane tension t_C reaches its lytic value t_L (rule 39) and, with the application of rule 43, we describe the membrane disruption (hence we substitute the semibracket symbol for the membrane with the new symbol †, which also stands for the death of the cell). In this case, the external and internal multisets are mixed in the environment. We also consider another possible transition (rule 40), which has very low associated probability, according to an applied suction of $p \gg 40$ mmHg, namely the one which causes the MscL to enter its expanded (but still closed) conformation.

Again, if the membrane tension is equal to t_C and the applied pressure has a value $p \ll 40$ mmHg, then the conformation of the MscL changes to the expanded state with very low probability (rule 2) because the applied suction is not enough to trigger the channel activation.

However, if the membrane tension is equal to t_C and the applied suction is between 0 and 40 mmHg (or close to the last value), then the closed conformation is more likely to change to the expanded conformation (rule 4). The next transition occurs only if the applied suction is kept around a constant value of 40 mmHg (rules 6→11). The four subsets of rules 13→17, 18→22, 23→27, 28→32 describe the flickering through subconducting open states when the applied suction is maintained around the value of 40 mmHg. All these rules consist of a coupling between an environmental rule and a communication rule; the communication of objects $x \subseteq M_{Reg}$ from the internal region to the environment is to be considered consistent with the conductance of the respective subconducting state (see Section 2.1).

Once the MscL is in an open substate (values $t_{SO1}, t_{SO2}, t_{SO3}, t_{SO4}$ of the membrane tension), if the pressure is maintained at a constant value $p \cong 40$ mmHg, then the MscL can reach its fully open conformation with different probabilities depending to the current conformation 17, 22, 27, or 32). The tension assumes the value t_O and the next subset of rules to be applied is 33→35. With the highest probability, a multiset x in M_{Reg} is communicated from the region to the environment and the channel returns to the initial closed conformation (rule 36), assuming that no more pressure is applied. The simulation of the cycle ends.

Starting from the open conformation, we also assume that some more possible transitions (rules 37, 38) can happen when the applied suction is increased, but we do not specify any particular probability value for these rules, since we would need a comparison with biological data that, to the best of our knowledge, is still missing.

We now give the formal definition of the *in vitro* model for MscL activity.

Definition 3. A P system modeling MscL *in vitro* is defined by the construct

$$\Pi_{invitro} = (V, \mu, Tension, M_{Env}, M_{Reg}, R),$$

where:

- $V = V_{chem} \cup \{w\}$ is the alphabet of the system, where $V_{chem} = \{a_1, \ldots, a_n\}$ is a finite set of symbols corresponding to the chemicals diluted in water. To denote a water molecule we use the different symbol $w \notin V_{chem}$;
- $\mu \in \{[_t\ ]_t, \dagger\}$ is the membrane structure consisting at most of a unique membrane labeled with the variable parameter t (which corresponds to the membrane tension value). If the membrane is no longer present in the system, because of its lysis, then we use the notation $\dagger$;
- $Tension = \{t_C, t_{CE}, t_{SO1}, t_{SO2}, t_{SO3}, t_{SO4}, t_O, t_L\}$ is a finite set of labels for the membrane structure $[_t\ ]_t$, that is a set of values for the membrane tension t;
- M_{Env}, M_{Reg} are the multisets present in the environment and in the region, respectively (we can assume that, initially, $M_{Env} = M_{Reg}$, because identical buffers are considered as *in vitro* experiments);
- R is the set of environmental rules and communication rules previously listed.

Remark 1. Typical "symmetrical buffers" used for in vitro experiments [44] contain, for instance, 200 mM KCl and 40 mM $MgCl_2$; hence we might define $M_{Env} = M_{Reg} = a_1^{200} a_2^{40} w^N$, where a_1=KCl, a_2=$MgCl_2$, and N is an integer much larger than the number of chemicals dissolved in the solution. The P system defined for the modeling of MscL activity during patch clamping experiments with symmetrical buffers can be used also for modeling experiments where nonsymmetrical buffers are used; that is the case when $M_{Reg} \neq M_{Env}$. See Section 7 for a comprehensive analysis, covering also the case of in vivo *model.*

A configuration of the system is determined by the state of the membrane, its tension value (in the case the membrane has not been broken) and the multisets present in the environment and inside the region. If the membrane has undergone a lytic rupture, then we denote it by the symbol $\dagger$; obviously, we do not associate any tension value to it and, in this case, only the environmental multiset (which is the union of the multisets initially present in *Env* and *Reg*) remains. Thus, a configuration C of $\Pi_{invitro}$ is a 4-tuple $([\ ], t, M_{Env}, M_{Reg})$ or a couple $(\dagger, M_{Env})$, where

$t \in \mathit{Tension}$, and M_{Env}, M_{Reg} are multisets over V. The initial configuration is $C_0 = ([\,], t_C, M_{Env}, M_{Reg})$, with $M_{Env} = M_{Reg}$; the final configuration C_f is a 4-tuple $([\,], t_C, M'_{Env}, M'_{Reg})$ or a couple $(\dagger, M''_{Env})$, with $M'_{Env}, M'_{Reg}, M''_{Env}$ such that $M'_{Env} \cup M'_{Reg} = M''_{Env} = M_{Env} \cup M_{Reg}$.

A transition from one configuration to the next one is a function τ_p of the states of each component:

$$\begin{aligned} \tau_p : (\{[\,]\} \times \mathit{Tension} \times \mathit{Env} \times \mathit{Reg}) \cup (\{\dagger\} \times \mathit{Env}) \longrightarrow \\ (\{[\,]\} \times \mathit{Tension} \times \mathit{Env} \times \mathit{Reg}) \cup (\{\dagger\} \times \mathit{Env}) \end{aligned}$$

such that

$$\begin{aligned} ([\,], t, M_{Env}, M_{Reg}) &\mapsto \{([\,], t', M'_{Env}, M'_{Reg}), (\dagger, M''_{Env})\}, \\ (\dagger, M_{Env}) &\mapsto (\dagger, M_{Env}). \end{aligned}$$

Given the current configuration C, the next configuration $C' = \tau_p(C)$ is obtained from C by applying the rules from R, according to the pressure values p in environmental rules and the arrows among tension values in Figure 2. At each step, the multisets M'_{Env}, M'_{Reg} in C' are such that $M'_{Env} \cup M'_{Reg} = M_{Env} \cup M_{Reg}$ holds (that is, objects are never modified, but only communicated).

We define a *cycle simulation* as a (finite) sequence of transitions starting from the initial configuration C_0 and ending in one of the final configurations C_f. Observe that, formally, there can also exist infinite sequences of transitions because rule 1, which describes the environmental condition where no suction is applied, could be applied forever (with the highest probability). Anyway, only a finite number of transitions is meaningful in modeling an MscL cycle. We refer to Section 8 for a theoretical investigation of subsequent activation cycles of MscL.

5 A P Model for *In Vivo* Functioning of MscL

In this section, we propose a model for the *in vivo* activity of MscL. By considering the knowledge accumulated from laboratory experiments as a basis for investigation, we make some hypotheses about the functioning of MscL which are subject to downshocks in natural or artificial environments. With respect to the *in vitro* model, we consider only a reduced number of transitions among channel conformations (see Figure 3), add a new membrane tension value reachable only after a cycle, and propose further possible transitions among membrane tension values which can occur in response to different environmental conditions. In particular, for the sake of simplicity we do not consider the channel subconducting open states used in the *in vitro* model. For the moment, we do not associate any probability with evolution rules because our first goal is to describe the general behavior of MscL when it is not

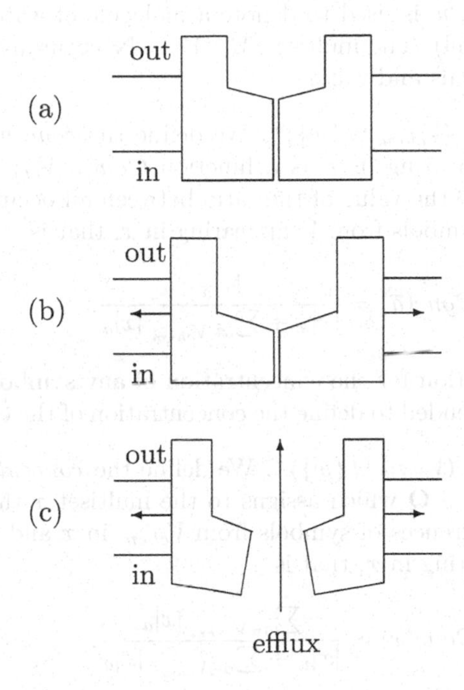

Fig. 3. States of the mechanosensitive channel during a cycle: (a) closed conformation; (b) expanded conformation; (c) open conformation.

studied with patch clamping. We also assume that after a closed-to-open cycle has been performed the membrane can be subject to a stress which causes its tension to be higher than it was before the channel activation. A deeper analysis of the latter situation, corresponding to the behavior of the channel and of membrane tension during subsequent activation cycles, will be given in Section 8.

We recall that a hypotonic shock occurs when a certain quantity of pure water is suddenly added to the environment, thus lowering the concentration of solutes. The water added enters the cell by osmosis very quickly and hence the osmotic pressure (and, consequently, the membrane tension) rises, causing MscL to open in order to prevent cell bursting.

From a formal point of view, for the *in vivo* model we have to consider not only the variable values of the membrane tension, but also the composition of the multisets present inside and outside the bacterial region. Actually, we will account for the *concentrations* of objects (in both the environment and the region), as well as for the addition of water which may occur at any time in the environment.

Let $V = V_{chem} \cup \{w\}$ be the alphabet consisting of a subset of symbols for solutes (V_{chem}) and a special symbol ($w \notin V_{chem}$) for the solvent composing a solution (namely, w is used to denote a molecule of water, the solvent considered in this model). The multiset $M : V \to \mathbf{N}$ naturally describes the multiplicities of chemicals and water.

Definition 4. Let $x \in (V_{chem} \cup \{w\})^+$. We define the *concentration* of any symbol $\overline{a} \in V_{chem}$ appearing in x as a function $Conc : V_{chem} \to \mathbf{Q}$ which assigns to the symbol $\overline{a}$ the value of the ratio between all occurrences of $\overline{a}$ in x and the sum of all symbols from V appearing in x, that is

$$Conc(\overline{a}) = \frac{|x|_{\overline{a}}}{|x|_w + \sum_{a \in V_{chem}} |x|_a}.$$

The previous definition for the concentration of any symbol appearing in the multiset can be extended to define the concentration of the entire multiset.

Definition 5. Let $x \in (V_{chem} \cup \{w\})^+$. We define the *concentration* of x as a function $Conc : V^+ \to \mathbf{Q}$ which assigns to the multiset x the value of the ratio between all occurrences of symbols from V_{chem} in x and the sum of all symbols from V appearing in x, that is

$$Conc(x) = \frac{\sum_{a \in V_{chem}} |x|_a}{|x|_w + \sum_{a \in V_{chem}} |x|_a}.$$

Remark 2. Given the definition of concentration for a multiset x, observe that if $|x|_w = 0$ or $\sum_{a \in V_{chem}} |x|_a = 0$, then x does not correspond to a realistic solution (that is, a mixture of solutes and solvent). Indeed, in the first case x would consist only of pure chemicals, while in the second case x would consist only of pure water. The two cases are not biologically relevant or plausible from the point of view of the activity we intend to model. Hence, to avoid these border situations, we will assume that

$$(alph(x) \cap \{w\} \neq \emptyset) \wedge (alph(x) \cap V_{chem} \neq \emptyset),$$

that is, the multiset x contains at least one symbol w and at least one symbol from V_{chem}.

In the following discussion, when it is not necessary to give a detailed description of the concentration in terms of all chemicals occurring in the solution, we will briefly denote by $Chem(x) = \sum_{a \in V_{chem}} |x|_a$ the multiset of symbols from V_{chem} present in x; hence the concentration of x will be given by

$$Conc(x) = \frac{Chem(x)}{|x|_w + Chem(x)}.$$

For a sound definition of the *in vivo* model, the system for MscL activity has to be defined in terms of three fundamental components: the external *environment*, the *membrane tension*, and the internal *region*.

The environment (Env) is made of solutes and water molecules, whose relative ratio (that is, concentration) may vary during hypotonic shocks, whenever some water is suddenly added to the environment. The internal region (Reg) consists of objects over the same alphabet of the environment ($V_{chem} \cup \{w\}$). This component is needed to account for the difference in concentration with that of the environment. We also emphasize that, in order to simplify the model, we are not considering any other processes occurring inside the region.

The variable tension assumes values from the finite set $Tension = \{t_{close_0}, t_{close_1}, t_{expanded}, t_{open}, t_{lysis}\}$. Each element in the set $Tension$ will be used to denote, as follows, the particular value (or the interval of values) of membrane tension and, hence, also a conformation of the channel:

- t_{close_0} is the initial value of the membrane tension, the setting where the cell grows in the medium, the channels are closed, and no events that might trigger the activation of MscL occur;
- t_{close_1} is the value at the end of the first cycle, when the channels return to be closed. In the following discussion, we will assume that t_{close_1} may be either equal to or different from t_{close_0};
- $t_{expanded}$ is the value that the membrane reaches during the phase of MscL expansion, while the channel still remains closed; its value is between t_{close_0} and t_{open};
- t_{open} is the value corresponding to the opening of MscL;
- t_{lysis} is the value corresponding to the lytic threshold of the membrane; when reached, the membrane will break and the cell will die.

The transitions among tension values are due to addition of water in the environment (see Figure 4). We particularly emphasize here the transitions from $t_{expanded}$ and t_{open} to t_{lysis}, which are assumed to be possible behaviors of MscL when, during a cycle, more water is added to the environment. We point out that, to the best of our knowledge, no biological investigations have been done to bring these events to light.

In the sequel, the addition of water will be described by means of an environmental rule which adds a fixed number of symbols w to the multiset already present in the environment. Obviously, this is a rough representation of biological reality, since it is difficult to count the exact number of water molecules intervening during a hypotonic shock. Nonetheless, P models of this type allow us to use specific values for symbols multiplicities and also to consider significant ratios between multiplicities.

We stress the fact that for the *in vivo* model both the environment and the region play a fundamental role in the description and application of rules (and hence in the modeling of the MscL activity), together with the membrane tension already present in the *in vitro* model. The environmental rules will be defined here with respect to water addition and, as will be clear soon, it will also be necessary to define a new class of rules which accounts for the differences between the multiset in the environment and the multiset in the region.

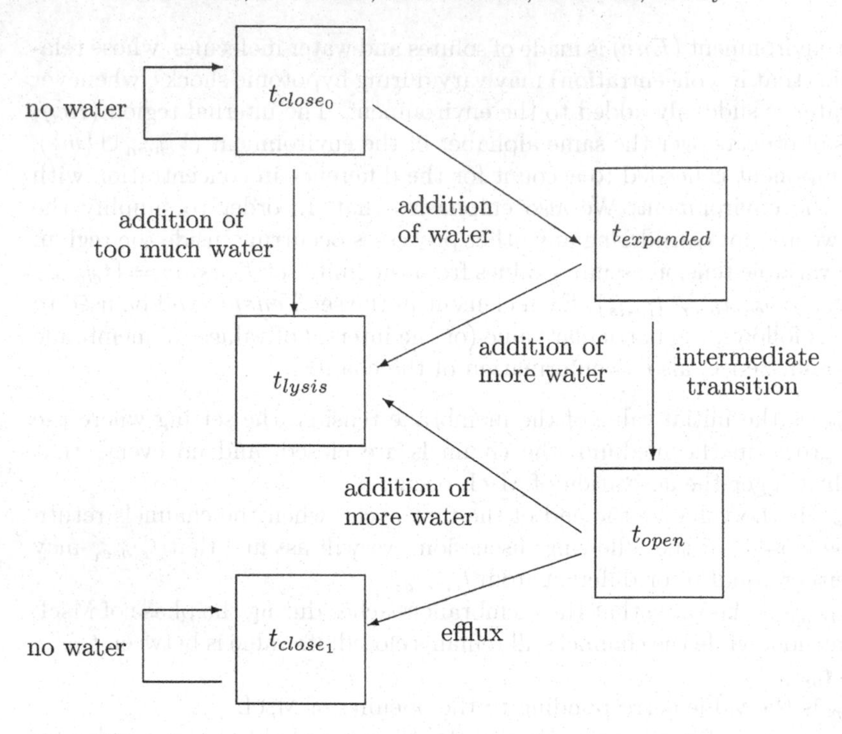

Fig. 4. Membrane tension transitions with respect to the addition of water.

Definition 6. An *in vivo environmental rule* is denoted as

$$\langle w^n, add \rangle [_t \rightarrow [_{t'}$$

for some $n \in \mathbf{N}, t, t' \in Tension$. It describes a (possible) change of the membrane tension value due to an addition of n water molecules to the environment, which can happen at any time and cannot be controlled by any component of the system.

We define now a new type of evolution rule whose application depends on the number of water molecules and chemicals which are present (at the time of its application) in both the environment and inside the region.

Definition 7. A *concentration-based evolution rule* is a rewriting rule of the form

$$x \; [_t \; y \xrightarrow{k_1 \leq C(x,y) \leq k_2} \; [_{t'},$$

where x is the environment multiset over the alphabet V, y is the region multiset over V, $t, t' \in Tension$ are membrane tension values, and $k_1 \leq C(x, y) \leq k_2$ is a condition depending on the environment and the region multisets x, y for some $k_1, k_2 \in \mathbf{R}$ (even null values), defined as

$$C(x,y) = \frac{Conc(y)}{Conc(x)}.$$

Environmental and concentration-based rules can be coupled to describe environmental changes in the quantity of water, which can have different consequent actions according to the condition $C(x,y)$ satisfied; that is, we can write

$$\langle w^n, add\rangle \; [_t \overset{k_1 \leq C(x,y) \leq k_2}{\longrightarrow} [_{t'}.$$

By Definition 7 and the coupling between environmental and concentration-based rules, we deduce that the ratio

$$\frac{Conc(y)}{Conc(x)} = \frac{Chem(y)}{Chem(x)} \cdot \frac{|x|_w + Chem(x)}{|y|_w + Chem(y)}$$

increases whenever some water addition occurs in the environment. This increase causes the influx of water into the region and the activation of MscL, but if the ratio suddenly increases too much, then the membrane can undergo a lytic rupture. Hence, in what follows we will consider two threshold values for the condition $C(x,y)$, namely k_a for triggering the channel activation (hence, $C(x,y) > k_a$ must be satisfied to begin the simulation of an MscL cycle) and k_l for the membrane lysis situation (hence, if $C(x,y) \geq k_l$ is satisfied, the cell dies).

The passage of objects from the environment into the region (and vice versa) will be described by means of communication rules with different target indications. Precisely, the classical target of type *out* will be used to denote the efflux of water and chemicals from the region to the environment, which happens if and only if the MscL is open. Instead, the influx of water from the environment to the region directly through the membrane (by osmosis) will be denoted with a new target indication called *through*. Formally, a communication rule of type *out* will be simply described by the couple (x, out), where $x \in (V_{chem} \cup \{w\})^*$, while a communication rule of type *through* will be described by the couple $(w^m, through)$ for some $m > 0$.

All necessary ingredients of the *in vivo* model have been introduced and it is now possible to list the rules for the simulation of MscL activity. In what follows, we assume that initially the bacterium is in an equilibrium state with the surrounding environment (see Section 7 for some more considerations about this assumption).

1. $[_{t_{close_0}} \longrightarrow [_{t_{close_0}}$,
2. $\langle w^n, add\rangle \; x[_{t_{close_0}} y \overset{C(x,y) < k_a}{\longrightarrow} (w^h, through) \; [_{t_{close_0}}$, for some h such that $0 \leq h \ll n$,
3. $\langle w^n, add\rangle \; x[_{t_{close_0}} y \overset{k_a \leq C(x,y) \ll k_l}{\longrightarrow} (w^h, through) \; [_{t_{expanded}}$, for some h such that $0 < h \leq n$,

4. $\langle w^n, add\rangle\ x[_{t_{close_0}} y \xrightarrow{k_l \leq C(x,y)} (w^h, through)\ [_{t_{lysis}}$, for some h such that $0 < h \leq n$,
5. $[_{t_{lysis}} \longrightarrow \dagger$,
6. $(w^h, through)\ [_{t_{expanded}} \longrightarrow [_{t_{open}}$, for some h such that $0 < h < |x|_w$,
7. $\langle w^n, add\rangle\ x[_{t_{expanded}} y \xrightarrow{k_a \ll C(x,y) \leq k_l} (w^h, through)\ [_{t_{lysis}}$, for some h such that $0 < h \leq n$,
8. $[_{t_{open}}\ (y, out) \longrightarrow [_{t_{close_1}}$, for some $y \subseteq M_{Reg}$,
9. $\langle w^n, add\rangle\ x[_{t_{open}} y \xrightarrow{k_a \ll C(x,y) \leq k_l} (w^h, through)\ [_{t_{lysis}}$, for some h such that $0 < h \leq n$,
10. $[_{t_{close_1}} \longrightarrow [_{t_{close_1}}$.

In the initial configuration of the system, the environment and the region are in osmotic equilibrium (see Section 8), and the membrane tension value is equal to t_{close_0}: no event is occurring in the environment, the cell is resting and MscL are closed. No changes are made in any of the components (rule 1).

If a water addition occurs but the quantity of added water is very small (i.e., the condition $C(x,y) < k_a$ is satisfied), some water molecules (h) may enter the region – by osmosis – but no transitions occur for the channel (rule 2). After the application of this rule, we have the multiset $x' = x \setminus w^h$ in the environment and the multiset $y' = y \cup w^h$ in the region. This evolution step corresponds to a small change in the concentration of the environment with respect to the concentration in the region. Since the model we are presenting is the first attempt to describe MscL activity and we are not considering multiple substates, we assume that this small change is not enough to cause a channel transition from the closed state to the expanded substate considered in this work. Further refinements of our model could include all transition substates and their flickering between each other.

A transition from the channel closed state to the expanded substate can instead be modeled by the addition of a greater quantity n of water molecules, such that the condition $k_a \leq C(x,y) \ll k_l$ is satisfied. The tension value reaches the value $t_{expanded}$ and, at the same time, water continues to enter the bacterium by osmosis (rule 3).

If the water added is too much and its influx causes a steep rise in turgor pressure (in this case $k_l \leq C(x,y)$ holds), then the membrane tension suddenly reaches its lytic threshold (rule 4), the membrane breaks, the multisets mix in the environment, and the cell dies (rule 5).

Let us now turn back to the channel expanded substate, described by the tension value $t_{expanded}$. In this condition, water continues to pass directly through the membrane until the turgor pressure is high enough to open the channel, and hence the tension value reaches the new value t_{open} (rule 6).

Once the channel is open, chemicals and water can exit the region and pass to the environment by crossing the open pore (rule 8); then the channel returns to the closed state (we assume that this transition can be modeled by a single evolution step). The multiset y in M_{Reg}, communicated from the region to the environment, is left unspecified, but it should correspond to the conductivity of the channel. In further developments of the model, as for the *in vitro* model, this data will be considered and the form of the communicated multiset will be specified.

When the tension value is equal to t_{close_1}, the cycle simulation ends since we are defining a model for one activation cycle only. We assume that no other water addition will occur and hence the cell rests in equilibrium with the environment (rule 10).

Two more rules have been added to the model in order to predict the possible behavior of MscL in case of further addition of water when the tension is equal to $t_{expanded}$ or t_{open}. In both cases (rules 7 and 9) we assume that consistent addition of water (with a quantity near the lytic one, such that the condition $k_a \ll C(x,y) \leq k_l$ is satisfied) will cause the membrane lysis (rule 5) and, hence, cell death.

Definition 8. A P system modeling MscL *in vivo* is defined by the construction

$$\Pi_{invivo} = (V, \mu, Tension, M_{Env}, M_{Reg}, R),$$

where:

- $V = V_{chem} \cup \{w\}$ is the alphabet of the system, where $V_{chem} = \{a_1, \ldots, a_n\}$ is a finite set of symbols corresponding to the chemicals diluted in water, for which we use the different symbol $w \notin V_{chem}$;
- $\mu \in \{[_t\ ]_t, \dagger\}$ is the membrane structure consisting at most of a unique membrane labeled with the variable parameter t (which corresponds to the membrane tension value). If the membrane is no longer present in the system because of its lysis, we use the notation $\dagger$;
- $Tension = \{t_{close_0}, t_{close_1}, t_{expanded}, t_{open}, t_{lysis}\}$ is a finite set of labels for $[_t\ ]_t$, that is, a set of values or intervals of values for the membrane tension t;
- M_{Env}, M_{Reg} are the multisets present in the environment and in the region, respectively;
- R is the set of environmental, communication, and concentration-based rules given above.

As for *in vitro* model, a configuration of the system is determined by the state of the membrane, its tension value (in the case it has not been broken), and the multisets present in the environment and inside the region. If the membrane has undergone a lytic rupture, we denote it by the symbol $\dagger$; obviously we do not associate any tension value to it and we consider only the remaining environmental multiset. Thus, a configuration C is a 4-tuple $([\ ], t, M_{Env}, M_{Reg})$ or a couple $(\dagger, M_{Env})$, with $t \in Tension$ and M_{Env}, M_{Reg} multisets over V. The initial configuration is $C_0 = ([\ ], t_{close_0}, M_{Env}, M_{Reg})$ and the final configuration C_f is of the form $([\ ], t_{close_0}, M'_{Env}, M'_{Reg}), ([\ ], t_{close_1}, M'_{Env}, M'_{Reg})$, or $(\dagger, M''_{Env})$.

A transition from one configuration to the next one is a function τ_w of the states of each component:

$$\tau_w : \{[\,]\} \times \mathit{Tension} \times \mathit{Env} \times \mathit{Reg} \cup \{\dagger\} \times \mathit{Env} \longrightarrow$$
$$\{[\,]\} \times \mathit{Tension} \times \mathit{Env} \times \mathit{Reg} \cup \{\dagger\} \times \mathit{Env}$$

such that

$$([\,], t, M_{Env}, M_{Reg}) \mapsto \{([\,], t', M'_{Env}, M'_{Reg}), (\dagger, M''_{Env})\},$$
$$(\dagger, M_{Env}) \mapsto (\dagger, M_{Env}).$$

Given the current configuration C, the next configuration $C' = \tau_w(C)$ is obtained from C according to the rules in R and the arrows among tension values in Figure 4. At each step, the multisets M'_{Env}, M'_{Reg} in C' are such that $M'_{Env} \cup M'_{Reg} = M_{Env} \cup M_{Reg}$ holds (that is, objects are never modified but only communicated). Moreover, since in the *in vivo* model water can pass through the membranes even when the channel is closed, it is possible that $\mathit{Conc}(M'_{Env}) \neq \mathit{Conc}(M_{Env})$ and $\mathit{Conc}(M'_{Reg}) \neq \mathit{Conc}(M_{Reg})$, that is, the distribution of symbols from V may be different in successive configurations.

We define a *cycle simulation* as a (finite) sequence of transitions starting from the initial configuration C_0 and ending in one of the final configurations C_f. Observe that there can also exist infinite sequences of transitions, because rules 1 and 10 could be formally applied forever. In any case, we still consider that only a finite number of transitions is meaningful in modeling a MscL cycle.

6 Simulations *In Silico*

It is problematic to reconstruct the behavior of a complex entity like an MscL using just mathematical models since it is difficult to capture all the factors at play in equations. Several models have been proposed and constructed to capture complex biological system dynamics. They include systems based on integration of differential equations and Petri-net models. The major drawback of these models is that they adopt a centralized approach and try to model the behavior of a complex system as a single atomic entity. As a consequence, the observable behavior is constrained by the explicit variable and rules programmed into the system.

A complex system, such as a biological one, consists of several subsystems and the observable behavior of the system emerges out of parallel interactions between its constituent parts, most of which may not have been scripted in advance. In order to gain some experimental evidence about the quality of the foregoing P models, they were used to conduct simulations *in silico* to obtain some predictions about observable data that could be used to test their soundness and validity. Important observables are the compound effects of complex chains of interactions over space and time based on the model, particularly as they occur in global emergent behavior observed in biological systems. This approach has been used in this section, where we describe some of these simulations in software and the corresponding results for the *in vitro* P model.

6.1 About Simulation Environments

Simulations conducted *in silico* have several advantages over actual experiments *in vitro* and sometimes even observations *in vivo*. Biological experiments need to be run several times in order to validate hypotheses. Each run may cost hundreds of dollars, adding up to a large cost. The costs are even higher if the runs get corrupted due to errors on part of the experimenter, e.g., by using a wrong concentration of a reagent. Although it must be acknowledged that simulations may lack physical realism and appear incomplete and unbelievable at first sight, as with most scientific models a suitable choice of granularity and relevant features may provide powerful hints and insights.

Simulations are more cost effective, reasonably faster, and provide a higher level of control than experiments *in vitro*. While the latter may take days from setting up to extraction of results, simulations take only a few hours. Once a simulation program is coded and debugged, it may be run as many times as needed (with different parameters, if required) with little additional cost. Simulations usually detect errors fairly early and they can be easily rectified. This has been the case with the simulation environment used for this chapter.

6.2 Virtual Test Tubes

The MscL experiments were implemented on *EdnaCo*, a complex systems simulator running on a cluster of 24 PCs. *EdnaCo* is a virtual test tube (VTT for short) that can be used as a distributed discrete-event simulation environment. It was originally developed to better understand reactions among DNA molecules for computational purposes [18, 17]. While a full description of the features and performance of *EdnaCo* can be found in [16], we summarize some of the features here in order to make this chapter self-contained.

EdnaCo follows the complex systems paradigm of entities (objects) and interactions, i.e., instead of programming their entire behavior over time, only entities (originally DNA molecules) and individual interactions between pairs of them are programmed by the user. The VTT is spatially arranged as a 3D coordinate system in which molecules (entities) reside. The tube may move entities to simulate motion (possibly Brownian, according to a predetermined schedule, or no motion at all) and they are allowed to interact freely. Entities could be homogenous or heterogeneous and may represent any complex biomolecules. Each molecule is located at a unique space coordinate at any given time. The VTT can also code for physico-chemical properties such as temperature, pressure, salinity, and pH that may vary across space and time and affect the way entities interact. Depending on the nature of the simulation, interactions between entities are programmed by the users. Multiple instances of an entity behave in the same manner.

All entities are capable of sensing the position of other entities up to a specific distance defined as the radius of interaction. If two or more entities come within interaction distance, they may interact with each other. An interaction between two entities may be viewed as a chemical or mechanical reaction between them. As result of interaction, existing entities may get consumed, their status may change, and/or new entities may get created. Moreover, the concentration of entities may be manipulated externally by adding or removing entities to or from the tube at any point in time. The running time of a simulation is divided into discrete time steps, or iterations. At any iteration, the state of the objects and the tube may change recursively, based on previous changes, to reflect the interaction rules among themselves and/or with their environment. Parallelism is implemented in *EdnaCo* by dividing the VTT into a number of discrete segments, each running on a different processor. This allows multiple interactions to take place at once. When entities move, they may either change positions within a segment or migrate across processor boundaries. The architecture of *EdnaCo* allows it to be scaled to an arbitrarily large number of processors. Further details of this simulation environment can be found in [16, 7].

6.3 Experimental Design and Emergent Behavior in Simulation

The MscL channels were simulated according to the local interactions described above in Section 4. The behavior of a single MscL may change under changing pressure conditions. *EdnaCo* is capable of simulating the behavior of multiple MscL interacting in parallel under changing environmental conditions. An MscL entity has a tension variable associated with it. As the external pressure is varied at each iteration, the tension value gets updated according to the probabilistic rules. The probabilistic rules are implemented using a state-of-the-art random number generator with a very large period and other desirable properties (see [12, 19] for details). For a typical run, the external pressure may be increased at a fixed or variable rate. The MscL tension value keeps on fluctuating until the pressure reaches the lytic threshold, upon which the cell lyses (dies).

While the above simulation of a single channel could be accomplished using a simple program, the major advantage of using *EdnaCo* is that a simulation requiring a large number of objects (such as multiple MscL and other cellular components in the extensions of the P models proposed below in Section 9) can be run in a relatively short time due to the massive parallelism present in *EdnaCo*.

The following are sample results from the simulation. The measured quantities, the tension (Figure 6), the conductance (Figure 7), and the current (Figure 8) are emergent observed quantities that were not explicitly programmed in the simulation. The figures show their behavior over time. An iteration is the time required for the objects to effect one interaction, i.e., to apply one set of rules simultaneously. These results appear to be in line with general biological phenomena and offer experimentalists the challenge of verifying by actual experiments *in vitro* and, perhaps with more difficulty, *in vivo*.

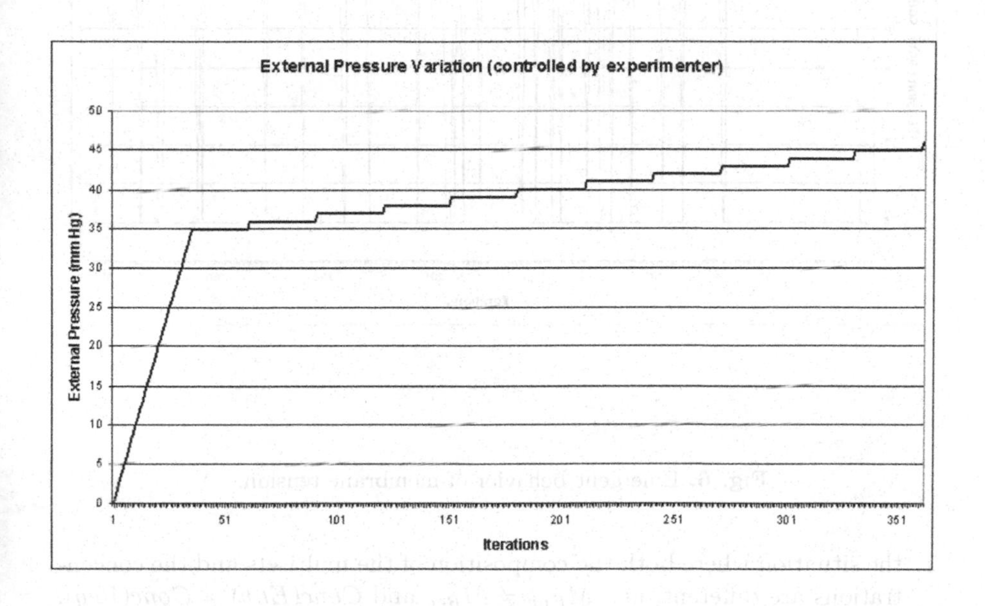

Fig. 5. Variation of applied pressure.

7 Some General Considerations About the Models

In the *in vivo* model, we have made the assumption that the initial configuration of the system corresponds to biological equilibrium, that is a situation where no activation of MscL can occur. A biological equilibrium can be formally described by considering the composition of the multisets in the environment and in the region and their respective concentrations. We list here all the possible combinations for multisets and concentrations, as well as the corresponding biological reality of different prokaryotes:

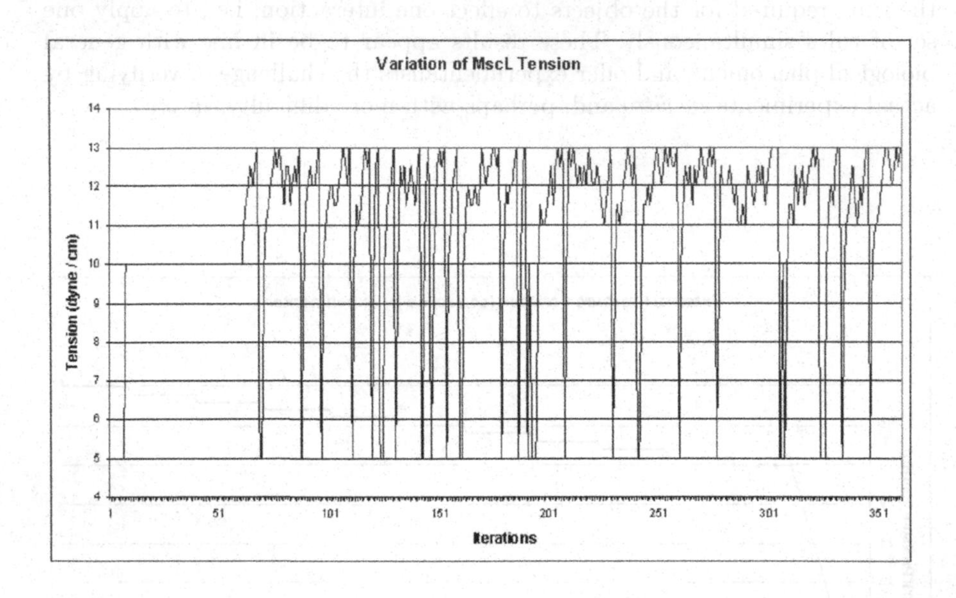

Fig. 6. Emergent behavior of membrane tension.

1. the situation where both the composition of the multisets and the concentrations are different, i.e., $M_{Env} \neq M_{Reg}$ and $Conc(Env) \neq Conc(Reg)$, corresponds to the natural habitat of *E. coli*;
2. the situation where the composition of the multisets is different but the concentrations are equal, i.e., $M_{Env} \neq M_{Reg}$ and $Conc(Env) = Conc(Reg)$, corresponds to the natural habitat of *Halobacter salinarum*;
3. the situation where both the composition of the multisets and the concentrations are equal, i.e., $M_{Env} = M_{Reg}$ and $Conc(Env) = Conc(Reg)$, corresponds to patch clamp experiments with the use of symmetrical buffers considered in the *in vitro* model.

Note that $M_{Env} = M_{Reg}, Conc(Env) \neq Conc(Reg)$ can never occur, because if the multisets are equal, then, by definition, the concentrations must be equal too; hence the previous list is exhaustive of all possibilities.

We claim that the *in vivo* model is general enough to cover all different biological possibilities, and similar considerations can be applicable also to the *in vitro* model (thus, it can be also used to describe patch clamping with nonsymmetrical buffers).

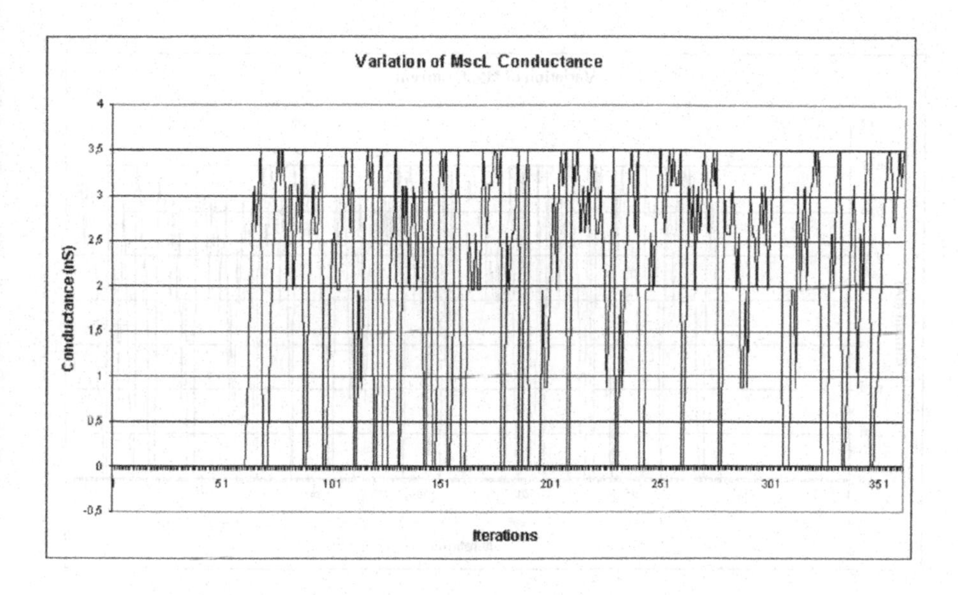

Fig. 7. Emergent behavior of conductance variation.

The above considerations are consistent with the situation of the prokaryotes belonging to the domain *Archaea* living in hypersaline environments (such as the Dead Sea). In halophilic *Archaea* such as *Halobacter salinarum* or *Haloferax volcanii* (in *Haloferax volcanii* two mechanosensitive channels were identified [27, 26]) there is practically no turgor pressure because the overall concentration of solutes inside the cell and outside the cell (in the growing medium) is practically the same. However, the quantity of each solute present inside and outside the cell is very different. For example, in the case of *Halobacter salinarum*, sodium ions are 4 mol/L in the growing medium, whereas inside the cell the concentration is much lower (1.37 M); for potassium ions the external concentration is very low (0.032 M), whereas the internal concentration is very high (4.57M). For more details see [34]. The model proposed in this chapter is appropriate for describing this situation too.

This is one strategy developed by halophilic prokaryotes, the so-called "salt-in" strategy [35], to cope with high sodium concentration outside the cell (that is, with hypersaline environments). This strategy requires a considerable number of changes to safeguard all regulatory and metabolic functions at high salinity as, e.g., salt-adapted enzymes [14, 34]. This strategy is relatively inexpensive in terms of energy, but requires far-reaching adaptations to the

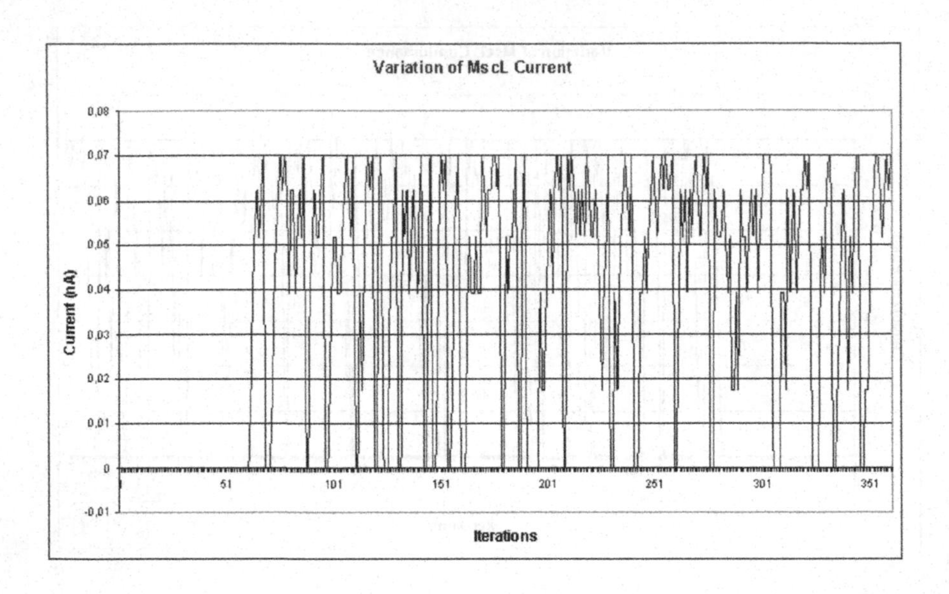

Fig. 8. Emergent behavior of current variation.

presence of high salt concentrations by the intracellular enzymatic machinery. Adaptability of the cells to change salt concentrations is limited [35]. The proposed model could be further modified to describe this strategy.

Another strategy used by other prokaryotes like *Escherichia coli* involves the presence inside the cell (in the cytoplasm) of high levels of organic chemicals (such as glutamate, sugars, glycinebetaine, etc.) that act as organic osmolites because they increase the osmotic pressure, thus decreasing the entry of sodium ions inside the cell [28]. This strategy is energetically expensive, the actual energetic cost depending on the type of organic solute synthesized. No major modification of the intracellular machinery is needed (in comparison to halophilic) and in most cases cells can rapidly adapt to changes in external salinity [35].

8 An Investigation of Subsequent Activation Cycles

In this section we suggest some theoretical ideas, to be further investigated, about some possible behavior of MscL during subsequent activation cycles, depending both on artificial and natural environmental conditions. Specifically, we assume different responses of the membrane tension after one activation cycle and the corresponding consequences of the responses.

Our main hypothesis is that the value assumed by the membrane tension after a closed-to-open cycle of MscL, namely t_{close_1}, might be higher than the value t_{close_0} it assumed before the activation of the channel. That is, the relation

$$t_{close_1} = t_{close_0} + l_0$$

holds for some $l_0 \in \mathbf{R}^+$, as already suggested in Section 5.

Thus, it is now natural to wonder about the behavior of the channel as time goes by and during subsequent cycles. We can speculate that, after some time the cellular membrane relaxes again and the tension value reached after the first cycle returns equal to the initial value, that is, $t_{close_1} = t_{close_0}$. In this case, a subsequent cycle – occurring after the time elapsed for the complete membrane relaxation – would have the same dynamic of the first cycle.

We may suppose that the membrane is not able to relax and neutralize the stress it suffered because of the turgor pressure. In this case, some different assumptions can be made for the behavior of the channel during successive cycles:

1. After each cycle the membrane tension value increases by a positive (or null) real value until a threshold tension value T is reached for some $T \in \mathbf{R}^+, T < t_{lysis}$. That is, there exists an index $i' \in \mathbf{N}, i' \geq 1$, such that $t_{close_i} = t_{close_{i'}}$ for all $i \geq i'$:

$$\begin{aligned} t_{close_1} &= t_{close_0} + l_1, \\ t_{close_2} &= t_{close_1} + l_2, \\ &\dots \\ t_{close_i} &= t_{close_{i-1}} + l_i, \\ t_{close_{i+1}} &= t_{close_i}, \end{aligned}$$

   where, following the notations above, $i' = i+1$. The increments $l_1, \dots, l_{i-1} \in \mathbf{R}^+, l_i \in \mathbf{R}$ are not all necessarily distinct; in any case, the succession of values is decreasing, that is, $l_1 \geq l_2 \geq \dots \geq l_i \geq 0$.

   In this case, the behavior of the channel could be quasi-periodic, since the tension finally reaches a threshold value for its gating. A similar dynamics could occur in the case where the tension of the membrane after each cycle would be equal to or slightly different from the value it had before the previous activation;
2. After each cycle the membrane tension value increases by a strictly positive real value; hence sooner or later it will be equal to the lytic tension value, that is, there exists an index $i'' \in \mathbf{N}, i'' \geq 1$, such that $t_{close_{i''}}$ is equal to t_{lysis}:

$$\begin{aligned} t_{close_1} &= t_{close_0} + l_1, \\ t_{close_2} &= t_{close_1} + l_2, \end{aligned}$$

$$\dots$$

$$t_{close_i} = t_{close_{i-1}} + l_i = t_{lysis},$$

where, following the notations above, $i'' = i$. Note that the finite sum $\sum_{1 \leq i < i''} l_i$ of all increments $l_1, l_2, \dots, l_i \in \mathbf{R}^+$ must be equal to the difference $L = t_{lysis} - t_{close_0}$.

In this case, no periodic behavior could probably be observed.

We stress the fact that the above considerations still have no underlying biological data, because, to the best of our knowledge, no experiments have ever been performed to analyze the behavior of Msc during subsequent activation cycles. However, we want to emphasize the relevance that an analysis of this type might have for biologists or biotechnologists, probably for the investigation of the life cycles of prokaryotes.

9 Final Remarks and Future Extensions

In this chapter we focused on the opening of MscL during patch clamping experiments and hypotonic shocks. However, when bacteria are exposed to either hyper or hypotonic shocks there are a lot of changes that occur in the cells in order to respond, at both the short term and long term levels, to these changes (see the review [47] and Section 9.4).

As future extensions to MscL models, we plan to consider also the response time of the MscL opening, as well as times when the MscL remains open, which would mean defining *timed evolution rules.* We intend to improve the connections between *in vitro* and *in vivo* models (see Section 9.1) by adding, for instance, all substate conformations and associated probabilities to the *in vivo* model.

Indeed, the design of the software simulator presented in Section 6 provides an easier way to check the effectiveness and the correctness of the models, and hopefully it could become a tool for biologists for the investigation *in silico* of MscL functioning.

9.1 Relations Between *In Vitro* and *In Vivo* Models

In vitro and *in vivo* P models might propose a platform for the integration of the data obtained on MscL in prokaryotes, with special emphasis on *E. coli.* The corroboration of both *in vitro* and *in vivo* results were reported in original empirical papers on MscL [9, 33, 32, 3]. Thus, we expect that the further refinement of our models (by means also of the software environment used to produce simulations) would accelerate the integration of *in vitro* and *in vivo* results. Moreover, with the explosion of molecular biology and the increase in the quantity and quality of data obtained by high throughput technologies, there is a trend nowadays in biology to move from the reductionistic approach to the integrative approach [36], either at the supermolecular level [22] or at the systemic level [36, 25]. Indeed, the reductionistic approach and the systemic, integrative approach are today on the same side of the barricade, a totally different position than four or five decades ago when both approaches started to flourish in biology [31].

In this perspective, the correlation between *in vitro* and *in vivo* results represents one of the most important trends in biological research, and the models proposed in this chapter could make an important contribution to it.

Moreover, they could be improved for describing the function of other mechanosensitive channels with a different conductance (MscS and MscM), working separately or together in a prokaryotic cell. In the absence of detailed biological results concerning *in vivo* cooperation between different mechanosensitive channels in the response to osmotic challenges, the P model could elaborate scenarios starting from the results separately obtained on each type of Msc by *in vitro* experiments. In this way, the improved model could be of further use for biology scientists working on Msc to see how their experimental results fit, not only with their own working hypothesis, but also with the *in silico* experiments.

9.2 Multiple Occurrences of MscL

The *in vitro* model simulates the activity of a single mechanosensitive channel; nonetheless it is known that about 50-100 occurrences of MscL may be found in a single bacterial cell, and many copies of MscL can be present in a membrane patch. The apparent excess of MscL protein in bacterial cells, besides the fact that one single MscL would suffice to dissipate osmotic gradients within 1 millisecond, deserves further attention in both biology and membrane systems.

We claim that the *in vitro* model can be easily extended to describe and simultaneously analyze multiple occurrences of MscL. In fact, one cannot be certain that all occurrences of MscL are in the same conformation in any fixed environmental condition: when a suction is applied on a patch membrane containing many channels, some may be completely open, some in a subconducting (open) state, and some still closed. To describe all possible reactions, it suffices to consider the same model for all channels but with different probability values associated with the evolution rules of each channel, or even different membrane tension values ranging in a known real interval.

For instance, consider a patch membrane containing two channels, and define the rules

$$\langle p, apply\rangle [_{t_C} \longrightarrow_{prob=0.8} [_{t_{CE}} \quad \text{for channel 1,}$$
$$\langle p, apply\rangle [_{t_C} \longrightarrow_{prob=0.3} [_{t_C} \quad \text{for channel 2,}$$

for the same value of $p \in (0, 40]$ (that is, the first channel progresses to the expanded conformation, while the second remains closed), and

$$\langle p', apply\rangle[t_{CE} \longrightarrow_{prob=0.7} [t_{SO1} \quad \text{for channel 1,}$$
$$\langle p', apply\rangle[t_{CE} \longrightarrow_{prob=0.5} [t_{CE} \quad \text{for channel 2,}$$

for the same value of $p' \cong 40$ (that is, the first channel progresses to the first subconducting conformation, while the second remains in the expanded conformation); and so on for all other rules.

For an analysis of this type the software simulator (supported by real data for multisets and channel conductivity) can correctly perform different simulations, with different input values assigned to all parameters, and can be used to follow all possible choices for next state transitions.

From the relation[5] $t = ap + b$ between the tension t and the pressure p, where a and b are values depending on the radius of the patch (see [44] for some picture of patches with different values for radius), we know that the tension which determines the status of the channel depends on its actual position on the patch membrane. For instance, we might observe different behaviors in the case in which the channel is close to the border of the patch, or in the case in which it is placed in the middle of the patch, and so on. Obviously, such differences are larger when the geometry of the patch is not uniform; specifically:

(*i*) if the patch is a perfect half sphere, then the tension of the channel will always be the same whatever the position of the channel on the patch, since the radius is always the same;

(*ii*) otherwise, the tension of the channel changes from place to place, since the tension increases as the radius increases, that is, when the channel is closer to the border of the patch (if the patch is flat) or when the channel is far from the border (if the patch is long and narrow).

[5] We derive the relation between the membrane tension and the patch pressure from a version of Laplace's law (in [21]):

$$t - t_{1/2} = \frac{r}{2} \cdot (p - p_{1/2}),$$

where r is the radius of curvature of the membrane patch (under the external suction applied to the patch pipette) and $p_{1/2}$ and $t_{1/2}$ are, respectively, the negative pressure and the membrane tension at which the channel is open 50% of the time (they can be assumed to remain constant for a membrane patch during an experiment). Since our aim is to define a discrete model, we may assume that r also remains nearly constant for the pressure at which the MscL is open ($r = r_{open}$) and for the pressure at which it is closed ($r = r_{close}$); we can also fix a value for the radius during substate transitions ($r = r_{substate}$). It follows that we can assume a linear dependence of the membrane tension with respect to the applied pressure, that is,

$$t = ap + b,$$

where $a = r/2, b = t_{1/2} - (r/2)p_{1/2}$, and r takes a fixed constant value in the set $\{r_{close}, r_{substate}, r_{open}\}$.

We can assume the same kind of behavior for many channels. Hence, when considering multiple channels in the same patch, we can have the following situations: if the patch is a perfect half sphere (case (*i*)), then the tension is equal for all channels (that is, the same setting of the *in vitro* model is valid for all channels); otherwise (case (*ii*)), the tension changes from channel to channel and hence different settings of the *in vitro* model must be used for each channel.

We believe that by determining the right set of values for probability for a single channel and by considering multiple channels (each with a specified set of parameters corresponding to a presumed radius) we can predict the positions (the distribution, more precisely) of the channels on the patch membrane, which is something unknown to biologists.

9.3 Effects of Inhibitors and Activators

The occurrence of inhibitors for MscL such as gadolinium ion, amiloride, and aminoglycoside antibiotics (e.g., streptomycin, gentamicyn, and neomycin) as well as of activators (such as amphipathic molecules, molecules that have both hydrophilic and hydrophobic groups, etc.) are not only very useful tools in biological experiments, but could also be used to refine the P model of mechanosensitive channels. For example, P features as promoter and inhibitor objects could be included in the models in order to describe the changes in the activation of MscL and in cell functioning.

Gadolinium ion is an inhibitor of mechanosensitive channels (not only for the ones of large conductance); its presence causes a decrease in the osmotic efflux of solutes from *E. coli* and other microorganisms during hypotonic shock (more details in [48]). Relatively low concentration inhibits MscL activity both *in situ* and in reconstituted liposomes, but increasing the pressure gradient can reactivate the blocked channels. Gadolinium ion may act directly on mechanosensitive channels, or its effect may be exerted via membrane lipids, rendering the bilayer less effective in transmitting the stretch force. Moreover, it was shown that Gadolinium ion is an inhibitor of MscL in the cyanobacterium *Synechocystis* PCC 6803 during a hyperosmotic shock [3].

Similar attention may be paid to simple peptides like gramicidin, which is used to study artificial mechanosensitive channels because it forms channels in bilayers. When the bilayer tension is increased, two molecules of gramicidin link to each other and thus form a dimer which is responsible for the formation of a channel. Gramicidin has provided the opportunity to analyze possible underlying molecular mechanisms in extremely well defined simple systems (for more details see [21]).

We believe that our P model could be used and adapted to predict the inhibition activity of gadolinium and the function of gramicidin channels, to hopefully derive relevant data for biologists.

9.4 Short and Long Term Responses to Osmotic Pressure

Mechanosensitive channels are not the only structures involved in osmoregulation and water balance in a prokaryote cell, either on a short or a long term basis. Other structures involved in these processes are porins, aquaporins, glycerol facilitators, aquaglyceroporins, K transporters, and osmoprotectant transporters, whose structure and functioning we briefly discuss.

Porins are proteins found only in the external membrane of Gram negative bacteria; three molecules belonging to the same type of porin together form a pore across the external membrane. This pore is filled with water and allows the passage of some ions and molecules according to the concentration gradient. Bacterial porins are not static, permanently open pores, but can switch between short-lived open states and closed conformations, and can also remain in an inactivated, non-ion conducting state for prolonged periods [11]. For instance, in *E. coli* several porins are known, such as OmpC, OmpF (where Omp stays for "outer membrane protein"), and so on. The total quantity of OmpC and OmpF are fairly constant, but their ratio changes with the osmolarity of the medium. The proportions of these porins can be different for cells grown under given conditions. For instance, in *E. coli* the increase in the osmolarity of the growing medium induces an increase in the synthesis of the porin OmpC (and a decrease in the synthesis of the porin OmpF). Thus, at high osmolarity the outer membrane in *E. coli* contains more OmpC and less OmpF [11].

Aquaporins are transmembrane water channel proteins, they are tetrameric assemblies of four subunits each containing its own aqueous pore (more details in [40]). Glycerol facilitators are channels permeable to glycerol or small uncharged molecules, whereas aquaglyceroporins are a new class of water channels also permeable to glycerol, but to a lesser degree than glycerol facilitators [13, 23].

K transporters act in this way: when *E. coli* is placed in a medium of high osmolarity (not produced by K ions), the cells respond by synthesizing a specific uptake system for potassium, whose specific and controlled entry into the cell will contribute to maintaining the internal osmotic pressure at optimum. In *E. coli* there are Trk transporter, Kdp transporter, and a sensor kinase (KdpD) that catalyze potassium uptake with different kinetic parameters during osmotic upshifts (for more details see [48]).

Osmoprotectant transporters become active also during osmotic upshifts, thus enabling the cell to accumulate osmoprotectants from the external medium. The best known osmoprotectant transporters are the ProP transporter in *E. coli* and the Betp transporter in *Corynebacterium glutamicum* [48].

So, the interplay between different structures involved in the response to osmotic challenges (either upshifts or downshifts) at short and long term levels, all hot topics in microbiology, could be positively affected by contributions from P systems. Moreover, it has to be stressed that during osmotic challenges other fundamental processes (e.g., respiration) inside the cell are also affected [3, 48]. The integration of these changes in a more generalized mathematical model could help in modeling biological processes occurring in the entire cell.

Acknowledgement. This work was partially supported by a contribution of the EU Commission under The Fifth Framework Programme, project "MolCoNet" IST-2001-32008.

References

1. I.I. Ardelean: Molecular Biology of Bacteria and its Relevance for P Systems. In [38], 1–18.
2. I.I. Ardelean, D. Besozzi: Mechanosensitive Channels, a Hot Topic in (Micro)Biology: Any Excitement for P Systems? In *Brainstorming Week on Membrane Computing*, Tarragona, February 2003 (M. Cavaliere, C. Martín-Vide, Gh. Păun, eds.), Technical Report No 26/03, URV Tarragona, 2003, 32–36.
3. I.I. Ardelean, S. Tunaru, M. Hagemann, M. Scharnagl, G. Zarnea: Mechanosensitive Channels are Involved in the Enhancement of Respiration After Osmotic or Saline Upshocks in Synechocystis SP.PCC 6803. *Proc. Rom. Acad. Series B.*, 2, 3 (2000), 227–232.
4. F. Bernardini, V. Manca: Dynamical Aspects of P Systems. *BioSystems*, 70, 2 (2003), 85–93.
5. F. Bernardini, V. Manca: P Systems with Boundary Rules. In [38], 107–118.
6. D. Besozzi, I.I. Ardelean, G. Mauri: The Potential of P Systems for Modeling the Activity of Mechanosensitive Channels in *E. coli*. In *Pre-Proceedings of Workshop on Membrane Computing–WMC03* (A. Alhazov, C. Martín-Vide, Gh. Păun, eds.), Technical Report No. 28/03, URV Tarragona, 2003, 84–102.
7. D. Blain, M. Garzon, S.Y. Shin, B.T. Zhang, S. Kashiwamura, M. Yamamoto, K. Kameda, A. Ohuchi: Development, Evaluation and Benchmarking of Simulation Software for Biomolecular Computing. *Journal of Natural Computing*, 3, 4 (2004), in press.
8. P. Blount, M.J. Schroeder, G. Kung: Mutation in a Bacterial Mechanosensitive Channel Changes the Cellular Response to Osmotic Stress. *J. Biol. Chem.*, 272 (1997), 32150–32157.
9. P. Blount, S. Sukharev, C. Kung: A Mechanosensitive Channel Protein and its Gene in *E. Coli*. *Gravitational and Space Biology Bulletin*, 10, 2 (June 1997).
10. L.N. Csonka: Physiological and Genetic Responses of Bacteria to Osmotic Stress. *Microbiol. Rev.*, 53, 1 (1989), 121–147.
11. A.H. Delcour: Function and Modulation of Bacterial Porins: Insight From Electrophysiology. *FEMS Microbiol. Lett.*, 151 (1997), 115–123.
12. L. Deng: Generalized Mersenne Prime Numbers and Its Application to Random Number Generation. In *Monte Carlo and Quasi-Monte Carlo Methods* (H. Niederreiter, ed.), Springer, Berlin, 2004, 167–180.
13. A. Froger, J.P. Rolland, P. Bron, V. Lagree, F. Le Caherec, S. Deschamps, J.F. Hubert, I. Pellerin, D. Thomas, C. Delamarche: Functional Characterization of a Microbial Aquaglyceroporin. *Microbiology*, 147 (2001), 1129–1135.

14. E.A. Galinsky, H.G. Truper: Microbial Behavior in Salt-Stressed Ecosystems. *FEMS Microbial. Rev.*, 15 (1994), 95–108.
15. M. Garzon, ed.: Biomolecular Machines and Artificial Evolution. *Genetic Programming and Evolvable Machines*, 4, 2 (2003), Kluwer Academic Publishers.
16. M. Garzon, D. Blain, A. Neel: Virtual Test Tubes for Biomolecular Computing. *Journal of Natural Computing*, 3, 4 (2004), in press.
17. M.H. Garzon, E. Drumwright, R.J. Deaton, D. Renault: Virtual Test Tubes: A New Methodology for Computing. *Proc. 7th Int. Symposium on String Processing and Information Retrieval*, A Coruña, Spain, IEEE Computer Society Press, 2000, 116–121.
18. M.H. Garzon, C. Oehmen: Biomolecular Computation in Virtual Test Tubes. *Proc. 7th Int. Workshop on DNA-Based Computers, DNA 2001* (Revised Papers), LNCS 2340, Springer, 2001, 117–128.
19. E. Gentle: *Random Number Generation and Monte Carlo Methods.* Springer, Berlin, 2003, p. 51.
20. O.P. Hamill, A. Marty, E. Neher, B. Sakman, F.J. Siqworth: Improved Patch Clamp Technique for High Resolution Currents Recording from Cells and Cell-Free Membranes Patches. *Pflugers Arc.*, 391 (1981), 85–100.
21. O.P. Hamill, B. Martinac: Molecular Basis of Mechanotransduction in Living Cells. *Physiological Reviews*, 81, 2 (April 2001).
22. L.H. Hartwell, J.L. Hopfield, S. Leibler, A.W. Murray: From Molecular to Modular Cell Biology. *Nature*, 402 (1999), C47–C52.
23. S. Hohmann, R. M. Bill, G. Kayingo, B. Prior: Microbial MIP Channels. *Trends Microbiol.*, 8 (2000), 33–38.
24. http://www.e-cell.org.
25. H. Kitano: Systems Biology – A Brief Overview. *Science*, 295 (2002), 1662–1664.
26. A. Kloda, B. Martinac: Mechanosensitive Channels in Archaea. *Cell. Biochem. Biophys.*, 34 (2001), 349–381.
27. A.C. Le Dain, N. Saint, A. Kloda, A. Ghazi, B. Martinac: Mechanosensitive Ion Channels of the Archaeon Haloferax Volcanii. *J. Biol. Chem.*, 273, 20 (1998), 1216–1219.
28. K. Lippert, E.A. Galinsky: Enzyme Stabilization by Ectoine-Type Compatible Solutes: Protection Against Heating, Freezing and Drying. *Appl. Microbiol. Biotechnol.*, 37 (1992), 61–65.
29. L. M. Loew, J. C. Schaff: The Virtual Cell – A Software Environment for Computational Cell Biology. *TRENDS in Biotechnology*, 19, 10 (2001), 401–406.
30. B. Martinac, E. Perozo: Physical Principles of Mechanosensitive Channel Gating by Bilayer Deformation Forces. *J. Physiol.*, 544.P (2002), 16–17.
31. E. Mayer: *This is Biology.* The Belknap Press of Harward University Press, 1998.
32. P.C. Moe, G. Levin, P. Blount: Correlating a Protein Structure with Function of a Bacterial Mechanosensitive Channel. *The Journal of Biological Chemistry*, 275, 40 (October 2000), 31121–31127.
33. Y. Nakamuru, Y.Takahasi, T. Unemoto, T. Nakamura: Mechanosensitive Channel Functions to Alleviate the Cell Lysis of Marine Bacterium Vibrio Alinilyticus, by Osmotic Downshock. *FEBS Lett.*, 444 (1999), 170–172.
34. A. Oren: *Halophilic Microorganisms and Their Environments.* Kluwer Academic Publishers, 2002.

35. A. Oren: The Bioenergetic Basis for the Decrease in Metabolic Diversity at Increasing Salt Concentrations: Implications for the Functioning of Salt Lake Ecosystems. *Hydrobiologia*, 466 (2001), 61–72.
36. B. Palsson: The Challenges of In Silico Biology. *Nature Biotechnology*, 18 (2000), 1147–1150.
37. Gh. Păun: *Membrane Computing. An Introduction.* Springer, Berlin, 2002.
38. Gh. Păun, G. Rozenberg, A. Salomaa, C. Zandron, eds.: *Membrane Computing. Proceedings of Membrane Computing International Workshop WMC–CdeA2002, Curtea de Argeş, Romania, August 2002.* LNCS 2597, Springer, Berlin, 2003,
39. J.C. Schaff: A General Computational Framework for Modeling Cellular Structure and Function. *Biophys. J.*, 73 (1997), 1135–1146.
40. S. Scheuring, P. Titmann, H. Stahlberg, P. Ringler, M. Borgnia, P. Agre, H. Gross, A. Engel: The Aquaporin Sidedness Revisited. *J.Molcc. Biol.*, 229 (2000), 1271–1278.
41. S.I. Sukharev: Mechanosensitive Channels in Bacteria as Membrane Tension Reporters. *The FSEB Journal*, 13 (Supplement 1999), 55–61.
42. S.I. Sukharev, M. Betanzos, C.S. Chiang, H.R. Guy: The Gating Mechanism of the Large Mechanosensitive Channel MscL. *Nature*, 409 (2001), 720–724.
43. S.I. Sukharev, P. Blount, B. Martinac, C. Kung: Mechanosensitive Channels of Escherichia Coli. The MscL Gene, Protein, and Activities. *Annu. Rev. Physiol.*, 59 (1997), 633–657.
44. S.I. Sukharev, W.J. Sigurdson, C. Kung, F. Sachs: Energetic and Spatial Parameters for Gating of the Bacterial Large Conductance Mechanosensitive Channel, MscL. *J. Gen. Physiol.*, 113 (1999), 525–539.
45. M. Tomita et al.: E-CELL – Software Enviroment for Whole-Cell Simulation. *Bioinformatics*, 15, 1 (1999), 72–84.
46. M. Tomita et al.: The E-CELL Project – Towards Integrative Simulation of Cellular Processes. *New Generation Computing*, 18 (2000), 1–12.
47. J.M. Ward: Patch-Clamping and Other Molecular Approaches for the Study of Plasma Membrane Transporters Demystified. *Plant Physiol.*, 114 (1997), 1151–1159.
48. J.M. Wood: Osmosensing by Bacteria Signals and Membrane-Based Sensors. *Microbiol. Mol. Biol. Rev.*, 63, 1 (1999), 230–262.

Chapter 3
P Systems for Biological Dynamics

Luca Bianco, Federico Fontana, Giuditta Franco, Vincenzo Manca

University of Verona
Department of Computer Science
Strada Le Grazie 15, 37134 Verona, Italy
{bianco,fontana,franco}@sci.univr.it, vincenzo.manca@univr.it

Summary. P systems have clear structural analogies with the cell. However, certain difficulties arise when one attempts to represent a biomolecular process using these systems. This chapter suggests some ways to overcome such difficulties and to provide P systems with further functionalities aimed at increasing their versatility in the modeling of biomolecular processes. Concepts from state transition dynamics are taken to put P systems in a general analysis framework for dynamical discrete systems. An explicit notion of environment is proposed to provide P systems with a regulatory and constraining agent, as real biomolecular processes must deal with. The chapter focuses on a new rewriting strategy inspired by biochemistry, in which reactivities play a central role in driving the rules as happens during biochemical reactions. Tests on an algorithm implementing rewriting with reactivities, realized on a simulator called *Psim*, show the capability of this algorithm to express several processes with precision, particularly those presenting oscillatory phenomena. Finally, an analysis of the process of leukocyte recruitment is also performed using *Psim*.

1 Introduction

We know that the simplicity of a computing device does not limit its power to solve problems, assuming that its design criteria follow certain specifications and enough space is provided to represent the data and to store the processing rules, and enough time is allocated to the device to compute the solution. This fact became clear after the Turing machine was designed and the proofs of universality and equivalence with other computing machines and formalisms (most of them being as simple as the architecture proposed by Alan Turing) were given. So, why is a modern computer so different in practice from a Turing machine, if it has the same computational power?

The answer is obvious: a modern computer architecture is much better interfaced with the external world than a Turing machine, and, consequently, a task can be more easily implemented on it. Algorithms and features that would need indefinite time to be organized to run on a Turing or a von Neumann machine, are rapidly implemented over a modern computer instead. On the other hand, Turing machines and equivalent "ideal" architectures have proved to be invaluable for crossing the bridge between computer theory and practice. In some sense, we can say that ideal machines cannot be avoided, although any practical application of them must move through the existence of more elaborate systems.

In 1998 *P systems* were presented as a new model of computation [27]. We argue that this model can be considered and developed in such a way as to become an analogue of Turing machines, playing the role of a mathematically idealized model for biological systems. The following discussion will provide specific arguments for this claim.

Before P systems, some other classes of rewriting systems had already shown the ability of expressing specific biological phenomena [32, 20, 13]. P systems move a step further: they have clear structural analogies with the cell; in particular they model several features of the biological membranes (for this reason they are often referred to as *membrane systems*). Moreover, the transitions occurring in these systems recall certain evolution processes that take place in a living cell.

From a formal viewpoint, P systems satisfy a result of universality even in their basic definition [27]. In this sense they have all the computational power needed to capture a biomolecular process – provided that we are able to arrange it into an algorithmic procedure. In addition to this, the similarities existing between P systems and (at least some aspects of) biological cells suggest that P systems are able to represent the same biological process in a meaningful way, that is, not only to compute it as any universal machine would, but also to provide insight into the biological mechanisms determining and controlling the process via the observation of the transitions of the system.

Unfortunately, for most of the classes of P systems considered so far, this is true only to some extent. Modeling specific biological activities inside a P system is not an easy task. Many alternative constructs derived from the basic definition of P systems have been proposed, sometimes capturing crucial aspects of the biology of cells such as *thickness*, *polarity*, *transport* via *symport* and *antiport* rules, *catalysis*, *dissolution*, *polarity*, *permeability*, *inhibition*, *promotion*, communication via *carriers*, and *energy* [29, 26, 25], sometimes importing paradigms from other formal systems also having biological implications, such as *splicing* and object structuring (in the form of strings) [30]. All these alternative constructs exhibit properties of universality; hence they represent a first, necessary attempt to move P systems closer to the world of biomolecules while preserving their computational power.

Nevertheless there are some aspects, crucial in almost any study of biomolecular processes, that the traditional formulations of P systems do not take into account (at least not so explicitly as to turn into versatile constructs for biological applications):

- *Dynamics* of biosystems. The halting of a P system indicates that a computation has terminated successfully, but the dynamical behavior of biomolecular processes has significant relevance in the description of the processes themselves. This means that two or more processes that terminate with identical configurations may move through completely different transitions. Thus, in the context of living organisms it is more appropriate to consider the dynamical patterns of the "life" evolving in a given environment. The knowledge and classification of these patterns is a preliminary task for understanding or influencing some behaviors (possibly harmfully) for specific purposes.
- *Environmental energy* and *resources.* The resources available in the environment play a major role in the control a biomolecular process. The existence in the environment of elements, which can act as catalysts or provide the energy needed for the biochemical elements to react, can radically change the nature of a process. In particular, an environment which periodically feeds the system with resources can transfer properties of periodicity to the system as well.
- *Asynchronous* system control. Biomolecular mechanisms are the result of many individual local reactions, each of them formed by processes whose extension is limited in time and space. These processes interact with each other by means of specific communication strategies, in a way that they finally exhibit a (sometimes surprising) overall coordination. In this sense, and despite the coordination, biomolecular processes are asynchronous.

Clearly, these aspects are closely related one another: shifting the focus on the system dynamics means that less attention is paid to the final configuration of the system; meanwhile, the continuous control of the resources needed by the process to evolve is critical to drive the system dynamics along a specific trajectory. The environment itself has the role of a "supervisor" in the process control, since it becomes responsible for a sort of external input whose effects in the system propagate via local reactions.

How do today's P systems deal with the points just outlined? About the first point, we know that P systems are intended to "consume" the available resources in a maximally parallel way during the rewriting of symbols. Imposing this property, all the symbols that are present in the system in a given configuration become potential resources: they are consumed as much as possible, and new symbols are produced as a consequence of that action. In other words, maximal parallelism constrains the system to consume all the available resources during a transition.

We know that we can regulate the system evolution by adding auxiliary symbols (and corresponding cooperative rules which use such symbols) or, alternatively, by providing the system with *priority* constraints on the rules to form (sometimes complex) relationships of precedence for the rewriting rules. But, this modification of the system structure is not guaranteed to have a corresponding biological counterpart. So, we are looking for an alternative, more biologically founded strategy for the regulation of parallelism.

About the second point, we know that nonterminating processes are of key importance in the study of *periodicity* and *quasi-periodicity*, two aspects whose detection is important for understanding many biological processes [33]. From this viewpoint, many existing types of P systems do not provide versatile tools to handle periodicity.

The question of resource availability (this issue leads us to the third point) is unavoidable in the study of biomolecular phenomena. P systems, in their native definition, do not take any energetic constraint into account. So, it is not unfeasible for them to use indefinitely large amounts of resources to perform a computation: clearly, this cannot happen in nature. Other types of P systems have been proposed in which energy is considered as a constraining factor, although in those systems the environment, intended as a place to exchange resources, does not play a central role [31, 12].

Finally, P systems are organized in a way such that their evolution is synchronous, i.e., a global clock triggers the production of new symbols inside all membranes. This limits their versatility in modeling asynchronous phenomena.

Coming back to our initial considerations on ideal and practical computing devices, it is our opinion that P systems are still at an early development stage. On the one hand, their simplicity does not limit their computational power and, in fact, this simplicity has allowed us to prove important results of universality. On the other hand, much research still has to be done to get P systems closer to the world (especially, but not only) of biomolecular applications while keeping them theoretically well-founded. Turing machines have found in modern PCs their applied counterpart: it is time, now, to look for ways that turn P systems into *real* biomolecular computing devices.

The considerable research aimed at proving the well-foundedness of the various types of P systems (see, e.g., [28]) is the background of the presentations in this chapter. However, we have focused our effort on addressing some novel theoretical and practical issues especially oriented to biomolecular computing – this applicative direction is followed also by other groups [6, 3, 1].

First, we consider a new perspective (for many aspects still in progress) according to which P systems are cast in a dynamical framework. In this perspective, we will characterize the transitions and the space state of a discrete system using *state transition dynamics* [23]. Then we will introduce *P systems with boundary rules* (PB systems) and *PB systems with environment* (PBE systems) [5, 4] as constructs in which the concept of *environment cycle* is proposed to represent cyclic biological behaviors; meanwhile the definitions of periodicity and quasi-periodicity of state transition dynamics of P systems apply at an operative level.

Next, we propose to observe the rules of a rewriting system from a different viewpoint: not only will they produce new symbols starting from existing ones, they will also be part of a system reproducing a reaction in which the application of every rule changes the relative amounts of reacting substances present before and after the production. Moreover, such relative amounts will influence the *reactivity* of the rules in a way that their application will be dependent on the substance concentration, as normally happens in biochemical phenomena. An application of this model to some known biochemical problems – described so far in terms of differential equations – is proposed, in particular in the simulation of the *Brusselator*, a simplified model of the Belousov-Zabotinskii (BZ) reaction having great relevance in biochemistry [33, 15, 37, 24].

Finally, we look at an extended version of a P system in action, aimed at simulating some mechanisms in the human immune system when it activates the *leukocyte selective recruitment* against an inflammatory process.

Closing this section, we want to stress once again the importance that the dynamical characterization and the environment have in our vision of a class of P systems especially intended to serve as biomolecular computers:

- the system dynamics, as a way to capture recurrent (or however determined) behaviors by "clues," that are left even by those biomolecular processes that cannot be decoded due to their apparently chaotic behavior;
- the environment, as an entity that regulates parallelism, alters an otherwise unavoidable terminal state, provides resources, and acts as a delocalized control for the system; as we have seen, these issues are intimately connected with the dynamics of our system.

Although this survey does not pretend to define a comprehensive application framework for membrane systems, the authors nevertheless hope that the issues proposed here will suggest possible points of investigation from where to carry on some more applied research on P systems. In the following sections of this chapter we will constantly refer to the basic definitions and notation of P systems and of *multisets* introduced by Păun [28].

2 The Dynamics of Discrete Systems

Continuous systems are often described in terms of differential equations. A common strategy to figure out such equations consists of writing down equilibrium conditions for *infinitely small* physical units such as time units, dt, and spatial volume units, d$\boldsymbol{s}$. From there, a classic approach to discrete system modeling consists of picking up the continuous phenomenon (i.e., the differential problem describing it) and then producing a discrete model of it according to a given discretization method. Finally a simulation is run, provided that the discrete model respects certain stability conditions. This approach is largely used in the practice of system modeling.

A discrete model differs from the continuous phenomenon it comes from. Sometimes this discrepancy can be arbitrarily reduced, that is, the model precision is proportional to the granularity with which the continuous phenomenon is reproduced in the discrete domain.

We are here interested in those physical phenomena whose characterization (that is, the information needed to describe them) is *inherently* discrete. In this case a discrete model can represent the physical phenomenon completely. This is the case of many biomolecular processes (think, for instance, of DNA replication [30]).

For the above reasons – particularly for the last one – we are especially interested in discrete systems regardless of any specific relationship with a continuous system, and of any prior argument on the precision of the discrete solution versus the continuous one. Furthermore, in most cases of biological interest the discrete paradigm can be extended even to the values the system assumes during its evolution, in a way that numerical values are conveniently substituted by symbols.

The *state transition dynamics* formalism considers a system defined in a discrete domain assuming discrete values. It studies properties such as state orbits and trajectories, periodicity, eventual periodicity and divergence, recurrence of states, attractors, and fixed points. By means of this analysis we are able not only to characterize such properties, but also to make important considerations about determinism vs. nondeterminism, and about regularity vs. *chaos*.

To give an idea of the characterization given by state transition dynamics, we report here the most important definitions and results (sometimes in a quite informal presentation). For further details, discussions, and mathematical insight we refer you to [23] where a general approach to discrete system dynamics has been investigated in its formal and computational aspects.

Definition 1. *A state transition dynamics is a pair* (S, q) *where* S *is a set of states and* q *is a function from* S *into its power set:*

$$q : S \to \mathcal{P}(S)\,.$$

By calling *quasi state* any subset $X \subseteq S$, and extending the application of q over quasi states, i.e.,

$$q(X) = \bigcup_{x \in X} q(x)\,,$$

we map quasi states to quasi states by means of q to form *orbits*, and characterize specific *trajectories* along these orbits by means of the following definitions.

Definition 2. *An* X-orbit *is a sequence* $\{X_i\}_{i\in\mathbb{N}}$ *of quasi states such that*

$$\begin{array}{ll} X_0 = X, & i = 0, \\ X_i \subseteq q(X_{i-1}), & i > 0. \end{array} \tag{1}$$

An x-trajectory *is a function* $\xi : \mathbb{N} \to S$ *such that*

$$\begin{array}{l} \xi(0) = x, \\ \xi(i) \in q\big(\xi(i-1)\big), \;\; i > 0. \end{array} \tag{2}$$

Let us denote by q^i the composition of q repeated i times, and let us define $q^*(x) = \bigcup_{i \in \mathbb{N}} q^i(x)$. We refer to the following special trajectories as *flights* and *blackholes*:

Definition 3. *An x-trajectory is an* x-flight *if it is an injective function on* $\mathbb{N}$. *An x-flight is an* x-blackhole *if $q^*(x) \subseteq \xi(\mathbb{N})$, where $\xi(\mathbb{N})$ is the image set of ξ.*

When S is made of symbolic values, the relation $y \in q(x)$ induced by q between two states, x and y, is conveniently expressed using the notation typical of rewriting systems: $x \rightarrow y$. Note that we can easily introduce non terminating computations as long as q is total.

It is clear that the notion of a dynamical system defined above is nondeterministic, because any state can transform into a set of possible states, though an equivalently expressive deterministic system where states are the quasi states of the original system can be figured out. The nondeterministic aspect is essential for the modeling of many phenomena.

We now give a characterization of the evolution in these systems.

Definition 4. *An X-orbit is* periodic *if $q^n(X) = X$ for some $n > 0$. An orbit is* eventually periodic *if $q^{n+k}(X) = q^k(X)$ for some $k, n > 0$. In this case k is called the* transient *and n the* period.

Definition 5. *An X-orbit is* $\Omega(f(n))$-divergent *with respect to a function $\mu : S \rightarrow \mathbb{N}$, called Ljapunov function, if $\mu(q^n(X))$ has order $\Omega(f(n))$. A similar definition holds for the order of divergence $O(f(n))$.*

Definition 6. *A state x is a fixed point if the transition relation transforms it into itself deterministically, that is, $q(x) = \{x\}$.*

Periodicity and eventual periodicity are properties with strong computational significance. It can be shown that, in a suitable computational framework where every machine finds a counterpart in a corresponding state transition dynamics, the periodicity decision problem turns out to be computationally equivalent to the termination problem [23]:

Proposition 1. *Given a computationally universal class of machines, the (eventual) periodicity of the related dynamical systems is not decidable.*

Affine to periodicity (but weaker) is recurrence:

Definition 7. *A state x is* recurrent *if $x \in q^n(x)$ for some $n > 0$. A state x is* eternally recurrent *if for all $n > 0$ such that $y \in q^n(x)$ there is $m > 0$ such that $x \in q^m(y)$.*

A system dynamics is ultimately characterized by its *attractors*, which in very first approximation can be seen as quasi states into which the system must fall in the end. First, we say that a Y-orbit is *included* in an X-orbit if the former sequence is contained in the latter sequence, and *eventually included* if the former sequence is included in the X-orbit except for a finite number of quasi states.

We call *basin* a set $B \subseteq S$ such that $q(x)$ is included in B for every state $x \in B$. Inside a basin we possibly find an *attracting set* A, i.e., a subset which eventually includes the x-orbit of every state $x \in B$. If A is minimal under set inclusion, i.e., no subsets (even those made up of a single state) can be removed from A without causing the loss of the attracting property, then our attracting set is an attractor.

A complete characterization of attractors requires more definitions than those recalled in this chapter; we refer to [23] for details. In particular, we have outlined here only the so-called *unavoidable* attracting sets and the corresponding attractors. It can also be shown (we omit here all the intermediate results, along with further definitions) that a state transition dynamics can have three different types of attractors:

1. *periodic attractors*, that is, periodic orbits (fixed point attractors are a special case);
2. *eternally recurrent blackholes*;
3. *complex attractors*, that is, a combination of the two previous cases.

The notion of an attractor opens a wider perspective on the classical notion of calculus, which seems to fit better with a computational interpretation of biological systems. In fact, these systems do not compute states that encode results (according to the Turing paradigm); rather, they "compute" attractors or stable regimes satisfying behavioral requirements that respect certain conditions for life.

Particularly interesting are chaotic attractors. Life chooses forms approaching chaos, while trying not to fall into it. While getting closer to this total freedom, simple cycles take the rich and complex forms featured by evolution and adaptation. The expression *at the edge of chaos* [19], in fact, expresses the typical condition in which biological systems explore the space of computable forms moving along a threshold that lies between biological *status quo* and chaotic evolution, both of them destructive for a species. However, at the edge of this threshold lies that constructive evolution life is constantly searching for.

The (nondeterministic) notion of orbit becomes useful also in an attempt to give a characterization of *discrete chaos*. Looking at chaos from this perspective gives further insight into the meaning of nondeterminism.

Chaotic dynamical systems are characterized by the following features:

- Global recurrence. In a chaotic dynamical system the set of all states is its own attractor, which is also called a *strange* attractor. In other words, a chaotic behavior is a global property that cannot be decomposed into distinct parts.
- Sensitivity to initial conditions. This requirement implies an exponential divergence of orbits where points that are "near" become exponentially

"far" in time. This aspect can be viewed as an explosion of orbits, or as an "informational drift" along the orbits, and it is relative to some Ljapunov function with respect to some measure of distance between states.
- Ubiquitous periodicity. This property refers to the erratic aspect of chaos: orbits are wandering everywhere and forever, that is, explosions of orbits are mixed with orbit implosions in such a way that the dynamics return periodically to themselves, according to their intrinsic recurrence, but these periods are endlessly overlapping each other.

Such features can be expressed in the context of state transition dynamics [23].

The definition of chaos expressed in terms of state transition dynamics allows us to consider chaos of dynamical systems very similarly to the deterministic chaos of continuous systems (logistic maps, Bernoulli shifts, Manneville maps [9, 7]), which, although ruled by very simple dynamics, represent evident chaotic behaviors. What makes these systems intrinsically chaotic is the essential role of quasi states in their descriptions. In fact, their dynamical systems are defined on states given by real numbers, but these numbers are always expressed by some of their finite approximations, that is, by rational numbers. In this sense, one may view an infinite set of states as a rational number, that is, a quasi state which comprises all the real numbers that, at some level of approximation, share the same rational number. Therefore, the sensitivity to initial conditions corresponds to the exponential growth of a Ljapunov function along the orbits associated with the finite approximations of the states of the system. Analogously, the overlapping of periods in these systems corresponds to the eventual intersections of their periodic orbits, when we consider the quasi states which correspond to the finite representations of their states.

In conclusion, what is called deterministic chaos does not differ from nondeterministic chaos. The difference is only a matter of the way orbits are defined. In deterministic chaos these are introduced by the intrinsic approximation of states; in the nondeterministic chaos of state transition dynamics, the orbits are defined by state transition relations that provide many possible states that can be reached from a single state. But it is very important to remark that neither is determinism synonymous to predictability, nor is nondeterminism synonymous to unpredictability. Indeed, a system that is deterministic but chaotic becomes unpredictable, and a nondeterministic system can be predictable in several aspects [14].

Similar behaviors have also been observed in other constructs, such as Kauffman networks and cellular automata [17, 38, 39]. These systems show that many relevant characteristics of their dynamical behavior are consequences of the relationships existing between the transition function and the state structure. Parameters like *connectivity*, *channeling*, *majority*, *input entropy*, and figures taken from Derrida plots might inspire the search for similar quantities in P systems. Also, many concepts of formal language theory can be revisited from the perspective of state transition dynamics: for instance, the languages generated by grammars or recognized by automata are special cases of attractors.

These considerations of nondeterministic chaos in discrete systems allow to gain insight into those biomolecular behaviors that must be classified as chaotic. In particular, such a notion of chaos might help in identifying orbits from apparently "unreadable" biomolecular process dynamics, as life cycles depict periodicities that are masked or blurred by nondeterministic shifts away from the main trajectories, but leave fingerprints of these trajectories along the way.

Conversely, it could enable us to specify richer and more "open" dynamics than those defined by other representations, for instance the dynamics provided by the (deterministic) solution of a differential problem. In Section 4 we will see some examples, both formulated in terms of a differential problem and analyzed using a discrete dynamical (string transition) system.

3 Resource Drawing from the Environment

The boundary of a P system with the external world is represented by the skin membrane. In most types of P systems every membrane limits the scope of the rules in such a way that, by definition, they can generate and/or consume symbols only in the region delimited by their own membrane, and the skin does not make an exception to this definition. The environment has no special roles other than providing the (possibly infinite) amount of resources needed by the system to evolve and receiving the (possibly indefinitely many) symbol-objects that, once properly decoded, give the result of the computation the system has performed.

PB systems, PBE systems, and PBE systems with resources [5, 4] enrich the P construct by giving a more active role to the boundary and to the environment. This idea has a strong foundation in some typical features often exhibited by biological systems, such as:

- *periodicity and quasi-periodicity.* Life is always related to temporal cycles where, even if some temporal irreversibility is intrinsic, many parameters change periodically and some basic rhythms are preserved;
- *stability and adaptability.* Biological systems tend, within some limits, to keep their "form" and their basic "behavior" even if their external world changes;
- *growth and degeneration.* A living organism is able to demonstrate a correct life cycle when it maintains some basic oscillating reactions in time.

In this sense, periodical behavior, resource availability, and influence of the environment to the system are in close relationship to each other.

Recalling the basic construct of a P system, and in particular the notion of a *configuration* as a string μ encoding the membrane topology and the multiset contained in every membrane [27], we define a *PB system* in the following way:

Definition 8. *A P system with boundary rules (PB system) is a construct*

$$\Pi = (V, \mu_0, R, i_O),$$

where:

(i) V *is an alphabet of symbols;*
(ii) μ_0 *is the initial configuration;*
(iii) R *is a finite set of rules of the following two forms:*
- $xx' \, [_i \, y'y \rightarrow xy' \, [_i \, x'y$, *for* $x, y, x', y' \in V^*$ *and* $1 \leq i \leq m$ (communication rules)*;*
- $[_i \, y \rightarrow [_i \, y'$, *for* $y, y' \in V^*$ *and* $1 \leq i \leq m$ (transformation rules)*;*

(iv) $i_O \in \{1, \ldots, m\}$ *is the label of the output membrane.*

In addition to basic P systems we have essentially the communication rules of the form $xx' \, [_i \, y'y \rightarrow xy' \, [_i \, x'y$. By means of these rules we can move objects through membranes: if the membrane i contains the multiset $y'y$ and the multiset xx' is present outside the membrane i, then the multiset x' moves into the membrane i and the multiset y' is sent out from it; clearly, some of these multisets may be empty. The salient fact in the action of the communication rules is that they can "see" the immediate outside of the membrane region they belong to. Indeed, their nature recalls the *antiport* rules from P systems with *linked transport* [26].

The computational universality of PB systems was proved by showing that they are able, using three membranes, to characterize the recursively enumerable sets of vectors of natural numbers [4].

We present the notion of *environment cycle of period* k as the infinite sequence where k multisets $\beta_0, \beta_1, \ldots, \beta_{k-1}$ occur periodically in time. Now we can define a *PBE system.*

Definition 9. *A PB System with Environment (PBE system) is a construct*

$$\Pi = (V, \mu_0, R, E, R_E, i_O),$$

where:

(i) V, μ_0, R, i_O *are as in Definition 8;*
(ii) E *is an environment cycle of period* k*;*
(iii) R_E *is a finite set of rewriting rules on multisets of the form* $x \rightarrow y$*, for* $x, y \in V^*$ (environment rules).

The system configuration at time j, i.e., η_j, is related to the previous configuration η_{j-1} by the following relation, in which we make use of the dynamics functions q_E and q_R, related to R_E and R (see Definition 1), respectively mapping the environment configuration $\beta_{j-1}\gamma_{j-1}$ and the (overall) system configuration η_{j-1} at time $j-1$ onto respective new multisets

$$\eta_j = \beta_j \gamma_j \mu_j = \beta_j \, q_E(\beta_{j-1}\gamma_{j-1}) \, q_R(\eta_{j-1}) \,, \tag{3}$$

where:

- $\beta_j \in E$ is the multiset produced by the environment at time j;
- $\gamma_j = q_E(\beta_{j-1}\gamma_{j-1})$ is the environment configuration at time j resulting from mapping the previous environment configuration, i.e., the multiset $\beta_{j-1}\gamma_{j-1}$, to the multiset γ_j by means of q_E;
- $\mu_j = q_R(\eta_{j-1}) = q_R(\beta_{j-1}\gamma_{j-1}\mu_{j-1})$ is the internal system configuration at time j, resulting from applying q_R to the previous configuration, η_{j-1}.

Relation (3) is initialized with β_0 (the first multiset of the environment cycle), μ_0 (the initial configuration), and $\gamma_0 = \emptyset$.

Since the environment cycle E is periodic, it is not difficult to see that the sequence of configurations $\eta_0, \eta_1, \ldots$ produced by the PBE system, read as a sequence of quasi states, is eventually periodic in the sense of Definition 4. The peculiar aspect of PBE systems is the assumption of a periodic behavior of the environment; the system behavior is obtained consequently.

We now add resources to a PBE system as symbols of a finite set $\{r_i\}_{1 \le i \le h}$, with $h > 0$, hence obtaining a *PBE system with resources.*

Definition 10. *A PBE system with resources is a construct*

$$\Pi = (V, \mu_0, R, E, R_E, i_O),$$

where:

(i) V*,* μ_0*,* E*,* R_E*,* i_O *are as in Definition 9;*
(ii) the rules in R *have the following form:*
- $xx' \,[_i\, y'y\, r_j^k \rightarrow xy' \,[_i\, x'y$*, for* $x, y, x', y' \in V^*$*,* $1 \le j \le h$*,* $k > 0$*, and* $1 \le i \le m$*;* (communication rules)*;*
- $[_i\, y\, r_j^k \rightarrow [_i\, y'$*, for* $y, y' \in V^*$*,* $1 \le j \le h$*,* $k > 0$*, and* $1 \le i \le m$ (transformation rules).

Along with resources, *waste objects* can be also introduced in the definition of PBE systems with resources. The interplay between resource and waste objects makes PBE systems with resources satisfy a condition of *noncreativity*: every object produced by some rule is consumed by some other rule. In other words, a noncreative system defines a cycle where no object is created or destroyed, and every object is transformed into another one. The only new objects that are introduced in the system are provided by the environment cycle [4].

In the following example we exhibit a simple periodic PBE system with resources.

Example 1. Consider the system

$$\Pi_1 = (V, \mu_0, R, E, R_E, i_O),$$

where:

$V = \{a, b, r_1, r_2\}, \quad \mu_O = [_1\, a\, [_2\,]_2\,]_1,$
$R = \{[_1 a r_1 \rightarrow [_1 ab, \ \ b[_2 r_2 \rightarrow [_2 b, \ \ r_1[_1 \rightarrow [_1 r_1, \ \ r_2[_1 \rightarrow [_1 r_2, \ \ r_2[_2 \rightarrow [_2 r_2\},$
$E = \{r_1, r_1, r_2^2, \lambda, \lambda\}, \quad R_E = \emptyset, \quad i_O = 1.$

The behavior of this PBE system is described by the following sequence of transitions:

$$\begin{array}{rll}
\eta_0 = & r_1[_1 a[_2\ ]_2]_1 & \rightarrow \\
& r_1[_1 r_1 a[_2\ ]_2]_1 & \rightarrow \\
& r_2^2[_1 r_1 ab[_2\ ]_2]_1 & \rightarrow \\
& [_1 r_2^2 ab^2[_2\ ]_2]_1 & \rightarrow \\
& [_1 ab^2[_2 r_2^2]_2]_1 & \rightarrow \\
\eta_5 = & r_1[_1 a[_2 b^2]_2]_1 & \rightarrow \\
& r_1[_1 r_1 a[_2 b^2]_2]_1 & \rightarrow \\
& r_2^2[_1 r_1 ab[_2 b^2]_2]_1 & \rightarrow \\
& [_1 r_2^2 ab^2[_2 b^2]_2]_1 & \rightarrow \\
& [_1 ab^2[_2 r_2^2 b^2]_2]_1 & \rightarrow \\
\eta_{10} = & r_1[_1 a[_2 b^4]_2]_1 & \rightarrow \\
& \ldots &
\end{array} \tag{4}$$

In configuration η_5 we note that the observable membrane again assumes the value taken during the initial configuration η_0. Since the environment has started a new cycle of production in η_5, the environment configurations also again assume the value taken during the initial configuration. Hence, the initial sequence of rules must repeat starting from η_5, and so on, at every new 5-step cycle.

In conclusion, the system continues to cyclically repeat the same sequence of transitions. Actually, it can be proved that the observable sequence $\{X_i\}$ generated by Π_1 is eventually periodic with transient $k_0 = 0$ and period $k = 5$. In fact, the finite observable subsequence read along two configurations, η_{5n} and $\eta_{5(n+1)}$, $n \geq 0$, is always equal to $\{a, a, ab, ab^2, ab^2\}$ [4].

This example shows an intriguing analogy with a fundamental result coming from linear system theory [16]. According to this result, the excitation of a *linear system* with a pure sinusoid always produces, at the system output, another (scaled and time-shifted) sinusoid having the same frequency as the incoming one. In practice a transient is always present before the system goes to a *stationary* condition. After this transient the output becomes purely sinusoidal.

More formally, if we look at the linear system as an operator $\mathcal{F}$ mapping time-domain functions into time-domain functions, or *signals*, then, if we inject into it a purely sinusoidal signal

$$x(t) = A_x \sin(\omega_x t + \phi_x)$$

having *amplitude* A_x, *frequency* ω_x, and *phase* ϕ_x, then the system in its turn responds with a sinusoidal signal having different amplitude and phase, respectively A_y and ϕ_y, but the same frequency:

$$A_y \sin(\omega_x t + \phi_y) = \mathcal{F}\big[A_x \sin(\omega_x t + \phi_x)\big] .$$

It can be shown for linear systems [16] that, in the case of purely sinusoidal excitation, the amplitude of the (sinusoidal) output signal depends on the resonance properties of the system: the closer the frequency ω of the sinusoid to the *resonance frequency* Ω, the larger the amplitude of the output signal.

Although formally provable, this result has an immediate interpretation in basic system dynamics. If we consider, for example, an extremely simple linear dynamical system such as the pendulum, then it is easy to show that this system has its own natural oscillation frequency that depends only on its structural parameters (i.e., size and mass). This oscillation frequency *is* the resonance frequency of the system and, in fact, making a pendulum oscillate out of its resonance frequency (i.e., forcing it to an unnatural oscillation by repeatedly moving it with the hand, corresponding to injecting a "sinusoidal" non-resonant signal into the system given the limit of our capability to reproduce a sinusoidal signal with our hand) becomes as harder as the forcing sinusoid is farther from the resonance frequency. In more technical detail, linear system dynamics tells us that in a second order linear system having natural resonance frequency Ω (as the pendulum is), A_y will be as much greater as the frequency of the input signal is closer to Ω. Equivalently, we say that the system resonates at frequency $\omega = \Omega$ [16].

The aforementioned properties also hold when the linear system is defined in the discrete-time domain. In this case it can be shown that a similar relationship involving discrete-time sinusoids exists between the system input and output:

$$A_y \sin(\omega_x n + \phi_y) = \mathcal{F}_D\big[A_x \sin(\omega_x n + \phi_x)\big] ,$$

where we have substituted the continuous-time operator $\mathcal{F}$ with the discrete-time operator $\mathcal{F}_D$, and the continuous-time variable t with the discrete-time variable n.

The environment cycle constantly feeds a PBE system with a sequence of multisets that can be seen as a *symbolic* input signal. This signal is by all means periodic (of period k) since it can be univocally encoded by a discrete-time sinusoid having the same period. A general question arises: if a PBE system is constantly fed by a periodic environment cycle, is the system output (i.e., the content of the observable membrane) periodic as well and, if so, has this periodical behavior any affinity with that of a linear system?

Membrane systems seem to have poor affinity to linear systems: it suffices to say that they have structural analogies with the cell, which is definitely a nonlinear system. Going into more technical details, we can straightforwardly design a P system that, for instance, sends to the environment a periodic sequence made of k_S different symbols repeating forever, after it has been triggered by injecting the symbol S inside the skin, independently of any further symbol injected into the system after the first one. This P system has a clearly nonlinear behavior, as the period of its output is completely determined by the initial symbol injected rather than by the symbolic input sequence. Despite this, Example 1 describes a likely linear behavior in which the system resonance is driven by the period of the environment cycle.

These clues, once put together, suggest a possible research direction aimed at studying the system dynamics (in all its aspects of periodicity, chaos, and so on) not only through the analysis of the trajectories drawn by the system, as state transition dynamics does, but also by means of an analysis of the *structural properties* of the system. Predicting at least to some extent the dynamical behavior (either close to linearity, nonlinearity, nondeterminism, etc.) of a membrane system by an evaluation of its structure would mark a step forward in the application of P systems as models of biomolecular processes.

4 Oscillatory Biochemical Systems

In this section we focus our attention on some biochemical processes that exhibit periodic behavior. These processes are modeled by means of membrane systems, whose objects are reinterpreted in terms of *concentrations* of biological or chemical elements. Such membrane models are finally implemented on a simulator, called *Psim*, running on a normal PC.

First, we will explain the structure of the simulator. Then, we will illustrate some details of the algorithm it implements. Finally, we will discuss some experimental results obtained by the simulation of three well known dynamical systems.

4.1 Structure of the Simulator

Psim simulates a P system using the algorithm explained in Section 4.2. Its structure is inspired from some existing software especially designed for the simulation of P systems: the application developed in Prolog by Malita [22], that emphasizes the execution speed; the simulator written in LISP by Suzuki and Tanaka [36], particularly useful in long simulations of relatively simple systems; the *membrane simulator* developed by Ciobanu and Paraschiv [8], that is able to provide a graphical representation of the whole system through time.

The main features of *Psim* are (1) a flexible definition of the membrane structure via an XML file, (2) a user-friendly interface provided with printing and graphical capabilities, and (3) the possibility to save and reload intermediate results.

The main screen of the simulator is depicted in Figure 1. It is divided into three frames. The top frame contains two text fields, in which the user types

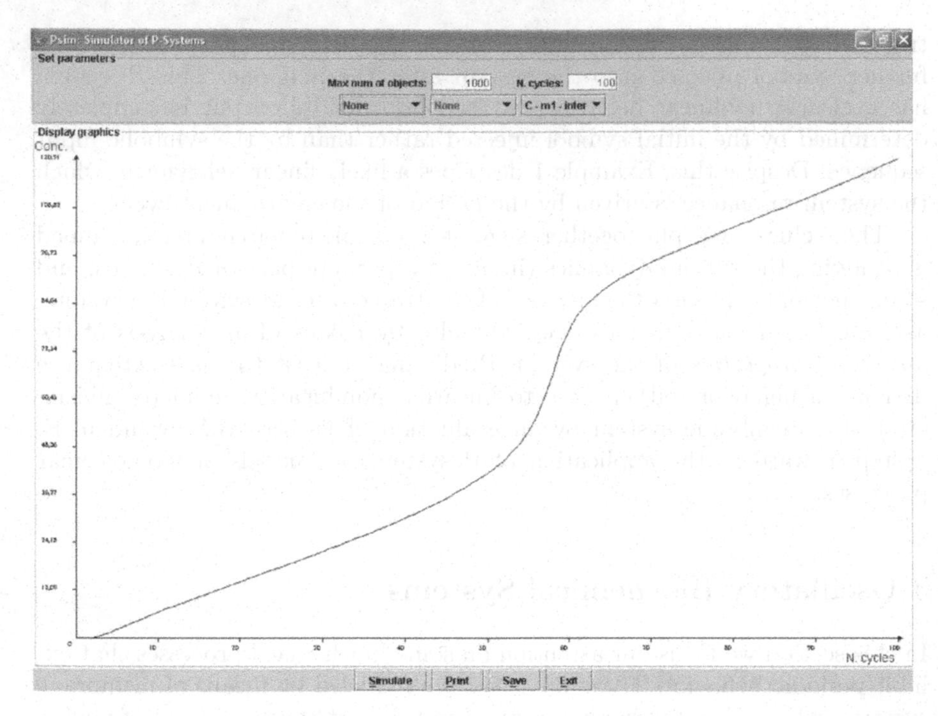

Fig. 1. Screenshot of the simulator.

the largest number of elements allowed in a membrane, and the number of simulation cycles; it also contains three drop-down boxes, allowing the user to select the type of output graph. The frame in the middle displays the computation results. The frame at the bottom contains four buttons, which tell *Psim* to start a simulation, to print results, to save the system state, or to exit.

The membrane structure is encoded into an XML file that is selected at launch time, at the command line, and then loaded by the simulator. A membrane is completely specified by a name, a position in the system, and a multiplicity. Each membrane is made of three different regions: an internal region called *in*, an external region called *out*, and, unlike previous implementations, a third region called *inter*. Similarly to the structure described in Section 5, this is the part of the membrane that can contain receptors: we can imagine it as an intermediate region located between the membrane inside and the outside, separated from these two regions in a way that both can see its content. This region, hence, can be used for interchanging objects and for communication and, thus, it can be opened or closed to allow or inhibit communication, respectively. Each of these parts can contain objects, as we will explain later, and also more membranes in such a way that the user can design more complex structures. In the end *Psim* models a multiset of membranes, each containing three (possibly empty) multisets of objects.

The information on the topological structure of the membranes is followed by the description of the objects that are initially present in the system. These objects are associated with a name, a multiplicity and a reference to a membrane containing them. The first two parameters are attributes of the tag *object* of the XML file, whereas the information about the outer membrane (thus, more generally, the information regarding the topology of the system) is encoded by nesting corresponding *object* tags.

The last part of the input file is the description of the rule set. The syntax describing such rules implements the notation of a PBE system (see Section 3). Each rule is associated with a single membrane, e.g., it describes its evolution by controlling the production of a subset of its objects, and is specified by a name, a reactivity factor that alters the uniform distribution of probability in the application of the rule, and a nonempty list of elements related to it.

These elements are of two kinds: *reactants* and *products*. Each is specified by a name, a position within the membrane structure, and a *reaction factor* (later on in this chapter we will talk about reactivities more extensively).

Intuitively a rule states that certain reactants take part in a reaction, with proportions given by their stoichiometric coefficients. In this way they generate corresponding products at a rate depending on some chemical and physical factors. According to these ideas every XML tag describing an element of a rule is composed of a reactant part and a product part. Each of these parts has one attribute identifying the types of objects it refers to, one attribute regarding the quantity of the objects, and a pair of attributes specifying the position of the element in the system in terms of the membrane and its relative region inside it. Together, these two parts indicate how a reactant transforms itself into a product in terms of types of objects, position within the system, and multiplicity.

This information, together with the reactivity coefficient of a rule, is the only information our algorithm needs to compute the contribution of a reaction in the production of an object over time. It is important to stress at this time that with such a syntax it is possible to implement both communication and transformation rules (according to the terminology used in Section 3).

Figure 2 contains a simple XML file that can be used as an input to the simulator. All previously discussed points can be recognized to take part in it.

From a theoretical point of view, and following the terminology proposed in [27], the system we have implemented can describe every family of P systems with $m \geq 1$ membranes, both with or without priorities between rules, catalysts, and various types of target indications. Features like *dissolution*, *electrical charge*, *thickness*, and *permeability* have not been implemented yet, but the flexible backbone of our system would eventually allow their introduction. Formally speaking, we can model systems of the form $P_m(Pri, Cat, i/o, n\pm, n\delta, n\tau)$, where $m > 0$.

The simulator output consists of a series of graphs representing the objects' multiplicity over time. The interaction between the user and the simulator is

```
<?xml version='1.0' encoding='utf-8'?>
<!DOCTYPE sistema SYSTEM "psim.dtd">

<simulator>
    <membrane id="m1" mult="1">
        <in>
            <object name="E" mult="1"/>
        </in>
    </membrane>

    <rule name="R1" rho="1"  membrane="m1">
        <elem reactant="A" r_mult="1" r_memb="m1" r_zone="out"
          product="C" p_mult="1" p_memb="m1" p_zone="inter"/>
    </rule>
    <rule name="R2" rho="1"  membrane="m1">
        <elem reactant="C" r_mult="1" r_memb="m1" r_zone="inter"
          product="D" p_mult="1" p_memb="m1" p_zone="inter"/>
    </rule>
    <rule name="R3" rho="1"  membrane="m1">
        <elem reactant="D" r_mult="1" r_memb="m1" r_zone="inter"
          product="E" p_mult="1" p_memb="m1" p_zone="in"/>
    </rule>
    <rule name="R5" rho="1"  membrane="m1">
        <elem reactant="E" r_mult="1" r_memb="m1" r_zone="in"
          product="A" p_mult="1" p_memb="m1" p_zone="out"/>
    </rule>
</simulator>
```

Fig. 2. Example of an XML file. It implements a simple set of impulses sent to membrane $m1$ from the outside in the form of concentration of an object E. This concentration goes repeatedly from 0 to its greatest allowed amount, then drops down to 0, with a period of four simulation steps.

mediated by a simple Graphical User Interface (GUI) that helps the user to write all the parameters needed by the simulation, and to select the graphs that will be displayed after the computations.

Both the simulator and the GUI are written in Java, and are executed by a Java virtual machine. This makes the simulator cross-platform. However, at this stage of development the system dynamics in our simulator includes only the production and the spatial movement of objects. We want to extend this functionality, allowing the topological structure to change in time.

4.2 The Metabolic Algorithm and Some Applications

The algorithm implemented by the simulator is inspired by a *chemical* reading of the rewriting rules. Due to the biological implications of this type of reading, we called the algorithm *metabolic.*

The reinterpretation of the rewriting rules in light of a specific application is not new: several researchers have applied rewriting systems to contexts different from a purely abstract one, giving alternative meanings to the rules [2, 35, 37]. In P systems every rule can be seen as a binary relation between strings, mapping the left argument to the right one. For instance, a rule r : $AB \to CD$ containing symbols defined over an alphabet V states that every occurrence of the object $A \in V$ in the system, once paired with $B \in V$, can be substituted with the new object pair $CD \in V^*$.

If we look at r as a chemical *reaction*, then the objects on the left have the role of *reactants* while those on the right are *products.* Following this interpretation, we propose to look at rules as descriptors of the changes in concentration of the reactants into products. In other words, r says that a number of objects of type A and B transforms into objects of type C and D. In this way we deal with populations rather than single objects.

This interpretation needs the introduction of some definitions. Consider a P system on an alphabet $V = \{A, B, C, \dots\}$, provided with a nonempty set R of rewriting rules. Every rule $r : \alpha \to \beta$, with $\alpha, \beta \in V^*$, is associated with a *reactivity coefficient* k_r whose role will be made clear in the following.

For each membrane M we give a maximum number of objects, $|M|$, that cannot be overcome. This parameter is related to the physical properties of M, and we will call it the *capacity* of M. We define a conventional *molarity unit*:

$$\mu = \nu\,|M|\,,$$

where ν is the *reaction factor*, taking values between 0 and 1 ($\nu = 0.01$ in our experiments), which defines a fraction μ of the membrane capacity as the reactive unit or, simply, *mole.*

Denoting with $|X|$ the number of elements of type X in M, we define the quantity

$$||X|| = \frac{|X|}{\mu} \tag{5}$$

as the number of *moles* of X inside M. This molar formulation for the quantities involved in a reaction leads to the *α-molar concentration*, defined as the product of the moles of every object in a string $\alpha = \alpha_1 \dots \alpha_{l(\alpha)}$, $l(\alpha)$ being the length of α [21]:

$$||\alpha|| = \prod_{i=1}^{l(\alpha)} ||\alpha_i||\,. \tag{6}$$

It is now possible to describe an algorithm that translates the rewriting rules into a set of equations defining the *molar variation*, $\Delta||X||$, of every element X as a consequence of the application of the rules.

A rule $r : \alpha \to \beta \in R$ acts on the left (i.e., reactant) and right (i.e., product) objects: the left part of r diminishes the concentration of the reactants, while the right part increases the concentration of the products. Hence, the change in the amount of an element X in M due to r is equal to

$$|\beta|_X - |\alpha|_X, \tag{7}$$

where $|\gamma|_S$ indicates the number of occurrences of S contained in γ. Note that this factor is independent of the concentrations; it "syntactically" ties the rule to the object. Note also that the final balance for X can be either negative or positive, depending on the specific reaction.

In chemical terms, r affects the concentration of every element appearing in it by a similar contribution, depending on the concentration of all the reactants at the time of application. The term $||\alpha||$ takes this aspect into account, according to equation (6). Thus, we can compute the effect $p(X, r)$ of a rule $r : \alpha \to \beta$ on the concentration of X as

$$p(X, r) = k_r \left(|\beta|_X - |\alpha|_X\right) ||\alpha|| \,, \tag{8}$$

where k_r is the *reactivity coefficient* of the rule. Therefore, $p(X, r)$ is the product of three factors: i) the reactivity k_r; ii) the quantity (7), which plays the role that stoichiometric coefficients have in chemical reactions; and iii) the molar concentration (6) of the reactants.

In general, an object is present in more than one rule. In order to compute the overall molar variation of an object X we have to take the contributions of all rules into account. This is done by summing up their effects on the concentration of X:

$$\Delta||X|| = \sum_{r \in R} p(X, r), \tag{9}$$

where R is the set of rules in our P system.

Hence, after the application of a set of rules our algorithm updates the number of moles of an object X according to the following assignment:

$$||X|| := ||X|| + \Delta||X|| \,. \tag{10}$$

The multiplicity of X is updated accordingly:

$$|X| := |X| + \mu \, \Delta||X|| \,. \tag{11}$$

Let us now see a concrete example of this translation from rewriting rules to *metabolic equations*. Consider the set of rules

$$\begin{aligned} r1 &: AC \;\; \to AB, \\ r2 &: BC \;\; \to A, \\ r3 &: BBB \to BC, \end{aligned} \tag{12}$$

each associated with a reactivity coefficient, respectively, k_{r1}, k_{r2}, and k_{r3}. We want to calculate the variation in the multiplicity of every object in the system caused by the rules.

If we apply equation (9) to each object, we obtain the following system of metabolic equations:

$$\begin{aligned}\Delta||A|| &= 0\cdot k_{r1}||AC|| + 1\cdot k_{r2}||BC|| + 0\cdot k_{r3}||BBB||,\\ \Delta||B|| &= +1\cdot k_{r1}||AC|| - 1\cdot k_{r2}||BC|| - 2\cdot k_{r3}||BBB||,\\ \Delta||C|| &= -1\cdot k_{r1}||AC|| - 1\cdot k_{r2}||BC|| + 1\cdot k_{r3}||BBB||,\end{aligned} \tag{13}$$

where k_{r1}, k_{r2}, and k_{r3} can be read as "rates" of application of $r1$, $r2$, and $r3$, respectively. As we can see from (13), where all contributions (including the null ones) are represented, it is always possible to figure out an equation for every object of the P system from the corresponding set of rewriting rules. Each of these equations gives the molar variation of the related element as time elapses.

By applying equation (5) we can figure out the finite *differentials* associated with the system (13):

$$\begin{aligned}\Delta a &= & & +\mu\cdot\tfrac{k_{r2}}{\mu^2}\cdot bc,\\ \Delta b &= +\mu\cdot\tfrac{k_{r1}}{\mu^2}\cdot ac & & -\mu\cdot\tfrac{k_{r2}}{\mu^2}\cdot bc \; -2\mu\cdot\tfrac{k_{r3}}{\mu^3}\cdot b^3,\\ \Delta c &= -\mu\cdot\tfrac{k_{r1}}{\mu^2}\cdot ac & & -\mu\cdot\tfrac{k_{r2}}{\mu^2}\cdot bc \; +\mu\cdot\tfrac{k_{r3}}{\mu^3}\cdot b^3,\end{aligned} \tag{14}$$

in which we have denoted numbers of elements with a, b, c instead of $|A|, |B|, |C|$, respectively. Note that the correspondence between rewriting rules and differential equations is not bidirectional: in general there is no unique way to translate a system of differentials into a set of rewriting rules, whereas the converse holds.

We want to emphasize some important facts about the coefficients k_r. We have seen that in the molar formulation of rewriting rules they are called reactivities, and their role is to weigh each rule's action. Reactivities take several things into account. They model many physical parameters of the reaction environment uniformly acting on the rules, such as pressure and temperature. They also model other chemical parameters, not acting uniformly on rules, such as the pH and the presence of catalysts or enzymes, supporting a reaction that could not effectively take place otherwise.

The application of the reactivity factors in a P system should account for the following:

- strategy of application;
- synchronization times;
- degree of parallelism;
- times and degrees of activation;
- reactants' partition;
- energy partition;
- reaction rate.

If we consider all the interconnections existing between the points introduced in the previous list, then it is easy to understand that the tuning of reactivity factors is very important. We think it needs further investigation, and our future work will proceed along this line. For instance, it is important to note that a change in the reactivity can be put in relation to the granularity with which we observe a reaction. This type of change has a clear analogy with the tuning of the temporal step that controls the resolution at which we observe the discrete approximation of the solution of a differential equation (obtained using the Euler's method, for example). Furthermore, the discovery of algorithms reproducing specific dynamics that are observable in nature might be useful in the tuning of the reactivities. This is something else we want to investigate.

As previously seen, the multiplicity of X is updated according to (10) after each system transition. Unfortunately it might happen that a rule is applied too many times with respect to the reactant allowance, due to a wrong choice of the reactivity coefficients. In other words, the system in principle can consume more reactants than those available at a given configuration. This violates the Principle of Mass Conservation.

To account for this, we add to our model a set of constraints that force the system to respect the Principle of Mass Conservation. One possibility is that for every object X, before calculating its molar variation $\Delta||X||$, check if the negative contribution on X due to the variation exceeds the amount $|X|$ calculated at the previous step; if so, then stop the computation, else go on. Another possible work-around to a violation of the previously discussed constraints is to decrease each reactivity value to a lower rate and then repeat the check.

To clarify this, it is useful to illustrate the constraint set with a concrete example. Consider the P system defined by the rule set (12). As discussed before we associate the following constraints with each reactant, A, B, and C – respectively, $\mathbf{C}_{|A|}$, $\mathbf{C}_{|B|}$, and $\mathbf{C}_{|C|}$:

$$\begin{aligned} \mathbf{C}_{|A|} &: k_{r1}||AC|| < |A| \\ \mathbf{C}_{|B|} &: k_{r2}||BC|| + k_{r3}||BBB|| < |B| \\ \mathbf{C}_{|C|} &: k_{r1}||AC|| + k_{r2}||BC|| < |C| \end{aligned} \tag{15}$$

One may think that the constraint on an object X can be equivalently calculated *after* the updating of $|X|$, by simply checking that it never assumes negative values. Once more, this is the wrong approach. In fact, even if the balance of positive and negative contributions results in an admissible variation, no one is able in this way to prevent the amount of X consumed by all the reactions (i.e., those reactions including it among their reactants) during a transition from exceeding its real amount. So, conditions such as (15) must be checked *before* every transition.

Once a constraint violation has been discovered, there are several ways to react. The investigation on how to do that is in progress. There are some open questions in our model, and our future work will try to answer them. One such question deals with the temporal variation of the reactivity parameters: we think that setting these parameters free to vary with time would have a strong impact on the system behavior, enabling it to simulate reactions more precisely.

At this stage the model is time invariant and the overall dynamical system behavior depends completely on a vector I, of size $n+m$, containing n normalized reactivities $k_1/k_M, \ldots, k_n/K_M$ (k_M being the largest reactivity coefficient), divided in turn by μ, plus m initial concentrations, $a_{01}, \ldots, a_{0m}$, of the elements $a_1, \ldots, a_m$, respectively:

$$I = \left(\frac{k_1}{\mu k_M}, \ldots, \frac{k_n}{\mu k_M}, a_{01}, \ldots, a_{0m} \right) . \tag{16}$$

In other words, every reactivity coefficient can be expressed as a fraction of the molarity unit as well as the largest reactivity, in a way that it is possible to think at every coefficient as the result of two normalizations: i) a *molar* normalization, accounting for the mass and time granularity of the reaction, and ii) a *kinetic* normalization, accounting for the relative strengths of rules.

We have tested the algorithm with *Psim*, by simulating some well known oscillating biochemical reactions. The first test has been conducted on the Brusselator. This reaction occurs when certain reactants such as sulphuric acid, malonic acid, ferroin, and bromate sodium are combined together in presence of a cerium catalyst [33, 15, 37, 24].

After a period of inactivity, the resulting compound starts producing a series of periodic changes in color ranging from red to blue. The corresponding chemical reaction, according to the formula given in [37], can be symbolically described in terms of the following rewriting rules:

$$\begin{aligned} r_1 &: A &&\rightarrow X \\ r_2 &: BX &&\rightarrow YD \\ r_3 &: XXY &&\rightarrow XXX \\ r_4 &: X &&\rightarrow C \end{aligned} \tag{17}$$

Usually, the assumption is made that the reaction is continuously fed by the external environment. To account for this we provide the set (17) with two further generative rules, whose role is to feed the system with reactants A and B:

$$\begin{aligned} r_5 &: \lambda \rightarrow A \\ r_6 &: \lambda \rightarrow B \end{aligned} \tag{18}$$

Starting from this extended set of rules, and assuming that every rule r has reactivity k_r, and then using the algorithm discussed previously, it is possible to obtain the following set of metabolic equations:

$$\begin{aligned} \Delta||A|| &= k_{r5} - k_{r1}||A|| \\ \Delta||B|| &= k_{r6} - k_{r1}||BX|| \\ \Delta||C|| &= k_{r4}||X|| \\ \Delta||D|| &= k_{r2}||BX|| \\ \Delta||X|| &= k_{r1}||A|| - k_{r2}||BX|| + k_{r3}||XXY|| - k_{r4}||X|| \\ \Delta||Y|| &= k_{r2}||BX|| - k_{r3}||XXY|| \end{aligned} \tag{19}$$

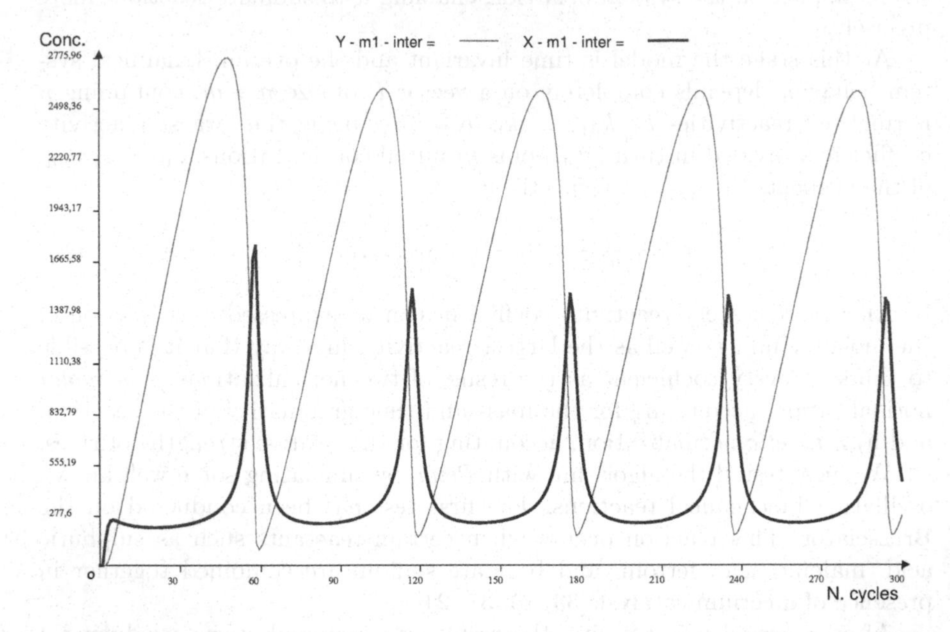

Fig. 3. Oscillations of Belousov-Zhabotinskii reaction model simulated by *Psim* with parameters $k_1 = 0.9$, $k_2 = 0.7$, $k_3 = 0.36$, $k_4 = 0.36$, $k_5 = 0.1$, $k_6 = 0.15$, and $\mu = 1000$ ($|M| = 100000$). Parameters can be rewritten as functions of k_1 as $k_2 = 0.78\,k_1$, $k_3 = 0.4\,k_1$, $k_4 = 0.4\,k_1$, $k_5 = 0.11\,k_1$, and $k_6 = 0.17\,k_1$.

each describing the evolution of the concentration of the respective element. The oscillating behavior of the BZ reaction turns, in the abstract system expressed by (17) and (18), into corresponding oscillations in the number of X and Y.

We have encoded rule set (19) into an XML input file, and fed this file to the simulator. The number of X and Y as functions of time is plotted in Figure 3: the oscillating behavior of these functions is clearly visible. According to the assumptions made in [35], initially all objects have multiplicity equal to zero. Note that it is possible to normalize all the reactivity coefficients by the largest reactivity (k_1 in the example of Figure 3).

Prigogine and Nicolis [24] have studied a simpler dynamics of the BZ reaction in terms of only objects X and Y. This formulation yields the following system of differential equations:

$$\begin{aligned} x' &= k_1 - k_2 x + k_3 x^2 y - k_4 x \\ y' &= k_2 x - k_3 x^2 y \end{aligned} \tag{20}$$

depending on four parameters, k_1, k_2, k_3, and k_4. By expressing (20) in terms of rewriting rules we obtain the following rewriting system, that can be viewed as a simpler form of the Brusselator:

$$\begin{array}{ll} r1: \lambda & \rightarrow X \\ r2: X & \rightarrow Y \\ r3: XXY & \rightarrow XXX \\ r4: X & \rightarrow \lambda \end{array} \tag{21}$$

The associated molar equations are:

$$\begin{array}{ll} \Delta||X|| & = k_{r1} - k_{r2}||X|| + k_{r3}||XXY|| - k_{r4}||X|| \\ \Delta||Y|| & = k_{r2}||X|| - k_{r3}||XXY|| \end{array} \tag{22}$$

where, as usual, every rule r has reactivity k_r.

Equations (20) produce interesting behaviors depending on the reactivity coefficients. As outlined in [24], when choosing in particular $k_1 = 100$, $k_2 = 3$, $k_3 = 10^{-4}$, and $k_4 = 1$, they originate the typical oscillating dynamics of the Brusselator.

These values must be normalized to equate the dynamics obtained with the differential approach with the dynamics obtained with the metabolic algorithm. In fact, let us denote with k_i the parameters characterizing the differential formulation, and with $k_i^{(m)}$ the corresponding parameters in the metabolic algorithm. The translation from the differential to the metabolic parameter – refer also to (14) – is done for every rule according to the following *molar normalization*:

$$k_i^{(m)} = \frac{k_i}{\mu^{1-l(\alpha)}}, \tag{23}$$

where $l(\alpha)$ is the number of reactants in the rule to which the formula applies.

As we can see from Figure 4, the simulation of this model of the Brusselator yields the same behavior described in [24] and [35] after molar normalization of the parameters given in [24]. Moreover, by solving (20) with a well known numerical integration method like Runge-Kutta [18] we have verified that the correspondence between the differential approach and our algorithm also holds when the parameters are chosen in a way such that no oscillations are observed. All these correspondences existing between the results from our simulations and those obtained using robust integration methods, like Runge-Kutta, suggest that the algorithm implemented by *Psim* can certainly be considered a reliable candidate for modeling this kind of system.

The second dynamical system we investigate is a simple predator-prey model [15]. The model is formed by two objects evolving in time: preys, X, and predators, Y. We make the following simplifying assumptions:

- the number of preys increases following a Malthusian model;
- the increase of preys reduces proportionally to the number of predators;
- predators extinguish in absence of preys since in that case they become preys in their turn;
- predators increase proportionally to the number of preys.

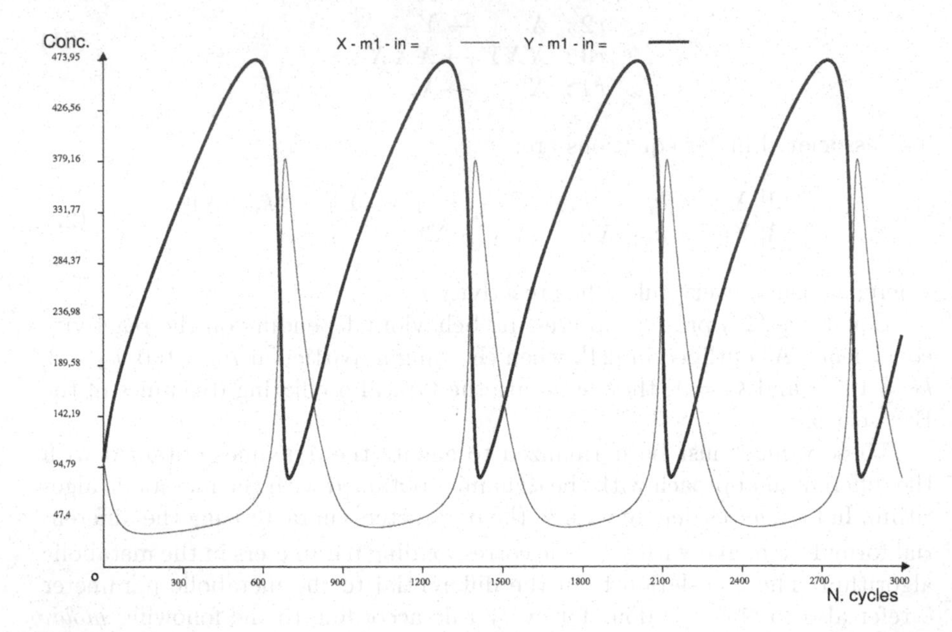

Fig. 4. Oscillations of the simplified Brusselator model simulated by *Psim* with reactivities scaled by a factor 100: $k_1 = 0.001$, $k_2 = 0.03$, $k_3 = 1$, $k_4 = 0.01$, and $\mu = 1000$ ($|M| = 100000$). The initial cardinality of X and Y is set to 100.

This model is described by the Lotka-Volterra differential equations:

$$\begin{aligned} x' &= ax - dxy \\ y' &= exy - by \end{aligned} \tag{24}$$

with initial conditions $x_0 > 0$ and $y_0 > 0$.

We can translate these differential equations into the following set of rewriting rules (recall that $x = ||X||$ and $y = ||Y||$):

$$\begin{aligned} r1 &: X \rightarrow XX \\ r2 &: XY \rightarrow YY \\ r3 &: Y \rightarrow \lambda \end{aligned} \tag{25}$$

In this way we obtain the corresponding metabolic equations:

$$\begin{aligned} \Delta||X|| &= k_{r1}||X|| - k_{r2}||XY|| \\ \Delta||Y|| &= -k_{r2}||XY|| - k_{r3}||Y|| \end{aligned} \tag{26}$$

The above rules and objects are contained into a system with just one membrane.

We have simulated this system using an initial amount of 100 preys and 20 predators. The simulation, as we can see from Figure 5, confirmed the

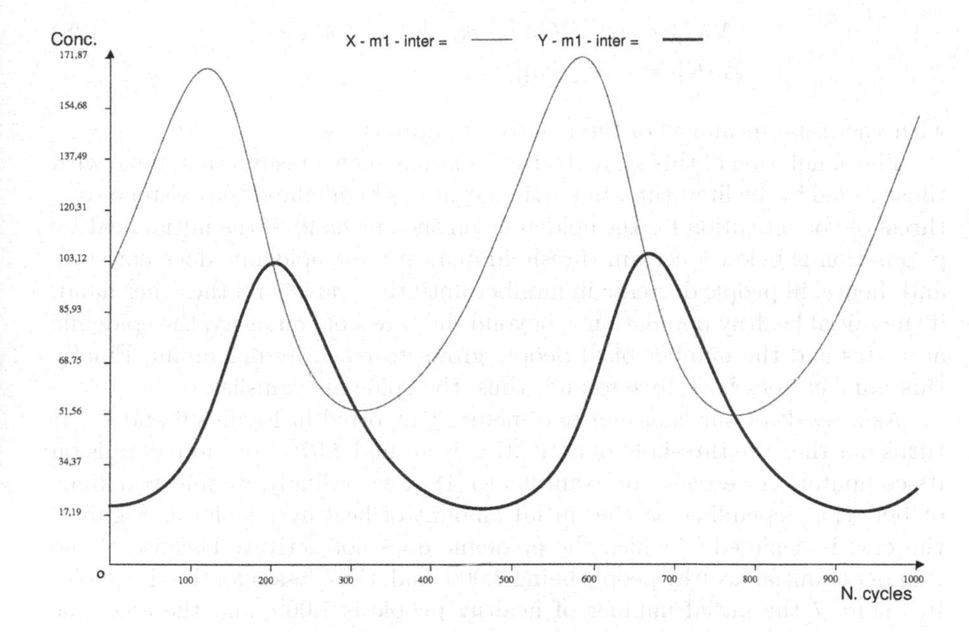

Fig. 5. Oscillations of the predator-prey model simulated by *Psim* with $k_1 = 0.01$, $k_2 = 0.02$, $k_3 = 0.02$, and $\mu = 100$ ($|M| = 10000$).

oscillating behavior of the number of preys and predators in the predator-prey model described by the Lotka-Volterra equation system.

The last model we discuss in this section is that of an infective disease that spreads over a population, causing death or permanent immunity to the infected. We make the simplifying assumption that the population is closed (e.g., no births, immigration, or emigration are allowed) in a way that the population can be partitioned into three different categories: healthy people, C, ill people, G, and immune people, K.

When a healthy person meets an ill one he (or she) gets ill with a probability expressed by the reactivity of the rule. An ill person has three possibilities: he dies, becomes immune forever, or survives indefinitely, although ill. On the other hand, a healthy individual maintains his state as long as he is not in contact with an ill one. This system can be expressed by the following set of rules:

$$\begin{aligned} r_1 &: CG \rightarrow GG \\ r_2 &: G \rightarrow K \\ r_3 &: G \rightarrow \lambda \end{aligned} \tag{27}$$

in which all the symbols have the meanings previously discussed.

It is now possible to write down the set of associated metabolic equations:

$$\begin{aligned}\Delta||C|| &= -\ k_{r1}||CG|| \\ \Delta||G|| &= k_{r1}||CG|| \ -\ k_{r2}||G|| \ -\ k_{r3}||G|| \\ \Delta||K|| &= -\ k_{r2}||G||\end{aligned} \tag{28}$$

with the usual meanings of the reactivity parameters.

The simulation of this system with *Psim* has shown results that agree with those found in the literature. In particular, it has highlighted the existence of a threshold of activation for the epidemic: on the one hand, if the initial healthy population is below a certain threshold quantity, the epidemic does not start and, hence, ill people decrease in number until they vanish; on the other hand, if the initial healthy population is beyond the threshold quantity, the epidemic activates and the number of ill people grows to reach its maximum. Finally this number goes back to zero and, thus, the epidemic vanishes.

As a result of our choice of parameters – reported in Figures 6 and 7 – it turns out that the threshold of activation is around 2,570 (for more details on its computation we refer, for example, to [18]); accordingly, we find two kinds of behavior depending on the initial amount of healthy people; in Figure 6 the case is depicted in which the epidemic does not activate because of the number of initial healthy people being 2,000, and, thus, less than the threshold; in Figure 7 the initial number of healthy people is 7,000, and the epidemic reaches a maximum, and then vanishes. In both cases the initial number of ill people is equal to 300.

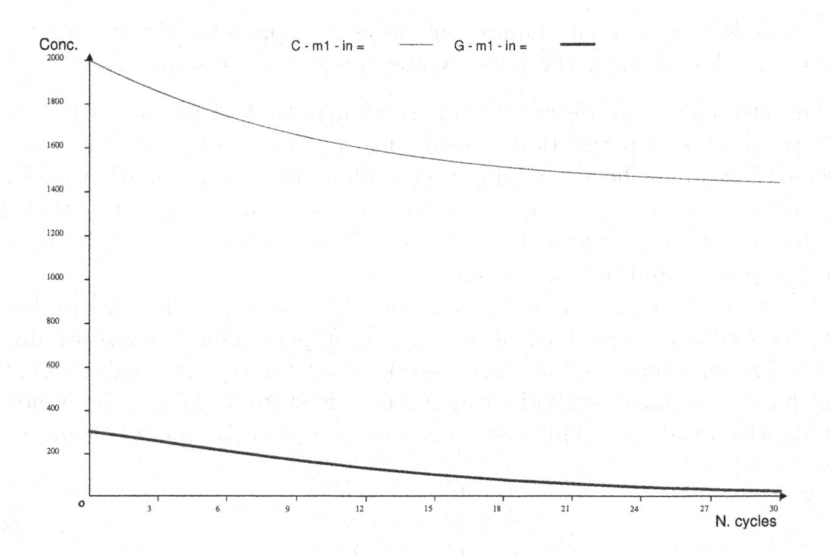

Fig. 6. Inactive epidemic model simulated by *Psim* with $k_1 = 0.3$, $k_2 = 0.1$, $k_3 = 0.12$, and $\mu = 3500$.

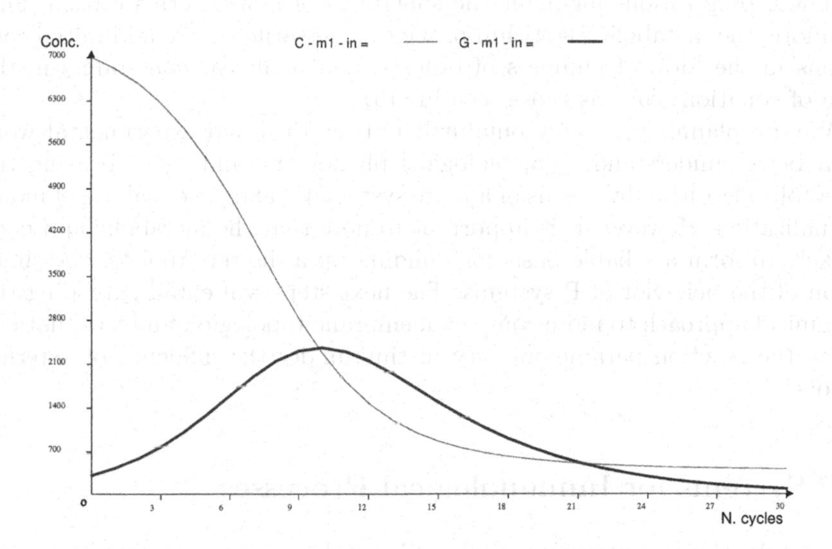

Fig. 7. Active epidemic model simulated by *Psim* with $k_1 = 0.3$, $k_2 = 0.1$, $k_3 = 0.12$ and $\mu = 3500$.

We end this section with some considerations that will guide our future work. In particular, we will try to investigate carefully the molar normalization, whose importance is emphasized by the following propositions.

Let E be a system of metabolic equations derived from a set of rewriting rules, and let $MA(E, \mu)$ be the dynamics we get by applying the metabolic algorithm MA using a molarity unit μ. Let us call E_μ the molar normalization of equations E obtained by replacing every reactivity parameter k_i in E with $k_i^{(m)}$. Finally, let us call dE the differential form obtained by replacing, in E, the finite difference operator Δ with the differential operator, d/dt, and the molar quantities with absolute quantities (that is, by setting $\mu = 1$). If *Euler* is the Euler's approximation method for solving differential equations, then the following proposition is easily proved.

Proposition 2. $MA(E, \mu) = Euler(dE_\mu)$.

Nevertheless, the scale factor could be essential for the correct observation of a system and, in computational terms, the scale factor should be relevant for the reliability of numerical simulations. This is the case with Euler's approximation method. In fact, in general we have: $Euler(d(E)) \neq Euler(d([E]_\mu))$. In the case of the oscillatory phenomena we studied – especially in the Brusselator as reported in [24] – we had the following experimental result.

Proposition 3. $MA(E, \mu) = Runge\text{–}Kutta(dE)$.

These propositions highlight the soundness of molar normalization. Furthermore, the metabolic algorithm provides a versatile way to add initial conditions in the form of numbers of objects, and nonlinear constraints in the form of equations such as those seen in (15).

We are planning to carry out further theoretical and experimental work for a better understanding of biological phenomena and for improving the metabolic algorithm by means of a more systematic and practical use of molar normalization. However, it is important to note that the metabolic approach is likely to form a reliable basis for building up a discrete tool for the simulation of the behavior of P systems. The next steps will entail extending the dynamical approach to more complex membrane topologies, and to situations where the reaction parameters vary in time under the influence of external factors.

5 P Systems for Immunological Processes

The immune system represents a case of complex adaptive system where the notion of cell membrane is essential. For example, in the description of phenomena such as *lysis* or *opsonization*, that are processes by means of which the bacterial membrane is destroyed and coated, respectively, we must be able to formalize the concept of cell surface. P systems provide a versatile means to describe the main processes happening in the immune system.

In spite of its complex nature and capability to rapidly adapt against the attack of infectious agents, the immune system can be considered a typical example of a distributed system [34], consisting of many locally interacting components that provide global protection without any need for centralized control. Moreover, the immune system is inherently dynamical, since the individual components circulating throughout the body are continually created and destroyed. A dynamical study of P systems [4] – for example, from the point of view of string transition dynamics, as reported in Section 2 – may help in better understanding the relevant features of immunological processes.

In this section we analyze the basic aspects of the innate immune system in order to discover structures of objects and types of membrane rules useful for formalizing, in terms of membrane systems, the fundamental steps of the primary immunological response. Then, a simplified version of the membrane system for leukocyte selective recruitment [11] is simulated using *Psim*. We will see that using certain parameters the experimental behaviors are well reproduced. A formal description of the adaptive immune system is left to forthcoming research.

A realistic description of the cell needs a richer structure than the one defined for traditional P systems and their variations [28]. For example, we need a representation of membranes that describes the recognition and pairing mechanism between cells by means of "receptor-receptor" bonds. The surface of many (in particular, immune) cells is covered with *receptors*, that are complex three-dimensional, electrically charged structures well visible from the outside. The more complementary the structure and charge of receptors belonging to two cells, the more likely the binding between such cells. The receptor of a pathogen is called *epitope*, and the bond strength between a receptor and an epitope is called *affinity*. Receptors are deemed *specific* because they tightly bind only to a few similar epitope structures or patterns.

The notion of a receptor, crucial in the cellular interplay occurring in the immune system, has not been considered yet in the basic model of P systems. In order to formally express this here, we use $\{\ [_i$ instead of $[_i$ to indicate a membrane, expressing with this notation the presence (between braces and square brackets) of an interstice, intended as a region belonging to the external surface where objects (receptors) are detectable from the outside and, in turn, can detect objects in the external world. This feature can be considered as an extension of the communication mechanism of PB systems [5] (refer to Section 3) and symport/antiport P systems [26]. In fact, in PB systems $x[_i y$ means that a membrane labeled i can see outside its boundary, in particular, it can see an object x close to its external surface. Here we allow *semi-internal* objects to be also are visible from the outside.

A second relevant matter is the introduction of rules that manage entire membranes (including their contents) as objects. This approach is suitable and realistic to express phenomena such as *phagocytosis* by macrophages and *diapedesis*. In both the processes, microorganisms are engulfed and consumed by immune system cells, and selected cells have to pass through tissues to attack the infection.

Moreover, in order to describe *adhesion* between cells, that is, the cellular complex obtained after bonds are created among respective receptors, we use the following rule:

$$\{p[_a\ ]_a\}\ \{q[_b\ ]_b\} \rightarrow \{p\ q[_{a+b}\ ]_{a+b}\}$$

where p represents the receptors of the membrane a, q represents the receptors of the membrane b, and $+$ is a special symbol for joining labels.

6 The Architecture of the Immune System

Some replicating pathogens constantly attack the body, and can be harmful if left unchecked. Since different antigens have to be destroyed in different ways, the problem faced by the immune system is to recognize them and to choose the right tools for destroying a particular kind of pathogen. We can associate all relevant biological properties that characterize the antigens with the symbols of an alphabet A, and all the possible forms of epitopes with the symbols of an alphabet E, so that a particular external microorganism can be represented by $\{r[_s\ ]_s\}^v$, where $s \in A^\star, r \in E^\star$, and $v \in \{0, +, -\}$. The charge associated with the membrane is neutral when the antigen is opsonized, positive when it is replicating, and negative when it is not replicating.

To protect multicellular organisms from foreign pathogens – especially those that replicate such as viruses, bacteria, or parasites – the immune system must be capable of distinguishing harmful foreign material from normal constituents of the organism, which we will indicate with a membrane $[_{self}\ ]^{v}_{self}$ where $v \in \{0, +, -\}$. The charge associated with this membrane is normally positive; it becomes negative when the *self* cell is infected, and neutral when it is immune to the attack of antigens.

The recognition of an antigen as foreign in the immune system can be seen as a problem of pattern recognition implemented by binding. For example, lymphocytes recognize pathogens by forming molecular bonds between pathogen fragments and receptors on the surface of the lymphocyte. The more complementary the molecular shape and electrostatic surface charge between pathogen and receptor, the stronger the bond (and the higher the affinity). When an immune system detector binds to an antigen, we say that the immune system has recognized the pattern encoded by the antigen.

To describe this kind of affinity between cells we define a function R_c that measures the bond strength between receptors. If we call R the set of elements representing all types of receptors (of self cells), we have:

$$R_c : \{(x, y) \mid x \in R, y \in E\} \to \mathbb{N}.$$

This definition can be extended to strings. In this case computing the value of R_c becomes a hard problem, as it depends on structural and topological cellular contraints. For this reason we avoid an explicit formulation of R_c for strings.

We also need to define an affinity function between molecules (identified with elements of a set *Mol*) and cell receptors, because a cytokine produces its actions by binding to specific high affinity cell surface receptors (typically in close proximity to where it is produced):

$$R_m : \{(x, y) \mid x \in R, y \in Mol\} \to \mathbb{N}\,.$$

This is a general definition which, in the context of cytokine molecules, we can restrict to a boolean function since *cytokine receptors* are associated with the structurally unique family of receptors, termed JAKs, expressed by many different cells. So, we can see R_m as the characteristic function on $JAK \times Y$ where $JAK \subset R$ is the subset of symbols associated with JAKs receptors.

The architecture of the immune system is multilayered, and it is provided with defenses on several levels. The most elementary one is the skin, which is the first barrier to infection. Another barrier is constituted by the physiological conditions, such as pH and temperature, that provide uncomfortable living conditions for foreign organisms [34]. We will describe these two levels as more internal filter membranes $[_p\ ]_p$ of a membrane system (see Figure 8).

Once pathogens have entered the body they encounter the *innate* immune system and, then, the *adaptive* immune system. Both systems consist of a multiplicity of cells and molecules that interact in a complex manner to *detect* and *eliminate* pathogens. The first system is that part of the immune system we are provided with since our birth. It does not change, or adapt, to specific pathogens and provides a rapid first line of defense to keep an early infection in check, giving the adaptive immune system enough time to prepare a more specific response. The term *adaptive* in fact refers to the part of the immune system that learns how to recognize specific kinds of pathogens, and retains memory of this for speeding up future responses.

Both detection and elimination depend on chemical bonds. The surfaces of immune system cells are covered with various receptors; some of them chemically bind to pathogens and some bind to other immune system cells or molecules.

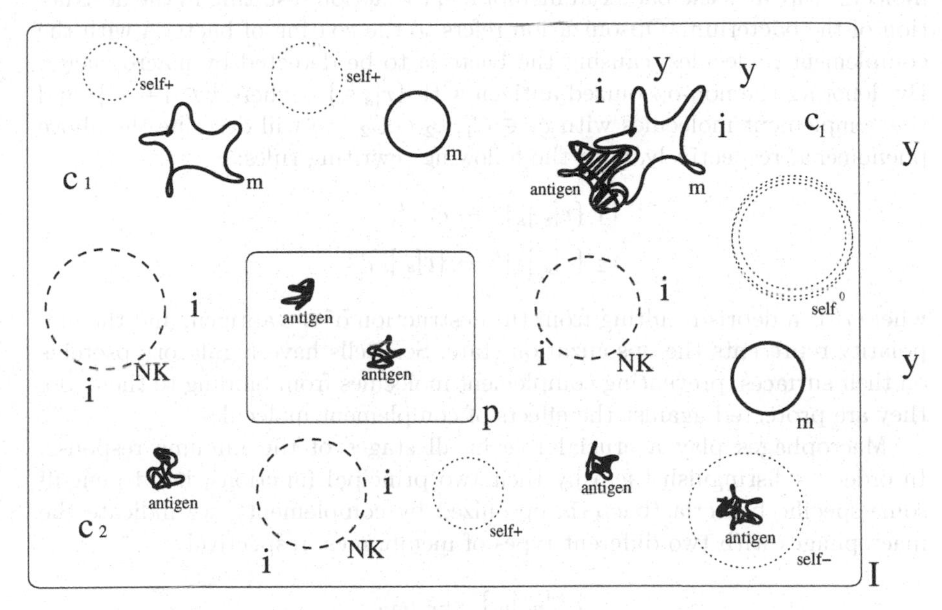

Fig. 8. A membrane system for the immune system.

The membrane system we propose describes the principal steps of first immune response. It contains two nested membranes representing the first two levels of defense before the adaptive part of the immune system, located in the environment: in Figure 8 the attack of antigens must be seen from the inside to the outside of the system, because the most internal region represents the external world, and the most external region represents the last, and more specific, body defense. This choice is motivated by the increasing complexity of the processes that must be described.

For the sake of simplicity, in the following discussion we do not close brackets indicating corresponding membranes if not necessary. The most internal membrane $[_p$ represents the body skin. It is a filter that selects, according to specific criteria, membranes that are labeled by strings in $A^\star$, and allows only membranes which have labels belonging to a language $L \subset A^\star$ to exit. The region of the most external membrane $[_I$ represents the phase during which the innate immune system is active. Only the pathogens that need a specific action for being destroyed can pass through the membrane labeled I. Now we see how the innate immune system attacks the antigens. We first introduce some notation.

Complement molecules, together with macrophage cells, are respectively the primary chemical and the phagocytic response of the immune system in the early stages of infection. We split the complement molecules into two sets, C_1 and C_2, because they are involved in two distinct phenomena, respectively, lysis and opsonization. Lysis is the process by means of which the complement molecule ruptures the bacterial membrane, the action resulting in the destruction of the bacterium. Opsonization refers to the coating of bacteria with the complement molecules, causing the bacteria to be detected by macrophages. By denoting the non-opsonized antigen with $\{r[_s\,]_s\}^v$, where $v \in \{+,-\}$, and the complement molecules with $c_1 \in C_1$, $c_2 \in C_2$, we will describe the above phenomena, respectively, with the following rewriting rules:

$$c_1\,\{r[_s\,]_s\}^v \to c_1\,r',$$

$$c_2\,\{r[_s\,]_s\}^v \to \{r[_s\,]_s\}^0,$$

where r' is a debris resulting from the destruction of the antigen, and the null polarity represents the opsonization state. Self cells have regulatory proteins on their surfaces, preventing complement molecules from binding to them. So, they are protected against the effects of complement molecules.

Macrophages play a crucial role in all stages of the immune response. In order to distinguish them by their two principal functions, i.e., to engulf some specific bacteria (bacteria opsonized by complement), we indicate the macrophages with two different types of membranes, respectively:

$$\{\,t\,[_m\,]_m\} \quad [_m\,]_m\,,$$

where $t \in R^\star$ is the string of cell receptors. Actually, macrophages have receptors both for certain kinds of bacteria and for complement molecules, but this abstract description of macrophages contains all relevant aspects needed by the dynamical system we are describing.

We associate the *cytokine* molecules with symbols belonging to Y. Cytokines are molecules that act as a variety of important signals, and their release activates the next phase of the host defense, called early induced response. They are produced not only by macrophages and other immune system cells, but also by some self cells (which are not part of the immune system) when they are damaged by pathogens. Their major effect is to induce an inflammatory response, associated with some physiological changes (fever) which reduce the activity of pathogens and reinforce the immune response by triggering the production of *acute phase proteins* (APPs), substances which bind to bacteria, thus activating macrophages or complement molecules.

When infected by viruses certain cells produce *interferons*, a family of cytokines, so-called because they inhibit viral replication. Moreover, they activate certain immune system cells called *Natural Killers* (NKs), that kill virus infected self cells. We classify the interferons as elements of a subset $\mathcal{I} \subset Y$, and the NK cells as membranes labeled *nk*:

$$[_{nk}\ ]_{nk}$$

NK cells bind to normal host cells, but they are normally not active because healthy cells express molecules that act as inhibitory signals. When virally infected cells cannot express these signals they are killed by activated NK cells that release special chemicals that trigger the *apoptosis* (programmed cell death) on an infected cell.

The overall validity of this representation must be verified in terms of its capability to explain immunological dynamics. *Psim* applies rewriting rules to objects taking into account their reactivity, i.e., the ability shown by such objects (reactants) to meet (react) together. In particular, the concept of concentration used by *Psim* helps in formalizing a sort of substance (or cellular) adjacency which is not defined in membrane systems. In fact, P systems are topological spaces without a metric on objects. The existence of such a metric can be important when dealing with cellular reactions.

Nevertheless, the notation used in membrane systems shows features that are essential in the representation of immunological phenomena. Cells and molecules in fact move around compartments; moreover, cells engulf objects and in turn are engulfed by other cells. Likewise, agents change their internal state, or their state of perception, with respect to the objects they meet, their place, and their state.

7 Innate Immune System with Membranes

In the following discussion we propose a membrane system dealing with the objects and membranes previously introduced, and provide them with suitable rules.

Consider the following P system [10]:

$$(V, \mu, R)$$

where V is the alphabet, μ is the initial configuration (where one can see also the initial membrane structure of the system), and R is the set of rewriting rules that extend the evolution and communication rules [28] by considering also membranes having receptors, and managing membranes as objects.

Our system has the following alphabet:

$$A \cup E \cup E' \cup R \cup C_1 \cup C_2 \cup Y \cup \{APP, d\},$$

where A contains the biological properties of the antigen, E contains the epitopes in all possible forms, E' contains the primed versions of the symbols in E (this is useful to indicate the epitopes of the bacterium after it has been processed by lysis), R contains the self cell receptors, C_1 and C_2 contain the complement, Y contains the cytokines, and, finally, APP and d are two special symbols needed to simulate the activation of macrophages and the programmed death, respectively.

The system starts from the initial configuration (see Figure 8):

$$[_I [_{self}]^{+}_{self}{}^{k_s} \; [_{nk}]_{nk}{}^{k_n} \; y \; [_m]_m{}^{k_m} \; \{t_1 [_m]_m\}^{k_1}$$
$$\cdots \{t_h [_m]_m\}^{k_h} \; [_p \{r_1 [_{s_1}]_{s_1}\} \cdots \{r_n [_{s_n}]_{s_n}\} \;]_p \;]_I,$$

where $k_s, k_n, k_m, k_1, \ldots, k_h$ are membrane multiplicities. Then, we have the following rules:

- the *starting rule* simulates the entry of some antigens through the body skin:

$$[_p \cdots \{r [_s]_s\}^+ \cdots]_p \rightarrow [_p]_p \; \{r [_s]_s\}^+ \quad r \in E^\star, s \in L \quad \text{entrance}$$

- once antigens have entered the body (from now on we will not write $r \in E^\star$, $r' \in E'^\star$, $s \in L$, $c_1 \in C_1$, and $c_2 \in C_2$ all the time) they induct the primary response by replicating themselves and by infecting self healthy cells (which switch from positive to negative polarity):

$$\{r [_s]_s\}^+ \rightarrow \{r [_s]_s\}^+ \; \{r [_s]_s\}^+ \qquad \text{replication}$$

$$[_{self}]^+_{self} \; \{r [_s]_s\}^v \rightarrow [_{self}]^-_{self} \quad v \in \{+, -\} \qquad \text{infection}$$

- complement molecules, which are present in the region I, can thus act by lysis and opsonization of some kinds of antigens, indicated by a string of the set $\hat{L} \subset L$:

$$c_1 \; \{r [_s]_s\}^v \rightarrow c_1 \; r' \quad s \in \hat{L}, v \in \{+, -\} \qquad \text{lysis}$$

$$c_2 \; \{r [_s]_s\}^v \rightarrow \{r [_s]_s\}^0 \quad s \in \hat{L}, v \in \{+, -\} \qquad \text{opsonization}$$

- macrophages act as scavenger cells (engulfing debris and opsonizing pathogens), as process antigens (engulfing and digesting antigens, and then presenting fragments of their proteins on the surface), and when activated by bindings between receptors, as signal emitters (via cytokines) to the immune system:

$$r' \, [_m \,]_m \rightarrow [_m \,]_m \quad \text{debris remotion}$$

$$\{r[_s \,]_s\}^0 \, [_m \,]_m \rightarrow [_m \,]_m \quad \text{opsonized antigen remotion}$$

$$\{t[_m \,]_m\} \, \{r[_s \,]_s\}^v \rightarrow \{t \, s[_{m+s} \,]_{m+s}\} \quad \text{antigen-macrophage complex}$$

$$\{t \, r[_{m+s} \,]_{m+s}\} \rightarrow \{t \, r[_m \,]_m\} \quad \text{functionality restoration (with trace)}$$

$$\{t \, r[_{m+s} \,]_{m+s}\} \rightarrow \{t \, r[_{ms} \,]_{ms}\} \, y \quad \text{cytokine production by macrophages}$$

where in the third rule $v \in \{+, -\}$ and $R_c(t, s) > 0$, and in the fifth rule $y \in Y$. In our notation one can distinguish whether the macrophages are activated or not by the presence of the symbol $+$ in their respective label;

- cytokines are probably the most important biologically active group of molecules to identify; with hundreds of known cytokine-like molecules, it is necessary here to restrict the discussion to a few key cytokines and their most important properties, which include starting and maintaining the inflammatory response. However it is clear that, being the common signaling system for cell growth, inflammation, immunity, differentiation, and tissue repair processes, cytokines are involved in many, if not all, physiological functions.

Interferons, and in general cytokines, are produced in the adaptive immune system by T cells and B cells, by activated macrophages, and by endothelial cells of inflamed tissue:

$$[_{self} \,]_{self}^- \rightarrow [_{self} \,]_{self}^- \, y \quad y \in Y \quad \text{cytokines production by infected cells}$$

Cytokines increase the number of macrophages and complement by production of APPs:

$$y \rightarrow y \, APP \quad \text{APP production by cytokines}$$

$$APP \rightarrow c_1 \, c_2 \quad \text{complement molecules increasing by APP}$$

$$APP \rightarrow [_m \,]_m \quad \text{macrophages increasing by APP}$$

They also increase the resistance of self cells against bacterial infection (especially mycobacteria and certain viruses):

$$y^k [_{self} \,]_{self}^+ \rightarrow [_{self} \,]_{self}^0 \quad \text{self cells immunization by cytokines}$$

where $k \in \mathbb{N}$ is a given number. Moreover, interferons (that are a special class of cytokines) inhibit virus replication and activate NK cells, respectively:

$$i\{r[_s \,]_s\}^+ \rightarrow \{r[_s \,]_s\}^- \, i \in \mathcal{I} \quad \text{virus replication/inhibition}$$

$$i[_{nk} \,]_{nk} \rightarrow [_{nk} \, i \,]_{nk} \quad i \in \mathcal{I} \quad \text{NK cell activation by interferons}$$

Interferons are also produced by activated NK cells:

$[_{nk} i\,]_{nk} \rightarrow j\ [_{nk} i\,]_{nk} \quad i, j \in \mathcal{I}$ interferon production by activated NK cells

Finally, activated NK cells induce programmed death in infected cells:

$$[_{nk}\, i\,]_{nk}\ [_{self}\,]^-_{self} \rightarrow [_{nk}\, i\,]_{nk}\ [_{self}\, d\,]^-_{self} \qquad \text{programmed death}$$

$$[_{self}\, d\,]^-_{self} \rightarrow d \qquad \text{self cell death}$$

$$d \rightarrow \lambda$$

These rules are not used as in P systems, but in the way of *Psim*, therefore the process dynamics and effects are regulated at every step by the actual quantities of reactants. Every dissolving membrane delivers its content to the immediately outer membrane [29], and every dividing membrane replicates its content inside the new membrane.

Due to its complexity, the description of the adaptive immune system and its interaction with the innate immune system is left for the future. Priorities between regions (and not between rules) might be established to simulate different macro steps, in a way that some processes of the innate immune system would activate with higher priority than those of the adaptive immune system. This regulation might be performed more accurately by using reactivities. Since cells and substances are sent from the adaptive system to the innate one as macrophages and antibodies acting as complement molecules, and as cytokines, respectively, from now on we will suppose that the environment periodically provides these materials [4].

8 Leukocyte Selective Recruitment

In an organism the first response against an inflammatory process consists of the activation of a tissue-specific recruitment of leukocytes. Activation relies on the complex functional interplay between the surface molecules that are designed for specialized functions. These molecules are differently expressed by leukocytes circulating in the blood, and by endothelial cells covering the blood vessel.

Leukocyte recruitment in tissues requires extravasation from the blood. Extravasation is made possible by a process of transendothelial migration, and three major steps have been identified in the process of leukocyte extravasation: (i) tethering-rolling of free-flowing white blood cells, (ii) activation of the same cells, and (iii) arrest of their movement due to their adherence to endothelial cells. After this arrest a phenomenon of *diapedesis* occurs; that is, leukocytes move from the blood beyond endothelial cells toward the tissue.

A leukocyte cell has some receptors put on its surface that bind with counter-receptors located on the surface of the endothelial cells. These bonds slow down the initial speed of leukocyte. Moreover, some molecules, called chemokines, are produced by the epithelium and by the bacteria that have activated the inflammation process. Chemokines can bind with the leukocyte receptors, producing signals inside it. Such signals generate on the leukocyte surface new and different receptors that, interacting with the endothelial receptors, strongly slow down the cell speed until it stops (see Figure 9).

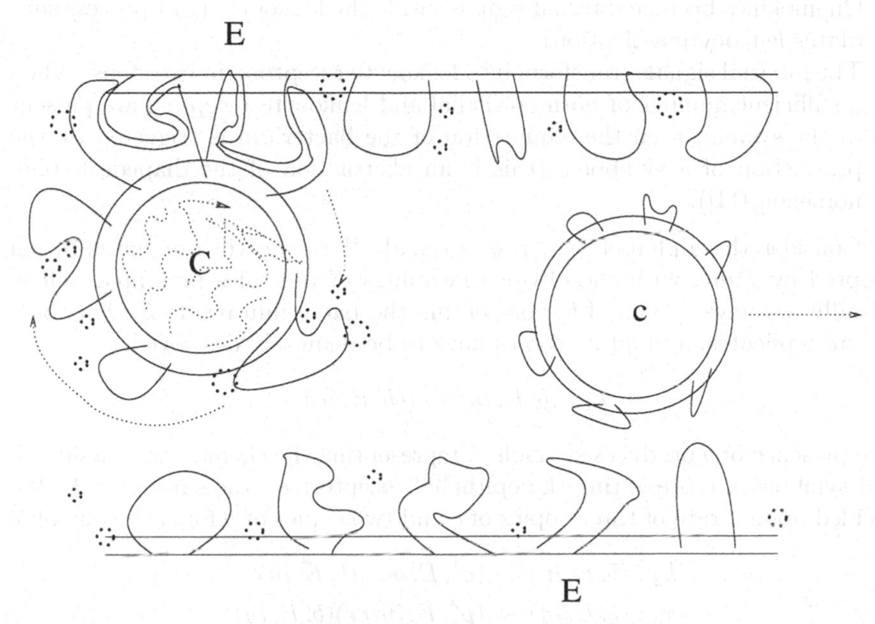

Fig. 9. Leukocyte cell attacked by chemokines and endothelial receptors. C: leukocytes; E: epithelium.

We call A the initial state with fast circulating leukocytes into the blood, B the rolling state, C the activation state, and D the final adhesion state. Therefore, the system evolves through three main phases:

1. $A \to B$, by means of some receptor-receptor interactions,
2. $B \to C$, by means of some chemokine-receptor interactions, and
3. $C \to D$, by means of some receptor-receptor interactions.

(refer to [11] for a description of a membrane system representing this immunological phenomenon; in that paper, ad-hoc notations have been introduced for adding more realism to the model).

Here we analyze, using *Psim*, a simplified model of the system where only a specific kind of leukocyte is present and the dynamics are ruled by only one production of symbols. Accordingly, in this model receptor-receptor bonds are not explicitly represented.

We use two membranes, labeled E and L: E represents the epithelium entered and infected by the bacterium; L represents the leukocyte which interacts with the epithelium by producing or transforming symbols. Initially we have one symbol b inside the membrane labeled E.

The process develops as follows:

- The antigen inside the epithelium produces chemokines and epithelial receptors externally, and copies of itself internally.
- Chemokines become internal signals inside the leukocyte (this process simulates leukocyte activation).
- The internal signals transform into leukocyte receptors, in two steps. When a sufficient number of both epithelial and leukocyte receptors are present in the system, then the elimination of the bacterium is triggered by the production of a symbol v (this is an abstraction of the diapedesis phenomenon [11]).

Consider the alphabet $\{b, c, p, q_1, q_2, s, v\}$. Here we will use the notation adopted by *Psim*, with the obvious meanings of the rules presented below. Initially, we have a symbol b representing the bacterium inside E. This bacterium replicates, and all its copies have to be definitely destroyed:

$$k_1 : (b, E, in) \rightarrow (bb, E, in)$$

The presence of b produces symbols c, representing the chemokines outside E, and symbols p, representing the epithelial receptors on the surface of E. We decided upon a rate of three copies of c and two copies of p for each copy of b:

$$k_2 : (b, E, in) \rightarrow (c^3, E, out)(b, E, in)$$
$$k_3 : (b, E, in) \rightarrow (p^2, E, inter)(b, E, in)$$

The chemokines c outside membranes become symbols s, representing internal signals inside membranes labeled with L. Moreover, s is 'slowly' transformed into q_2, representing leukocyte receptors, by moving trough a transformation into five copies of q_1:

$$k_4 : (c, E, out) \rightarrow (s, L, in)$$
$$k_5 : (s, L, in) \rightarrow (q_1^5, L, inter)$$
$$k_6 : (q_1, L, inter) \rightarrow (q_2, L, inter)$$

Finally, the presence of both p and q_2 – respectively representing epithelial and leukocyte receptors – activates the production of a symbol v inside E, where each copy of b is deleted by the presence of v.

The problem is to regulate the reactivity values $k_1, \ldots, k_8$ to correctly modulate the application of rules, i.e., to neutralize the infection. As one can see in Figures 10, 11, and 12, setting $k_1 = 1$, $k_2 = 1$, $k_3 = 0.8$, $k_4 = 0.7$, $k_5 = 1$, $k_6 = 0.2$, $k_7 = 0.8$, and $k_8 = 1$ lets the immune response work soundly, as it destroys all copies of replicating b.

Note that the infection decay (e.g., the decreasing amount of b inside E) corresponds to the increasing activation of leukocytes (e.g., amount of s

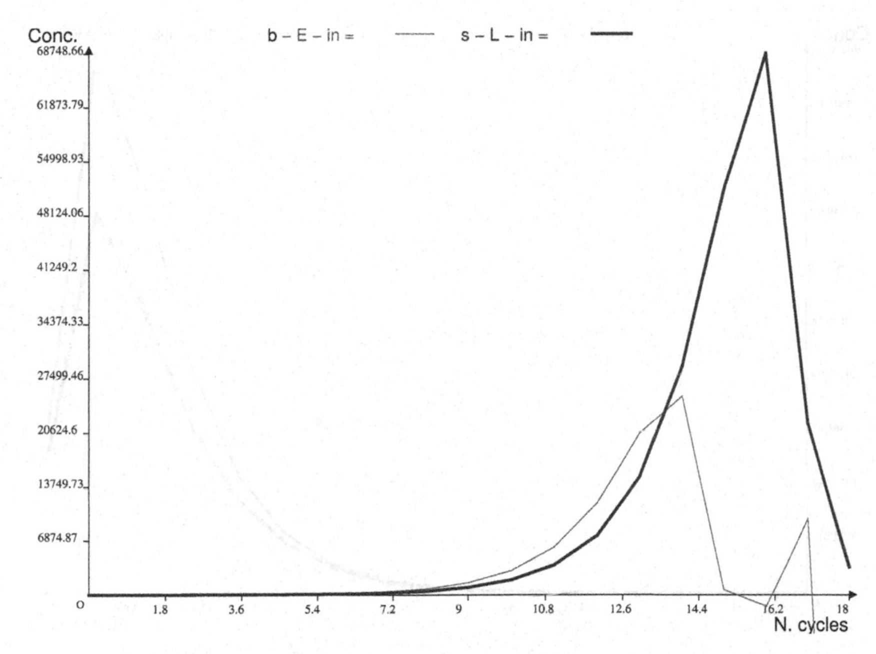

Fig. 10. The amount of b (representing the quantity of infection) inside E (the inflamed epithelial tissue) and s (internal signals as immune responses) inside L (leukocyte cell) after 18 applications of the rules; the membrane maximum capacity is equal to 10^6.

inside L), as is well known from immunological studies [34]. Also, when the infection disappears, the production of s slows down quickly, as can be seen in Figure 10. Moreover, the production of epithelial receptors is higher than the production of chemokines, as in real immunological processes, and the chemokines' behavior follows the amount of b plotted in Figures 11 and 12.

The application of the rules at each computational step is dictated by the principles outlined in Section 4. In this case we have the following differentials:

$$\begin{aligned}
\Delta b &= |b| - \frac{|b||v|}{\mu}, \\
\Delta c &= 3|b| - 0.7|c|, \\
\Delta p &= 2 * 0.8|b| - 0.8\frac{|p||q_2|}{\mu}, \\
\Delta s &= 0.7|c| - |s|, \\
\Delta q_1 &= 5|q_1| - 0.2|q_1|, \\
\Delta q_2 &= 0.2|q_1| - 0.8\frac{|p||q_2|}{\mu}, \\
\Delta v &= 0.8\frac{|p||q_2|}{\mu}.
\end{aligned}$$

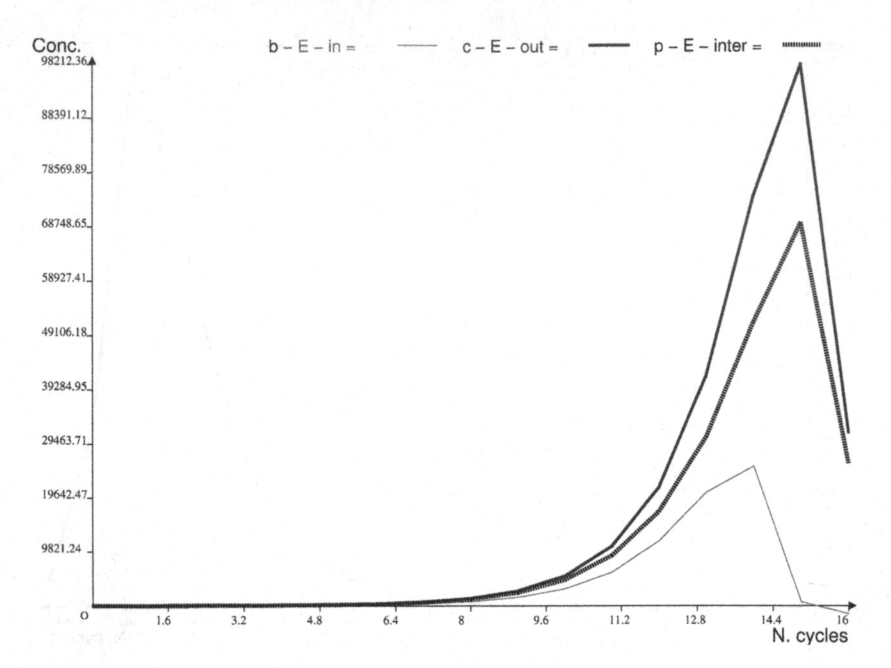

Fig. 11. The amounts of b (antigens), c (chemokines), p (epithelial receptors) after 16 applications of the rules; the membrane maximum capacity is equal to 10^6.

9 Conclusion and Future Directions

This chapter describes theoretical facts as well as experiments and applications we have recently dealt with during our investigation of the dynamics of P systems. The leading idea of this investigation has been that of looking at P systems not as machines which in front of an input produce a corresponding output (provided they reach a final configuration), but, rather, as systems capable in principle of reproducing a wide spectrum of dynamics, including those pertaining to molecular processes.

If we resolve, for example, the Brusselator reaction expressed by (17) using the *metabolic graph* drawn in Figure 13, then we somehow emphasize in the associated P system a sort of "neuron-like" membrane structure (according to Păun's terminology).

Such a "neural" representation of a P system can be immediately extended to any membrane structure by making a proper use of labels. In this way any membrane system takes on the aspect of a *dynamical network*, i.e., a graph whose nodes have a state that at every temporal step depends on the state of other nodes, and whose nodes and edges can be added and/or removed dynamically.

It is easy to discover that a dynamics can present oscillatory behaviors only if the associated graph contains cycles. In general, finding parameters that

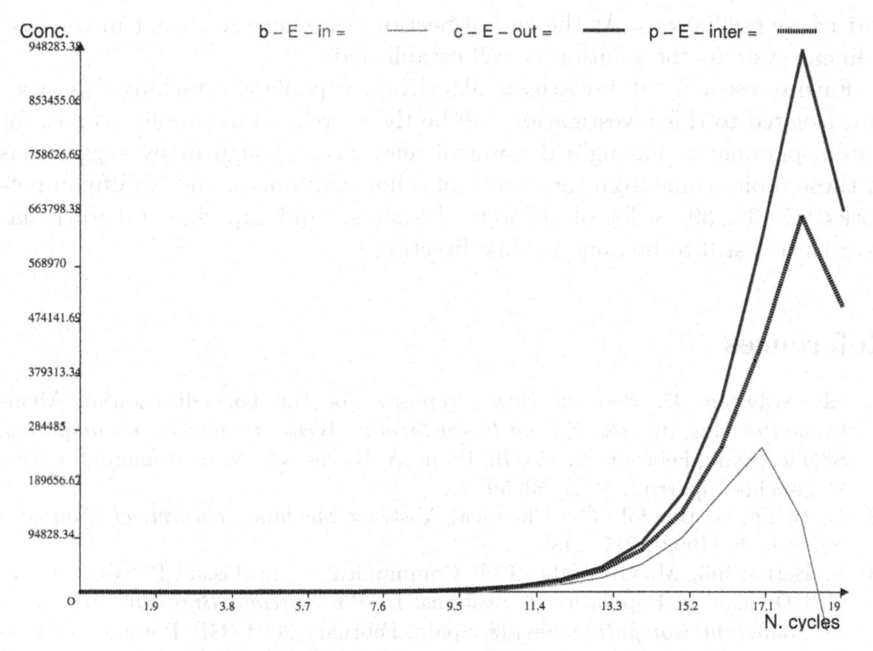

Fig. 12. The quantities of b (antigens), c (chemokines), and p (epithelial receptors) after 19 applications of the rules; the membrane maximum capacity is equal to 10^7.

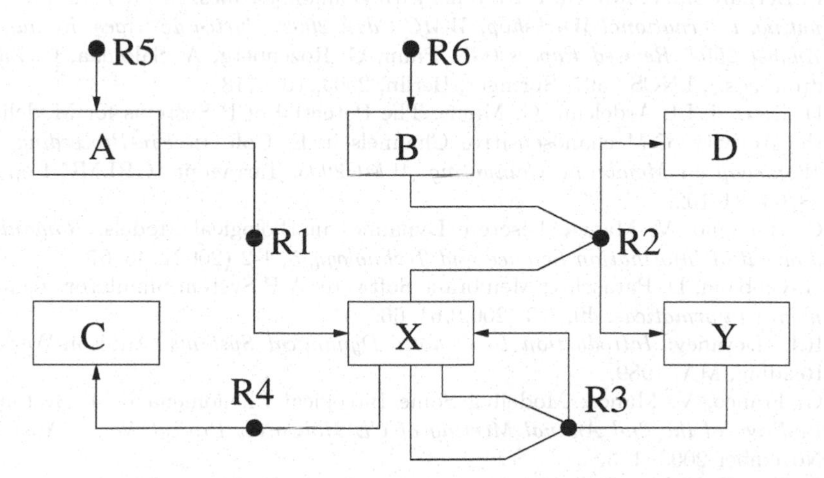

Fig. 13. Brusselator's *Metabolic Graph.*

translate into oscillations is not easy. The *inverse oscillation problem* can be stated in the following way: given a metabolic graph, find initial concentrations and reactivity parameters (i.e., a collection of values for I as defined by (16)) that cause oscillations. At the end of Section 3 we have seen that in the case of linear systems the solution is well established.

Future research will investigate algorithms capable of attacking this problem. Related to this investigation will be the search, on metabolic graphs, for system parameters having a dynamical relevance. Though many suggestions on these topics come from the theory of cellular automata and Kauffman networks [17, 38, 39], a lot of theoretical analysis and experimental work has nevertheless still to be done in this direction.

References

1. I.I. Ardelean, D. Besozzi: New Proposals for the Formalization of Membrane Proteins. In *Proc. Second Brainstorming Week on Membrane Computing*, Seville, Spain, February 2004 (Gh. Păun, A. Riscos-Núñez, A. Romero-Jiménez, F. Sancho-Caparrini, eds.), 53–59.
2. G. Bellin, G. Boudol: The Chemical Abstract Machine. *Theoretical Computer Science*, 96 (1992), 217–248.
3. F. Bernardini, M. Gheorghe: Cell Communication in Tissue P Systems and Cell Division in Population P Systems. In *Proc. Second Brainstorming Week on Membrane Computing*, Seville, Spain, February 2004 (Gh. Păun, A. Riscos-Núñez, A. Romero-Jiménez, F. Sancho-Caparrini, eds.), 74–91.
4. F. Bernardini, V. Manca: Dynamical Aspects of P Systems. *BioSystems*, 70 (2002), 85–93.
5. F. Bernardini, V. Manca: P Systems with Boundary Rules. In *Membrane Computing, International Workshop, WMC-CdeA 2002, Curtea de Argeş, Romania, August 2002, Revised Papers* (Gh. Păun, G. Rozenberg, A. Salomaa, C. Zandron, eds.), LNCS 2597, Springer, Berlin, 2003, 107–118.
6. D. Besozzi, I.I. Ardelean, G. Mauri: The Potential of P Systems for Modeling the Activity of Mechanosensitive Channels in E. Coli. In *Pre-Proceedings of Workshop on Membrane Computing, WMC2003*, Tarragona, GRLMC Report 28/03, 84–102.
7. C. Bonanno, V. Manca: Discrete Dynamics in Biological Models. *Romanian Journal of Information Scicnce and Technology*, 5, 1-2 (2002), 45–67.
8. G. Ciobanu, D. Paraschiv: Membrane Software. A P System Simulator. *Fundamenta Informaticae*, 49, 1-3 (2002),61–66.
9. R.L. Devaney: *Introduction to Chaotic Dynamical Systems*. Addison-Wesley, Reading, MA, 1989.
10. G. Franco, V. Manca: Modeling Some Biological Phenomena by P Systems. In *Proc. of the 2nd Annual Meeting of CE MolCoNet Project*, Wien, Austria, November 2003, 1–3.
11. G. Franco, V. Manca: A Membrane System for the Leukocyte Selective Recruitment. In *Membrane Computing, International Workshop, WMC2003, Tarragona, Spain, July 2003, Revised Papers* (C. Martín-Vide, Gh. Păun, G. Rozenberg, A. Salomaa, eds.), LNCS 2933, Springer, Berlin, 2004, 181–190.
12. R. Freund: Energy-Controlled P Systems. In *Membrane Computing, International Workshop, WMC-CdeA 2002, Curtea de Argeş, Romania, August 2002, Revised Papers* (Gh. Păun, G. Rozenberg, A. Salomaa, C. Zandron, eds.), LNCS 2597, Springer, Berlin, 2003, 247–260.

13. T. Head: Formal Language Theory and DNA: An Analysis of the Generative Capacity of Specific Recombinant Behaviours. *Bull. Mathematical Biology*, 49 (1987), 737–759.
14. R.C. Hilborn: *Chaos and Nonlinear Dynamics*. Oxford University Press, Oxford, UK, 2000.
15. D.S. Jones, B.D. Sleeman: *Differential Equations and Mathematical Biology*. Chapman & Hall/CRC, London, UK, 2003.
16. T. Kailath: *Linear Systems*. Prentice-Hall, Englewood Cliffs, 1980.
17. S.A. Kauffman: *The Origins of Order*. Oxford University Press, New York, NY, 1993.
18. J.D. Lambert: *Computational Methods in Ordinary Differential Equations*. J. Wiley & Sons, New York, NY, 1973.
19. C.G. Langton: Computation at the Edge of Chaos: Phase Transitions and Emergent Computation. *Physica D*, 42, 12 (1990).
20. A. Lindenmayer: Mathematical Models for Cellular Interaction in Development. *J. of Theoretical Biology*, 18 (1968), 280–315.
21. B.H. Mahan: *University Chemistry*. Addison Wesley, Reading, MA, 1967.
22. M. Malita: Membrane Computing in Prolog. In *Pre-Proceedings of the Workshop on Multiset Processing*, Curtea de Argeş, Romania, 2000, 159–175.
23. V. Manca, G. Franco, G. Scollo: String Transition Dynamics – Basic Concepts and Molecular Computing Perspectives. In *Molecular Computational Models – Unconventional Approaches* (M. Gheorghe, ed.), Idea Group, London, 2004, 32–55.
24. G. Nicolis, I. Prigogine: *Exploring Complexity. An Introduction*. Freeman and Company, San Francisco, CA, 1989.
25. A. Păun: On P Systems with Membrane Division. In *Unconventional Models of Computation* (I. Antoniou, C.S. Calude, M.J. Dinneen, eds.), Springer, London, UK, 2000, 187–201.
26. A. Păun, Gh. Păun: The Power of Communication: P Systems with Symport/Antiport. *New Generation Computing*, 20, 3 (2002), 295–306.
27. Gh. Păun: Computing with Membranes. *J. Comput. System Sci.*, 61, 1 (2000), 108–143.
28. Gh. Păun: *Membrane Computing. An Introduction*. Springer, Berlin, 2002.
29. Gh. Păun, G. Rozenberg: A Guide to Membrane Computing. *Theoretical Computer Science*, 287 (2002), 73–100.
30. Gh. Păun, G. Rozenberg, A. Salomaa: *DNA Computing – New Computing Paradigms*. Springer, Berlin, 1998.
31. Gh. Păun, Y. Suzuki, H. Tanaka: P Systems with Energy Accounting. *Int. J. Computer Math.*, 78, 3 (2001), 343–364.
32. G. Rozenberg, A. Salomaa: *Handbook of Formal Languages*. Springer, Berlin, Germany, 1997.
33. K.S. Scott: *Chemical Chaos*. Oxford University Press, Oxford, UK, 1991.
34. L.A. Segel, I.R. Cohen: *Design Principles for the Immune System and Other Distributed Autonomous System*. Oxford University Press, Oxford, UK, 2001.
35. Y. Suzuki, H. Tanaka: Chemical Oscillation in Symbolic Chemical Systems and Its Behavioral Pattern. In *Proc. International Conference on Complex Systems* (Y. Bar-Yam, ed.), Nashua, NH, September 1997.
36. Y. Suzuki, H. Tanaka: On a Lisp Implementation of a Class of P Systems. *Romanian Journal of Information Science and Technology*, 3, 2 (2000), 173–186.

37. Y. Suzuki, H. Tanaka: Abstract Rewriting Systems on Multisets and Their Application for Modeling Complex Behaviours. In *Proc. of Brainstorming Week on Membrane Computing*, Tarragona, Spain, February 2003, 313–331.
38. S. Wolfram: *Theory and Application of Cellular Automata.* Addison-Wesley, Reading, MA, 1986.
39. A. Wuensche: Basins of Attraction in Network Dynamics: A Conceptual Framework for Biomolecular Networks. In *Modularity in Development and Evolution* (G. Schlosser, G.P. Wagner, eds.), Chicago University Press, Chicago, MA, 2003.

Chapter 4
Modeling Respiration in Bacteria and Respiration/Photosynthesis Interaction in Cyanobacteria Using a P System Simulator

Matteo Cavaliere[1], Ioan I. Ardelean[2]

[1] Department of Computer Science and Artificial Intelligence
University of Sevilla
Avda. Reina Mercedes s/n, 41012 Sevilla, Spain
martew@inwind.it

[2] Centre of Microbiology,
Institute of Biology of the Romanian Academy,
Splaiul Independentei 296,
PO Box 56-53, Bucharest 79651, Romania
ioan.ardelean@ibiol.ro

Summary. We present a probabilistic simulator of P systems that implements the evolution-communication model proposed in [8] enriched with some probabilistic parameters inspired by cell biology. After describing the software and its working, we compare the mathematical model with the biological reality of the cell. Then, we present some biological applications showing how one can use this software to simulate simple but interesting biological phenomena, related to respiration in *Escherichia coli* and the interaction between respiration and photosynthesis in cyanobacteria. The present chapter is an extension of the work presented in [5].

1 Introduction

In this chapter we present a probabilistic software simulator for a P system model and we show how it can be used to approach some biological processes and to study new mathematical problems.

Membrane computing is a branch of natural computing which starts from the idea that a formal computing device can be abstracted from cell functioning. Since this model has been introduced, [15], many papers have investigated the computational and mathematical aspects of membrane systems (also called P systems, from the name of its creator). For an up-to-date bibliography of P systems the reader can consult the Web page [27].

In this chapter, we address a different class of questions: we try to compare the mathematical model with the biological reality, indicating how one can use the P system framework (and in this case the software simulator) to model very important processes that occur in cells.

In Section 2 we give an overview of evolution-communication (EC) P systems, recalling the idea of the model and the main definitions.

In Section 3 we present, in a general way, the software which simulates (probabilistic) EC P systems, together with a short description of its working.

In Section 4 we present the simulation of a very simple probabilistic P system, while Section 5 is dedicated to a comparison between the features of the mathematical model and the biological reality. In this way we try to establish a link between the mathematical framework, the simulator realized, and the biological reality, introducing new concepts in the P system area and, at the same time, comparing the standard concepts of membrane computing with what we have in cell biology.

Finally, in Section 6 we show how it is possible to use the mathematical model, and therefore the software, to simulate different situations of important biological processes; in particular, we simulate the process of respiration in *Escherichia coli* and *Synechocystis PCC*6803, the corresponding proton pumping by cytochrome *c* oxydase in *Anacystis nidulans*, and the interplay between oxygen consumption and oxygen production by photosynthesis II ($PSII$) in *Synechocystis PCC*6803 in the presence of a specific synthetic inhibitor of $PSII$. We also show how to interpret the results obtained in a way as to infer useful results for biologists.

This work has been done in the framework of P systems because we believe that the emergence of P systems, together with its collaboration with biologists, could be as important and fruitful for biology as the introduction of physics and chemistry almost two centuries ago.

2 Definitions

For an introduction to P systems we refer the reader to the monograph [16], and to the Web page [27]; for the essential cell biology elements we refer the reader to [1]; we also use some (very) basic elements of formal languages here: the reader can consult the classic book [26] for more details.

We recall here the definition of the particular model (evolution-communication) of P systems, for which we have also realized a simulation software package (for the probabilistic case). This model has been proposed in [8] and its motivation is rooted in the idea to separate the evolutive mechanism of the cell (chemical reactions described by rewriting rules) from the communicative mechanism (passage of chemical objects through the membranes of the cell described by the so-called symport/antiport rules). In an evolution-communication P system the evolutive mechanism and the communicative mechanism can work simultaneously and independently.

The formal definition of an evolution-communication P system is as follows:

Definition 2.1 *An evolution-communication P system (in short, an EC P system) of degree $m \geq 1$ is a construct*

$$\Pi = (\Sigma, \mu, w_1, \cdots, w_m, R_1, \cdots, R_m, R'_1, \cdots, R'_m, i_0),$$

where Σ is the alphabet of symbol objects, μ is a membrane structure with m regions, labeled $1, \cdots, m$, and $i_0 \in \{0, \cdots, m\}$ is the output region (the environment if $i_0 = 0$).
Every region $i \in \{1, \cdots, m\}$ has:

- $w_i \in \Sigma^*$, *a string representing a multiset of symbol objects from Σ;*
- R_i, *a finite set of evolution rules over Σ of the form $u \to v$, for $u \in \Sigma^+$ and $v \in \Sigma^*$;*
- R'_i, *a finite set of symport/antiport rules over Σ of the forms (a, in), (b, out), and $(b, out; a, in)$, for $a, b \in \Sigma$.*

In other words, we can say that the evolution-communication P systems consist of a membrane structure μ (that follows the fluid-mosaic model proposed in [24]) composed of several membranes hierarchically embedded in a main membrane called the skin membrane. The membranes delimit regions and each region can contain symbol objects (that represent chemical objects), and each symbol object can be present in many occurrences (i.e., many molecules). For this reason a multiset of symbol objects is associated with each region (a multiset is a set where each element is associated with its number of occurrences, described by a natural number).

Each occurrence of the symbol objects evolves according to given evolution rules (that represent chemical reactions, i.e., biochemical transformations) associated with the regions and described by rewriting rules.

At the same time, in the evolution-communication P systems we have the communication rules represented by symport/antiport rules that simulate some of the biochemical transport mechanisms present in the cell.

The transport of molecules and chemicals (symbol objects) across membranes is one of the fundamental functions of a cell and its study in bacteria (as well as in other living cells) is under strong development (for more information the reader is advised to consult [7, 13, 23]). The transport can be passive or active. The transport is *passive* when molecules (symbol objects) pass across the membrane from the compartment with a higher concentration to that with a lower concentration, and, in this case, there is no metabolical energy used for the transport. An example of passive transport is the entry of oxygen molecules by diffusion into the cell of *Escherichia coli* bacteria or the exit of carbon dioxide outside the bacterium. These two passive processes are both important for the aerobic respiration in *Escherichia coli*, as we will show later. The transport is *active* when molecules pass across the membrane from a compartment with a lower concentration to one with a higher concentration. In this case, it is necessary to expend some metabolical energy to accomplish the transport.

For example, during the process of respiration in *Escherichia coli*, the protons are transported across the cell membrane, from the inner space (determined by the cell membrane) to the periplasmic space (external space). In such a way a difference in the concentration of protons across the cell membrane is established (more protons in the periplasmic space and less in the inner space of the cell) and this is a chemical form of energy (so-called proton motive force, described in [13]). These protons are used for symport and antiport transport of chemicals.

The symport of different substances needed for bacterial growth is very well documented in [10]. For example, *Escherichia coli* uses the symport of protons with either lactose, arabinose, or galactose. A classical example of antiport mechanism is the proton/sodium antiporter found in many bacteria, where the main function of this antiporter is the maintaining of a (quite) constant concentration of either protons or sodium ions inside the cell, [14].

The application of the rules of either type (evolution and symport/antiport rules) of an EC P system is made in a nondeterministic way (the rules to be applied are chosen randomly) and in a maximally parallel way (at each step all objects that can be moved or evolve must do so). The process is synchronized (there is a global clock marking the time units and it is common to all the regions of the system). In this way we get a sequence of *configurations* (an instantaneous description of the system) that defines a *computation* of an evolution-communication P system. From a mathematical point of view it is interesting to study the *halting* computations, that is, the computations which reach a configuration where no rule (neither an evolution rule nor a symport/antiport rule) is applicable.

Here, we consider any computation (halting or non-halting) of an evolution-communication P system as corresponding to the biochemical transformations and transport processes in a living cell; on the other hand, in our case, the presence of an output region of a P system is not relevant.

3 The Software

The software we have realized is able to simulate the behaviour of an evolution-communication P system as described earlier. Moreover, the simulator, using the *weak priority* approach, solves the conflicts that can appear when the rules choose the symbol objects; the weak priority can be seen as a competition of the rules for each single occurrence of the objects.

In what follows we explain the software in more detail; in the next section we will discuss, from a biological point of view, the implementation choices. The simulator takes as an input the rules of the P system to simulate (evolution rules and symport/antiport rules), the structure of the model (actually, the structure is not limited to a tree, as customary in the P systems field, but is permitted to be a graph), and the occurrences of the symbol objects present at the beginning of the computation in the regions of the P system.

Also, we must specify for each rule r two kinds of probabilities: *the probability to be available* (we call it Pav_r) and the *probability to win a conflict* (we call it $Pwin_r$); actually, as we will see below, the simulator computes this second probability using an integer coefficient, $Cwin_r$, fixed for each rule r.

The simulation takes place in the following way. At each step, the simulator decides which rules are available, and this decision is taken using the probability Pav_r fixed for each rule r. These probabilities are independent of each other and they express the probability of each rule being *available* for the application at a step.

After this choice, the simulator solves conflicts present among *available* rules, using the probability $Pwin_r$ that indicates the probability of the rule r winning a conflict over a symbol object with other rules; in this case the probability $Pwin_r$ is calculated for each conflict and depends on the coefficient $Cwin_r$ associated with r and on the coefficients of the other rules involved in the same conflict: in this way, for each conflict, a probability distribution among the rules involved in that conflict is produced. If $r_i, i = 1, \cdots, k$, are the rules involved in a conflict over a symbol object, then $Pwin$ is calculated, for each $r_i, i = 1, \cdots, k$, as follows:

$$Pwin_{r_i} = Cwin_{r_i}/(Cwin_{r_1} + \cdots + Cwin_{r_k}),$$

for $i = 1, \cdots, k$.

Finally, when every conflict has been resolved, the rules are applied in parallel in each region, as is usual with P systems.

3.1 How the Software Works: A Simple Example

The best way to explain the working of the software is to discuss a simple example.

Suppose we have a simple P system composed of one membrane (labeled 1) and in this region there are the evolution rules $r_1 : a \to aa$, $r_2 : b \to a$, and $r_3 : aa \to a$; The first rule r_1 says that one occurrence of the symbol object a is replaced by two occurrences of the same symbol object, r_3 is exactly the opposite, and the second rule means that one occurrence of the symbol object b is transformed into one occurrence of the symbol object a. In region 1 (at the beginning of the computation) there are 10 occurrences of the symbol object a and one occurrence of the symbol object b.

Moreover, we suppose that $Pav_{r_1} = 0.8, Pav_{r_2} = 0.2$, and $Pav_{r_3} = 0.6$, and, $Cwin_{r_1} = Cwin_{r_3} = 100$, and $Cwin_{r_2} = 40$.

At each step, the simulator chooses the available rules in region 1 and the choice is made using the probabilities $Papp_{r_i}, i = 1, 2, 3$. These probabilities imply that at some steps the available rules might be different from those at the previous steps, and there might be rules that are chosen more often than others (their Pav is bigger than the probabilities of the other rules). After constructing the list of the available rules, the simulator must solve the conflicts that, at this point, might be present in the region.

Suppose that, at some step, the list of available rules is composed of $r_1 : a \to aa$ and $r_3 : aa \to a$.

At this point, the simulator searches for possible conflicts on the occurrences of the symbol objects present in region 1.

In our case, there are conflicts because both rules can use the symbol object a. Then, the simulator starts to assign each occurrence of a in region 1 to one of the two rules involved in the conflict.

For this goal one uses the probability $Pwin_{r_i}$ calculated for each rule from the coefficient $Cwin_{r_i}$ associated with the rule r_i; actually, in our case, the probabilities are computed in the following way:

$$Pwin_{r_i} = Cwin_{r_i}/(Cwin_{r_1} + Cwin_{r_3}), i \in \{1, 3\}.$$

In this example, the probability is the same (0.5) for both rules involved in the conflict. Using such probabilities, each occurrence of object a is assigned. When a rule *wins* the other, it takes the number of occurrences of objects that it needs (two for r_3 and one for r_1). In this way, when there are no more conflicts, the rules can be applied in parallel. For example, in our case suppose that r_3 wins the first conflict against r_1; then r_3 can take the two occurrences of a that it needs; in this way, the occurrences which remain to assign are eight. Now, there are still conflicts to solve, so the process continues in the same way.

Suppose that, at the end, six occurrences of a are assigned to r_1 and four to r_3; after applying the rules in parallel, the new number of occurrences of a in region 1 is 14. To clarify the idea of the working of the simulator, we discuss the previous example step by step, starting from the initial configuration. The P system is

$$\Pi = (\{a, b\}, [_1\ ]_1, a^{10}b, R_1), \text{ with}$$
$$R_1 = \{r_1 : a \to aa, \quad r_2 : b \to a, \quad r_3 : aa \to a\}.$$

Moreover, we have that

$$Pav_{r_1} = 0.8, \quad Pav_{r_2} = 0.2, \quad Pav_{r_3} = 0.6,$$
$$Cwin_{r_1} = Cwin_{r_3} = 100, \quad Cwin_{r_2} = 40.$$

Step 1: *List of rules available*

The simulator decides which are the available rules *in this step*. Suppose that, using the probabilities Pav, the simulator chooses the rules r_1 and r_3. Then, the list of available rules in this step of computation is $L = \{r_1, r_3\}$.

Step 2: *Searching for conflicts in region 1*

The number of occurrences of a is 10. There are conflicts among the rules in the list L (in this case, both rules can use the symbol object a).

Step 3: *Solving the conflicts*

The simulator solves the conflicts by assigning, using the probabilities $Pwin$, the occurrences of a to the rules involved in the conflicts.

In this case, $Cwin_{r_1} = Cwin_{r_3} = 100$ and hence $Pwin_{r_1} = Pwin_{r_3} = 100/200 = 0.5$; therefore, in this case the two rules have the same probability of winning a conflict in which they are involved.

Suppose that, for example, the rule r_3 wins the first conflict; therefore, two free occurrences of a are assigned to this rule. Now, there are eight more occurrences of object a to assign. There are still conflicts, so the simulator repeats, in a similar way, step 3 until all the conflicts are solved.

Step 4: *All conflicts in region 1 have been solved*

Suppose that all the conflicts in region 1 have been solved and suppose, for example, that six occurrences of a are assigned to r_1 and four to r_3.

Step 5: *Execute the rules*

When the objects have been assigned the simulator executes the rules present in the membrane, in parallel, according to the assigned objects. In particular, after the execution of the rules, we will have 14 occurrences of object a and one occurrence of the object b.

Step 6: *Repeat the same process* for a new step of computation (it starts again from step 1, where a new list L of available rules is created).

In this example, we had only one membrane, but if the P system to simulate has many membranes, then the same algorithm is applied to each membrane of the system, and only when each occurrence of each symbol object is assigned (or, at least, the simulator has tried to assign it) are the rules executed in parallel in each region, as is usual in P systems.

4 A First Probabilistic Simulation: Is Life Unpredictable?

In this section, we show the results of a first experiment, consisting of the simulation of a very simple EC P system. This simple experiment clarifies the working of our simulator and it is interesting because of the *unexpected* results and because it reminds other systems (in different fields) where similar phenomena appear.

The P system that we have simulated is:

$$\Pi = (\{a\}, [_1\ ]_1, a, R_1), \text{ with}$$
$$R_1 = \{r_1 : a \to aa, \quad r_2 : aa \to a\}.$$

This P system is composed of one region (with label 1) and of two simple evolution rules, r_1 and r_2. The first rule duplicates the occurrences of symbol objects a, while the second rule does the opposite: it reduces to half the number of occurrences of a.

We fix Pav_{r_1} and $Pav_{r_2} = 0.8$; this means that r_1 and r_2 have the same probability of being available at each step of the simulation. Also, we take $Cwin_{r_1}$ and $Cwin_{r_2} = 100$; this means that, if there is a conflict between r_1 and r_2, both the rules have the same probability of winning the conflict.

The two rules of the system can be considered as describing *associations* ($r_1 : a \to aa$) and *dissociations* ($r_1 : aa \to a$) of molecules, and this reminds us of the importance of these phenomena for the origin of life on earth [12, 25]. Thus, the models and software package as considered above can also be related to the important subject of origin of life (whatever "life" means in this framework). Moreover, from a *formal language* point of view, it is natural to ask about the link between systems as above and some kinds of *Lindenmayer systems* (formal language devices inspired by biology, [22]) with probabilistic behavior.

Even if the system constructed could seem very *stable* (all the probabilities are equal), the results (the number of occurrences of symbol object a present in the system during the evolution – we can say, metaphorically, during the *life* of the system) of this experiment are quite surprising. The behaviour of the system seems very *unstable*, and we notice large variations of the number of occurrences of a during the computation, with many large jumps (see Figure 1). Roughly speaking, we can think that, for this simple system, with these coefficients and probabilities, we can expect any kind of behavior; this makes it very hard to make predictions on the behavior. The picture in Figure 1 very well represents the controversial results of this experiment (the number of copies of a is indicated on the y axis and the number of the steps on the x axis). Moreover, we have noticed that, the lower we fix the probability for the two rules to be available,the more unstable (with larger jumps) the system becomes. We think that the study of this (simple) system could be a good starting point to understand more about the dynamics of probabilistic P systems.

Some simple questions arise here: is it possible to make some kind of prediction about the behaviour of this system? Are there other systems (also in different fields like biology, economics, etc.) that behave in a similar way and that can be modeled by a similar model?

5 Mathematics and Biology: A Comparison

In this section we try to clarify the link between the software mentioned before and some basic elements of cell biology. Moreover we show in which way many fundamental concepts present in biology have been *translated* to the underlying mathematical model, and, therefore, introduced in the software.

We will also discuss several concepts not (yet) introduced in the P system area, but clearly related to the biological processes studied here.

We start this comparison speaking about the model of P systems implemented. This is based on the evolution-communication model, where we do not have indications for communications, but the transport of chemical substances is done using the symport/antiport rules (that can simulate the biochemical transport mechanisms present in the cell). In this model we can have symbol objects and particular objects called *catalysts* already well studied in the P

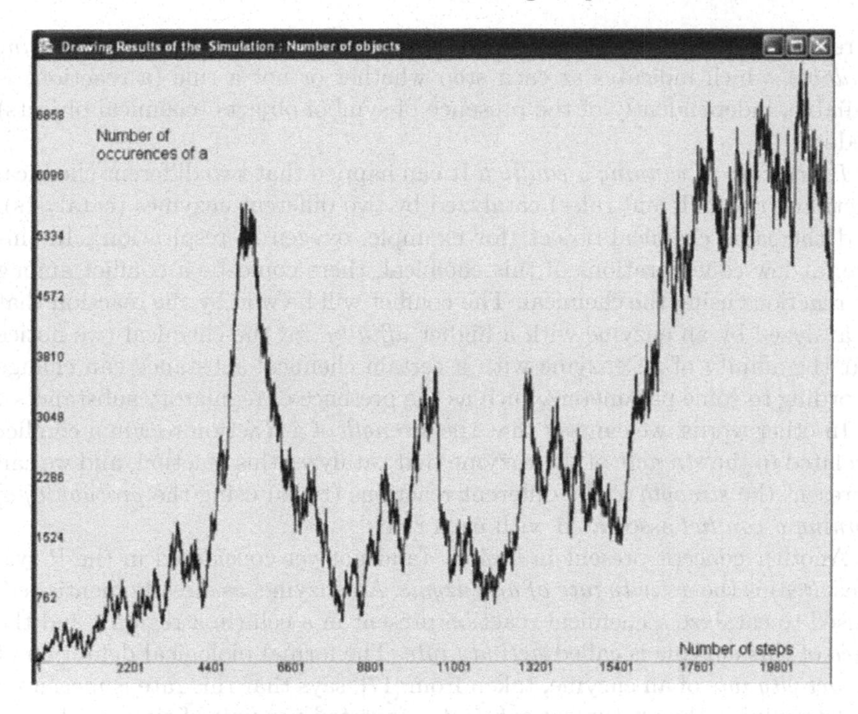

Fig. 1. The simulation of a simple probabilistic EC P system.

systems area [16]: in particular, a catalyst corresponds to what in bioloy is usually called *enzyme*, and it is used to *catalyze* chemical reactions (in other words, to permit the chemical reactions to occur).

Moreover, in the software, the conflicts among the rules (chemical reactions) when using some common symbol objects (chemical substances) are solved using the *weak priority approach*. Using this approach all the chemicals present are assigned to the available reactions (according to some biological parameters) and, if a reaction that is *stronger* than another cannot be applied (for many reasons, the main one being the lack of some necessary chemicals), then a reaction with a *lower* priority can be applied.

Now, we discuss the biological relevance of the two probabilistic parameters used in the software: the *probability of being available* and the *probability of winning a conflict* assigned to each rule.

Probability of being available: In a cell, it can happen that all the chemical substances needed for a chemical reaction are present but the reaction does not take place. Why? This fact is very important in biology and it is related to many possible factors. For example, in case of the process of respiration of bacteria, the factors that stop a chemical reaction are the inhibition of synthesis of some enzymes, the presence of inhibitors like potassium cyanide, the quantity of oxygen or bound hydrogen, etc. We need a probabilistic device to represent this phenomenon. The idea is to use the *probability of being available*, which indicates at each step whether or not a rule (a reaction) is available, independently of the presence of symbol objects (chemical objects) needed.

Probability of winning a conflict: It can happen that two different chemical reactions (two different rules) catalyzed by two different enzymes (catalysts), need the same chemical object (for example, oxygen in respiration). In this case, at low concentrations of this chemical, there could be a conflict among the reactions using the chemical. The conflict will be won by the reaction that is catalyzed by an enzyme with a higher *affinity* for the chemical (we notice that the affinity of an enzyme with a certain chemical substance can change according to some parameters, such as the presence of regulatory substances).

In other words, we can say that the *strength* of a reaction to win a conflict is related to the *strength* of the enzyme that catalyzes this reaction, and we can represent the *strength* of the different reactions (rules) using the *probability of winning a conflict* associated with each rule.

Another concept present in biology (and not yet considered in the P systems area) is the *activity rate of an enzyme*. An enzyme, as already mentioned, is used to catalyze a chemical reaction present in a cell (in a region), and the *speed* of this enzyme is called *activity rate*. The formal biological definition of the *activity rate* of an enzyme, taken from [17], says that this rate is measured by determining the amount of *substrate* converted per unit of time, under exactly defined and strictly controlled conditions. In other words, the activity of an enzyme is the number of reactions that it can catalyze in a fixed unit of time (the chemicals consumed by such reactions are called *substrate*). Every type of enzyme has its own activity rate, and this can change in time according to biological parameters dependent on the particular chemical reaction considered.

Usually, in biology the activity rate of an enzyme is calculated for the total quantity and not for single occurrences (single molecules). The international unit of enzyme activity (named *katal*) is the quantity of enzyme that converts one mol (the quantity in grams numerically equally to the molecular weight of the molecule) of substrate in one second. In practice one utilizes the milikatal or the nanokatal, for the quantity of enzyme that converts one milimol or one nanomol of substrate (for details, see [17]).

In the P systems area, it is usually assumed that each symbol object has the same *velocity* (we can say the same "activity rate": at each *step*, each occurrence is used, if possible, in one reaction); if we consider enzymes (catalysts in the P systems area) with different activity rates, then we can have a different mathematical model; in this case not all the objects proceed at the same velocity, but, at the same step different kinds of catalysts can be used in a different number of reactions (rules). This means that, in general, for each object o, it is important to consider the number of rules (chemical reactions) where it can be used at one single step, and we will call it s_o (speed of an object); different kinds of objects can have different speeds.

Another feature present in biology (and which could be introduced in the mathematical model) is the *activity* of an enzyme: in fact, a molecule (an occurrence) of an enzyme can be present but not *active* (this means that the molecule of the enzyme is present in the cell but it cannot be used to catalyze any reactions). Then, by activity we mean the capacity of an enzyme to help the catalyzed reaction to occur.

We can look at the activity of an enzyme of a certain kind in this way: we have a fixed number of occurrences of such enzyme in the cell (for example k molecules) and part of these occurrences are active and the remaining occurrences are not active.

Only the occurrences of an enzyme that are active can be used to catalyze the reaction, and the number of active occurrences change according to some biological parameters. For example, enzymes are inhibited, reversibly or irreversibly. The effect of irreversible inhibitors is to reduce the amount of enzymes available for the reaction; their effect cannot be overcome by simple physical techniques such as *washing out*. On the other hand, reversible inhibitors combine with the enzyme in such a way that in general they can be easily removed.

It is easy to observe that, changing the number of active occurrences of an enzyme, it is possible to control the general *activity rate* of the enzyme itself, and, therefore, the rate of the catalyzed reaction.

Finally, we would like to introduce the concept of *closed* and *open* systems. Open systems can exchange energy and chemical objects with the environment while closed systems can exchange only energy and not chemicals. Being open systems, living organisms continuously obtain energy and materials from the external environment and eliminate the end products (for example, carbon dioxide produced by the respiration process) of their metabolism [11].

Sometimes, to study various processes occurring in living cells, the cells are closed in some measuring device; for example, to study the oxygen consumption by a suspension of cells, the cells are closed in an oxygen electrode chamber and the decrease in the concentration of the molecular oxygen (as a result of the respiration) is monitored by a so-called Clark-type electrode.

6 Modeling Some Biological Processes

In this section we describe how some important biological processes, such as the respiration process in *Escherichia coli* and the interactions between respiration and photosynthesis in cyanobacteria, can be modeled and translated in the P systems framework, and, how, in this way, using the simulator, it is possible to obtain results that can be useful for biologists. Also, we intend to illustrate with a realistic example what we have discussed in the previous section.

6.1 Modeling the Respiration in Bacteria

Respiration is the biological process that allows cells (from bacteria to humans) to obtain energy. In short, respiration promotes a flux of electrons from electron donors to a final electron acceptor, which in most cases is the molecular oxygen.

In *Escherichia coli*, as well as in other bacteria, the cell ability to consume molecular oxygen during the respiration is determined by the presence of two different enzymes that catalyze the final step of respiration: the reduction of molecular oxygen with protons and electrons.

In *Escherichia coli*, these two terminal oxydases (enzymes) – called *terminal* because they are the last components of the respiratory electron transport pathways – are *cytochrome bd* and *cytochrome bo*. The occurrence of multiple (two or more) terminal oxydases enables the cell to modulate its respiration, in accordance with its energy requirements and the availability of chemicals in the environment [21].

For example, in *Escherichia coli* cytochrome *bd* has high affinity for oxygen and is involved in energy conversion with medium efficiency: more precisely, for every electron (passed through the cytochrome *bd* to molecular oxygen) one proton (one atom of bound hydrogen without its electron) is transported from the inside to the outside the cell.

Thus, because of its higher affinity for oxygen, the cytochrome *bd* "works more" at a relatively low oxygen concentration in the growing medium.

On the other hand, the cytochrome *bo* oxidase has a lower affinity for oxygen. Thus, cytochrome *bo* works at a higher oxygen concentration in the growing medium; however, it has greater efficiency in energy conversion.

Recently, Alexeeva et al. [2, 3] studied the relationship between the oxygen concentration in the growing medium and the *activity rate* of the two terminal oxidases, including the flux of electrons to molecular oxygen through each of the two pathways.

Simply, but correctly (for more details the reader can consult the papers cited above), we can say that, at low oxygen concentration in the growing medium (lower than about 40% of oxygen saturation) the cytochrome *bd* oxydase is responsible for the entire respiratory activity of the cells; in other words, the flux of electrons to molecular oxygen proceeds 100% through the cytochrome *bd* oxydase. At high oxygen concentration in the growing medium (between 90% and 100% of oxygen saturation), the cytochrome *bo* oxydase is responsible for almost the entire respiratory activity of the cells. Furthermore, between 40% and 90% of oxygen saturation, the two types of terminal oxidases contribute together to the respiration of the cell.

Now, we show how to translate this biological reality into the P systems framework so that it is possible to use the simulator.

We know that in *Escherichia coli* the consumption of molecular oxygen is made using two different chemical reactions catalyzed by two different kinds of enzymes: *bo* and *bd* (for simplicity of notation we call *bo* E'' and call *bd*

E'). The activity rate of these two enzymes is different, according to the percentage of saturation of molecular oxygen: this means that the activity rate of the enzymes determines how many chemical reactions catalyzed by such enzymes occur in a fixed time unit.

The two reactions can be represented as

$$E'AB \rightarrow CE', \quad E''AB \rightarrow CE'',$$

where E'' and E' are the two enzymes that catalyze the two reactions, B is the molecular oxygen (consumed by the two reactions), A is the hydrogen, and C is the molecular water produced.

As we can see, both the chemical reactions consume oxygen and hydrogen but the activity rate of the two reactions will be different, according to the concentration of oxygen (substrate of the reaction) in the cell, because it depends on the activity rate of the enzymes E' and E'' (which, in turn, depends on the concentration of oxygen in the cell, as discussed earlier). Moreover, only the occurrences of an enzyme that are active can be used to catalyze the reaction, and the number of active occurrences change according to some parameters (in this case, the concentration of oxygen).

It is possible to *control* the activity rate of the enzymes E' and E'' by controlling the number of *active* occurrences of the enzymes in the cell. In other words, it is possible to simulate the *increasing* (*decreasing*) of the activity rate of an enzyme by increasing (decreasing) the number of *active* occurrences of the enzyme in the cell.

In reality, this is only a biological hypothesis and the situation is not very clear; in fact, using intact cells it is very difficult to measure simultaneously the activity of an enzyme and its quantity. For example, we do not know if the changes in the activity rate of an enzyme are related to changes in the quantity of active enzymes (this means that there is a larger quantity of active enzymes, but the activity rate of the single occurrence remains the same) or happen in the activity rate of every single enzyme, but the quantity of active enzymes remains the same.

However, for our purpose at this stage, the biological hypothesis can be taken as true. Then, by using experimental and mathematical results given by the software, we can check the truth of this biological hypothesis.

Finally, we have to observe that at a low concentration of oxygen there could be conflicts among the two reactions using this chemical; these conflicts must be resolved by using the different affinities of the two enzymes with respect to the oxygen, and this means using the *probability of winning* associated with the two rules, as discussed earlier.

6.2 Simulating the Respiration Process in Bacteria

In this subsection we show how one can use the simulator to model the process of respiration in *Escherichia coli* described in the previous subsection and how, in this way, we can have some useful results for biologists. Initially we consider only one of the two reactions that take place in *Escherichia coli* during respiration, and we try to detail and to simulate it using the software.

In particular we consider a system where only the first enzyme e' (*cytochrome bd*) is present. It is known (see, e.g., [17]) that the activity of an enzyme can be expressed as the activity of each *unit* enzyme and the quantity of substances involved in the reactions can be given in *nanomols*. Then, using units of enzyme and nanomols, the chemical reaction introduced before,

$$E'AB \to CE',$$

must be rewritten in a more precise way

$$E'A^kB^j \to C^iE'. \tag{1}$$

This means that one unit of the enzyme E' (*cytochrome bd*) catalyzes a reaction that consumes k nanomols of hydrogen (A) and j nanomols of oxygen (B), producing i nanomols of water (C). Actually, for the purpose of this experiment, we are interested only in the consumption of oxygen, and we can see the reaction in the following simpler way:

$$E'B^j \to E'.$$

From [17] we know that one unit of enzyme E' consumes (more precisely, we should say, "catalyzes the reaction that consumes") 42 nanomols of oxygen in one hour. We must specify that this is the activity rate of the enzyme when the concentration of oxygen (substrate) is between 40% and 60% of saturation, and we suppose this as true for our experiment. Moreover, we suppose that our system is closed (no oxygen is introduced from the environment). Using these data we can rewrite the chemical reaction in the following way:

$$E'B^{42} \to E'.$$

This way we have made the translation from biological reality to mathematical model: now each occurrence of the symbol object E' represents one unit of enzyme *cytochrome bd*, each occurrence of the symbol object B represents one nanomol of oxygen, and each step of the P system is one hour (the time unit of the activity rate of the enzyme).

Now we can simulate the consumption of oxygen in the respiration process of *Escherichia coli* using the software on the following simple P system:

$$\begin{aligned} \Pi &= (\{E', B\}, [_1\ ]_1, (E')^kB^j, R_1), \text{ with} \\ R_1 &= \{E'B^{42} \to E'\}, \end{aligned} \tag{2}$$

where *Papp* is 1.0 for the rule in R_1 (we suppose the reaction is always available and that *Pwin* is not relevant here because there is only one rule).

In the formal definition of the P system, we have left as parameters the number of occurrences of nanomols of oxygen (j) and the number of unit enzymes (k) present at the beginning of the experiment; changing the values of this parameters, we can run different experiments for different biological situations.

An immediate application for our simulator is to have diagrams of the consumption of oxygen, changing the quantity k (expressed in units) of enzyme *cytochrome bd* used. In other words, we are interested in looking at how much "faster" the oxygen decreases using different quantities of enzymes. This way we can check in a real experiment what quantity of enzymes we have used by looking at what extent the consumption obtained in laboratory is "similar" to the simulation. Of course, the verification of the "similarity" can be automated, using the software and checking online the quantity of enzymes used in an experiment. We illustrate this application with a simple practical example.

In Figure 2 is shown the simulation of the P system described before, where the consumption of oxygen is reported when the quantity of enzyme used is 1/10 of aunit of enzyme (1), 1/6 of a unit of enzyme (2), 1/2 of a unit of enzyme (3) and exactly 1 unit of enzyme (4).

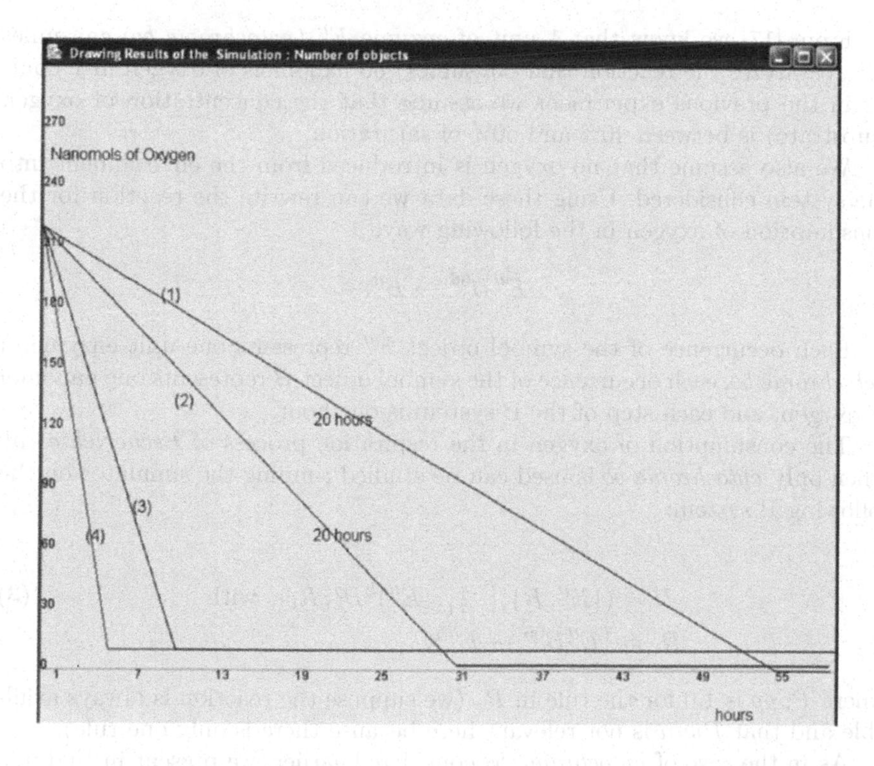

Fig. 2. Consumption of oxygen in Escherichia coli using only cytochrome *bd*.

On the x axis is given the time (in hours) of the experiment, while on the y axis is given the quantity (in nanomols) of oxygen present in the cell; the initial quantity of oxygen is 220 nanomols, and this corresponds to saturation of the cell; for this first step and for the goal of our experiment we can, as an approximation, assume that the activity rate of the enzymes used is constant in the time (we have to remark that, in practice, the activity rate of the enzyme changes according to the concentration of oxygen).

Now we want to consider the consumption of oxygen in *Escherichia coli*, where only the second enzyme (*cytochrome bo*) is present.

Using units to express the quantity of enzyme *cytochrome bo* and nanomols for the quantity of oxygen and hydrogen, the reaction for the consumption of oxygen is:

$$E''A^kB^j \rightarrow C^iE''.$$

This formula is interpreted exactly as the formula described in (1) except for the fact that the enzyme used here is E'', indicating the *cytochrome bo*.

Again, because we are interested only in the consumption of oxygen, the reaction can be simplified and written as:

$$E''B^j \rightarrow E''.$$

From [17] we know that 1 unit of enzyme E'' (*cytochrome bo*) consumes (i.e., catalyzes the reaction that consumes) 66 nanomols of oxygen in 1 hour. As in the previous experiment we assume that the concentration of oxygen (substrate) is between 40% and 60% of saturation.

We also assume that no oxygen is introduced from the environment into the system considered. Using these data we can rewrite the reaction for the consumption of oxygen in the following way:

$$E''B^{66} \rightarrow E''.$$

Each occurrence of the symbol object E'' represents one unit enzyme of *cytochrome bo*, each occurrence of the symbol object B represents one nanomol of oxygen, and each step of the P system is one hour.

The consumption of oxygen in the respiration process of *Escherichia coli* when only *cytochrome bo* is used can be studied running the simulator on the following P system:

$$\begin{aligned} \Pi &= (\{E'', B\}, [_1\ ]_1, (E'')^kB^j, R_1), \text{ with} \qquad (3)\\ R_1 &= \{E''B^{66} \rightarrow E''\}, \end{aligned}$$

where *Papp* is 1.0 for the rule in R_1 (we suppose the reaction is always available and that *Pwin* is not relevant here because there is only one rule).

As in the case of *cytochrome bd* considered earlier, we present in Figure 3 the diagram representing the consumption of oxygen when the quantity of

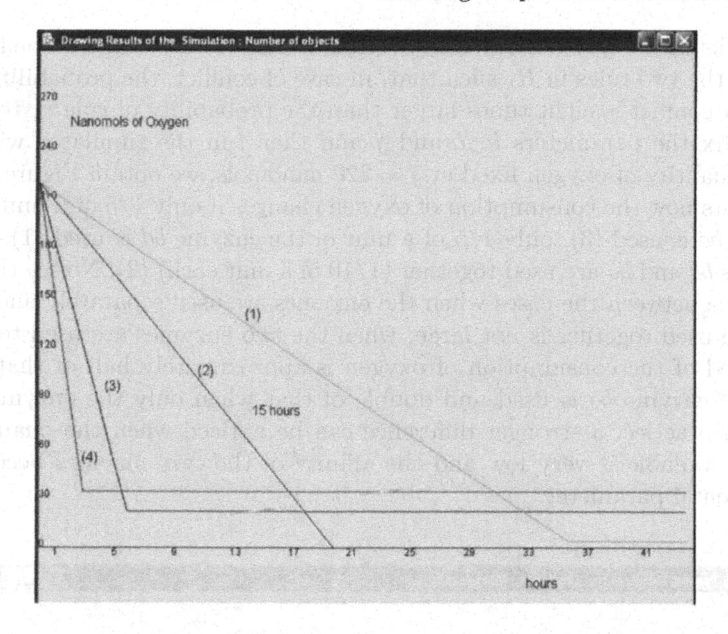

Fig. 3. Consumption of oxygen in Escherichia coli using only cytochrome *bo*.

(the unique) enzyme *cytochrome bo* used is 1/10 of a unit of enzyme (1), 1/6 of a unit of enzyme (2), 1/2 of a unit of enzyme (3) and exactly 1 unit of enzyme (4), starting with an initial amount of oxygen of 220 nanomols.

Now we consider the consumption of oxygen in *Escherichia coli* when both the enzymes cytochrome *bo* and cytochrome *bd* oxydase are present. In this case the two enzymes are somehow "concurrent": both the enzymes catalyze the consumption of oxygen. Because the oxygen is a common resource for the two enzymes, the ability of an enzyme to catalyze the consumption of oxygen is related to its affinity, as discussed in Section 6.1.

From [2, 3] we find out that when the oxygen present is 220 nanomols, the affinity of the enzyme *bo* with respect to oxygen is almost eight times larger than the affinity of the enzyme *bd*.

Now we can simulate the consumption of oxygen in the respiration process of *Escherichia coli* using the software on the following simple P system obtained by "composing" the two P systems previously considered in (2) and (3).

$$\begin{aligned} \Pi &= (\{E', E'', B\}, [_1\ ]_1, (E')^k (E'')^p B^j, R_1), \\ R_1 &= \{r_1 : E'B^{42} \to E',\ r_2 : E''B^{66} \to E''\}, \end{aligned} \tag{4}$$

where *Papp* is 1.0 for the rules in R_1 (we suppose the reactions are always available).

The two rules r_1 and r_2 compete for the use of the oxygen B. Such conflicts are solved using the *probability to win* that simulates the affinity of an enzyme and works in the way described in Section 3. Therefore we fix the coefficients C_{win} of the two rules in R_1 such that, in case of conflict, the probability of r_2 to win a conflict is eight times larger than the probability of rule r_1 to win.

We fix the parameters k, p, and j and then run the simulator with the initial quantity of oxygen fixed at $j = 220$ nanomols; we obtain Figure 4 that represents how the consumption of oxygen changes if only 1/5 of a unit of the enzyme *bo* is used (3), only 1/5 of a unit of the enzyme *bd* is used (1), or the enzymes *bd* and *bo* are used together (1/10 of a unit each) (2). Notice that the difference between the cases when the enzymes are used separately and when they are used together is not large; when the two enzymes are used together the speed of the consumption of oxygen is approximately half of that when only the enzyme *bo* is used and double of that when only the enzyme *bd* is used. In practice, a stronger difference can be noticed when the quantity of oxygen available is very low and the affinity of the two enzymes becomes a fundamental parameter.

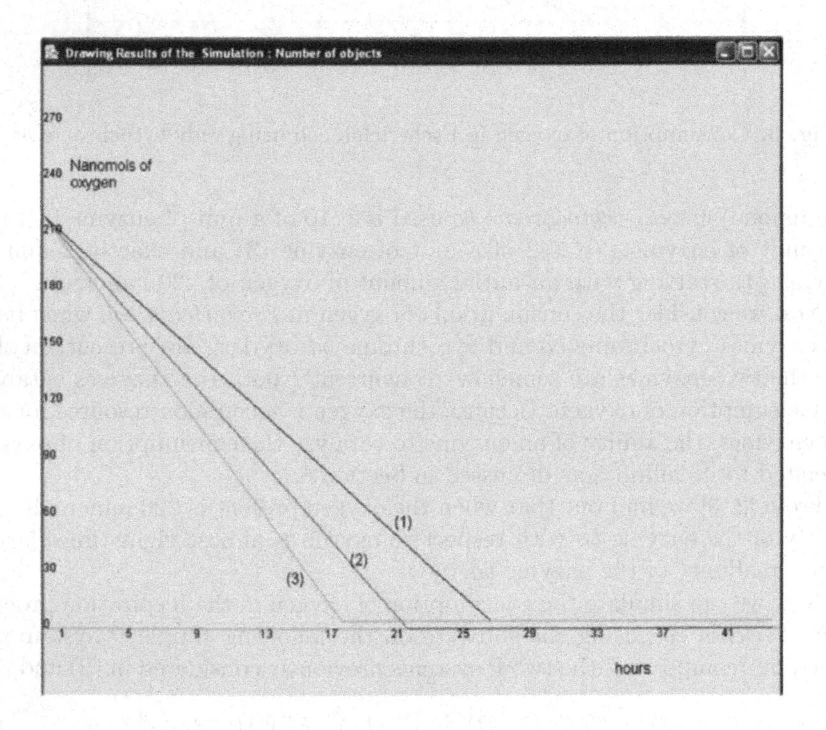

Fig. 4. Consumption of oxygen in Escherichia coli using cytochrome *bo* and *bd*.

6.3 Pumps in *Escherichia coli*

As already discussed in Sections 2 and 5, proton pumping is fundamental for biological energy conservation.

In *Escherichia coli* the consumption of molecular oxygen is related to the translocation (pumping out) of protons outside the cell.

Specifically, cytochrome *bd* oxydase translocates one proton for every electron transported to molecular oxygen. Following the data presented in [2, 3], we conclude that 42 nanomols of protons are pumped out, using one unit of enzyme *bd* oxydase.

In the case of cytochrome *bo* oxidase, two protons are translocated for every electron transported to molecular oxygen. This means that, from the data presented in [2, 3], 132 nanomols of protons are pumped out in one hour using one unit of enzyme *bo* oxidase.

Formally the following P system represents the situation where only consumption of molecular oxygen and translocation of protons are taken into consideration and where only the enzyme cytochrome *bo* oxidase is present.

$$\begin{aligned} \Pi &= (\{E'', O, P\}, [_1\ ]_1, E''^k O^j, R_1, R_1'), \text{ with} \qquad (5) \\ R_1 &= \{E'' O^{66} \rightarrow E'' P^{132}\}, \\ R_1' &= \{(P, out)\}, \end{aligned}$$

where $Papp = 1$ for both rules (the reactions are always available; $Pwin$ is not relevant here because there are no conflicts between the rules); each step of this system corresponds to 60 minutes. Each occurrence of the symbol object O corresponds to one nanomol of molecular oxygen; each occurrence of the symbol object P corresponds to one nanomol of protons, and each occurrence of the symbol object E'' corresponds to one unit of enzyme cytochrome *bo* oxidase. The consumption of molecular oxygen is simulated using the evolution rule in R_1 while the translocation of protons toward the environment is simulated using the symport rule present in R_1'.

Figure 5 represents the accumulation of protons in the environment when 1/10 of a unit of enzyme is used and 220 nanomols of molecular oxygen are initially present in the cell.

The diagrams from Figures 2, 3, 4, and 5 can be used to obtain biologically significant results. For instance, suppose that for an experiment we take exactly 220 nanomols and a certain quantity of the enzyme cytochrome *bo*. Suppose we observe that after 20 hours of respiration the *Escherichia coli* still contains (around) 140 nanomols of oxygen (this can be measured in a lab). Then, looking at the diagram represented in Figure 2 we can understand that the quantity of enzymes present in the cell is around 1/10 of a unit (the same idea can be applied to the other diagrams obtained).

In this case the calculation is quite easy, but the same idea, together with an improved version of the simulator, can be adapted easily to calculate, online, the quantity of enzymes used when the activity rate of each enzyme

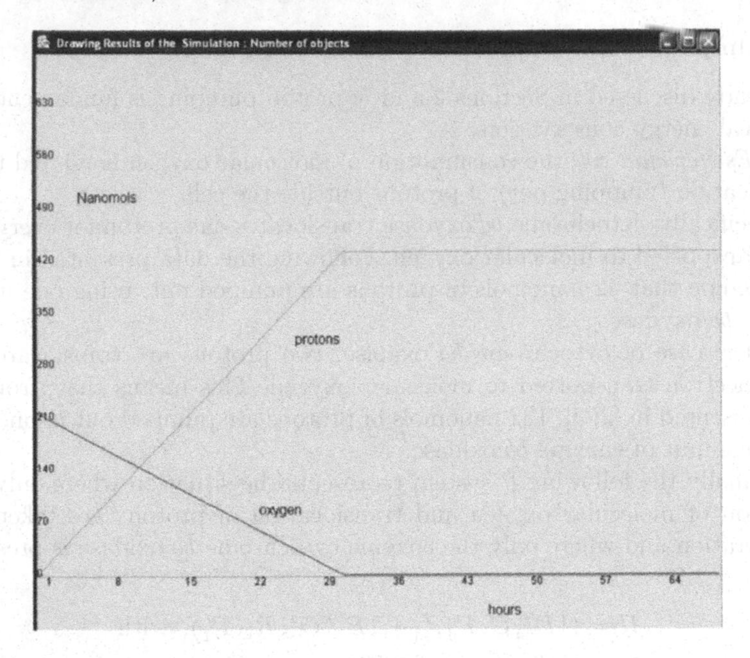

Fig. 5. Translocation of protons in Escherichia coli using cytochrome *bd*.

changes during the experiment according to the concentration of the oxygen substrate.

Moreover, in future investigations one can consider processes where the enzymes *cytochrome bo* and *cytochrome bd* oxydase are involved in the following interesting situations:

(i) *Decreasing of oxygen concentration:* The oxygen concentration decreases from saturation toward zero. In this case, both the reactions (rules) can be applied; both enzymes are present in the cell, with different activity rates and different affinities that change according to the oxygen concentration.

(ii) *Increasing of oxygen concentration:* This is dual to case (i): the oxygen concentration increases from zero toward saturation.

(iii) *Constant oxygen concentration:* In this case, the oxygen concentration is maintained constant (at a chosen value, between 10% and 100%). Moreover, the activity of the two enzymes is fixed, according to the chosen concentration of oxygen, and the system is not *closed* (as in the cases considered in our experiments) but is fed with oxygen.

Finally, an interesting experiment could be the measurement, using the simulator, of the activity and the quantity used of each enzyme and, based on previous experimental knowledge, the determination of either oxygen consumption and/or oxygen concentration in the growing medium. This last application could especially be of real practical interest both in academic studies and in bioindustrial activities, because, while oxygen measurements are usually carried out (both in academia and industry), enzyme activity (or enzyme quantity) cannot be measured in intact cells. These measurements are very laborious and time consuming (and for this reason the online control of bioindustrial processes is practically impossible at this level). The online knowledge of enzyme activity and enzyme quantity in bioindustrial processes can enable the online optimization (with respect to cost, production of useful chemicals, etc.) of such processes.

6.4 Modeling Respiration-Photosynthesis Interaction in Cyanobacteria

Cyanobacteria are the largest and most diversified important group of prokaryotes [19], defined by the ability to carry out both oxygenic photosynthesis (within the thylakoid membranes) and respiration (within the plasma membranes and thylakoid membranes) [19], [20]. In this section and in Section 6.5 we will consider for our simulations the cyanobacterium *Synechocystis PCC 6803*.

In brief, the overall process of photosynthesis consists of using electrons from water to ultimately reduce carbon dioxide, producing some chemicals (for example, carbohydrates). This process is essential for life on earth, generating the main food source for almost all living cells and the only source of the molecular oxygen needed for respiration.

The first reaction in photosynthesis is the splitting of water (at the expense of light energy, not presented here for simplicity) to produce molecular oxygen, protons, and electrons [9].

The last reaction of the respiration process is the opposite of the first reaction that occurs in photosynthesis; this means that in respiration water is produced by the consumption of oxygen during its combination with protons and electrons, as in *Escherichia coli*.

In cyanobacteria there is strong interaction between respiration and photosynthesis; for example, the oxygen produced by photosynthesis in the inner membrane (thylakoid membrane) is used in the cell membrane for the process of respiration, [18, 20], and, at the same time, the carbon dioxide produced by respiration in the cell membrane is used (recycled) in the inner membrane for the process of photosynthesis. For simplicity, we can say that oxygen production (by the photosynthesis process) is catalyzed by only one type of enzyme (actually, this is called oxygen-evolving complex, [9]) and the last step of respiratory oxygen consumption in cyanobacteria is catalyzed by the enzyme cytochrome *c* oxydase [20].

From the data presented in [6], the photosynthetic oxygen production is of 26 nanomols of oxygen per microgram of chlorophyll in 10 minutes, while oxygen consumption in respiration is of five nanomols of oxygen per microgram chlorophyll in 10 minutes. The carbon dioxide production in respiration is of five nanomols per microgram chlorophyll (present in the cyanobacteria) in

10 minutes, while the carbon dioxide consumption in photosynthesis is of 26 nanomols of carbon dioxide per microgram chlorophyll in 10 minutes. We have to notice that the more detailed the data, nearer the simulation results to biological reality; for our purpose, in this case, the data expressed in nanomols and micrograms have good precision. The system considered here is closed: there is no exchange of chemicals with the environment, and this means that oxygen and carbon dioxide are not introduced in the system. We suppose, for simplicity, that we have inexhaustible quantities of water and light.

Then, in a formal simplified way, we can represent the process of respiration and photosynthesis in cyanobacteria with the following EC P system:

$$\begin{aligned} &\Pi = (\{c, O, D\}, [_1[_2\ ]_2]_1, c^k O^j, c^l D^m, R_1, R_2, R'_1, R'_2), \text{ with} \qquad (6)\\ &R_1 = \{cO^5 \to cD^5\},\\ &R_2 = \{cD^{26} \to cO^{26}\},\\ &R'_2 = \{(D, in), (O, out)\}, \end{aligned}$$

where $Papp = 1$ for each rule in R_1, R_2, R'_2 (we suppose the reaction is always available and that $Pwin$ is not relevant here because there are no conflicts between the rules) and each step of the system corresponds to 10 minutes.

The system is composed of two membranes labeled 1 and 2 that represent the cell membrane and the thylakoid membrane, respectively. In the two regions there are the symbol object c (each occurrence of c represents one microgram of chlorophyll), the symbol object D (each occurrence of D represents one nanomol of carbon dioxide), and the symbol object O (each occurrence of O represents one nanomol of oxygen).

In region 1, between membranes 1 and 2, we have the simple evolution rule $cO^5 \to cD^5$ that represents the last step of the respiration that occurs in the cell membrane. In particular, the rule says that, in one step, five occurrences (nanomols) of O (oxygen) are consumed, using one occurrence (microgram) of chlorophyll c, and five occurrences (nanomols) of D (carbon dioxide) are produced. In region 2, there is the simple evolution rule $cD^{26} \to cO^{26}$ that represents the oxygen production by photosynthesis that occurs in the thylakoid membrane. In particular, the rule says that in one step, 26 occurrences (nanomols) of D (carbon dioxide) are consumed, using one occurrence (microgram) of chlorophyll c, and 26 occurrences (nanomols) of O (oxygen) are produced.

In our general model we did not fix the quantity of O (oxygen), D (carbon dioxide), and c (chlorophyll) that is present in the cell at the beginning of the simulation, and we can apply the same model for different experimental data.

In the first simulation we take at the beginning 1 microgram of chlorophyll, 300 nanomols of oxygen in the cell membrane and 0 nanomols of oxygen in the thylakoid membrane, and 500 nanomols of carbon dioxide in the thylakoid membrane and 0 nanomols of carbon dioxide in the cell membrane. The result of the simulation of this system is given in Figure 6.

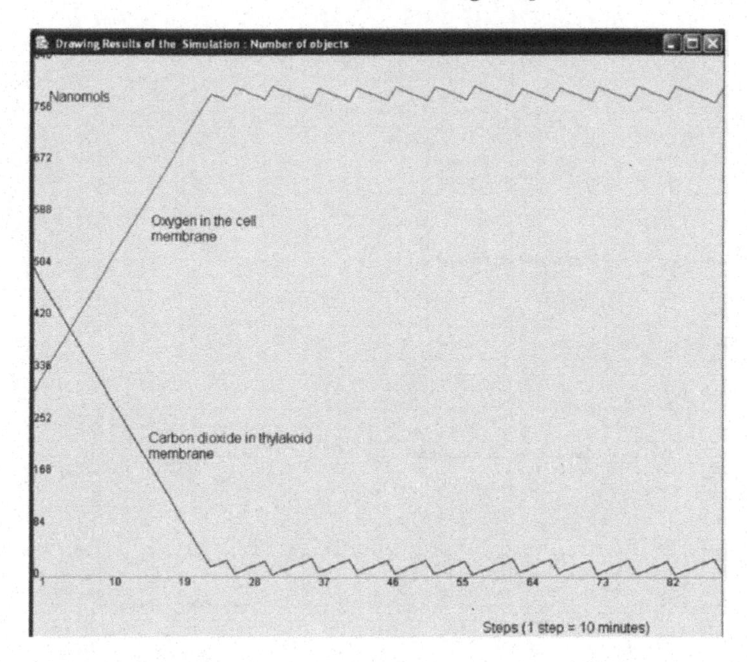

Fig. 6. Oxygen (cell membrane) and carbon dioxide (thylakoid membrane) in cyanobacteria (first simulation).

We can observe that after some time (in this case after 23 steps, or 230 minutes) the oxygen stops to accumulate at around 770 nanomols; after about 80 minutes we have the same amount of oxygen in the cell membrane as carbon dioxide in the thylakoid membrane.

In the second simulation we consider a system that does not contain any oxygen at the beginning, and we want see how it is able to produce oxygen and when the accumulation of oxygen stops. In particular, we take at the beginning one microgram of chlorophyll, 0 nanomols of oxygen in the cell membrane and 0 nanomols of oxygen in the thylakoid membrane, 450 nanomols of carbon dioxide in the thylakoid membrane and 0 nanomols of carbon dioxide in the cell membrane.

The result of the simulation of this system is given in Figure 7.

We can observe that after some time (about 21 steps, or 210 minutes) the accumulation of oxygen stops at around 440 nanomols; after about 110 minutes we have the same amount of oxygen in the cell membrane as carbon dioxide in the thylakoid membrane.

The simulation presented in Figure 7 corresponds to the situation where the cyanobacteria are in a medium without molecular oxygen, and this situation is of practical interest because it easily occurs in laboratory or natural environments in conditions of prolonged darkness. When the light is turned on (step 0) then the photosynthetic activity starts in the cyanobacteria. The

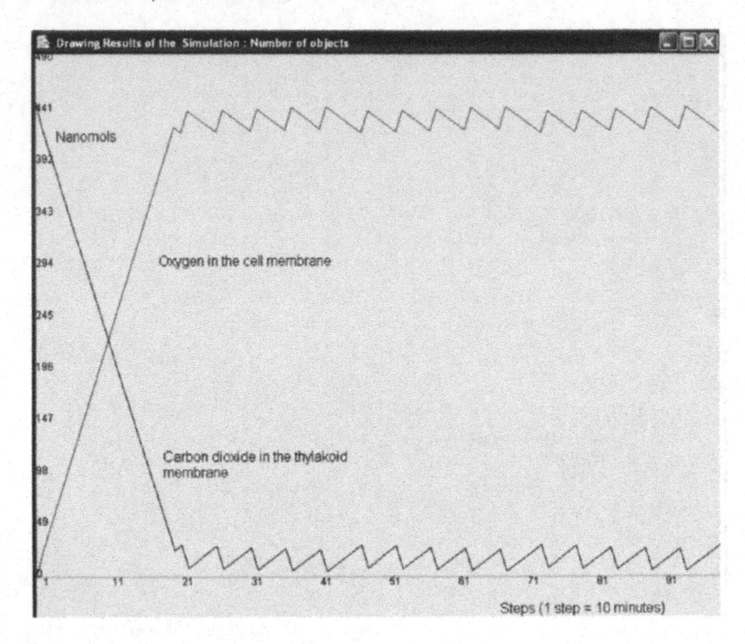

Fig. 7. Oxygen (cell membrane) and carbon dioxide (thylakoid membrane) in cyanobacteria (second simulation).

results presented in Figure 7 permit us to predict the time after which the oxygen stops to accumulate (with no oxygen at the beginning) and the quantity of molecular oxygen in the cell.

We want to stress again the fact that the systems considered in the two previous experiments are closed and oxygen is accumulated both in the liquid and the gaseous phase (for simplicity, in our simulation we did not take into account the well known inhibition of photosynthesis by the increasing of oxygen concentration).

For the future, we plan to add more details to the system, considering also other kinds of interactions existing between the two membranes, and, at the same time, the presence of regulatory mechanisms that are able to change the activity rate of the photosynthesis and respiration processes.

Moreover, we also plan to verify the stability of the system when it becomes *open* to interaction with the environment (to the introduction of oxygen or carbon dioxide from the environment) and to express the activity of the reactions in more specific and detailed terms; for example, the respiration activity can be expressed in micrograms of the specific enzyme used by cyanobacteria (cytochrome *c* oxydase).

6.5 Using an Inhibitor

In cyanobacteria it is possible to inhibit the production of oxygen during photosynthesis by adding a synthetic chemical inhibitor called *diuron* commonly used as herbicide. In particular, using diuron in low concentration it is possible to decrease the production of oxygen by 50%, while the consumption of oxygen is not modified.

That means that we can rewrite the P system considered in (6) in the following way:

$$\begin{aligned} \Pi &= (\{c, O, D\}, [_1[_2\ ]_2]_1, c^k O^j, c^l D^m, R_1, R_2, R_1', R_2'), \text{ with} \qquad (7)\\ R_1 &= \{cO^5 \to cD^5\},\\ R_2 &= \{cD^{13} \to cO^{13}\},\\ R_2' &= \{(D, in), (O, out)\}, \end{aligned}$$

where $Papp = 1$ for each rule in R_1, R_2, R_2' (we suppose the reaction is always available and the $Pwin$ is not relevant here because there are no conflicts between the rules) and each step of this system corresponds to 10 minutes. The components of the P systems can be interpreted exactly as in the case of the P system represented in (6); we can observe that in this case the production of oxygen by photosynthesis is decreased by half because of the addition of the inhibitor. It is interesting then to repeat the simulations described in Section 6.4 (whose results are presented in Figures 6 and 7) using this new P system.

In the first simulation we take, at the beginning, 1 microgram of chlorophyll, 300 nanomols of oxygen in the cell membrane and 0 nanomols of exygen in the thylakoid membrane, and 500 nanomols of carbon dioxide in the thylakoid membrane and 0 nanomols of carbon dioxide in the cell membrane. The result of the simulation of this system is given in Figure 8.

We can notice that after about 12 steps (i.e., after about 120 minutes) the quantity of oxygen present in the cell membrane is the same as the amount of the carbon dioxide present in thylakoid membrane (without using the inhibitor diuron this situation occurs after about 80 minutes; see Figure 6). On the other hand, we can see that after 600 minutes the oxygen stops to accumulate in the cell membrane with an amount of 760 nanomols (the reader can compare this data with that obtained in Figure 6).

Now we want to consider a system that does not contain any oxygen at the beginning, and we want to see how this system produces oxygen when *diuron* is added to the system. In particular, we take at the beginning 1 microgram of chlorophyll, 0 nanomols of oxygen in the cell membrane and 0 nanomols of oxygen in the thylakoid membrane, and 450 nanomols of carbon dioxide in the thylakoid membrane and 0 nanomols of carbon dioxide in the cell membrane. The result of this simulation is reported in Figure 9 (the reader can compare this with the equivalent experiment reported in Figure 7, where the inhibitor is not present).

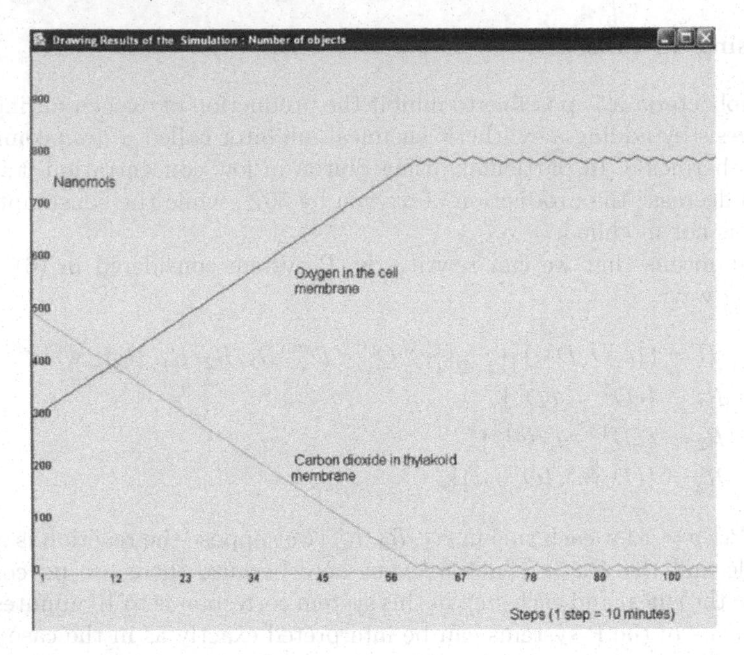

Fig. 8. Oxygen (cell membrane) and carbon dioxide (thylakoid membrane) in cyanobacteria (using inhibitor – first simulation).

We can see that after about 270 minutes the amount of oxygen in the cell membrane is the same as the amount of carbon dioxide in the thylakoid membrane; the accumulation of oxygen in the cell membrane stops after (about) 560 minutes with 480 nanomols of oxygen.

6.6 Pumps in Cyanobacteria

Now we use our simulator to study the process of *proton translocation* in the case of another cyanobacteria, *Anacystis nidulans.* The proton translocation consists in the pumping of protons, outside the cell, when oxygen is consumed. Therefore the decrease of oxygen concentration inside the cell corresponds to an increase of proton concentration outside the cell.

From [18] we can get the experimental data relating oxygen consumption and proton translocation. In particular considering 1 mg of dry weight of *Anacystis nidulans*, using one unit of enzyme; in two minutes 5 nanomols of oxygen are consumed and, at the same time, 21 nanomols of protons are sent out (translocated) from the cell and accumulated in the environment.

We can simulate this mechanism using our software.

In a formal simplified way, considering only the reaction involving oxygen and protons, we can represent the process of pumps in *Anacystis nidulans* with the following EC P system:

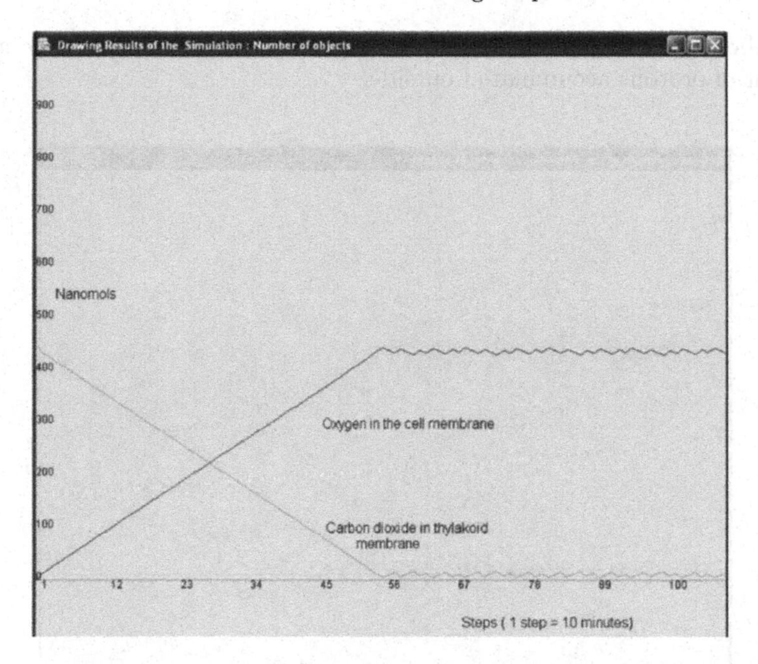

Fig. 9. Oxygen (cell membrane) and carbon dioxide (thylakoid membrane) in cyanobacteria (using inhibitor – second simulation).

$$\Pi = (\{E, O, P\}, [_1[_2\ ]_2]_1, E^k O^j, \lambda, R_1, R_2, R_1', R_2'), \text{ with} \tag{8}$$
$$R_1 = \emptyset,$$
$$R_2 = \{EO^5 \to EP^{21}\},$$
$$R_2' = \{(P, out)\},$$

where $Papp = 1$ for each rule in R_1, R_2, R_2' (the reaction is always available and the $Pwin$ is not relevant here because there are no conflicts between the rules) and each step of this system corresponds to two minutes.

The system is composed of two membranes, labeled 1 and 2, that represent the *Anacystis nidulans* cell and the environment, respectively.

In the two regions there are the symbol object E (each occurrence of E represents one unit of enzyme), the symbol object O (each occurrence of O represents one nanomol of oxygen), and the symbol object P (each occurrence of P represents one nanomol of protons).

The consumption of oxygen takes place in region 2 and using the symport rule present in R_2' the protons are sent out and accumulated in region 2.

Running the simulator over the P system described in (8), with $k = 1$ and $j = 220$, we obtain the diagram of Figure 10 describing the quantity (in nanomols) of oxygen contained in the cell (2), and the quantity (in nanomols) of protons accumulated in the environment (1). It is possible to observe that after about 15 minutes the amount of oxygen in the cell is the same as the amount of protons accumulated outside.

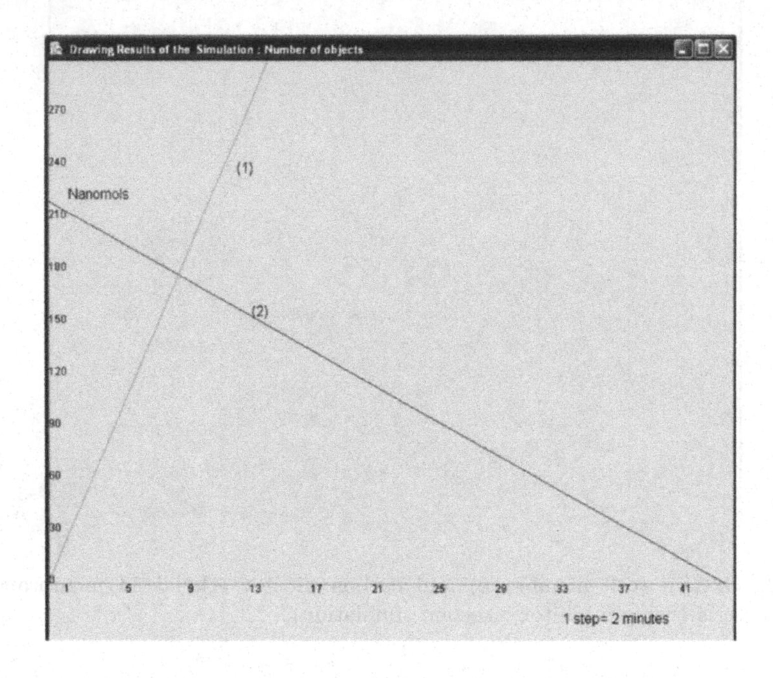

Fig. 10. Consumption of oxygen and the translocation of protons in *Anacystis nidulans*.

7 Concluding Remarks

In this chapter we have presented a probabilistic simulator of P systems recently implemented and have shown how various cellular processes can be modeled in the P systems framework and then simulated using the software. We have presented the main features of the simulator, showing its working and presenting the (strange and interesting) result of the simulation of a very simple probabilistic P system. In this respect, we have suggested the existence of a possible link between the *emergence of life* studied in biology and some kind of probabilistic P system. We also believe that a possibly fruitful research topic is the study of the fields (human life, economy, cell biology, etc.) that can modeled by (probabilistic) P systems and then simulated by a simulator; in this chapter we have discussed only biological applications, but we believe that the P systems framework can be applied to many other different fields.

In Sections 2 and 5 we have presented a comparison between the mathematical model (and the associated software) and the *real world*, discussing the way in which biological concepts have been implemented in software and introducing, in the P system area, new concepts, such as the *availability* of a chemical reaction, the *activity rate* of a catalyst, and the possibility of a catalyst to be at the same time *active* and *not active*.

Finally, in Section 6 we have shown how one can model some biological processes in the P systems framework and, in particular, we have studied two important biological processes: the (final step of) respiration in *Escherichia coli* bacteria and the interaction between respiration and photosynthesis in cyanobacteria, considering the consumption of oxygen, the accumulation of protons in the environment, and the influence of a synthetic inhibitor. We have shown how to translate these biological processes in the mathematical framework of P systems and how to obtain results relevant from a biological point of view, by using the simulator presented in Section 3.

In the near future, we plan to model other biological processes with more biological details considering the concept of *affinity* introduced in our simulator by the so-called probability of winning and the concept of *availability of a rule*, modeled in the software by the probability of a rule being available; in particular it would be very useful to add to the simulator the ability to change, at runtime, the values of some of the biological parameters considered (such as the *affinity* of an enzyme, that, for example, changes in *Escherichia coli* according to the concentration of molecular oxygen in the substrate); in this way, we can at the same time improve the simulator and the mathematical model presented, as well as make possible new applications to microbiology.

References

1. B. Alberts et al.: *Essential Cell Biology. An Introduction to the Molecular Biology of the Cell.* Garland Publ. Inc., New York, London, 1998.
2. S. Alexeeva, K.J. Hellingwerf, M.J. Teixeira de Mattos: Quantitative Assessment of Oxygen Availability: Perceived Aerobiosis and Its Effect on Flux Distribution in the Respiratory Chain of Escherichia Coli. *Journal of Bacteriology*, 184 (2002), 1402–1406.
3. S. Alexeeva, B. de Kort, G. Sawers, K.J. Hellingwerf, M.J. Teixeira de Mattos: Effects of Limited Aeration and of the ArcAB System on Intermediary Pyruvate Catabolism in Escherichia Coli. *Journal of Bacteriology*, 182 (2000), 4934–4940.
4. I. Ardelean: Molecular Biology of Bacteria and Its Relevance for P Systems. In *Membrane Computing, International Workshop, WMC-CdeA 2002, Curtea de Argeş, Romania, August 2002, Revised Papers* (Gh. Păun, G. Rozenberg, A. Salomaa, C. Zandron, eds.), LNCS 2597, Springer, Berlin, 2003, 1–18.
5. I. Ardelean, M. Cavaliere: Modeling Biological Processes by Using a Probabilistic P System Software. *Natural Computing*, 2, 2 (2003), 173–197.
6. I. Ardelean, S. Tunaru, M.L. Flonta, G. Teodosiu, E. Madalin, L.Dumitru, G. Zarnea: Increased Respiratory Activity in Light in Salt Stressed Synechocystis. In *The Phototrophic Prokaryotes* (G.A. Peschek, W. Loffelhardt, G. Schmetterer, eds.), Plenum Publisher, New York, 1999, 403–409.
7. I.R. Booth: *Bacterial Energy Transduction.* Academic Press, London, 1988.

8. M. Cavaliere: Evolution-Communication P Systems. In *Membrane Computing, International Workshop, WMC-CdeA 2002, Curtea de Argeş, Romania, August 2002, Revised Papers* (Gh. Păun, G. Rozenberg, A. Salomaa, C. Zandron, eds.), LNCS 2597, Springer, Berlin, 2003, 134–145.
9. D.O. Hall, K.K. Rao: *Photosynthesis.* Cambridge University Press, 1994.
10. H. Jung: Towards the Molecular Mechanism of Na/Solute Symport in Prokaryotes. *Biochem. Biophys. Acta*, 1505 (2001), 131–143.
11. E. Mayer: *This is Biology – The Science of the Living World.* The Belknap Press of the Harward University Press, Cambridge, Massachusetts, 1998.
12. S.L. Miller, L.E. Orgel: *The Origin of Life on Earth.* Englewood Clifs, Prentice Hall, 1973.
13. D.G. Nicholls, S.J. Ferguson: *Bioenergetics.* Academic Press, London, 2002.
14. E. Padan, M. Venturi, Y. Gercham, N. Dover: Na/H Antiporters. *Biochemical Biophisica Acta*, 1505 (2001), 144–157.
15. Gh. Păun: Computing with Membranes. *Journal of Computer and System Sciences*, 61, 1 (2000), 108–143.
16. Gh. Păun: *Membrane Computing. An Introduction.* Springer, Berlin, 2002.
17. J. Pelmont: *Catalyseurs du monde vivant.* Press Universitaires de Grenoble, Grenoble, 1995.
18. G.A. Peschek: Respiratory Electron Transport. In *The Cyanobacteria* (P. Fay, C. Van Baalen, eds.), Elsevier Science Publishers, Amsterdam, 1987, 119–161.
19. G.A. Peschek, C. Obinger, S. Fromwald, B. Bergman: Correlation Between Immuno-Gold Based and Activities of the Cytochrome-c Oxidase (aa_3-type) in Membranes of Salt Stressed Cyanobacteria. *FEMS – Microbiology Letters*, 124 (1994), 431–438.
20. G.A. Peschek, R. Zoder: Temperature Stress and Basic Bioenergetic Strategies for Stress Defence. In *Algal Adaptation to Environmental Stress* (L.C. Rai, J.P. Gaur, eds.), Springer, Berlin, 2001, 203–258.
21. A. Puustinen, M. Finel, T. Haltia, R.B. Gennis, M. Wikstrom: Properties of the Two Terminal Oxidades of Escherichia Coli. *Biochemistry*, 30 (1991), 3936–3942.
22. G. Rozenberg, A. Salomaa: *The Mathematical Theory of L Systems.* Academic Press, New York, 1980.
23. M.H. Saier: Genome Archaeology Leading to the Characterization and Classification of Transport Proteins. *Current Opinion Microbiology*, 2 (1999), 555–561.
24. S.J. Singer, G.L. Nicolson: The Fluid Mosaic Model of the Structure of Cell Membranes. *Science*, 175 (1972), 720–731.
25. P. Sowerby: Scanning Tunnelling Microscopy and Molecular Modeling of Xanthine Monolayers Self-Assembled at the Solid-Liquid Interface: Relevance to the Origin of Life. *Orig. Life Evol. Biosph.*, 29 (1999), 597–614.
26. J.E. Hopcroft, J.D. Ullman: *Introduction to Automata Theory, Languages, and Computation.* Addison-Wesley, Reading, Mass., 1979.
27. The P Systems Web Page: `http://psystems.disco.unimib.it/`

Chapter 5
Modeling Cell-Mediated Immunity by Means of P Systems

Gabriel Ciobanu

Romanian Academy, Institute of Computer Science, Iaşi, and
Research Institute "e-Austria", Timişoara, Romania
gabriel@iit.tuiasi.ro

Summary. The immune system represents the natural defense of an organism. It comprises a network of cells, molecules, and organs whose primary tasks are to defend the organism from pathogens, and to maintain its integrity. Since our knowledge of the immune system is still incomplete, formal modeling can help provide a better understanding of its underlying principles and organization. In this chapter we provide a brief introduction to the biology of the immune system, recalling several approaches used in the modeling of the immune system, and then describe a model based on P systems. Starting from a variant of P systems called client-server P systems, we use an abstract simulator as a useful intermediate step from a formal theory suitable for theoretical results to a software implementation of a molecular network. Finally, our approach leads to novel software able to provide new insights into the interactions influencing T cell behavior with the use of statistical correlations of the software experiments' results.

1 Introduction

The immune system is a complex network of cells and molecules whose primary tasks are to defend an organism from potentially dangerous foreign agents, and to maintain the integrity of the organism. Foreign agents include toxins, bacteria, fungi, parasites, viruses, various environmental and self-produced antigens. In this section we present some basic immune system concepts; further details can be found in immunology textbooks, for instance, [2] and [24]. The most important function of the immune system is the self/nonself recognition that enables an organism to distinguish between the harmless self and the potentially dangerous nonself. The mechanism of self/nonself recognition is mediated through major histocompatibility complex (MHC) molecules binding short peptides intracellularly, and transporting them to the cell surface for recognition by the T cells of the immune system. These peptides act as markers; cells presenting self-peptides are tolerated, while those presenting foreign peptides are subject to various immune responses.

We have two fluid systems in the body: blood and lymph. The blood and the lymphatic systems are responsible for transporting the agents of the immune system across the organism. Lymphocytes are the most important cells for adaptive immunity. They circulate in both the blood and lymphatic systems, and make their home in lymphoid organs. The lymph nodes are the usual places where antigens are presented to the cells of the immune system.

We have two main functionalities of the immune system: innate immunity, and adaptive immunity. We are born with a functional innate immunity system which is nonspecific; all antigens are attacked pretty much equally. This innate immunity represents the first defense of the organism, a defense achieved by many actions and components, such as surface barriers and mucosal immunity, chemical barriers, and normal flora (microbes living inside and on the surface of the body). The cells involved in the innate immune system bind antigens using hundreds of pattern recognition receptors. These receptors are encoded in the germ lines of each person; this immunity is passed from generation to generation. In this chapter we concentrate on adaptive immunity, which is more interesting for formal modeling.

Adaptive Immunity

Adaptive immunity (or acquired immunity) is a function of the immune system given by the fact the immune system has to learn the specific antigens before it can actually remove them from the organism. Adaptive immunity is developed and modified throughout the life of the host organism. The adaptive immunity appears to be a distributed system with a sort of coordination control, able to perform its complex task in an effective and efficient way. The most important components of adaptive immunity are the major types of lymphocytes: T cells, and B cells. Peripheral blood contains 20%-50% of circulating lymphocytes; the rest move in the lymph system. Roughly 80% of them are T cells, 15% are B cells, and the remaining are null or undifferentiated cells. Their total mass is about the same as that of the brain. B cells are produced from the *stem cells* in bone marrow; B cells produce antibodies and oversee humoral immunity. T cells are nonantibody-producing lymphocytes which are produced in the bone marrow, but sensitized in the thymus. Parts of the immune system are changeable and can adapt to better attack the invading antigen. There are two fundamental adaptive mechanisms: humoral immunity, and cell-mediated immunity. Humoral immunity is mediated by serum antibodies, which are proteins secreted by the B cell compartment of the immune response. Cell-mediated immunity consists of the T cells. Each T cell has many identical antigen receptors which interact with antigens.

Cell-Mediated Immunity

Phagocytes are cells able to attract, adhere to, engulf, and ingest foreign bodies. Promonocytes are made in the bone marrow, then released into blood, and called circulating monocytes which mature into macrophages. Once a macrophage phagocytizes a cell, it places portions of its proteins, called T cell epitopes, on the macrophage surface. These surface markers serve as an alarm to other immune cells which then infer the form of the invader. Macrophages engulf antigens, process them internally, and display parts of them on the cell surface together with MHC molecules. This mechanism sensitizes T cells to recognize the antigens. All cells are coated with various molecules and receptors. CD stands for *cluster of differentiation*, and there are more than one hundred and sixty clusters, each of which is a different molecule that coats the surface. Every T cell has about 10^5 molecules on its surface; T cells have CD2, CD3, CD4, CD28, CD45R, and other non-CD molecules on their surfaces. This large number of molecules residing on the surfaces of lymphocytes produce huge receptor variability. They produce random configurations on the lymphocytes surfaces; there exist around 10^{18} structurally different receptors. An antigen is likely to find a near-perfect fit with a very small number of lymphocytes.

T cells are primed in the thymus, where they undergo two selection processes. The first positive selection process weeds out only those T cells with the correct set of receptors that can recognize self-peptides presented by the MHC molecules. Then a negative selection process begins whereby T cells that can recognize MHC molecules complexed with foreign peptides are allowed to pass out of the thymus. Cytotoxic or killer T cells (CD8+) do their work by releasing lymphotoxins, which cause cell lysis. Helper T cells (CD4+) serve as regulators, and trigger various immune responses. They secrete chemicals called lymphokines that stimulate cytotoxic T and B cells to grow and divide, attract neutrophils, and enhance the ability of macrophages to engulf and destroy microbes. Suppressor T cells inhibit production of cytotoxic T cells, providing a mechanism for preventing the self-damage of healthy cells by the immune responses. Memory T cells are programmed to recognize and respond to a pathogen previously encountered by the organism.

2 Modeling in Immunology

In this section we briefly present some continuous and discrete models of the immune system.

The main problem of the immune system is to distinguish between self and nonself. We can say that the success of the immune system is dependent on its ability to distinguish between harmful nonself and everything else. This problem is difficult because there are $\sim 10^{16}$ patterns in nonself, and they have to be distinguished from $\sim 10^6$ self patterns. Moreover, the environment is highly distributed, resources are deficient in quantity compared with the demand, and the host organism must continue to work all the time. The immune system solves this problem by using a multilayered architecture of barriers, namely a physical barrier (the skin), physiological barriers (e.g., the pH values), and cells and molecules of the innate and adaptive immune response. The resulting system is flexible, scalable, robust, and resilient to subversion.

The immune system models are mostly based on two biological theories of the immune system, namely the clonal selection theory, and the idiotypic network theory. The *clonal selection theory* is described by Nobel Prize winner F. Burnet in [3], following the track first highlighted by P. Ehrlich at the beginning of the twentieth century. The theory of the clonal selection states that the immune response is the result of a selection of the right antibody by the antigen, much like the best adapted individual is selected by the environment in the theory of natural selection. The selected subsets of B cells and T cells grow and differentiate; then they turn off when the antigen concentration falls below some threshold. The *idiotypic network theory* is formulated by Nobel Prize winner N.K. Jerne in [14, 15]. According to the idiotypic network theory, the immune system is a regulated network of molecules and cells able to recognize one another even in the absence of antigens. The idiotypic network hypothesis is based on the concept that lymphocytes are not isolated, but communicate with each other. As a consequence, the identification of antigens is not done by a single recognizing set, but by several sets connected by antigen-antibody reactions as a network. Jerne suggested that during an immune response, antigens directly elicit the production of a first set of antibodies, Ab_1; then these antibodies act as antigens, and elicit the production of a second set of "anti-idiotypic" antibodies Ab_2 which recognize idiotypes on Ab_1 antibodies, and so forth. The immunologists consider these two independent theories as being consistent and complementary with each other [27]. The clonal selection theory is considered to be important for a global understanding of the immune system. The idiotypic network theory is useful for understanding the existence of anti-idiotypic reactions, and the immune responses. Most continuous models have been formulated using the framework of both immunological theories [21, 22], while the discrete models are mainly based on Jerne's theory.

The systemic models of immune responses have mainly been devoted to collective actions of various immune system components. These models do not study single cells or single molecules, but rather focus on cell interactions and collective behavior in activation, control, and supply of the immune responses.

Continuous Models

We briefly point out the main ideas, together with the advantages and disadvantages of some continuous models of immune systems. A good survey of the continuous models can be found in [22]. Continuous models describe the time evolution of concentrations of cellular and molecular entities which play a significant part in the immune system. Each of these models works with continuous functions defining the concentration (number of elements/volume) of cellular and molecular entities. These models use systems of nonlinear ordinary differential equations, mainly representing conservation laws, in which unknowns represent the concentrations. Some models assume that concentrations do not depend on space variables, and interactions between entities occur with uniformly random collisions. When concentrations do depend on the space variables, the problem becomes mathematically more difficult, and partial differential equations are needed.

The advantages of this approach are given by fact that the mathematical methods have a well known and theoretically established background, and the behavior of the solution can be described by qualitative and asymptotic analysis. Moreover, if a numerical solution is needed, then well known numerical methods are available. However, there are some biological disadvantages. The approximations necessary to keep the equations tractable may not be biologically evident, and so the model may move away from real biology. Worse, most of these models fail when the concentrations of some entities decrease drastically. The models are not compositional, and inserting new biological details may change the mathematical structure of the model.

The general framework of the continuous models is given by nonlinear equations describing generic iteration systems. They have a general form $\partial_t x = G(x) - L(x)$, where $x = \{x_1, \ldots, x_n\}$ is a vector describing concentrations, and the vectors $G(x)$ and $L(x)$ represent the gain and loss terms, respectively. The solution of these equations is represented by a curve in the n-dimensional state space describing the time evolution of the system starting from the initial condition $x^*(t = 0)$. The qualitative behavior of the solutions is usually investigated, including stationary points (fixed points), local and global stability, attractors, limiting cycles describing periodic behavior, and strange attractors describing chaotic behavior. In these models it is crucial to properly describe the affinity between cells and molecules. The simplest affinity function is a bilinear form; more detailed models need more complex nonlinear terms. In order to make them more realistic, there are some possible improvements of the continuous models. For instance, considering a time delay for the interactions between entities, the behavior of the delayed systems may be qualitatively different (e.g., the stability of fixed points can change). Usually new cells come into the system either from bone marrow or following some hypermutations. This could be modeled using a stochastic source term in the equations; the overall behavior can be very different.

Discrete Models

Discrete models consider individual entities as primitives, and the whole system dynamics arises from their collective actions. These models use various mathematical techniques such as the generalized Boltzmann equation, cellular automata, and lattice gas. Most of these approaches are widely used to study complex systems, and are based on computer simulations. Good surveys of the discrete models can be found in [22, 27, 9].

The approach based on generalized Boltzmann equation is described in [1]. Models based on cellular automata and lattice gas have been developed over the last twenty years, producing interesting results. Various mathematical models, particularly models based on cellular automata, are presented in [17]. Automata-based models are used to investigate the logic of interactions among a number of different cell types and their outcomes in terms of immune responses [16, 26]. The rules modeling the dynamic evolution of these automata-based models are expressed by logical operations. Application of the rules is iterated over discrete time. Some of the discrete models bring into the field the experience of computer scientists. The guiding line of these approaches is a deeper comprehension of immune system by describing and using immune system information processing in applications (see [9]). Some applications of the artificial immune system are described in [8].

A real advantage of these models is that they can be built using biological language, with biologically relevant approximations. New biological details are easy to insert without changing the mathematical structure of the model. On the other hand, qualitative and asymptotic analyses are no longer possible, or quite difficult.

Modeling Cell-Mediated Immunity

T cells play a central role in the cell-mediated immunity. They orchestrate the immune responses to foreign aggressors. When a T cell recognizes a foreign antigen, it initiates several signaling pathways, and the cell activates. The recognition of a foreign antigen is an extremely *sensitive*, *specific*, and *reliable* process; the models for T cell signaling network can help us understand how these features arise and work. So far, the study of T cell activation has benefited from the use of some mathematical models [4]. A key event for T cell activation is an appropriate interaction between the T cells armed with T cell antigen receptors (TCRs) and the professional antigen presenting cells (APCs). TCR recognizes only the foreign antigen in the form of short peptides presented in the groove of a molecule on the surface of the APC known as the major histocompatibility complex MHC. The physical interaction of TCRs with the peptide complexes is unique among signaling systems by its taking place over a continuous binding value process. The recognition of antigen initiates signal transduction. This can be broken down into series of discrete steps related to various molecular events (interactions, state transitions) within the signaling pathways. These are shown in Figure 1, reprinted from [18].

T cell responses show a hierarchical organization depending on the extent of TCR occupancy, the duration of antigen binding, the timing of encounters, and the engagement of costimulatory receptors. TCR is a very complex structure composed of a minimum of eight strongly associated chains. The actual arrangement and stoichiometry of CD3 and TCR chains within TCR complex is not entirely known. We refer in this chapter to a part of the signaling network, namely the activation of Zap70 and the phosphorylation of the adapters LAT, which are essential for connecting to the major intracellular signaling pathways Ca^{+2}/calcineurin and Ras/Rac/MAPK kinases. Although

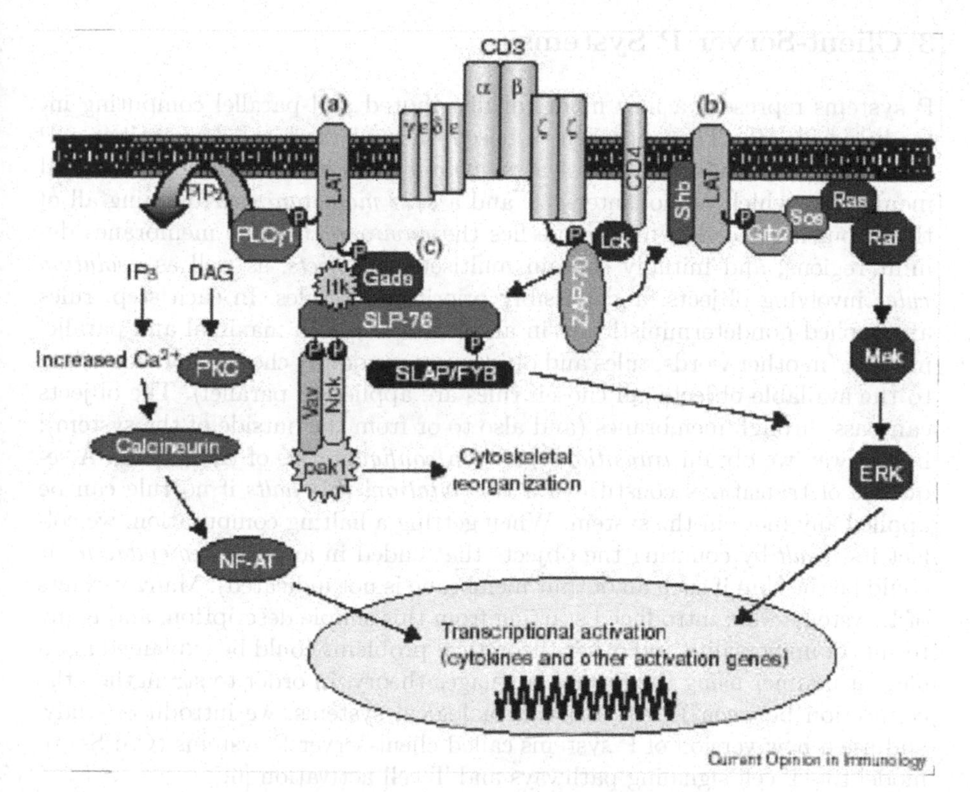

Fig. 1. TCR signaling pathways.

many other receptor interactions may contribute positively or negatively to the quality and the quantity of the T cell immune responses, TCR signaling upon antigen recognition determines a certain response (see [5]).

In this chapter we present a model of the T cell signaling network by a distributed version of P systems. Before presenting this version, let us emphasize *why the P systems are suitable for modeling the immune systems.* The immune system has more subsystems, each with its own rules. This structure can be represented faithfully by a P system where each subsystem is modeled by a membrane. The P system application of rules in a maximally parallel manner expresses the natural competition for scarce resources in the immune system. Communication and coordination is essential, and thus we consider P systems with symport/antiport rules of communication among membranes. Since the immune system environment is highly distributed, we introduce a P system using the well known client-server paradigm used in computer networks and distributed systems.

3 Client-Server P Systems

P systems represent a new model of distributed and parallel computing introduced in [19]. The approach is based on a hierarchical description: a P system is basically composed of a *membrane structure*, consisting of several membranes which do not intersect, and a *skin membrane*, surrounding all of them; outside the skin membrane lies the *environment*. The membranes delimit regions, and initially contain multisets of *objects*, as well as *evolution rules* involving objects, and possibly priorities for rules. In each step, rules are applied nondeterministically in all membranes in a maximal and parallel manner (in other words, rules and objects are randomly chosen, and according to the available objects, all chosen rules are applied in parallel). The objects can pass through membranes (and also to or from the outside of the system); in this way we obtain *transitions* between *configurations* of the system. A sequence of transitions constitutes a *computation*; this *halts* if no rule can be applied any more in the system. When getting a halting computation, we collect its *result* by counting the objects that ended in an *output membrane* (it could be the skin if such an output membrane is not indicated). Many variants of P systems were introduced starting from this simple description, and many results of universality and other theoretical problems could be explained in an elegant manner using the formal languages theory. In order to strengthen the connection between P systems and biological systems, we introduce, study, and use a new version of P systems called client-server P systems (CSPSs) to model the T cell signaling pathways and T cell activation [6].

Formally, we start from a particular variant of P systems, namely *P systems with symport/antiport rules*. The specificity of this type of P systems lies in the form of their rules, which can be one of:

- $(ab, in), (ab, out)$: objects a and b can pass through a membrane only together, in the same direction (*symport rules*), and
- $(a, out; b, in)$: objects a and b can pass through a membrane only together, but in different directions (*antiport rules*).

Theoretical results regarding this type of P system can be found in [20]. Generally, these results take into consideration the number of membranes and the *weight* of the port (i.e., the number of objects involved in a symport or antiport rule).

We formalize a client-server model according to the following description. The clients are characterized by their states, while the server stores the current states of clients and interaction rules defined over the states. When two clients can interact, the server notifies them, supplying at that time the corresponding rule. The clients interact and send their new states to the server, thus making the model consistent.

A *client-server P system* (CSPS) is a P system composed of elementary membranes (except the skin), with state objects modeling the states of the clients, and rule objects modeling the communication between clients. The communication is of symport type. In formal notation, the CSPS contains the skin membrane (numbered 1), together with m membranes representing clients (numbered 2 to $m+1$), and a membrane for the server (numbered $(m+2)$), all arranged inside the skin membrane. In the original approach of [6], a rule object $\eta_{\alpha_1\alpha_2\alpha_3\alpha_4\alpha_5}$ has the following meaning: two clients defined by states α_1 and α_3 can interact and pass to states α_2 and α_4, respectively; at the same time, it is possible to get supplementary information α_5.

Formally, a CSPS is a construct

$$\Pi = (V, \mu, w_1, \ldots, w_{m+2}, w_e, M_e, R_1, \ldots, R_{m+2}, m+2),$$

where:

1. $V = A \cup B$, with A, B disjoint sets such that:
   - $A = \bigcup_{i=2}^{m+1} S_i$, where S_i represents the states of client i;
   - $B = \{\eta_{\alpha_1\alpha_2\alpha_3\alpha_4\alpha_5} \mid \alpha_5 \in M_e \cup \{\lambda\}, \alpha_i \in A \cup M_e,\ 1 \le i \le 4$, where $\alpha_1 + \alpha_2 \Rightarrow \alpha_3 + \alpha_4 + \alpha_5$ is an interaction rule$\}$,
2. $\mu = [_1\, [_2\,]_2 \ldots [_{m+2}\,]_{m+2}\,]_1$,
3. $w_1 = \emptyset$, $w_{m+2} = B \cup S_{initial}$, where the initial state of the server is $S_{initial} = \{s_2, s_3, \ldots, s_{m+1}\}$, $s_i \in S_i$, $2 \le i \le m+1$ (the s_i represent the initial states of the clients), $w_i = S_i \setminus \{s_i\}$, $s_i \in S_{initial}$, $2 \le i \le m+1$,
4. $M_e = A$,
5. $R_1 = \{(\alpha_j\, \alpha_k\, \eta_{\alpha_1\alpha_2\alpha_3\alpha_4\alpha_5}, out) \mid j \in \{3,4\}, k \in \{1,2\}, j-k \ne 2, \alpha_k, \alpha_{k+2}, \alpha_5 \in M_e, \alpha_j, \alpha_{j-2} \in A\} \cup \{(\alpha_3\, \alpha_4\, \eta_{\alpha_1\alpha_2\alpha_3\alpha_4\alpha_5}, out) \mid \alpha_i \in A,\ 1 \le i \le 4,\ \alpha_5 \in M_e\} \cup \{(\alpha_3\alpha_4\alpha_5\eta_{\alpha_1\alpha_2\alpha_3\alpha_4\alpha_5}, in) \mid \alpha_i \in A \cup M_e,\ 1 \le i \le 5\}$,
   $R_{m+2} = \{(\alpha_1\, \alpha_2\, \eta_{\alpha_1\alpha_2\alpha_3\alpha_4\alpha_5}, out), (\alpha_3\, \alpha_4\, \alpha_5\, \eta_{\alpha_1\alpha_2\alpha_3\alpha_4\alpha_5}, in) \mid \alpha_i \in A \cup M_e,\ 1 \le i \le 4,\ \alpha_5 \in M_e \cup \{\lambda\}\}$,
   $R_i = \{(\alpha_j\, \eta_{\alpha_1\alpha_2\alpha_3\alpha_4\alpha_5}, in), (\alpha_{j+2}\, \eta_{\alpha_1\alpha_2\alpha_3\alpha_4\alpha_5}, out) \mid j \in \{1,2\}, \alpha_j, \alpha_{j+2} \in S_i\}$, $2 \le i \le m+1$.

Inside the server membrane (the one with label $m+2$) there are state objects (representing the current states of the clients) and rule objects. When two state objects can be combined according to a rule given by a rule object, the server membrane gives a "signal" to the respective client membranes.

The meaning of the rule $(\alpha_1\, \alpha_2\, \eta_{\alpha_1\alpha_2\alpha_3\alpha_4\alpha_5}, out) \in R_{m+2}$ is the following: the clients represented by membranes p and q, where $\alpha_1 \in S_p$ and $\alpha_2 \in S_q$, can interact according to the rule described by the rule object $\eta_{\alpha_1\alpha_2\alpha_3\alpha_4\alpha_5}$; as a result, these three objects (the current states and the rule object) exit the server region. The client membranes p and q involved absorb their own state objects and the rule object by means of their corresponding rules $(\alpha_1\, \eta_{\alpha_1\alpha_2\alpha_3\alpha_4\alpha_5}, in) \in R_p$ or $(\alpha_2\, \eta_{\alpha_1\alpha_2\alpha_3\alpha_4\alpha_5}, in) \in R_p$ (and similarly for membrane q). Then they release for further use their new states and the rule object into the skin membrane, by $(\alpha_3\, \eta_{\alpha_1\alpha_2\alpha_3\alpha_4\alpha_5}, out) \in R_p$ or $(\alpha_4\, \eta_{\alpha_1\alpha_2\alpha_3\alpha_4\alpha_5}, out) \in R_p$ (and similarly for membrane q). If $\alpha_5 \ne \lambda$, the supplementary information is brought in from the environment with rules from R_1. We emphasize the fact that the notifications of clients, and the interactions between them take place in a parallel manner.

Client-server P systems were theoretically investigated in [6], where it is proved that CSPSs of degree at most 4 and using symport rules of weight at most 4 are computationally universal. However, our goal here is to emphasize the use of P systems in modeling molecular biology, particularly in adaptive immunity. In order to make the theory able to describe the T cell signaling network, we adopt the following refined approach from abstract models to software experiments. We consider that the process of modeling and simulation implies three basic objects: the real system, the abstract model, and the simulator, together with two main relations: the modeling relation, which ties the real system to the model, and the simulation relation, which connects the model and the abstract simulator. Finally, the abstract simulator helps us design a faithful computer implementation able to ensure useful software experiments. These steps are described in the following picture:

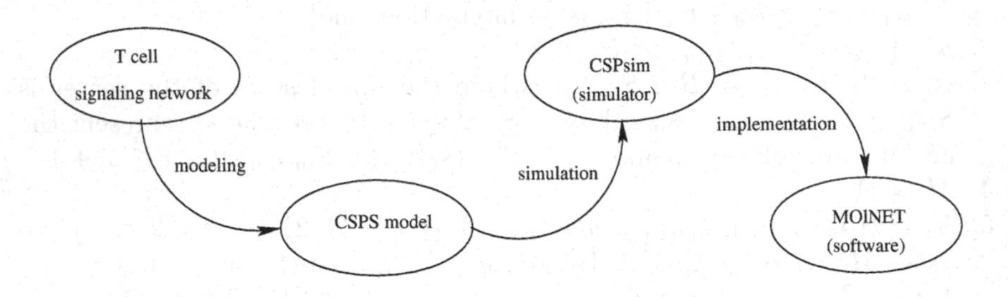

4 Client-Server P Simulators

Aiming to add both qualitative and quantitative features of the T cell signaling network, starting from CSPS, we define a client-server P simulator (CSPsim for short) as a set of communicating automata, together with appropriate internal transitions for each component, and communication steps between components. This approach leads to a novel software tool composed of a data server and its clients, together with a graphical tool for a visual representation of a molecular network. It is designed to work over computer networks, using the power of several processors and systems. It is platform-independent, and able to work over heterogeneous networks.

For each membrane of a CSPS, we define an automaton consisting of ports carrying input and output values, states, internal and external transitions, and timings. The structure of a CSPsim is defined as a set of automata together with associated interaction partners. Each membrane of a CSPS corresponds to an automaton in CSPsim. A state of the CSPsim has a set of rules (reactions) involving various membranes. We emphasize the existence of a coordinator (server) that controls a certain number of client membranes, and recomputes the new structure and rules of the system based on the least putative times of reactions. When a reaction occurs, the state of the simulator changes discretely, step by step. This means that the nonreactive collisions are ignored. It is possible to use a single automaton for a certain type of membrane, and therefore simulating a real system becomes more tractable. For each membrane i we define an automaton $M = (X, S, Y, \delta^{int}, \delta^{ext}, \lambda, \tau)$, where

- $X = \{(p, v) \mid p \in IPorts,\ v \in X_p\}$, where $IPorts$ is a set of input ports, and X_p is a set of possible input values;
- $Y = \{(p, v) \mid p \in OPorts,\ v \in Y_p\}$, where $OPorts$ is a set of output ports, and Y_p is a set of possible output values;
- S is the *set of states*;
- $\delta^{int} : S \to S$ is the *internal transition* function;
- $\delta^{ext} : Q \times X \to S$ is the *external transition* function, where $Q = \{(s, e) \mid s \in S, 0 \leq e \leq \tau(s)\}$, and e is the elapsed time since the last transition;
- $\lambda : S \to Y$ is the *output function*;
- $\tau : S \to \mathbb{R}^+$ is the *time advance function.*

Let us describe how a CSPsim is essentially simulating a molecular interaction. In biology, the interaction rules are given by $\alpha_1 A + \alpha_2 B \to \sum \beta_i C_i$, where α_1, α_2, and β_i represent multiplicities. The executive has specific input and output ports for each membrane client. In a certain configuration, the executive selects a rule that could be applied, i.e., check if the members of the left side of the rule are available in the current configuration. Then the executive sends to the clients involved in this rule a message describing the rule, by using its transition function δ. Each client receiving such a message uses its own transition function, and then send an acknowledgement to the executive. After receiving the acknowledgements from the clients, the executive performs a transition, and changes to a new configuration. And so on.

The executive receives from clients the necessary information regarding the reactions that took place, or unsuccessful attempts to react. When quantitative changes appear in a CSPsim, the executive recomputes the putative times for each reaction according to Gibson's algorithm, and modifies the clients membranes such that for the next reaction the client selected to participate is the one with the least putative time. The results yielded by the abstract simulation of a molecular network are strongly dependent on the algorithm used for choosing the performed reactions. Several algorithms were proposed to simulate the behavior of the biological systems. Our algorithm is based on widely accepted stochastic mesoscopic algorithms in biology. The mesoscopic view counts particles, but does not keep track of their position or momentum.

Several interaction algorithms regarding the mesoscopic view of the physical biology were proposed. Gillespie developed two such algorithms, the *Direct Method* [11] and *First Reaction Method* [12]. At each iteration these algorithms generate some putative times for each reaction according to an exponential distribution. The reaction with the least putative time will be executed and the system will be updated accordingly. Both algorithms have the time complexity of $\mathcal{O}(r)$ for each iteration, where r is the number of reactions. Gibson improved the *Direct Method* algorithm giving the *Next Reaction Method* [10]. This new algorithm uses Markov chains for choosing what reaction will be performed next. One major improvement consists of generating only one random number for each reaction. It also computes the putative times only for the reactions affected by the last executed reaction. The complexity of this algorithm for each iteration is $\mathcal{O}(\log(r))$, where r is the number of reactions. This algorithm is obviously more efficient than Gillespie's algorithms.

A client-server P simulator with two clients has the same computational power as a Turing machine. It is also possible to prove some important properties for our abstract CSP simulator, such as hierarchical, modular composition, universality, and uniqueness. These properties support the development of simulation environments. Thus an abstract CSPsim and its implementation provide support for building models in a hierarchical and modular manner. Moreover, within the framework presented for modeling and simulation we can prove, up to a specific modeling relation, that the simulations reflect faithfully the behavior of the real system.

5 Implementing T Cell Signaling Networks

The last step from modeling to software experiments is represented by the implementation of the abstract simulators. We consider generally the molecular network, and in particular the signaling network that grounds T cell activation.

Molecular networks and computer networks look and behave similarly. In this context, a suitable approach to implement the abstract CSPsim for molecular networks is to develop a software system running over computer networks. We developed a novel software system called MOlNET as an implementation of our model for T cell molecular networks. MOlNET has two main entities: the data server and the clients. The data server is the implementation of the executive, and the clients implement the basic CSPsim structure simulating the CSPS membranes. The formal frameworks of the CSPS model and CSPsim ensure model validity, as well as simulator correctness. A simulator performs the implicit operations of the model. An implementation is able to perform the simulations described by CSPsim. We can assume that every membrane is implemented by software clients, the executive is implemented by the MOlNET server, and the communication between membranes is implemented by MOlNET communication protocols.

The entire architecture is shown in Figure 2. The software has a modular architecture which allows us to easily integrate other tools, or even to use various interaction algorithms:

- *Graphical server*: it ensures a user-friendly graphical interface. The user is provided with multiple facilities such as for saving and loading certain

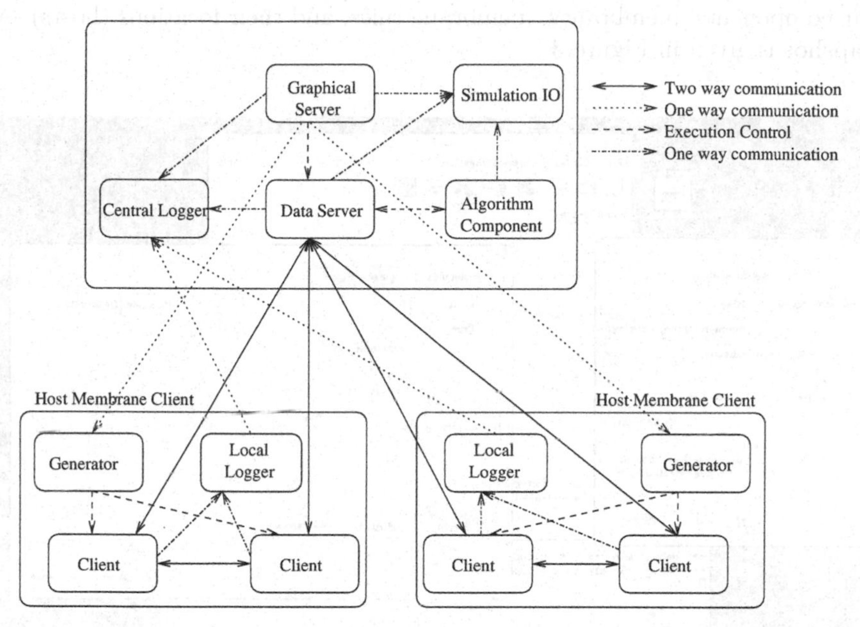

Fig. 2. MOlNET architecture.

simulations, for viewing both input and output data, for modifying the system data while software experiments run, or for defining tracers providing charts for the quantitative evolution of different clients of the system. The graphic server is also responsible for distributing the processes over the available hosts by communicating with the client generator through a specific protocol.

- *Data server*: it is the correspondent of the executive; its main role is to provide the clients with data regarding their interaction partners.
- *Client generator*: it is a process responsible only for starting other processes (i.e., the client processes) at a specific host.
- *Clients*: they correspond to client membranes of CSPS. According to the interaction algorithm, each client initiates reactions with its interaction partners, and participates in the reaction results. In this way we avoid getting any other client involved in a specific interaction. The clients keep track of the quantities involved in reactions. For each component which appears in a tracer, its client informs the graphical server about the quantitative modifications.

The output data consists of *tracers*. Each tracer is given for groups of clients. It provides both an output file and a visual representation (chart) of the evolution in time of the given clients. The user is provided with an intuitive graphical interface for introducing the input data for a simulation. The input data window includes several sections corresponding to various molecular components, membranes, membrane rules, and their locations (hosts). A snapshot is given in Figure 3.

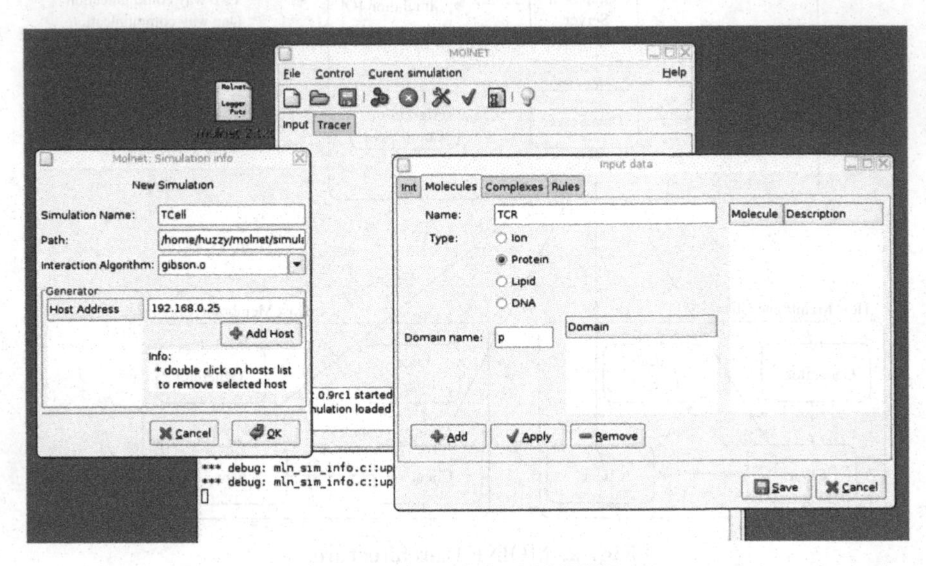

Fig. 3. A screenshot of the MOlNET software.

For implementation we use the C programming language, BSD socket interface for network communication, and GTK 2.0 for the graphical user interface. This implementation uses computer networks for distributing the computation, having the advantage of executing simulations over a large number of membranes, and so providing valuable and relevant data about the behavior of particular molecular networks.

Let us now return to modeling the behavior of the T cell signaling network. T cell activation is initiated by the interaction between the TCR arms of T cells and the professional antigen presenting cells. TCR recognizes the foreign antigen in the form of short peptides presented in the groove of a molecule on the surface of the antigen presenting cells. The recognition of antigen initiates signal transduction. This can be broken down into series of discrete steps that are related to various molecular events within the signaling pathways. In order to illustrate the evolution of the model, we consider only a part of the signaling network, namely the reaction between Zap70 and *LAT*, followed by the binding of $GADS \oplus SLP$-76 to phosphorylated *LAT*. Zap70 and *LAT* correspond to CSPS membranes, and CSPsim clients. First there is a communication between Zap70 and *LAT*, followed by the rules of these two components; they wait in their states for the time indicated by their reaction, then both send a message to that executive. The executive modifies the system so as to reflect the quantitative aspect of the reaction. The executive also modifies the transition rules from the initial state for these two components, allowing the binding of the complex $GADS \oplus SLP$-76 to phosphorylated LAT.

We simulate and analyze the interactions that drive the T cell behavior by using our approach based on client-server P systems and simulators. The software experiments provide data that can be interpreted with statistical methods. During these experiments we systematically perturb and monitor the signaling network outcomes, by adding or deleting new components, modifying the quantities and rates, establishing new interactions between components, or removing existing ones due to certain mutations. In this way, we can represent various factors (the number of triggered TCRs, the presence or absence of costimulation) which play a role in determining the outcome of the T cell. Software experiments can contribute to explaining how these factors determine differences in the formation and composition of the TCR signaling complexes, and how they drive various biological consequences of T cell signaling networks.

6 Software Experiments and Their Biological Relevance

Biological behavior is strongly influenced by the ability of molecules to communicate specifically with each other within large molecular networks. Crucial for the T cell behavior is the signaling network that could engage various cell responses due to potentially different signal types, quantities, and durations. We describe the network behavior both qualitatively and quantitatively. Many studies on T cell biology have revealed different types of functional responses (activation, proliferation, anergy, cell death) to different TCR stimulations. It is known that TCR engagement under some circumstances leads to proliferation and effector function, while under other conditions TCR stimulation leads to anergy. The factors that shape the response to antigen are the concentration of antigen, the duration of antigen binding, the timing of encounters, and the engagement of other receptors (such as CD28 or CTLA4). These T cell responses can be broken down into a series of discrete steps that are related to molecular events within a larger molecular network. Recent data suggest that unbalanced activation of NFAT relative to its cooperating AP-1 imposes the genetic program of T cell anergy that opposes the program of activation-proliferation mediated by NFAT-AP1 complex. Based on these results, we simulate the molecular network that drives T cell activation or T cell anergy.

The results obtained from experiments are statistically processed and interpreted. One of the main goals of statistics is inferring conclusions based solely on a finite number of observations about events likely to happen an indefinite (infinite) number of times. Nonetheless, the strength of conclusions yielded depends on the sample size on which the analysis is based. In fact, in many cases a contradiction appears between the minimum size required by statistics in order to make methods applicable, and the size that biological wet experiments can provide. This is why faithful computer simulators for biological processes are needed: the use of such tools overcomes the problems of budgeting, since the cost per software experiment is low in comparison with the biological lab experiments. Therefore, data sets of desired size can be obtained, thus allowing for correct statistical inferences and hypothesis testing.

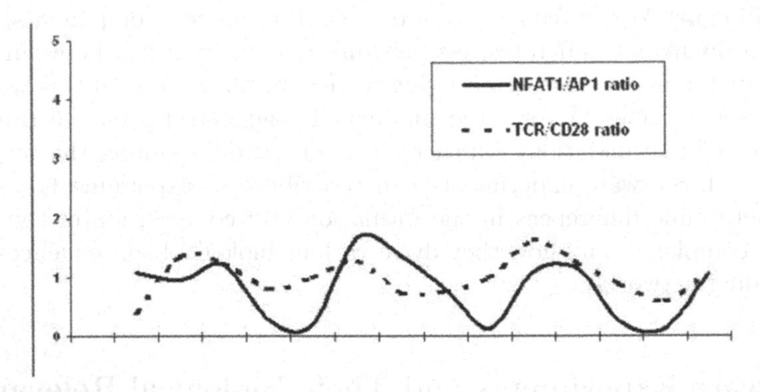

Fig. 4. MOlNET experiments regarding the T cell activation: ratios TCR/CD28 and NFAT/AP1 have similar trends.

We ran some experiments with our MOlNET software, using different input amounts for TCR and CD28 so as to vary their ratio across an interval. The output data of interest was the ratio between NFAT and AP1, the proteins with the main role in deciding the T cell behavior. In Figure 4 a representation of data yielded by experiments can be seen. The number of molecules were varied over the interval 10^3 to 10^5 for each component, and 10^{-3} to 10^{-5} for the Michaelis-Menten constant of enzymatic reactions. However, these may not always hold true, and further restrictions may be imposed. On the other hand, they could reflect the diversity of molecular environments throughout T cell population. To be rigorous, each reaction may be varied in terms of number of molecules that participate, and in terms of kinetic rates. Varying these parameters for all reactions in the network produces a huge number of software experiments.

We have applied the bootstrap method in the case of the ratio between quantities of NFAT and AP1. After generating 500,000 bootstrap samples from the original ratio sample, we obtained an estimate of the mean equal to 0.8023, as compared to the sample average of 0.8071. The full range of the bootstrap estimates was (0.2140, 2.4179), and from it we were able to indicate a 95% confidence interval for the mean (0.7071, 1.9003). For comparison, the 95% confidence interval obtained by using the classical t-test was (0.7220, 0.8921), which shows that the population of ratios has a distribution very close to normal. Moreover, by applying the bootstrap method we obtain a similar conclusion regarding the ratio TCR/CD28.

Other software experiments we made were related to the Lck recruitment model, that involves successive activations and inactivations of Zap70, Lck, and phosphatase. We ran several tests regarding the quantitative aspects of this particular interaction network. A chart representing the distribution of the quantitative modification corresponding to each of substances involved is presented below (we used 1.2 for a 20% increase).

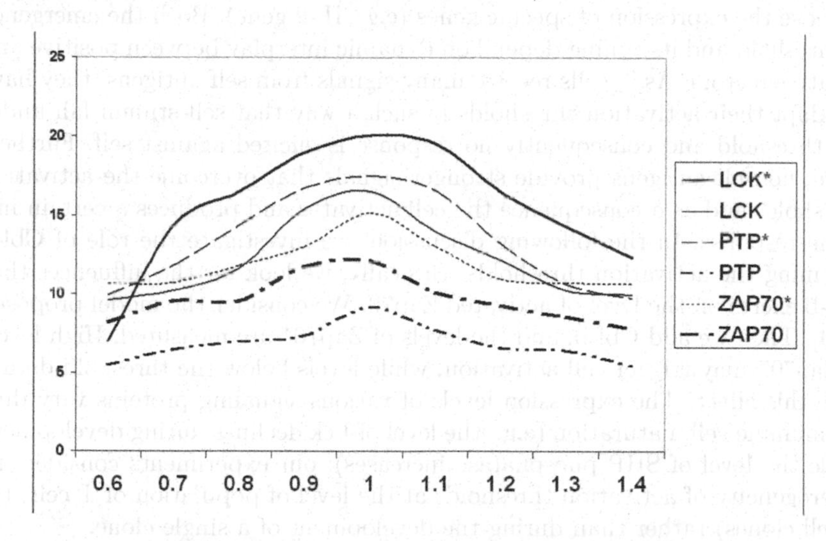

It can be seen that the samples follow a normal distribution, with the mean falling on either side of 1.0 mark, depending on the evolution of the system – for example, the mean for inactivated Lck is clearly less than 1 while the mean for its counterpart activated Lck is greater than 1. Furthermore, based on the data obtained, we could predict the 95% confidence interval for the mean of activated Lck modification as the interval (0.91; 1.14). This way, the data yielded by using MOlNET can be analyzed to find statistical correlations of the mechanisms inside the cell; it is also possible to make other predictions regarding the T cell signaling network.

Tuning the T Cell Activation Thresholds

T cells exert important control over the immune system. Therefore the fine-tuning of T cell activity can have great consequences on the responses that the immune system triggers against viruses or bacteria, as well as on the development of autoimmune diseases. It is very useful to see how a specific molecule type, namely Cbl-b, could tune the threshold required for T cell activation. More complex molecular networks that trigger two qualitatively different cell responses (full activation and anergy) were considered [5]. These results, together with other wet lab data, may open new perspectives in pharmacological manipulation of immune responses. Drugs may trigger, enhance, diminish, or stop the ways in which T cells respond, adjusting the expression level or activity level of the signaling proteins. We assume that simulations of T cell signaling mechanisms could reveal useful information on immunodeficiencies, autoimmune disorders, vaccine design, as well as the function of healthy immune system.

T cell activation is a threshold phenomenon that is dynamically modulated during cell maturation [13]. It reflects the signal intensity that is necessary to increase the expression of specific genes (e.g., IL-2 gene). Both the emergence of threshold and its tuning depend on dynamic interplay between positive and negative factors. As T cells receive many signals from self antigens, they have to adapt their activation thresholds in such a way that self-stimuli fall under the threshold and consequently no response is elicited against self. Furthermore, nonself antigens provide stronger signals that overcome the activation threshold, and as a consequence the cell activates and produces a certain immune response. In the following discussion, we investigate the role of Cbl-b in tuning the activation thresholds. Basically, we look for the influences that Cbl-b exerts on the level of activated Zap70. We consider the model proposed in [4]. Then we add Cbl-b, and the levels of Zap70* are measured. High levels of Zap70* may trigger cell activation, while levels below the threshold do not have this effect. The expression levels of various signaling proteins vary during immune cell maturation (e.g., the level of Lck declines during development while the level of SHP phosphatase increases); our experiments consider the heterogeneity of activation thresholds at the level of population of T cells (or T cell clones) rather than during the development of a single clone.

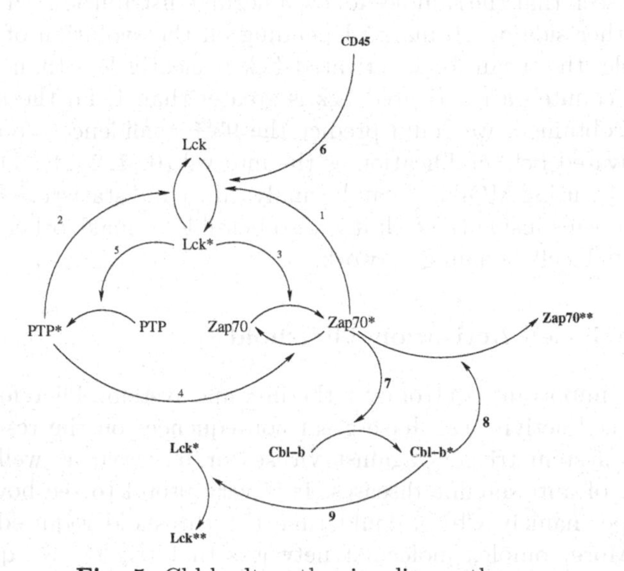

Fig. 5. Cbl-b alters the signaling pathways.

Recent reports highlight that Cbl-b is a key regulator of activation thresholds in T cells. Many proteins are associated with Cbl-b, including Lck* and

Zap70*. Cbl-b mediates chemical modification (ubiquitination) of these activated kinases that target them for degradation [23] (reactions 8 and 9 in Figure 5). The specific chemical modifications due to ubiquitination are denoted by "**". Degradation of active kinases results in the reduction of the activation of downstream signaling proteins. Furthermore, degradation of Lck can reduce the activation of Zap70, as shown in Figure 5. These events raise the threshold requirements for cell activation and prevents the development of autoimmunity [25]. Moreover, following TCR ligation, Zap70* activates Cbl [18] (reaction 7). Additionally, CD45 activates Lck (reaction 6). All these molecular events finely tune the signal intensity in such a way that it draws nearer to, or deviates from the activation threshold. The changes in the Lck*/Lck-total ratio, as well as in the Zap70*/Zap70-total ratio are shown in Figure 6. If the amounts of Cbl-b (and Cbl-b*) vary, and amounts of Lck (and Lck*) vary as well, then Zap70*/Zap70-total ratio still has slight variations (as in Figure 6). But when the amounts of Cbl-b and Cbl-b* equal 500,000 molecules, for an activation threshold set around 0.45 (that is, Zap70*/Zap70-total = 0.45, a thin horizontal line in our picture), the cell could either activate or not (during experiment 2, the signal intensity represented by a gray continuous curve is below the threshold, while in experiment 3 the threshold is overcome). These outcomes are produced by differentially regulating the amount of Lck* within the cell. In other words, Cbl-b fine-tunes T cell reactivity, and this also depends on the amount of Lck*.

The software experiments are represented by numbers from 1 to 4 along the horizontal coordinate in Figure 6. The input values of Lck and Lck* varied from 10,000 to 100,000, and from 500,000 to 1,000,000 molecules, while the input values of Zap70 and Zap70* were kept constant: 100,000. Cbl-b and Cbl-b* were first set to 10,000 (dark lines), and were then set to 500,000. The kinetic constants associated to the corresponding reactions of Figure 5 were $k_1 = 0.001, k_2 = k_3 = k_4 = k_5 = k_6 = 1, k_7 = 0.1$, and $k_8 = k_9 = 0.01$. Lck-total = Lck+Lck* and Zap70-total = Zap70 + Zap70*.

7 Conclusions

P systems were not initially created to model biological systems, although similarities can be observed. Despite many results of universality and several formal language problems which could be explained in an easier and elegant manner, it is useful and desirable to have more connections with applied computer science and molecular biology. Trying to strengthen the connections between P systems and molecular biology, we present a new version of P systems related to the client-server model used in computer networks. We use this version of P systems called client-server P systems to model the signaling network of the T cell. Then we introduce the client-server P simulators; these abstract simulators represent a useful intermediate step from a formal theory suitable for theoretical results to a software implementation of a molecular

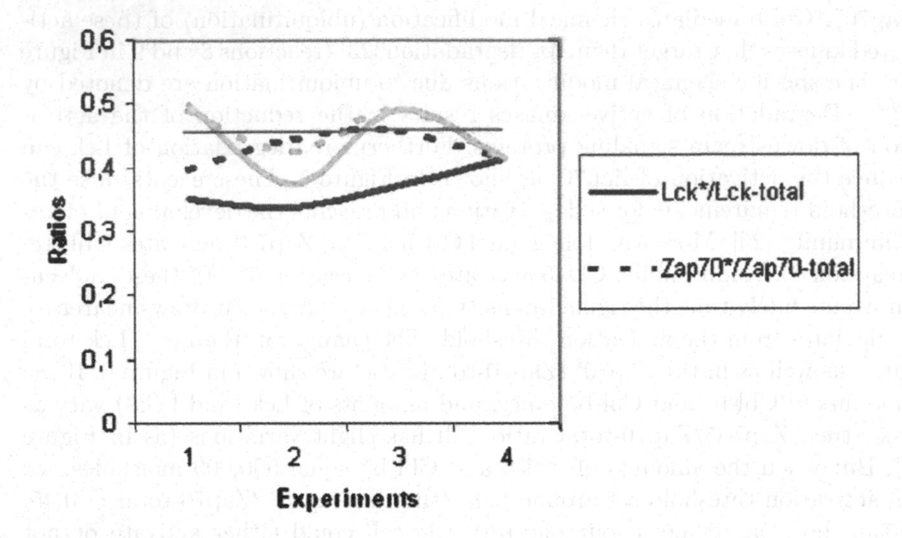

Fig. 6. Lck*/Lck-total and Zap70*/Zap70-total levels after TCR triggering and Cbl-b activation.

network. Using the abstract simulators called CSPsim, we design a software environment called MOlNET. Various software experiments (e.g., for tuning the activation thresholds) take into consideration both qualitative and quantitative aspects. In this way we get relevant biological information on T cell behavior, particularly on T cell responses. The simulations explain how various factors play a certain role in determining the T cell response, relating input and output values of T cell mechanisms. In particular, we see how a Cbl-b could tune the threshold required for T cell activation.

As far as we know, the existence of a server (an executive) for the molecular interaction is a new feature; the existence of a coordinator in molecular processes was recently advanced by some biologists. Therefore it is difficult to compare the architecture of our system with other systems. Regarding its functionality, we can mention its flexibility, modularity, and expressive power. The current approach may serve as a platform for further experimental and theoretical investigations.

Acknowledgements

Part of this work was supported by the academic research grant *Molecular Modeling* (R-252-000-120-112) from the National University of Singapore (NUS). Some of my former students were involved in this research grant, and they have contributed to this approach. Bogdan Tanasă has contributed importantly to the biological aspects related to T cell signaling networks, Dorin

Huzum was involved in various translations from P systems to software implementation, and Daniel Dumitriu was active in statistical experiments and interpretations.

References

1. N. Bellomo, M. Lo Schiavo: *Lecture Notes on the Generalized Boltzmann Equation.* World Scientific, Singapore, 2000.
2. E. Benjamini, R. Coico, G. Sunshine: *Immunology: A Short Course.* Wiley, 2000.
3. F. Burnet: *The Clonal Selection Theory of Acquired Immunity.* Vanderbilt University, Nashville, 1959.
4. C. Chan: *Modeling T Cell Activation.* PhD thesis, Center for Nonlinear Dynamics and its Applications, University College London, 2002.
5. G. Ciobanu, B. Tanasă, D. Dumitriu, D. Huzum, G. Moruz: Simulation and Prediction of T Cell Responses. *Proc. 3rd Conf. on Systems Biology* ICSB'02, Stockholm, 2002, 88–89.
6. G. Ciobanu, D. Dumitriu, D. Huzum, G. Moruz, B. Tanasă: Client-Server P Systems in Modeling Molecular Interaction. In *Membrane Computing, WMC-CdeA 2002, Curtea de Argeş, Romania, Revised Papers* (Gh.Păun, G.Rozenberg, A.Salomaa, C.Zandron, eds.), LNCS 2597, Springer, 2003, 203–218.
7. G. Ciobanu, D. Huzum: Discrete Event Systems and Client-Server Model for Signaling Mechanisms. In *Computational Methods in Systems Biology* (C. Priami, ed.), LNCS 2602, Springer, Berlin, 2003, 175–177.
8. D. Dasgupta, ed.: *Artificial Immune Systems and Their Applications.* Springer, 1999.
9. S. Forrest, S.A. Hofmeyr: Immunology as Information Processing. In *Design Principles for the Immune System and Other Distributed Autonomous Systems* (L.A. Segel, I. Cohen, eds.), Oxford University Press, 2001.
10. M.A. Gibson: *Computational Methods for Stochastic Biological Systems.* PhD thesis, California Institute of Technology, 2000.
11. D.T. Gillespie: Exact Simulation of Coupled Chemical Reactions. *J. Physical Chemistry*, 1977.
12. D.T. Gillespie: A General Method for Numerically Simulating the Stochastic Time Evolution of Coupled Chemical Reactions. *J. Computational Physics*, 1977.
13. Z. Grossman, A. Singer: Tuning of Activation Thresholds Explains Flexibility in the Selection and Development of T Cells in the Thymus. *PNAS*, 93 (1996), 14747–14752.
14. N.K. Jerne: The Immune System. *Sci. Am.*, 229, 1 (1973), 52–60.
15. N.K. Jerne: Towards a Network Theory of the Immune System. *Ann. Immunol.* (Inst. Pasteur), 125C (1974), 373–389.
16. M. Kaufman, J. Urbain, R. Thomas: Towards a Logical Analysis of the Immune Response. *J. Theor. Biol.* 114 (1985), 527.
17. S. Motta, V. Brusic: Mathematical Modeling of the Immune System. In *Modelling in Molecular Biology* (G. Ciobanu, G. Rozenberg, eds.), Springer, Berlin, 2004, 193–218.
18. P. Myung, N. Boerthe, G. Koretzky: Adapter Proteins in Lymphocyte Antigen-Receptor Signaling. *Current Opinion in Immunology*, 12 (2000), 256–266.

19. Gh. Păun: Computing with Membranes. *Journal of Computer and System Sciences*, 61 (2000), 108–143.
20. Gh. Păun: *Membrane Computing. An Introduction.* Springer, Berlin, 2002.
21. A.S. Perelson, Ed.: *Theoretical Immunology.* SFI Studies in the Sciences of Complexity, Addison-Wesley, Boston, 1988.
22. A.S. Perelson, G. Weisbuch: Immunology for Physicists. *Rev. Mod. Phys.*, 69 (1997), 1219–1267.
23. N. Rao, I. Dodge, H. Band: The Cbl Family of Ubiquitin Ligases: Critical Negative Regulators of Tyrosine Kinase Signaling in the Immune System. *Journal of Leukocyte Biology*, 71 (2002), 753–763.
24. I. Roitt, J. Brostoff, D. Male: *Immunology*, 6th edn. Harcourt, 2001.
25. C. Rudd, H. Schneider: Cbl Sets the Threshold for Autoimmunity. *Current Biology*, 10 (2000), 344–347.
26. G. Weisbuch, H. Atlan: Control of the Immune Response. *J. Phys. A*, 21 (1988), 189–192.
27. R.M. Zorzenon Dos Santos: Immune Responses: Getting Close to Experimental Results with Cellular Automata Models. In *Annual Reviews of Computational Physics* (D. Stauffer, ed.), Vol. V, World Scientific, Singapore, 1999, 159–202.

Chapter 6
A Membrane Computing Model of Photosynthesis

Taishin Yasunobu Nishida

Faculty of Engineering, Toyama Prefectural University
Kosugi-machi, Toyama 939-0398, Japan
nishida@pu-toyama.ac.jp

Summary. A model of light reactions taking place in photosynthesis is constructed using a variant of P systems. Behaviors of the model under various combinations of parameters are tested on a computer. Computer simulations show that the model explains in a good way many phenomena of photosynthesis, including photoinhibition mechanisms. A dynamical system using differential equations for photosynthesis is compared with the P system model. The comparison makes it clear that P systems are much better tools for dealing with biological phenomena than models based on differential equations.

1 Introduction

Living cells, especially eukaryotic cells, have many membrane-enclosed structures called organelles. There are many enzymes embedded in the membranes of organelles which act as catalysts of most biochemical reactions taking place in a cell. By enclosing organelles in membranes, living cells obtain many advantages; for instance, they can generate the electric and chemical potentials observed in mitochondrias and chloroplasts and isolate dangerous enzymes such as protease in lysosomes. This is one of the reasons why membrane computing is a promising framework for modeling issues related to cell biology.

Recently, our knowledge of biology has rapidly increased. However, we still do not have general methods for explaining biological phenomena. For example, we cannot explain some macroscopic phenomena, such as emotions, starting from the microscopic phenomena, such as impulses and neurotransmitters. We need a theory or a “language” describing the intermediate realm between macro and microscopic phenomena. It is our belief that membrane computing is an important step toward such a theory.

In this chapter we introduce two extensions of P systems. The first extension concerns the P systems with inner regions of membranes. Because the membrane of a living cell is not a thin film but a complex structure, it is natural to consider the inner region of a membrane, in between the two layers of phospholipidic molecules. Then, we incorporate nonintegral multiplicities, in fact, multiplicities given as real numbers, and operations (e.g., multiplications) among multiplicities. By making use of the latter extension, we can treat ratios, probabilities, concentrations, and so on. The membrane computing systems using nonintegral multiplicities are called $\mathbb{R}$-*subset transforming systems with membrane*.

After preparing these theoretical tools, we concentrate on the photosynthesis of plants, a process which occurs in chloroplasts. A chloroplast catches light energy, converts the energy into chemical energy, and makes starch from CO_2, H_2O, and the chemical energy. The first two reactions are called the light reactions. We consider information processing in the light reactions. The light reactions behave differently according to the strength of the light. We construct an $\mathbb{R}$-subset transforming system which is a model of the light reactions and simulate the behavior of this system on a computer.

In Section 2, we discuss P systems with inner regions of membranes. A rough sketch of photosynthesis will be found in Section 3. In Section 4, $\mathbb{R}$-subset transforming systems with membranes are defined. Then we construct a model of the light reactions. Section 5 is devoted to computer simulations of the model defined in Section 4 under various combinations of parameters and initial conditions. In Section 6, we construct a traditional differential equation based dynamical system as a model of photosynthesis and analyze it. This unveils the fact that P systems or $\mathbb{R}$-subset transforming systems are much more useful than systems of differential equations.

2 P Systems with Inner Regions of Membranes

In this section, we consider the inner region of a cytoplasmic membrane and define and analyze P systems having such regions, besides usual regions delimited by membranes. We assume that the reader is familiar with the basic notions about P systems, e.g., from [6] and [7].

The cytoplasmic membrane of a living cell consists of a double layer of phosphoric lipids and various proteins embedded in the layer (Figure 1). The inner region of a membrane is hydrophobic because there are hydrocarbon chains of lipids inside the membrane and phosphoric acids outside. The structure of the membrane induces a selective permeability for molecules. Small molecules, such as H_2O, O_2, and CO_2, easily go through membranes. On the other hand, large molecules (larger than about 100 Dalton, depending on molecular structure) and *all* ions cannot diffuse across membranes. However, the proteins embedded in the membrane control transportation of ions and molecules. The roles of such proteins is classified into three cases (details are found in textbooks on molecular biology, e.g., [1]):

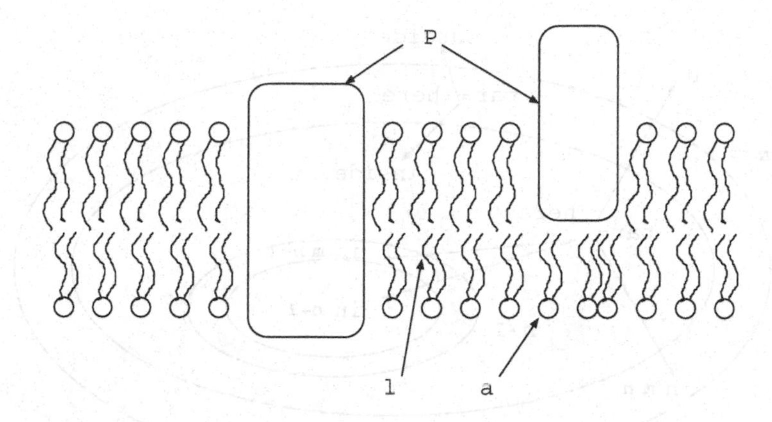

Fig. 1. Schematic diagram of a cytoplasmic membrane; 'a', 'l', and 'P' stand for phosphoric acids, hydrocarbon chains of lipids, and proteins, respectively.

1. Make channels through which ions move passively. A channel is specific to an ion, that is, there are different channels for different ions, e.g., sodium channels, calcium channels, and so on.
2. Provide pumps which transport ions (or molecules) against the gradient of the concentrations, that is, active transportation. Pumps are also specific to ions and molecules.
3. Catch molecules (or ions), change them if necessary, and release them on the same or the other side of the membrane.

We also note that membranes of living cells and organelles in cells discriminate between the inside of the membrane and the outside. The discrimination is the very origin of living organisms.

Now we introduce the inner regions of membranes using P system notations as follows (see Figure 2):

1. The inner region of a membrane which separates regions n and $n+1$ is denoted by $\mathbf{m}_n$ where the letter $\mathbf{m}$ stands for "membrane."
2. For an evolution rule $X \rightarrow (u, tar)$ from a normal region (not from an inner region), *tar* may also be $in_{\mathbf{m}_n}$ (in addition to *here*, in_n, and *out*), which implies that u is sent to the inner region $\mathbf{m}_n$. The target membrane must be adjacent to the region where the rule is applied.
3. For an evolution rule $X \rightarrow (u, tar)$ from an inner region of a membrane, *tar* may be one of *here, inside*, and *outside*.

In terms of computability, the modifications above bring nothing new, in the sense that the computing power of the family of P systems with inner regions of membranes is identical to that of standard P systems; the proof of this result can be found in [5].

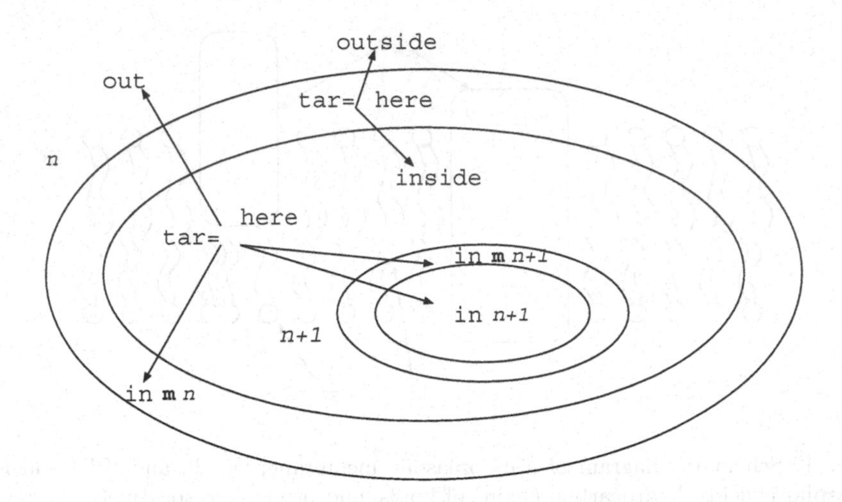

Fig. 2. Modified targets of evolution rules when inner regions of membranes are considered.

In order to treat reactions of photosynthesis, because a reaction occurs with some probability and the total number of reacting molecules is the product of the probability and the numbers of molecules, we have to introduce real numbers and operations with real numbers into P systems. That is why we will use here $\mathbb{R}$-subsets, which are instances of K-subsets as defined in [3].

Let X be a set of objects. An $\mathbb{R}$-subset A of X is a function from X to $\mathbb{R}$, where $\mathbb{R}$ is the set of all real numbers. A value $A(x)$ of this function for some element x in X is a "real valued multiplicity" of x in the "subset A." This notion can be considered a paraphrase of concentration of molecules, that is, x is a name of a molecule, e.g., CO_2, and $A(x)$ is the concentration of x. That is why we denote the multiplicity of x as $[x]_A$ instead of $A(x)$. Because the concentrations of molecules differ in various regions, there is a distinct $\mathbb{R}$-subset in each region, and this makes possible the identification of an $\mathbb{R}$-subset by the name of the corresponding region.

For two elements $[x]_A$ and $[y]_A$ in an $\mathbb{R}$-subset A and any real numbers p and q, the addition and multiplication operations are defined by

$$p[x]_A + q[y]_A \quad \text{and} \quad p[x]_A[y]_A,$$

which are nothing other than operations on real numbers.

Then, we extend P systems to $\mathbb{R}$-*subset transforming systems with membranes* by using evolution rules of the form

$$X \rightarrow (mul_1Y_1, tar_1), (mul_2Y_2, tar_2), \ldots, (mul_lY_l, tar_l),$$

where mul_i is a formula on an $\mathbb{R}$-subset, Y_i is an object, and tar_i is a modified target, considering as above inner regions of membranes. As in P systems, *tar* is omitted if $tar = here$. When the rule is applied in a region, the multiplicity of X in the region goes to 0 and the multiplicity of Y_i in the region indicated by tar_i is increased by adding mul_i for every $i = 1, \ldots, l$.

3 Photosynthetic Reactions Modeled by a Standard P System

Photosynthesis of higher level plants occurs in *chloroplasts*. In a chloroplast there are many membrane-surrounded structures called *thylakoids*. The region inside the thylakoid membrane is called *lumen*. The space between the chloroplast envelope and thylakoids is called *stroma*.

Photosynthetic reactions are classified into two groups: *light reactions* and *dark reactions*. Light reactions separate water into O_2 and H^+, reduce NADP to NADPH, and synthesize ATP. The enzymes which act as catalysts of light reactions are embedded in the thylakoid membrane. Dark reactions make starch from CO_2 and H_2O using the reduction power of NADPH and the chemical energy of ATP. Dark reactions occur in stroma.

Light reactions are carried out by two photosystems, photosystem I and photosystem II, PSI and PSII, respectively. PSI and PSII consist of many proteins and pigments such as chlorophyll and carotenoid. Pigments catch light energy and give it to the proteins in PSI and PSII. Using the energy, PSII splits water on the lumen side of the thylakoid membrane and transports H^+ from stroma to lumen, while PSI reduces NADP on the stroma side of the thylakoid membrane (see Figure 3).

Considering net reactions, light reactions are summarized as follows:

$$2\mathrm{H_2O}(L) \xrightarrow{\mathrm{PS},\gamma} \mathrm{O_2}(L) + 4\mathrm{H}^+(L), \tag{1}$$

$$2\mathrm{NADP}(S) + 2\mathrm{H}^+(S) \xrightarrow{\mathrm{PS},\gamma} 2\mathrm{NADPH}(S), \tag{2}$$

$$4\mathrm{H}^+(S) \xrightarrow{\mathrm{PS},\gamma} 4\mathrm{H}^+(L), \tag{3}$$

where γ represents photons and (L) and (S) represent molecules in lumen and stroma, respectively. We omit electrons from the chemical formulas above. The electric potential generated by electrons is the motive power of the active transport of H^+. So reactions (1) to (3) link each other and, as a consequence of these reactions, the concentration of H^+ in lumen increases or the pH in lumen decreases.

In the thylakoid membrane there is another structure which synthesizes ATP using the chemical potential of H^+ as follows

$$\mathrm{ADP}(S) + \mathrm{P}(S) + n\mathrm{H}^+(L) \rightarrow \mathrm{ATP}(S) + n\mathrm{H}^+(S), \tag{4}$$

where P stands for phosphoric acid and n is the number of H^+s which are required for the ATP synthesis. The number n depends upon the difference

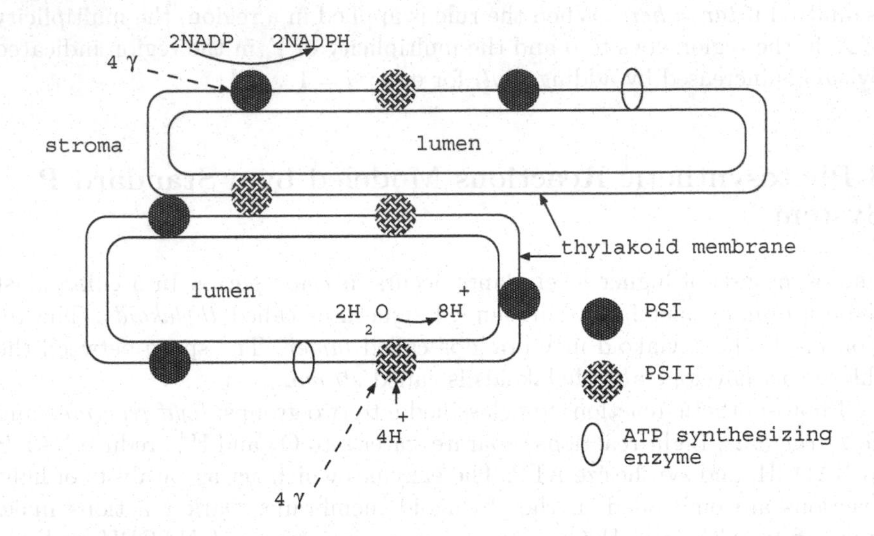

Fig. 3. The structures of the thylakoid membrane and light reactions; γ stands for a photon.

in pH between lumen and stroma; usually, $n = 3$. The dark reactions are essentially the reverse of (2),

$$CO_2 + 4NADPH \to CH_2O + 4NADP + H_2O, \tag{5}$$

which can be assumed a composition of the following four reactions:

$$\begin{aligned} 4NADPH &\to 4NADP + 4H^+ + 4e^-, \\ CO_2 + 4e^- &\to CO^{2-} + O^{2-}, \\ CO^{2-} + 2H^+ &\to CH_2O, \\ O^{2-} + 2H^+ &\to H_2O. \end{aligned}$$

In moderate luminosity, the products of light reactions are all consumed by dark reactions. But, if the light is strong or dark reactions stop for whatever reason, the products, NADPH at stroma and H^+ at lumen, become harmful to the structure of chloroplasts; they may destroy proteins, pigments, lipids, and so on. Consequently, plants have mechanisms to depress light reactions in high luminosity. The mechanisms are called *photoinhibition*. The details of photoinhibition are unknown. We consider two suggested photoinhibition mechanisms. The first consists of reactions which occur when there is no NADP in stroma and the pH in lumen is less than 5 [4]. The reactions are summarized as follows:

$$2H_2O(L) \xrightarrow{PS,\gamma} O_2(L) + 4H^+(L), \tag{6}$$

$$O_2(S) + 4H^+(S) \xrightarrow{PS,\gamma} 2H_2O(S), \quad (7)$$
$$4H^+(L) + O_2(L) \longrightarrow 2H_2O(L), \quad \text{and} \quad (8)$$
$$2NADPH(S) + 2H^+(S) + O_2(S) \longrightarrow 2H_2O(S) + 2NADP(S). \quad (9)$$

The second mechanism decreases the activity of PSII. PSII has many chlorophyll-protein complexes (called LHCII) as antennas to capture photons. PSII releases LHCIIs in strong light and decreases its light collecting activity. In other words, there are two type of PSII, PSIIh with high activity and PSIIl with low activity. Strong light conditions convert PSIIh to PSIIl and the total photosynthesis rate decreases [2].

4 Modeling of Light Reactions and Photoinhibition by an $\mathbb{R}$-subset Transforming System

In this section, probabilities of photosynthetic reactions are first considered. Then, an $\mathbb{R}$-subset transforming system called *Photo*, which is a model of light reactions and photoinhibitions, is constructed.

4.1 Reacting Probabilities of Photosynthesis

First let us consider the probability that a chemical reaction occurs in an ensemble of molecules. For a reaction

$$X \xrightarrow{C} Y,$$

where C is the catalyst of the reaction, the individual molecule X will be caught with probability p by the catalyst C in a time unit. Then, the probability for a molecule being caught by one of the n_C catalysts is given by

$$\begin{aligned} 1-(1-p)^{n_C} &= 1-((1+\frac{-1}{p^{-1}})^{p^{-1}})^{pn_C} \\ &\approx 1-e^{-pn_C}, \end{aligned}$$

because $p \ll 1$ and $(1+\frac{a}{x})^x \approx e^a$ for $x \gg 1$. Then we define

$$\Pr(n_C) = 1 - e^{-pn_C}.$$

If the catalyst C is activated by combining with some additional factor, say D, the probability will be described by

$$\Pr(n_C, n_D) = 1 - e^{-pn_C n_D},$$

where n_D is the multiplicity of factor D which activates the catalyst. So, after reactions taking place in a time unit, the multiplicities of molecules X and Y, denoted by n'_X and n'_Y respectively, are given by

$$n'_X = n_X(1 - \Pr(n_C)),$$
$$n'_Y = n_Y + n_X \Pr(n_C).$$

We note that we count molecules in the region in which the reaction occurs and that the "numbers" need not necessarily be the actual numbers of molecules but may be ratios of molecules to some constant[1], in which case we take concentrations of molecules into account.[2]

Then we compute the probabilities that reactions (1) to (9) in Section 3 take place. Because reactions (1) to (3) link to each other, they occur with the same probabilities P_1 and P_2 which are dependent on the photosystems PSI and PSII, respectively. The photon ν is the additive factor to activate PSI and PSII. Thus,

$$P_1 = 1 - e^{-p_1[\nu][\mathrm{PSII}]},$$
$$P_2 = 1 - e^{-p_1[\nu][\mathrm{PSI}]}.$$

Reactions (4) and (5) occur at a constant rate, which is adopted for normal light condition ν_0, and occur only if the concentration of H^+ in lumen is sufficiently larger than that in stroma, because reaction (4) is derived from the difference in concentrations of H^+ on both sides of the thylakoid membrane. So the probability of (4) and (5) have an identical value P_5, which is given by

$$P_5 = \begin{cases} 1 - e^{-p_1[\nu_0][\mathrm{PSI}]}, & \text{if } \frac{[\mathrm{H}_L^+]}{[\mathrm{H}_S^+]} \geq \theta_{\mathrm{ATP}}, \\ 0, & \text{otherwise.} \end{cases}$$

[1] The unit "mol" is the ratio of the numbers of molecules to Avogadro's number.

[2] By the Taylor expansion we have

$$\Pr(n_C) = -\sum_{i=1}^{\infty} \frac{1}{i!}(-pn_C)^i.$$

If $pn_C \ll 1$, we have $\Pr(n_C) \approx pn_C$. Then

$$n_X \Pr(n_C) = pn_C n_X,$$

which coincides with the reaction velocity $k[\mathrm{X}][\mathrm{C}]$ under the replacements $p \leftrightarrow k$, $n_X \leftrightarrow [\mathrm{X}]$, and $n_C \leftrightarrow [\mathrm{C}]$. Moreover, if there is a reverse reaction $\mathrm{X} \overset{\mathrm{C}}{\leftarrow} \mathrm{Y}$ with probability p' of occurring, then the number of X created from Y in a time unit is given by $n_Y \Pr'(n_C)$, where $Pr'(n_C) = 1 - e^{-p' n_C}$. The assumption that the numbers n_X and n_Y do not change leads to the equation

$$n_X \Pr(n_C) = pn_X n_C = n_Y \Pr'(n_C) = p' n_Y n_C,$$

or

$$\frac{n_X}{n_Y} = \frac{p'}{p},$$

which coincides with the equation assumed by the chemical equilibrium theory.

Because reactions (6) to (9) are photoinhibition reactions, they occur only if NADP in stroma is less than a threshold θ_{NADP} or if the pH in lumen is less than 5, i.e., the number of H^+ in lumen is more than a threshold θ_{H^+}. Reactions (6) and (7) are coupled and they occur with the same probability

$$P_6 = \begin{cases} 0, & \text{if } [\mathrm{NADP}] \geq \theta_{\mathrm{NADP}}, \\ p_1, & \text{otherwise.} \end{cases}$$

Also reactions (8) and (9) are coupled and are regulated by the difference in the numbers of H^+ in lumen and stroma. Therefore, we have

$$P_8 = \begin{cases} 0, & \text{if } [\mathrm{H}^+] \leq \theta_{\mathrm{H}^+}, \\ 1 - e^{(-p_8([\mathrm{H}_L^+]-[\mathrm{H}_S^+]))}, & \text{otherwise.} \end{cases}$$

4.2 A Model for Photosynthesis

After the previous preparations, we construct an $\mathbb{R}$-subset transforming system called *Photo*, which simulates photosynthesis and photoinhibition. *Photo* has three regions: the stroma which is denoted by S, the inner region of the thylakoid membrane which is denoted by $\mathbf{m}_L$, and the lumen which is denoted by L. That is, the envelope membrane of chloroplast is the skin membrane of *Photo*. The inner region $\mathbf{m}_L$ of the thylakoid membrane plays an important role in *Photo*. There are many thylakoids in a chloroplast, so we might use many regions of $\mathbf{m}_L$ and L. Also, there is experimental evidence which suggests differences between stacked thylakoids and unstacked thylakoids. But, as a first step of modeling, we use here the simplest membrane structure, as described above.

The set of objects V of *Photo* is given by

$$V = \{\mathrm{H}^+, \mathrm{NADP}, \mathrm{NADPH}, \mathrm{PSI}, \mathrm{PSIIh}, \mathrm{PSIIl}\}.$$

These objects are changed by the sets of evolution rules R_S, $R_{\mathbf{m}_L}$, and R_L which correspond to the regions S, $\mathbf{m}_L$, and L, respectively. These sets of rules are the following:

$$\begin{aligned} R_S &= \{R_S1, R_S2, R_S3\}, \\ R_{\mathbf{m}_L} &= \{R_{\mathbf{m}_L}1, R_{\mathbf{m}_L}2\}, \text{ and} \\ R_L &= \{R_L1\}, \end{aligned}$$

with the individual rules as given below:

R_S1 (2): $\mathrm{NADP} \longrightarrow (1-2r_2)[\mathrm{NADP}]_S\ \mathrm{NADP}\ 2r_2[\mathrm{NADP}]_S\ \mathrm{NADPH}$

R_S2 (5,9): $\mathrm{NADPH} \longrightarrow (1-r_5-2r_9)[\mathrm{NADPH}]_S\ \mathrm{NADPH}$
$(r_5+2r_9)[\mathrm{NADPH}]_S\ \mathrm{NADP}$

R_S3 (2,7,9): $\mathrm{H}^+ \longrightarrow (1-6r_1-4r_7-2r_9)[\mathrm{H}^+]_S\ \mathrm{H}^+,\ (4r_1[\mathrm{H}^+]_S\mathrm{H}^+\ \mathrm{in}_L)$

$R_{\mathbf{m}_L}1$: $\mathrm{PSIIh} \longrightarrow ((1-r_{PS})[\mathrm{PSIIh}]_{\mathbf{m}_L})\ \mathrm{PSIIh}\ r_{PS}[\mathrm{PSIIh}]_{\mathbf{m}_L}\ \mathrm{PSIIl}$

$R_{\mathbf{m}_L}2$: $\mathrm{PSIIl} \longrightarrow (1-r_{hk})[\mathrm{PSIIl}]_{\mathbf{m}_L}\mathrm{PSIIl}\ r_{hk}[\mathrm{PSIIl}]_{\mathbf{m}_L}\ \mathrm{PSIIh}$

R_L1 (1,4,6,8): $\mathrm{H}^+ \longrightarrow ((1-r_8)[\mathrm{H}^+]_L+(r_1+r_6)C_L-r_5C_{ATP})\ \mathrm{H}^+$
$(r_5C_{ATP}\mathrm{H}^+, \mathrm{out})$.

The numbers in the parentheses correspond to the reaction numbers explained in Section 3. The constants C_{PS}, C_L, and C_{ATP} correspond to the "housekeeping" PSIIh recovering, the maximum water splitting in lumen, and the maximum ATP synthesis in the thylacoid membrane, respectively. The reaction rates r_x are given by:

$$
\begin{aligned}
r_1 &= 1-\exp(-p_1\nu[\mathrm{PSIIh}]_{\mathbf{m}_L})+1-\exp(-p_1'\nu[\mathrm{PSIIl}]_{\mathbf{m}_L}),\\
r_2 &= 1-\exp(-p_1\nu[\mathrm{PSI}]_{\mathbf{m}_L}),\\
r_5 &= \begin{cases} 2(1-\exp(-p_1\nu_0[\mathrm{PSI}]_{\mathbf{m}_L})), & \text{if } \frac{[\mathrm{H}^+]_L}{[\mathrm{H}^+]_S} \geq \theta_{\mathrm{ATP}},\\ 0, & \text{otherwise,}\end{cases}\\
r_6 = r_7 &= \begin{cases} 0, & \text{if } [\mathrm{NADP}]_S \geq \theta_{\mathrm{NADP}},\\ r_1, & \text{otherwise,}\end{cases}\\
r_8 = r_9 &= \begin{cases} 0, & \text{if } [\mathrm{H}^+]_L \leq \theta_{\mathrm{H}^+},\\ \frac{1-\exp(-p_8([\mathrm{H}^+]_L-[\mathrm{H}^+]_S))}{4}, & \text{otherwise,}\end{cases}\\
r_{PS} &= 1-\exp(-p_{PS}\nu[\mathrm{PSIIh}]_{\mathbf{m}_L}),
\end{aligned}
$$

where p_{PS} stands for the probability of the photosystem II decreasing reaction. The above rules are easily derived from reaction probabilities obtained in subsection 4.1 and the appropriate number of molecules which jointly participate in the reactions.

Initially, the $\mathbb{R}$-subsets S_0, $\mathbf{m}_{L_0}$, and L_0 are placed in stroma, the inner region of the thylakoid membrane, and in lumen, respectively. Because the transition of *Photo* is assumed to continue forever, there is no specific output region. This means that all multiplicities (or numbers) of objects (or molecules) should be monitored forever.

5 Computer Simulations

We simulate the behaviour of *Photo* on a computer with various parameter values and various initial $\mathbb{R}$-subsets.

Conditions of Simulations. We examine seven cases, as shown in Table 1.

In Table 1, "photosystem decreasing activity disable" means that photosystem II always has high activity or there are no PSIIls, which is realized by setting $p_{PS} = 0$.

Table 1. Combinations of light conditions and photosystem decreasing activity.

light condition	photosystem decreasing activity enable	disable
dark ($\nu = 0$)	case Dark	
normal ($\nu = 1000$)	case EN	case DN
strong ($\nu = 10000$)	case ES	case DS
very strong ($\nu = 10^5$)	case EV	case DV

Tables 2 and 3 show standard parameter values and the initial $\mathbb{R}$-subsets[3].

Table 2. Parameters for the simulation.

constants	$r_{hk} = 1 - \exp(-p_{hk})$, $p_{hk} = 0.01$	$C_L = 30000$	$C_{ATP} = 500$
probabilities	$p_1 = 10^{-8}$	$p_1' = 10^{-9}$	$p_8 = 10^{-5}$
ν_0	1000		
thresholds	$\theta_{\mathrm{NADP}} = 200$	$\theta_{\mathrm{H+}} = 30000$	$\theta_{\mathrm{ATP}} = 10$

Table 3. The initial values for the simulation.

x	$[x]_{S_0}$	$[x]_{\mathrm{m}_{L_0}}$	$[x]_{L_0}$
H^+	3000	0	3000
NADP	1000	0	0
NADPH	0	0	0
PSI	0	1000	0
PSIIh	0	1000	0
PSIIl	0	0	0

Many simulations were done under different parameter and initial $\mathbb{R}$-subset values. They are listed in Table 4.

During simulations under very strong light conditions, the multiplicity of $[H^+]_S$ goes to unrealistic, i.e., negative, values. The simulation program resets these values to biologically reasonable values, i.e., 30, which corresponds to pH 8 under an assumed volume of stroma. We call such a phenomenon "lock-in." Lock-in occurs when reaction rates, such as r_1, r_2, etc. become too large in comparison with the number of the molecules. This will be avoided by

[3] The "standard" values are selected to fit observed figures of chloroplasts. For example, the numbers of H^+ ions correspond to the concentration of H^+ for pH 6. But "luminosity" of light is not an observed value, but it is selected to adjust all other parameters and the initial $\mathbb{R}$-subsets.

using small reaction rates, in other words, small probabilities p_1, p'_1, etc., and increasing iterations of rules in the simulation. But, because lock-in occurs in quite a few cases, we have not changed the model *Photo* and the simulation program.

Table 4. Parameters and initial $\mathbb{R}$-subset values under which simulations were done. We indicate here only the values which are different from the standard values.

Name of condition	value 1	value 2
small θ_{NADP}	$\theta_{\mathrm{NADP}} = 133$	
large θ_{NADP}	$\theta_{\mathrm{NADP}} = 300$	
small θ^+_{H} 1	$\theta^+_{\mathrm{H}} = 20000$	
small θ^+_{H} 2	$\theta^+_{\mathrm{H}} = 15000$	
large θ^+_{H}	$\theta^+_{\mathrm{H}} = 45000$	
small p'_1	$p'_1 = 5.0 \times 10^{-10}$	
large p'_1	$p'_1 = 2.0 \times 10^{-9}$	
initial low pH in lumen	$[\mathrm{H}^+]_{L_0} = 5700$	$[\mathrm{H}^+]_{S_0} = 300$
no photoinhibition	$\theta_{\mathrm{NADP}} = 0$	$\theta_{\mathrm{H}+} = 10^{14}$
extremely small p_1 & p'_1	$p_1 = 10^{-10}$	$p'_1 = 10^{-11}$
extremely large p_1 & p'_1	$p_1 = 10^{-6}$	$p'_1 = 10^{-7}$

Results of Simulations. In the case Dark, obviously no photosynthesis occurs and the initial $\mathbb{R}$-subset remains unchanged in all conditions. Simulations of the case Dark show soundness of *Photo* in an extreme condition, that when no light is present.

Figures 4 to 9 display changes of values of pH in stroma and lumen, numbers of NADPH and PSIIh for the cases EN, ES, EV, DN, DS, and DV, respectively, during simulations under the standard conditions. One can see that the values go to equilibria after 50–100 iterations, but some of them fluctuate around equilibria. The same tendency is observed everywhere, but not when using extremely small p_1 and p'_1. The average values of the last 100 iterations from the total of 210 iterations are shown in the following tables as equilibrium values.

Tables 5 and 6 show averages of values in the last 100 iterations of a total of 210 iterations under various conditions. In the tables, "standard" represents the same values as the standard condition and "←" indicates the same values as in the left column. The quadruple in Table 5 represents ($[\mathrm{H}^+]_L$, $[\mathrm{H}^+]_S$, [NADPH], [PSIIh]) and the triple in Table 6 represents ($[\mathrm{H}^+]_L$, $[\mathrm{H}^+]_S$, [NADPH]). We omit [PSIIh] from Table 6 because [PSIIh] keeps the initial value 1000 if the photosystem decreasing activity is disabled. For the same reason, we omit the conditions of small and large p'_1 from Table 6, because p'_1 affects only the photosystem decreasing activity.

Table 5. Results of simulations with photosystem decreasing activity enabled. The quadruples represent ($[H^+]_L$, $[H^+]_S$, [NADPH], [PSIIh]).

condition	case EN	case ES	case EV
standard	(29221, 194, 388, 628)	(30230, 36, 683, 271)	(52507, 7.8, 459, 97)
small θ_{H^+} 1	(19747, 166, 333, 628)	(21246, 33, 622, 271)	standard
small θ_{H^+} 2	(14935, 150, 301, 628)	(21246, 33, 622, 271)	standard
large θ_{H^+}	(39067, 241, 482, 628)	(43797, 39, 735, 271)	standard
small θ_{NADP}	standard	standard	standard
large θ_{NADP}	standard	(30303, 33, 661, 271)	standard
small p_1'	(29201, 200, 392, 628)	(29894, 41, 699, 271)	(45121, 9.9, 486, 97)
large p_1'	(29247, 182, 380, 628)	(30250, 30, 650, 271)	(53968, 7.5, 455, 97)
initial low pH in lumen	(12993, 78, 198, 628)	(30751, 21, 556, 271)	(78760, 14*, 394, 97)
nophotoinhibition	(40172, 249, 499, 628)	(189032, 48, 905, 271)	(851349, 9.3, 905, 97)
extremely small p_1 & p_1'	(3498, 2792, 32, 628)	(5498, 2065, 275, 271)	(20590, 194, 946, 97)
extremely large p_1 & p_1'	(42665, 658, 117, 628)	(53115, 469, 115, 271)	(53115, 469, 115, 97)

* Lock-in occurs.

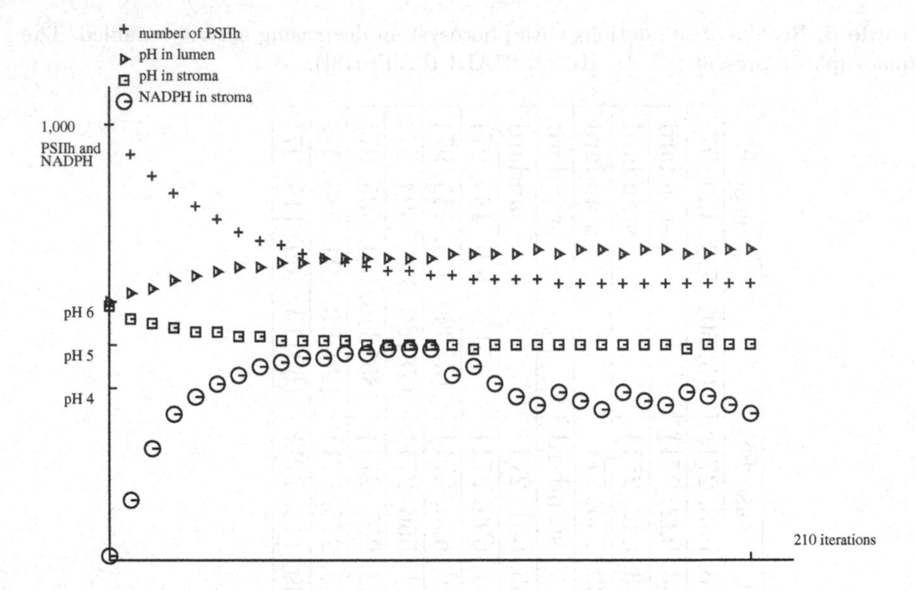

Fig. 4. Simulation of case EN under the standard condition.

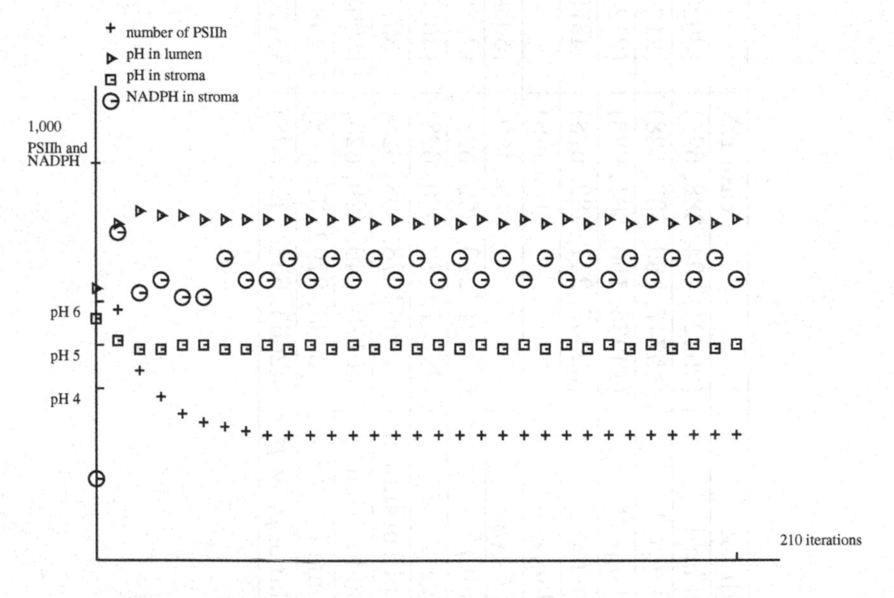

Fig. 5. Simulation of case ES under the standard condition.

One more condition is interesting and simulated: long iterations of extremely small p_1 and p'_1. The results are shown in Table 7 and displayed in Figures 10 and 11. The results in Table 7 tell us that there is a kind of trade-

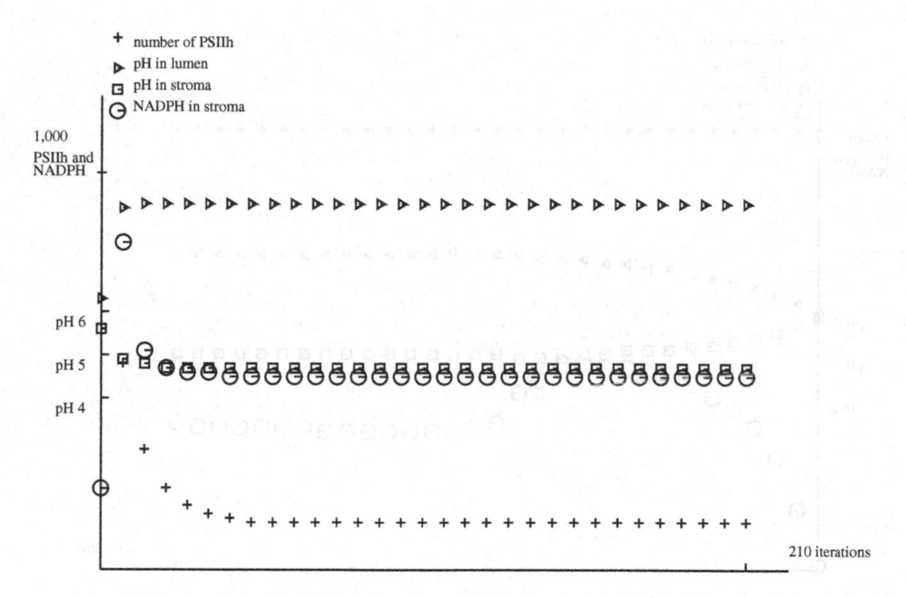

Fig. 6. Simulation of case EV under the standard condition.

Table 6. Results of simulations with photosystem decreasing activity disabled. The triple represents ($[H^+]_L$, $[H^+]_S$, [NADPH]).

condition	DN	DS	DV
standard	(29338, 125, 334)	(36931, 14, 522)	←
small θ_{H+} 1	(19824, 111, 285)	standard	standard
small θ_{H+} 2	(15029, 101, 251)	standard	standard
large θ_{H+}	(42780, 145, 415)	(45539, 14, 564)	←
small θ_{NADP}	standard	standard	standard
large θ_{NADP}	standard	standard	standard
initial low pH in lumen	(15998, 57, 175)	(54386, 8.6, 453)	←
nophotoinhibition	(52625, 167, 499)	(465522, 17, 905)	←
extremely small p_1 & p_1'	(3669, 2723, 32)	(9074, 1153, 275)	(29316, 1.4, 503)
extremely large p_1 & p_1'	(35218, 883, 119)	←	←

off between probabilities of reactions and the number of iterations, which validates the consideration about the lock-in.

Summary of Simulations. The simulations make it clear that *Photo* is a good model for light reactions and photoinhibitions of photosynthesis. One can observe the following:

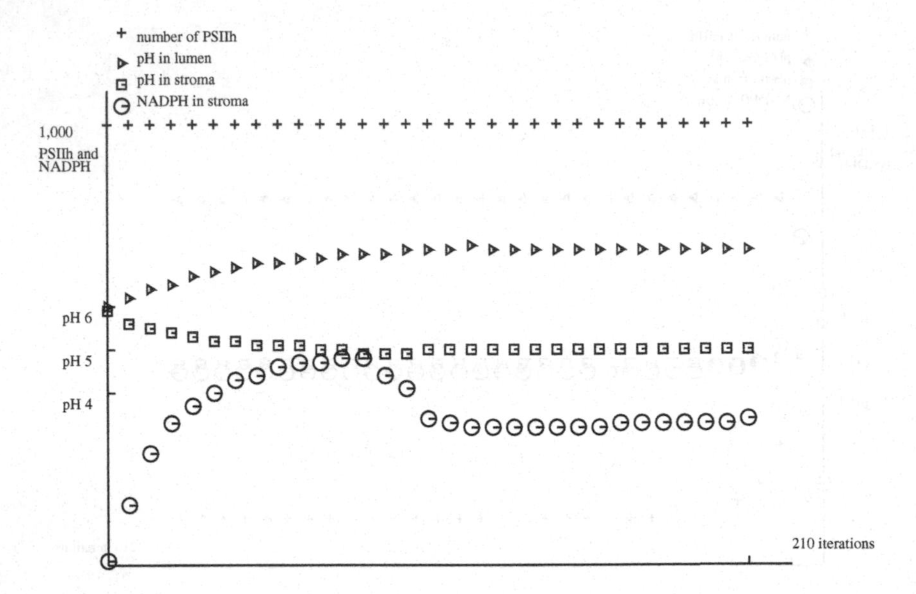

Fig. 7. Simulation of case DN under the standard condition.

Table 7. Results of 20000 iterations under extremely small p_1 and p'_1 condition.

case EN	case ES	case EV
(29593, 197, 389, 618)	(29229, 36, 691, 271)	(29000, 5.2, 783, 97)
case DN	case DS	case DV
(29067, 124, 330)	(29126, 12, 464)	(29313, 1.2, 485)

- The photosystem decreasing activity is effective in preventing damages caused by a low pH under strong light conditions.
- Photoinhibition reactions (6) to (9) are very important. If there were no such reactions, the cloroplasts would have fatal damage.
- *Photo* is sensitive to the threshold of pH (θ_{H+}), which may suggest that different θ_{H+} correspond to different plants which grow under different light conditions.
- *Photo* is insensitive to the threshold of NADP (θ_{NADP}) for photoinhibition. This may suggest that photoinhibition is triggered mainly by pH.

6 Comparison with a Dynamical System Model

In this section, a model of photosynthesis using a conventional method based on differential equations is analyzed. Then the advantage of using a P system becomes clear.

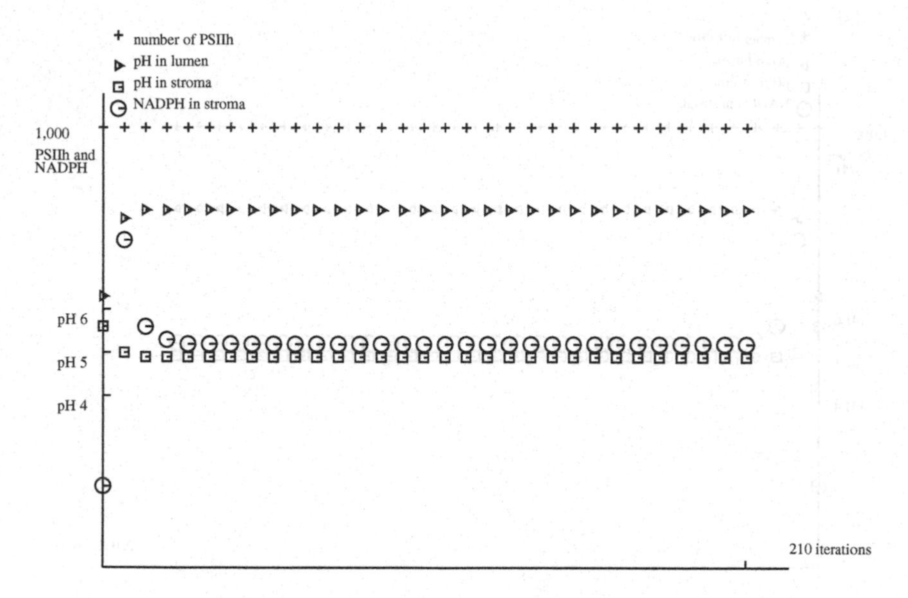

Fig. 8. Simulation of case DS under the standard condition.

The photosynthetic reactions without the photosystem decreasing activity, which were discussed in Section 3 are as follows:

$$
\begin{aligned}
2\mathrm{H_2O}(L) &\xrightarrow{\mathrm{PS},\gamma} \mathrm{O_2}(L) + 4\mathrm{H^+}(L), && (10)\\
2\mathrm{NADP}(S) + 2\mathrm{H^+}(S) &\xrightarrow{\mathrm{PS},\gamma} 2\mathrm{NADPH}(S), && (11)\\
4\mathrm{H^+}(S) &\xrightarrow{\mathrm{PS},\gamma} 4\mathrm{H^+}(L), && (12)\\
\mathrm{ADP}(S) + \mathrm{P}(S) + n\mathrm{H^+}(L) &\rightarrow \mathrm{ATP}(S) + n\mathrm{H^+}(S), && (13)\\
\mathrm{CO_2} + 4\mathrm{NADPH} &\rightarrow \mathrm{CH_2O} + 4\mathrm{NADP} + \mathrm{H_2O}, && (14)\\
2\mathrm{H_2O}(L) &\xrightarrow{\mathrm{PS},\gamma} \mathrm{O_2}(L) + 4\mathrm{H^+}(L), && (15)\\
\mathrm{O_2}(S) + 4\mathrm{H^+}(S) &\xrightarrow{\mathrm{PS},\gamma} 2\mathrm{H_2O}(S), && (16)\\
4\mathrm{H^+}(L) + \mathrm{O_2}(L) &\rightarrow 2\mathrm{H_2O}(L), \quad \text{and} && (17)\\
2\mathrm{NADPH}(S) + 2\mathrm{H^+}(S) + \mathrm{O_2}(S) &\rightarrow 2\mathrm{H_2O}(S) + 2\mathrm{NADP}(S). && (18)
\end{aligned}
$$

In the above reactions, the number of molecules of H_2O, O_2, ADP, ATP, and CO_2 can be considered to be constant because they are supplied from air or stabilized by other mechanisms. Thus, the number of NADP, H^+ in stroma, and H^+ in lumen are variable. The number of NADPH is dependent on NADP since the total number of NADP and NADPH is constant.

Then we can write a dynamical system with three differential equations:

$$\frac{dN}{dt} = (1 - 2r_2)N + (r_5 + 2r_9)(T_N - N), \tag{19}$$

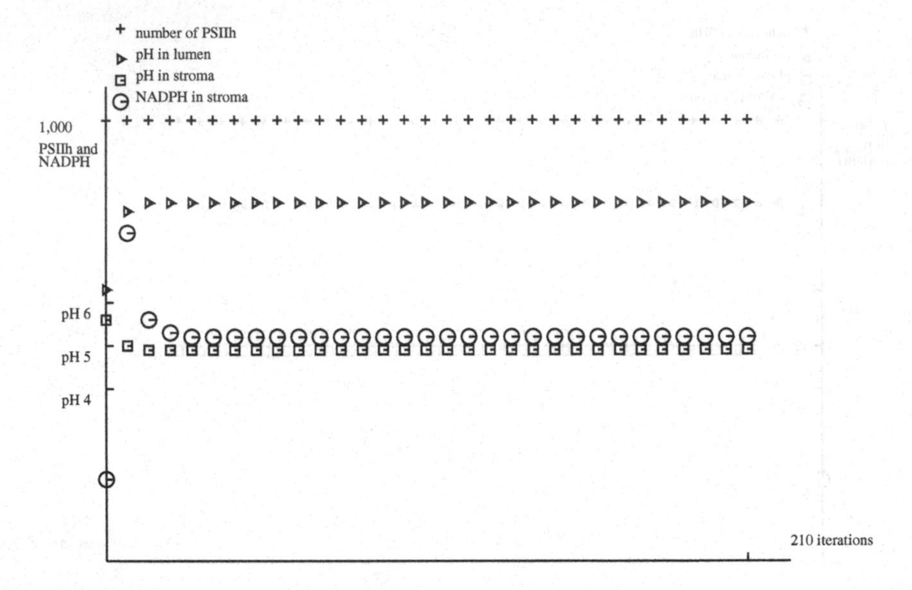

Fig. 9. Simulation of case DV under the standard condition.

$$\frac{dH_S}{dt} = (1 - 6r_1 - 4r_7 - 2r_9)H_S + r_5 C_{ATP}, \tag{20}$$

$$\frac{dH_L}{dt} = (1 - r_8)H_L + 4r_1 H_S + (r_1 + r_6)C_L - r_5 C_{ATP}, \tag{21}$$

where N, H_S, and H_L are the number of NADP, H^+ in stroma, and H^+ in lumen, T_N is the total number of NADP and NADPH, and the reaction rates r_x are given by

$$\begin{aligned} r_1 &= p_1 \nu \text{PSII}, \\ r_2 &= p_1 \nu \text{PSI}, \\ r_5 &= 2p_1 \nu_0 \text{PSI}, \\ r_6 = r_7 &= \begin{cases} 0, & N \geq \theta_{\text{NADP}}, \\ r_1, & \text{otherwise}, \end{cases} \\ r_8 = r_9 &= \begin{cases} 0, & H_L \leq \theta_{\text{H}^+}, \\ \frac{1 - \exp -p_8(H_L - H_S)}{4}, & \text{otherwise}. \end{cases} \end{aligned}$$

Equations (19) to (21) are highly nonlinear and cannot be solved analytically.

As a first step to treat the dynamical system, we consider the case where there are no photoinhibition reactions (reactions (15) to (18)). Then coefficients r_6, r_7, r_8, and r_9 become 0. So equations (19) to (21) look like

$$\frac{dN}{dt} = (1 - 2r_2)N + r_5(T_N - N), \tag{22}$$

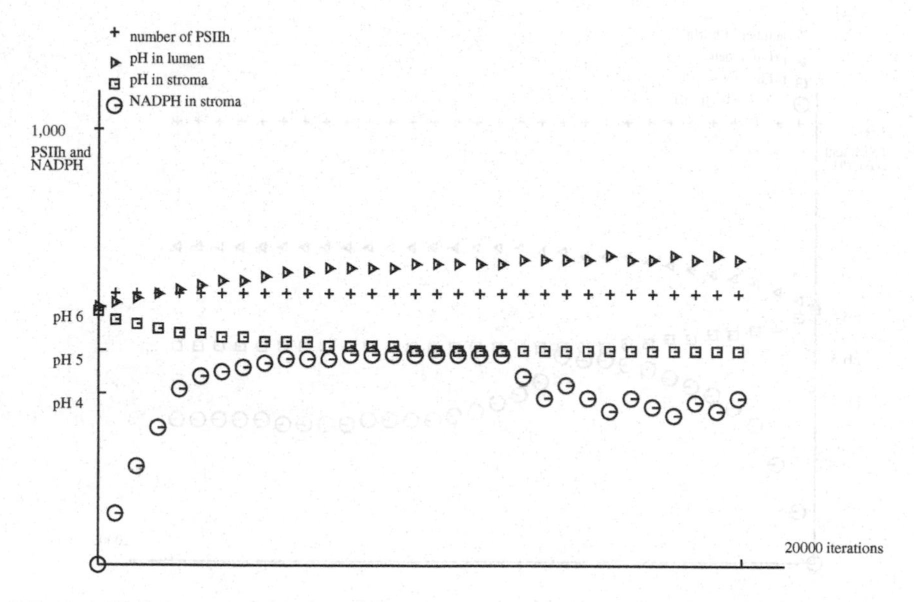

Fig. 10. Changes of values under extremely small p_1 and p_1' for case EN with 20000 iterations.

$$\frac{dH_S}{dt} = (1 - 6r_1)H_S + r_5 C_{ATP}, \tag{23}$$

$$\frac{dH_L}{dt} = H_L + 4r_1 H_S + r_1 C_L - r_5 C_{ATP}, \tag{24}$$

that is, they become linear differential equations. The solutions are

$$N = A_k e^{k_N t} - \frac{r_5 T_N}{k_N}, \tag{25}$$

$$H_S = A_S e^{k_S t} - \frac{r_5 C_{ATP}}{k_S}, \tag{26}$$

$$H_L = \frac{4r_1 A_S}{k_S - 1} e^{k_S t} + C + C_0, \tag{27}$$

where A_k, A_S, and C_0 are integral constants which should be determined by the initial conditions. k_N, k_S, and C are given by

$$k_N = 1 - 2r_1 - r_5,$$
$$k_S = 1 - 6r_1,$$
$$C = -r_1 C_L + r_5 C_{ATP} - \frac{4r_1 r_5 C_{ATP}}{k_S}.$$

One can easily see that solutions (25) to (27) are by no means realistic. The values N, H_S, and H_L tend to ∞ or 0 because they are dominated by the

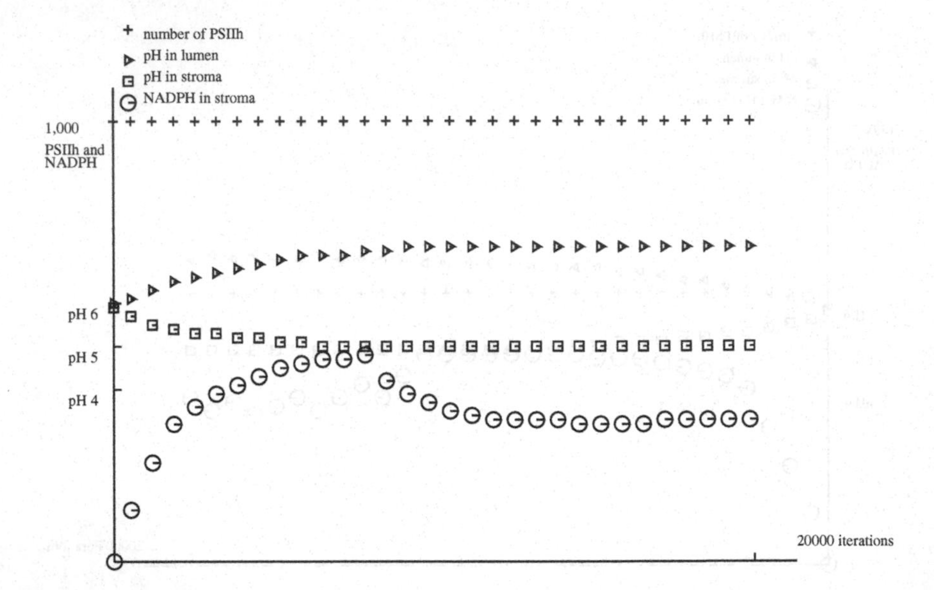

Fig. 11. Changes of values under extremely small p_1 and p_1' for case DN with 20000 iterations.

term e^{kt} with a real valued constant k. The observation tells us that the photoinhibition has an essential role in photosynthesis.

Now we assume $\theta_{\text{NADP}} = \infty$ and $\theta_{\text{H+}} = 0$; in other words, we assume that photoinhibition always occurs. Then equations (19) to (21) become

$$\frac{dN}{dt} = (1 - 2r_2)N + (r_5 + \frac{1}{2}p_8(H_L - H_S))(T_N - N), \tag{28}$$

$$\frac{dH_S}{dt} = (1 - 10r_1 - \frac{1}{2}p_8(H_L - H_S))H_S + r_5 C_{ATP}, \tag{29}$$

$$\frac{dH_L}{dt} = (1 - \frac{1}{4}p_8(H_L - H_S))H_L + 4r_1 H_S + 2r_1 C_L - r_5 C_{ATP}. \tag{30}$$

Equations (28) to (30) are again nonlinear. We compute equilibrium values by putting $\frac{dN}{dt} = \frac{dH_S}{dt} = \frac{dH_L}{dt} = 0$:

$$0 = (1 - 2r_2)N + (r_5 + \frac{1}{2}p_8(H_L - H_S))(T_N - N),$$

$$0 = (1 - 10r_1 - \frac{1}{2}p_8(H_L - H_S))H_S + r_5 C_{ATP},$$

$$0 = (1 - \frac{1}{4}p_8(H_L - H_S))H_L + 4r_1 H_S + 2r_1 C_L - r_5 C_{ATP}.$$

The equations are of the third degree in H_S and H_L, and general solutions are very complex. That is why we solve them with the same numerical parameters with which computer simulations of *Photo* are done. The parameter values are shown below.

p_1	ν_0	C_{PS}	T_N	C_{ATP}	C_L	p_8
10^{-8}	1000	1000	1000	500	30000	10^{-5}

Table 8 shows the equilibrium values of differential equations (28) to (30) for $\nu = 1000$ and $\nu = 10000$.

Table 8. Numerical values of equilibrium of equations (28) to (30).

	$\nu = 1000$	$\nu = 10000$
N	211.5	140.5
H_S	9.1	3.8
H_L	399789	526320

The equilibrium values resemble to some extent the results of simulations of *Photo* without the photosystem decrease activity of PSII. However, they also have unrealistic values in H_L. Moreover, we cannot know the changes of variables. Of course, numerical integration on a computer would give the behavior of the variables of the differential equations (28) to (30), but it is much simpler to construct a model using P systems and to simulate it on a computer than to construct a system of differential equations and then integrate it numerically.

7 Conclusion

We have discussed the effects of membrane-embedded objects or inner regions of membranes on P systems and, as an application, we have built a P system-like model called *Photo* which simulates the photosynthesis of plants. Since these attempts are in a preliminary state, we have addressed only a few problems and many issues are left for future research.

Results of simulations of *Photo* fit into several biological situations: no light, moderate light, photoinhibition under strong light, and a suggested mechanism of photoinhibition is supported. The results offer some hypotheses:

1. Photoinhibition is sensitive to pH.
2. Photoinhibition is insensitive to concentration of NADP.
3. There is a trade-off between probabilities of occurring reactions and the number of iterations of rules.

Two of these hypotheses (1 and 2) should be compared to the observed data about real plants. The third (3) should be checked by further theoretical and experimental investigations of $\mathbb{R}$-subset transforming systems.

Further efforts for improving *Photo* by incorporating biological facts are necessary. Photosynthesis is a set of complex reactions. There are important reactions, for instance, the dark reactions, which are not treated by the current version of *Photo*. However, *Photo* will be able to handle the dark reactions, because it is easy to add more rules to $\mathbb{R}$-subset transforming systems, and then we can analyze them on a computer.

As a conclusion, we again stress the fact that P systems and their variants are much more suitable as a tool for modeling and analyzing biological phenomena than traditional systems of differential equations.

References

1. B. Albers, D. Bray, J. Lewis, M. Raff, K. Robers, J.D. Watson: *Molecular Biology of the Cell*. Garland Publishing, New York, 1994.
2. B. Andersson: Thylakoid Membrane Dynamics in Relation to Light Stress and Photoinhibition. In *Trends in Photosynthesis Research* (J. Barber, M.G. Guerrero, H. Medran, eds.), Intercept, Andover, 1992, 71–86.
3. S. Eilenberg: *Automata, Languages, and Machines Volume A*. Academic Press, New York, 1974.
4. H.W. Heldt: *Plant Biochemistry & Molecular Biology*. Oxford University Press, Oxford, 1997.
5. T.Y. Nishida: Simulations of Photosynthesis by a K–Subset Transforming System with Membranes. *Fundamenta Informaticae* 49 (2002), 249–259.
6. Gh. Păun: Computing with Membranes. *Journal of Computer and System Sciences*, 61 (2000), 108–143.
7. Gh. Păun: *Membrane Computing: An Introduction*. Springer, Berlin, 2002.

Chapter 7
Modeling p53 Signaling Pathways by Using Multiset Processing

Yasuhiro Suzuki, Hiroshi Tanaka

Bioinformatics, Medical Research Institute
Tokyo Medical and Dental University
Yushima 1-5-45, Bunkyo, Tokyo 113, Japan
suzuki@bioinfo.tmd.ac.jp

Summary. We present an approach to modeling and simulating the protein p53 signaling pathways by means of a particular class of P systems, called ARMS (Abstract Rewriting Systems on Multisets). The results of the computer simulations are presented; they agree with the biological observations.

1 Abstract Rewriting Systems on Multisets

Living systems can be seen as huge chemical systems, where numerous substances interact with each other. Abstract Rewriting Systems on Multisets (ARMS), a class of P systems based on multiset processing but with a simple membrane structure, was introduced with the aim of modeling such chemical systems [23].

In ARMS, substances interact with each other according to reaction rules, which change the amounts of the substances. Reaction rules are of the form, customarily used in chemistry, $u \to v$, where u and v are multisets of molecules/reactants. Reactions occur at rates which depend on the amount of the substances from u.

The state of the system is represented by the amounts $[x]$ of the chemical substances (for generality, we use "amount" rather than "concentration" or "number of molecules"). When the reaction $u \to v$ occurs, the amount of substances x from u decreases from $[x]$ to $[x]-1$ and the amount of substances y from v increases from $[y]$ to $[y]+1$.

ARMS is a stochastic model, where rules are applied probabilistically. The probabilities are given by basic concepts of stoichiometric chemistry (for completeness, we present in Appendix A some basic elements of stoichiometry).

Various types of ARMS have been proposed and used for modeling a protocell [22], investigating the symbiotic mechanisms of an ecological system [20], considering a novel theory of evolution [18], and so on.

2 Modeling p53 Signaling Pathways by Using ARMS with Membranes

The p53 signaling network plays a major role in cell survival, and it safeguards against genetic instability, which leads to tumor formation. However, the complicated structure hampers modeling with the ordinary rate equation model.

The p53 signaling network has been studied intensively because over 50%-55% of all human cancers are reported to p53 gene mutations [8].

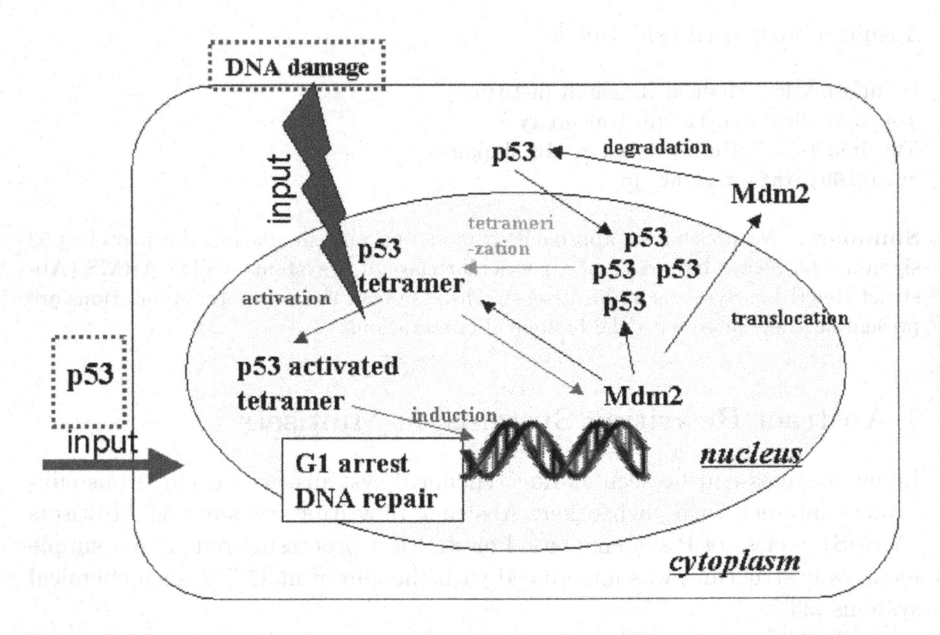

Fig. 1. The p53 signaling network

The p53 protein is a transcription factor which plays a major role in regulating the response of mammalian cells to stresses and damage, mainly through the transcriptional activation of genes involved in cell cycle control (G1 arrest), DNA repair, senescence, and apoptosis [12]. In normal cells, p53 is a short-lived and non-abundant protein because of its rapid degradation [12], and probably exists in a latent, inactive form [11], [9]. However, once a cell has a DNA damage, p53 transforms itself from latent to active conformation through tetramerization and translocates from the cytoplasm to the nucleus [16]. p53 has two levels of activation, depending on the level of DNA damage. The weakly activated p53 prevents damaged cells from proceeding in the cell division cycle and promotes DNA repair. The highly activated p53 induces apoptosis and eliminates mutated or irrevocably DNA damaged cells.

To delay the p53-induced apoptosis and permit cells that are not irretrievably damaged or mutated to survive, p53 forms the autoregulatory negative feedback loop with MDM2 oncoproteins. In addition, the survival factor promotes activation of MDM2 through PI3K-PDK1-Akt signaling and translocation of MDM2 from cytoplasm to nucleus, which downregulates the activity of p53. Moreover, the growth factor inhibits MDM2 activation through Arf protein, which upregulates the activity of p53. Then, p53, which forms an active conformation, inhibits MDM2 activation through PTEN protein and caspase activation, whereas p53 induces MDM2, which provides DNA damaged cells the opportunity for DNA repair. Subsequently, p53 induces PTEN, which then induces the death of mutated or irrevocably DNA damaged cells. p53 and MDM2 are networked with a high complexity to keep the cell normal and eliminate mutated cells [13].

In order to model these processes, we need a P system with two membranes, one representing the nucleus and one enclosing the cytoplasm; that is why we label them n and c, respectively. Therefore, the membrane structure is $[_c [_n]_n]_c$.

The reaction rules of the p53 signaling network are given in Table 1, in the multiset rewriting form customarily used in membrane computing, with the substances from their right hand side having target indications *in* or *out* in the case in which they are moved from one region of the system to another (the substances of each multiset are given by their names, and separated by a space; instead of the empty multiset, we write "vanish" for a better readability).

Rules in cytoplasm (the set R_c):

p53 → vanish,
p53 → (p53, in),

Rules in nucleus (the set R_n):

p53 p53 p53 p53 → p53-tetramer,
p53-tetramer → p53-tetramer MDM2,
p53 MDM2 → vanish,
MDM2 → (MDM2, out),
p53-tetramer MDM2 → (p53 p53 p53 p53, out),
p53-tetramer DNA-damage → p53-tetramer+ DNA-damage,
p53-tetramer+ DNA-damage → p53-tetramer.

Table 1. Reaction rules of the p53 signaling network.

The evolution of the P system described above was simulated on computer; the results are shown in Figure 2 and they indicate how p53 induces the MDM2 oncoprotein.

Biologically, MDM2 is phosphorylated by survival signaling through the PI3-kinase-PDK1-Akt pathway, which promotes rapid p53 degradation. We have summarized these interactions in the rule p53 MDM2 → vanish

from R_n. Thus, p53 and MDM2 form an autoregulatory negative feedback loop.

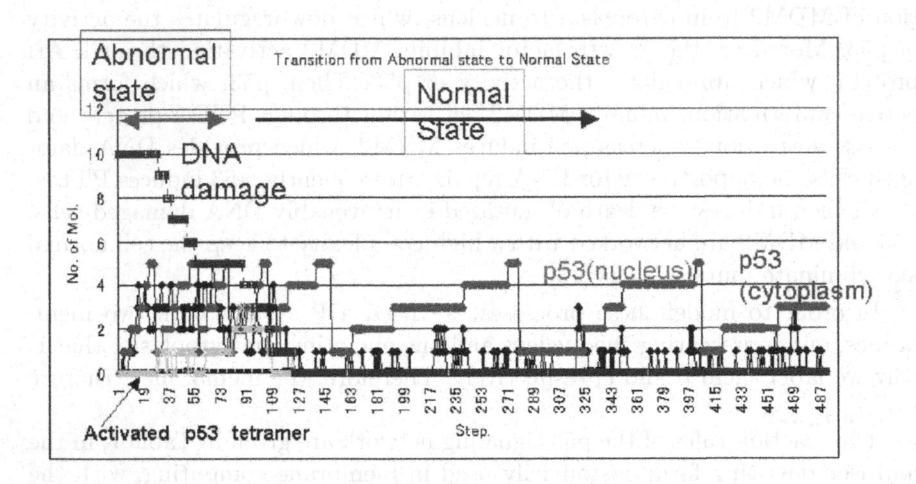

Fig. 2. The results of the simulation.

Once DNA damage increases, p53 is activated and translocated from the cytoplasm to the nucleus. The activated p53 complex induces PTEN protein and caspase activation that inhibit MDM2 activation. We have summarized these interactions in the rule

p53-tetramer DNA-damage → p53-tetramer+ DNA-damage

from R_n, where + indicates that p53-tetramer was activated. It forms the positive feedback loop to accelerate the p53 activation. In brief, p53 induces MDM2, which provides DNA damaged cells the opportunity for DNA repair.

The results of the simulation agree with biological knowledge: when the DNA is damaged (the "abnormal state" in Figure 2), p53-tetramer is activated, it repairs DNA damage, and after the damage is repaired p53-tetramer returns to the normal state and is degraded by MDM2 into single p53. Then these p53s and MDM2 form the autoregulatory negative feedback loop ("normal state" in the figure).

The feedback loop of p53 and MDM2 was predicted by a model based on differential equations [2] and then verified by biological experiments.

Here we have modeled the p53-MDM2 system in terms of multiset processing, a framework completely different from that in [2], but the results are similar.

3 Conservation Analysis of the p53 Signaling Network

Since ARMS is a model of stoichiometric chemistry, we can investigate the signaling network by using conservation analysis, which is a method for analyzing stoichiometric networks (in Appendix B we briefly present this method).

The stoichiometric matrix of the p53 signaling network is as follows:

$$M = \begin{bmatrix} & a & b & c & d & e & f & g & h & i \\ \text{MDM2 in } c & 0 & 0 & 0 & 0 & 0 & 1 & 0 & 0 & 0 \\ \text{p53 in } c & -1 & -1 & 0 & 0 & 0 & 0 & 4 & 0 & 0 \\ \text{p53 in } n & 0 & 1 & -4 & 0 & -1 & 0 & 0 & 0 & 0 \\ \text{p53-tetramer in } n & 0 & 0 & 1 & 0 & 0 & 0 & -1 & -1 & 1 \\ \text{p53-tetramer+ in } n & 0 & 0 & 0 & 0 & 0 & 0 & 0 & 1 & -1 \\ \text{MDM2 in } n & 0 & 0 & 0 & 1 & -1 & -1 & -1 & 0 & 0 \end{bmatrix}.$$

To examine the flux balance of the system, we solved the matrix equation $Nv = 0$ and obtained the following conditions for the rate vector v: $a = f, c = d = g, e + f = 0, 4g - (b + f) = 0, h = i$. If v satisfies these conditions, the system goes to a steady state, $Nv = 0$. However, the condition $e + f = 0$ means that one of e and f must be negative. Thus, since we do not assume reversible reactions, the system cannot reach the stoichiometric steady state. As for the dependencies of reactions,

$$\text{p53-tetramer MDM2} \to (\text{p53 p53 p53 p53, out}) \in R_n$$

depends on the reactions

$$\text{p53} \to (\text{p53, in}) \in R_c,$$
$$\text{MDM2} \to (\text{MDM2, out}) \in R_n.$$

This indicates that the degradation of the p53 tetramer in the nucleus depends on the fluxes of translocation of *p*53 (from the cytoplasm to the nucleus) and MDM2 (from the nucleus to the cytoplasm).

Next we examined the moiety conservation; since $rank(M) = 6$, all substances are independent. This indicated that there are no structural conserved cycles in the network.

However, in the experimental results of the simulation we can see several cycles in subspaces of the system. The conditions of generating cycles were given by the flux balance analysis, the conditions of rate vector ($a = f, c = d = g, e + f = 0, 4g - (b + f) = 0, h = i$). For example, if the reaction rate of the rule

$$\text{p53-tetramer MDM2} \to (\text{p53 p53 p53 p53, out}) \in R_n$$

is four times larger than the sum of the reaction rates of the rules

$$\text{p53} \to \text{vanish} \in R_c, \quad \text{p53} \to (\text{p53, in}) \in R_c,$$

then p53 in the cytoplasm shows oscillation. In turn, if the reaction rate of

$$\text{p53 p53 p53 p53} \to \text{p53-tetramer} \in R_n$$

is 1/4 times larger than the subtraction of the reaction rate of

$$\text{p53} \rightarrow (\text{p53, in}) \in R_c, \quad \text{p53 MDM2} \rightarrow \text{vanish} \in R_n,$$

then the p53 in the nucleus shows oscillation.

4 Final Remarks

Biological processes in the cell are the result of the interactions among a limited number of molecules in a limited space. Moreover, the number of molecules is not a continuous quantity but a discrete quantity.

Multiset processing is a simple technique, easy to be used by biologists, which contrasts with most continuous models and simulation systems. In particular, ARMS is based on stoichiometric chemistry, and if the number of elements in the system is large, then the behavior of the system is similar to the behavior of models based on differential equations or Markov models (the master equation [19]). Thus, when we consider the dynamical properties of ARMS, we can apply mathematical methods developed for analyzing differential equations or for Markovian analysis and we can also use conservation analysis.

A living system is a huge chemical system, so multiset processing in the framework of membrane computing can help biologists analyze the complicated biological processes taking place in living systems.

Appendix A (Stoichiometric Chemistry)

We start with an example: the net reaction for the formation of nitrogen dioxide (NO_2) from nitrous oxide (NO) and oxygen (O_2) is

$$2NO + O_2 \rightarrow 2NO_2, \tag{1}$$

which indicates that two molecules of nitrous oxide combine with one molecule of oxygen to form two molecules of nitrogen dioxide.

Chemists use this type of notation in describing any chemical reaction. In general, any chemical reaction can be described in the form

$$aA + bB \rightarrow cC + dD$$

where the symbols A and B stand for the reactants and the symbols C and D stand for the products of the reaction. The constants a, b, c, and d, which indicate the proportions in which the reactants combine and the products are formed, are called stoichiometric coefficients.

Unfortunately, stoichiometric descriptions of reactions do not tell the whole story. They tell us only the net result of a reaction, without telling us how the reaction takes place. Most chemical reactions actually have a mechanism involving the formation of intermediate products. For example, the net reaction (1) for the formation of nitrogen dioxide actually consists of three subreactions, all of which occur simultaneously. A proposed mechanism for this reaction is

$$2NO \rightarrow N_2O_2 \text{ (}NO\text{ reacts with itself to form } N_2O_2\text{)},$$
$$N_2O_2 \rightarrow 2NO \text{ (}NO\text{ is reformed from some of the } N_2O_2\text{)},$$
$$N_2O_2 + O_2 \rightarrow 2NO_2 \text{ (some of the } N_2O_2 \text{ reacts with } O_2 \text{ to form } NO_2\text{)}.$$

The intermediate product of dinitrogen dioxide (N_2O_2) does not appear in the net reaction (1), but it is involved in the mechanism of the reaction and, hence, a mathematical model of the reaction must take the presence of N_2O_2 into account.

Units of Measurement and Notations. Since molecules are very small, quantities of molecules are measured in units of moles. One mole of molecules is an Avogadro's number of molecules. Avogadro's number is approximately 6.022. Hence, for example, two moles is the same as 12.044 molecules. Concentrations of molecules in a solution are measured in units of molarities (M). One molarity is one mole of solute per liter of solution. For example, a 2M aqueous solution of sodium chloride ($NaCl$) is a solution consisting of two moles of $NaCl$ per each liter of solution. The notation $[A]$ denotes the concentration (in molarities) of a molecule A in solution. Thus, if we write $[NaCl] = 2M$, we mean that we have a solution with a 2M concentration of sodium chloride.

Rates of Reactions. Consider a simple chemical reaction of the form

$$2A + B \rightarrow C.$$

This reaction involves the combination of two molecules of A and one molecule of B to form one molecule of C. Because we are assuming that the reaction is simple (i.e., there are no intermediate steps), the concentration of A decreases at twice the rate that the concentration of B decreases. Thus, we have

$$\frac{d[A]}{dt} = 2\frac{d[B]}{dt}.$$

Also, the concentration of C increases at the same rate that the concentration of B decreases, so

$$\frac{d[C]}{dt} = -d[B]dt.$$

We can summarize the above two equations by writing

$$-\frac{1}{2}\frac{d[A]}{dt} = -\frac{d[B]}{dt} = \frac{d[C]}{dt}.$$

In general, for a simple chemical reaction of the form

$$aA + bB \to cC + dD,$$

we have

$$-\frac{1}{a}\frac{d[A]}{dt} = -\frac{1}{b}\frac{d[B]}{dt} = \frac{1}{c}\frac{d[C]}{dt} = \frac{1}{d}\frac{d[D]}{dt}$$

and we define the rate of the reaction, $v(t)$, as this common value. That is,

$$v = -\frac{1}{a}\frac{d[A]}{dt} = -\frac{1}{b}\frac{d[B]}{dt} = \frac{1}{c}\frac{d[C]}{dt} = \frac{1}{d}\frac{d[D]}{dt}.$$

Since the stoichiometric coefficients a, b, c, and d have no units, v has units of M/time. We stress that the reaction rate v is a function of time (t) because the reaction slows as the reactants are used up during the course of the reaction. The units of measurement of time for a particular reaction depend on the speed of the reaction. For a reaction that proceeds very quickly, it might be appropriate to measure time in seconds or milliseconds, whereas for a very slow reaction, such as the decay of some radioactive compounds, time is measured in years.

The Rate Law. An important concept of chemistry that is crucial to the development of mathematical models of chemical reactions is the rate law. For the homogeneous reaction

$$aA + bB \to cC + dD$$

with reaction rate v, the Rate Law gives the equation

$$v(t) = k[A]^{\alpha}[B]^{\beta} \tag{2}$$

Equation (2) is the framework on which mathematical models of chemical reactions are built. In this equation, the constant of proportionality, k, is called the rate constant of the reaction, and the constants α and β are called the order of the reaction with respect to the reactants A and B, respectively. The constants k, α, and β can be determined only by actual chemical experiments. In general, except possibly by coincidence, a and b are not related to the stoichiometric coefficients α and β.

Also, α and β have no units of measurement and the units of k are determined by the values of α and β in the following way. Recalling that v has units of M/time and $[A]$ and $[B]$ both have units of M, by writing equation (2) in terms of units, we obtain

$$\frac{M}{time} = (\text{units of } k) \cdot (M)^{\alpha} \cdot (M)^{\beta} = (\text{units of } k) \cdot M^{\alpha+\beta}.$$

From this we obtain

$$\text{units of } k = \tfrac{M}{time \cdot M^{1-\alpha-\beta}} = M^{1-\alpha-\beta} \cdot times^{-1}.$$

As an example of the Rate Law, suppose that we have a reaction

$$2A + B \rightarrow C + 2D$$

which is first order with respect to A and B (that is, $\alpha = \beta = 1$), and has rate constant $k = 0.02M^{-1} \cdot min^{-1}$. Suppose also that this reaction takes place in a container of fixed volume. The Rate Law then gives us the rate of this reaction as

$$v = 0.02[A]^{-1}[B]^{-1} \quad (3)$$

Note that the stoichiometric coefficients play no role in the rate equation (3). However, if we wish to rewrite the rate equation in terms of the rate of consumption of one of the reactants (or the rate of formation of one of the products), then the stoichiometric coefficients do play a role. For example, using the fact that

$$v = \frac{1}{2}\frac{d[A]}{dt},$$

we obtain the rate equation

$$v = -\frac{1}{2}\frac{d[A]}{dt} = 0.02[A][B]$$

or equivalently

$$\frac{d[A]}{dt} = -0.04[A][B] \quad (4)$$

for the rate of consumption of reactant A. Equation (4) is an example of a differential equation. In this equation, $[A]$ and $[B]$ are "unknown" functions of time and $d[A]/dt$ is the derivative of $[A]$ with respect to t. Likewise, we obtain the following differential equations for the concentrations of B, C, and D:

$$\frac{d[B]}{dt} = -0.02[A][B],$$

$$\frac{d[C]}{dt} = 0.02[A][B],$$

$$\frac{d[D]}{dt} = 0.04[A][B].$$

With respect to the definition of the Rate Law, for the homogeneous reaction $aA + bB \rightarrow cC + dD$ with reaction rate v, the rate law in ARMS is defined as the same equation, $v(t) = k[A]^a[B]^b$. In a single step of rewriting, the rules are applied randomly with probabilities depending on the rate law. The probability of $r_i \equiv aA + bB \rightarrow cC + dD$ is defined as

$$Prob(r_i) = \frac{k_i[A]^a[B]^b}{R},$$

where k_i is the rate constant and R is a coefficient for normalizing the probabilities ($\Sigma_i Prob(r_i) = 1$).

In the limit, from systems of a large size one obtains a set of deterministic differential equations.

Appendix B (Conservation Analysis)

The analysis of stoichiometric networks was undertaken in the 1960s in the chemical engineering community [1].

Stoichiometry Matrix. The analysis of stoichiometric network starts by considering the network topology. This information is embodied in the stoichiometry matrix, N. The columns of this matrix correspond to the distinct chemical reactions in the network, the rows to the chemical substances, one row per species. Thus the intersection of a row and column in the matrix indicates whether or not a certain species takes part in a particular reaction and, if so, according to the sign of the element, whether it is a reactant or a product, and, by the magnitude, the relative quantity of substance that takes part in that reaction. Stoichiometry is thus concerned with the relative mole amounts of chemical species that react in a particular reaction; it is not concerned with the rate of reaction. The stoichiometry matrix is therefore constant (in biological applications, it is determined by the genetic constitution of the organism).

The stoichiometric matrix represents a compact mathematical representation of a biochemical network. If a given network is composed of m substances involved in n reactants, then the stoichiometric matrix is an $m \times n$ matrix. Only those substances which evolve through the dynamics of the system are included in this count. Any source and sink substances needed to sustain a steady state (non-equilibrium in the thermodynamic sense) are set at a constant level and therefore do not have corresponding entries in the stoichiometry matrix.

System Equation. In order to fully characterize a system one needs to consider the kinetics of the individual reactions as well as the network topology. Modeling the reactions with differential equations, we arrive at a system of equations which involves both the stoichiometry matrix and the reaction vector

$$\frac{dS}{dt} = Nv,$$

where N is the $m \times n$ matrix and v is the n dimensional rate vector, whose ith component gives the rate of reaction i as a function of the concentration of substances.

Structural Properties. There are two key structural properties which can be derived from the stoichiometry matrix. These are *flux balance* constraints and *mass conservation* constraints, both of which are derived from the law of conservation of mass.

The flux balance constraints are valid only when the system is in a steady state, where each molecular concentration reaches a steady state value. As a result, the sum of all fluxes into a species pool must equal the sum of all fluxes out of the pool:

$$\Sigma \text{ Fluxes in} = \Sigma \text{ Fluxes out.}$$

For each species, this is equivalent to the statement that $N_i v = 0$, where N_i is the ith row of N. We arrive then at the matrix equation $Nv = 0$.

This constraint is imposed on the distributions of steady state fluxes at all nodes in the network, and is in direct analogy with Kirchhoff's first law of current flow. Thus to characterize the steady state flow through this system, we need only measure two of the fluxes, as the third is then determined. In general, one can divide the set of the fluxes into two subsets, the independent fluxes, and the dependent fluxes. The independent flux can take any values, while the values of the dependent fluxes are fixed by the independent fluxes and the flux constraints. This dependence is the essence of flux balance analysis.

Moiety Conservation Analysis. Molecular subgroups which are conserved during the evolution of a network are termed *conserved moieties* [17]. The total amount of a particular moiety in a network is time invariant and is determined solely by the initial conditions imposed on the system.

Moiety conservation analysis plays a dual role to the analysis of flux balance. Whereas flux balance is concerned with the conservation of mass as it flows into and out of species nodes, moiety conservation analysis is concerned with the conservation of mass as it moves around closed loops in the network. Again we see a direct analogy to electrical engineering since moiety conservation analysis is related to Kirchhoff's second law of potentials. Conserved moieties in the network reveal themselves as linear dependencies in the rows of the stoichiometry matrix. The question then arises: how, given a stoichiometry matrix, can we identify linear dependencies and determine the corresponding conserved moieties? Whenever the network exhibits conserved moieties, there will be dependencies among the rows of N, and the rank of N ($rank(N)$) will be less than m, the number of rows of N. The rows of N can be rearranged so that the first $rank(N)$ rows are linearly independent. The metabolites which correspond to these rows can then be defined as the *independent species* (S_i). The remainining $m - rank(N)$ are the *dependent species* (S_d). Note that if there are no structural conserved cycles in the network, then m is equal to $rank(N)$.

Acknowledgements. This research was supported by Grants-in-Aid for Scientific Research No. 14085205 and RCAST (at Doshisha University) from the Ministry of Education, Science and Culture of Japan.

References

1. R. Aris: Prolegomena to the Rational Analysis of Systems of Chemical Reactions. *Arch. Rational Mech. Anal.*, 19, 1 (1965), 81–99.
2. R.V. Bar-or, R. Maya, L.A. Segel, U. Alon, A.J. Levine, M. Oren: Generation of Oscillations by the p53-MDM2 Feedback Loop: A Theoretical and Experimental Study. *PNAS*, 97, 21 (2000), 11250–11255.

3. B.A. Barshop, R.F. Wrenn, C. Frieden: Analysis of Numerical Methods for Computer Simulation of Kinetic Processes: Development of KINSIM – A Flexible, Portable System. *Anal. Biochem.*, 130 (1983), 134–145.
4. P. Dittrich, J. Ziegler, W. Banzhaf: Artificial Chemistries – A Review. *Artificial Life*, 7, 3 (2001), 225–275.
5. M. Ehlde, G. Zacchi: MIST: A User-Friendly Metabolic Simulator. *Comput. Applic. Biosci.*, 11 (1995), 201–207.
6. C.W. Gardiner: *Handbook of Stochastic Methods*, 2nd ed., Springer, Berlin, 1985.
7. M.S. Herbert, B. Ingalls: *Conservation Analysis in Biochemical Networks: Computational Issues for Software Writers.* Keck Graduate Institute, 2003.
8. M. Hollstein, K. Rice, M.S. Greenblatt, T. Soussi, R. Fuchs, T. Sorlie, E. Hovig, B. Smith-Sorensen, R. Montesano, C.C. Harris: Database of p53 Gene Somatic Mutations in Human Tumors and Cell Lines. *Nucleic Acids Research*, 17 (1994), 3551–3555.
9. T.R. Hupp: Regulation of p53 Protein Function Through Alterations in Protein-Folding Pathways. *Cell Mollecular Life Science*, 55, 1 (1999), 88–95.
10. N.G. van Kampen: *Stochastic Processes in Physics and Chemistry.* North-Holland, 1981.
11. D. Lane: Awakening Angels. *Nature*, 394 (1998), 616–617.
12. A.J. Levine: p53, the Cellular Gatekeeper for Growth and Division. *Cell*, 88, 3 (1997), 323–331.
13. L.D. Mayo, D.B. Donner: The PTEN, MDM2, p53 Tumor Suppressor-Oncoprotein Network. *Trends in Biochemical Science*, 27, 9 (2002), 462–467.
14. P. Mendes: Biochemistry by Numbers: Simulation of Biochemical Pathways with Gepasi 3. *Trends in Biochemical Science*, 22, 9 (1997), 361–363.
15. Gh. Păun: Computing with Membranes. *Journal of Computer and System Sciences*, 61, 1 (2000), 108–143, and *Turku Center for Computer Science, TUCS Report* No 208, 1998 (www.tucs.fi).
16. C. Prives, P.A. Hall: The p53 Pathway. *Journal of Patholgy*, 187, 1 (1999), 112–126.
17. J.G. Reich, E.F. Selkov: *Energy Metabolism the Cell.* Academic Press, 1981.
18. Y. Suzuki, P. Davis, H. Tanaka: Emergence of Auto-Catalytic Structure in Stochastic Self-Reinforcing Reaction Networks. *J. of Artificial Life and Robotics*, 7 (2004).
19. Y. Suzuki, Y. Fujiwara, H. Tanaka, J. Takabayashi: Artificial Life Applications of a Class of P Systems: Abstract Rewriting Systems on Multisets. In *Multiset Processing. Mathematical, Computer Science, and Molecular Computing Points of View* (C.S. Calude, Gh. Păun, G. Rozenberg, A. Salomaa, eds.), LNCS 2235, Springer, Berlin, 2001, 299–346.
20. Y. Suzuki, J. Takabayashi, H. Tanaka: Investigation of Tritrophic System in Ecological Systems by Using an Artificial Chemistry. *J. of Artificial Life and Robotics*, 6 (2002), 129–132.
21. Y. Suzuki, H. Tanaka: A Symbolic Chemical System Based on an Abstract Rewriting System and Its Behavior Pattern. *J. of Artificial Life and Robotics*, 1 (1997), 211–219.
22. Y. Suzuki, H. Tanaka: Chemical Evolution Among Artificial Proto-Cells. *Artificial Life VII*, MIT Press, 2000, 54–64.
23. Y. Suzuki, S. Tsumoto, H. Tanaka: Analysis of Cycles in Symbolic Chemical Systems Based on Abstract Rewriting Systems on Multisets. *Artificial Life V*, MIT Press, 1996, 522–528.

Chapter 8
Static Sorting P Systems

Artiom Alhazov[1,2], Dragoş Sburlan[3,4]

[1] Research Group on Mathematical Linguistics
Rovira i Virgili University, Tarragona, Spain
[2] Institute of Mathematics and Computer Science
Academy of Sciences of Moldova, Chişinău, Moldova
artiom@math.md
[3] Department of Informatics and Numerical Methods
Ovidius University of Constantza, Romania
dsburlan@univ-ovidius.ro
[4] Department of Computer Science and Artificial Intelligence
University of Sevilla, Spain

Summary. This chapter deals with the application of P systems to sorting problems. Traditional studies of sorting assume constant time for comparing two numbers and compute the time complexity with respect to the number of components of a vector to be sorted. Here, we assume the number of components to be a fixed number k, and study various algorithms based on different models of P systems and their time complexities with respect to the maximal number or to the sum of the numbers. Massively parallel computations that can be realized within the framework of P systems may lead to major improvements in solving the classical integer sorting problems. Despite this important characteristic, we will see that, depending on the model used, the massive parallelism feature cannot be always used, and so some results will have complexities "comparable" with the classical integer sorting algorithms. Still, computing a word (ordered) from a multiset (unordered) can be a goal not only for computer science, but also, e.g., for biosynthesis (separating mixed objects according to some characteristics). Here, we will move from ranking algorithms that, starting with numbers represented as multisets, produce symbols in an order, to effective sorting algorithms.

1 Introduction

The sorting problem is an important one in computer science and many algorithms, both sequential and parallel, were developed for solving it. However, the time complexity remains at least $O(n\ log(n))$ for the sequential case and $O(log^2 n)$ for the parallel case.

Studying sorting within the P systems framework is a challenging task not only because it can produce better results (in some respects) than the classical sequential case, but also because we can compute a string (ordered) from a multiset (unordered) of objects. Moreover, one can remark that in the case of cooperative rules (in P systems with symbol objects and rewriting-like rules) the order of symbols on the left/right hand side of a production does not count. So, we deal with two "types of disorder" and can still compute an "order."

One of the first approaches to sorting with P systems was taken in [3] by considering a bead sort algorithm. There, the sorting procedure was constructed on a tissue P system with symport/antiport rules such that the objects were considered beads and the membranes were considered rods. The idea of the algorithm was that beads start to slide in the membrane structure to their appropriate places. The time complexity for this case was linear, which constitutes an improvement over the classical sequential sorting algorithm. However, the tradeoff for this was that it required a number of membranes proportional to $m \times k$, where m is the maximal number from the vector that we want to sort and k is the dimension of the vector.

Another study on this topic was done in [5], where the P systems with inhibitors/promoters and symport/antiport rules were used to develop comparators and then to organize them in a sorting network. As in the previous case, the input data was placed in different membranes and the computation started operating on elements already dissociated. The result was not obtained in a halting configuration but in a stable one, meaning that there were rules still applicable, but their application did not change the string/object contents of the membranes or the membrane structure. The time complexity was linear with respect to the number of components, and the number of membranes used in the computation was proportional to the number of components.

In [2], other methods and algorithms for the sorting problem are proposed by considering several variants of P systems. The feature shared by many algorithms from [2] is that the input components are placed in an initial input membrane and the computation dissociates this input according to the relation order among the multiplicities of components. In this way, we interpret the sorting as the order of elimination of the objects. The idea behind many of these algorithms (also presented below) is to consume objects from all components at once and, when one component is exhausted, to trigger a signal to find the next component to be eliminated. In other algorithms, we developed a comparator, using very weak "ingredients," which can be used practically in any sorting network design. For most of the algorithms the time complexity will be also linear, while for others it will depend on the largest multiplicity.

Finally, a sorting algorithm based on the existing ranking algorithm was proposed in [10].

The chapter is structured into three main parts corresponding to the type of problem that is solved: the strong sorting, the weak sorting, and the ranking. All these concepts will be defined below. In addition, the P system models considered in the paper are introduced in the following section, where a brief introduction to the framework is made. The last section is dedicated to conclusions and open problems.

2 Preliminaries

2.1 P Systems

The reader is supposed to be familiar with basic elements of membrane computing, e.g., from [9], so that here we specify only some notations and terminology used below.

A P system (of degree $m \geq 1$) with symbol objects and rewriting evolution rules is a construct

$$\Pi = (O, C, \mu, w_1, \ldots, w_m, (R_1, \rho_1), \ldots, (R_m, \rho_m), I, J),$$

where:

- O is the alphabet of Π; its elements are called *objects*;
- $C \subseteq O$ is a set of catalysts;
- μ is a membrane structure consisting of m membranes usually labeled 1, $2, \ldots, m$;
- w_i, $1 \leq i \leq m$, specify the multisets of objects present in the corresponding regions $1, 2, \ldots, m$, at the beginning of a computation;
- R_i, $1 \leq i \leq m$, are finite sets of evolution rules over O associated with regions $1, 2, \ldots, m$ of μ; the rules are called (1) *non-cooperative* if they are of the form $a \to v$ where a is an object from $O - C$ and v is a string over $\{a_{here}, a_{out}, a_{in} \mid a \in O - C\}$, and (2) *catalytic* if they are of the form $ca \to cv$, where a is an object from $O - C$ and v is a string over $\{a_{here}, a_{out}, a_{in} \mid a \in O - C, 1 \leq j \leq m\}$ and $c \in C$;
- I and J are the sets of numbers between 0 and m specifying the input regions and output regions of Π, respectively (in case of 0 the environment is used for the output).

Many other features were added to this basic definition of P systems; we refer you to [9] for details. In this work we will consider the following variants of P systems:

- P systems with weak/strong priorities. In each case, a partial order relation ρ_i is given over elements of R_i, $1 \leq i \leq m$. Weak priorities means that, in one step of computation the rules are applied according to the priority relation in a nondeterministic, maximally parallel manner such that a rule can be applied only if objects remain unused by rules with a higher priority. Strong priorities means that in a computation step, only the rules with the highest priority among those which can be applied are applied, in a maximally parallel manner (that is, even if unused objects remain, a lower priority rule cannot be used).

- P systems with target indications where the right hand string of an evolution rule is defined over $\{a_{here}, a_{out}, a_{in_j} \mid a \in O - C, 1 \leq j \leq m\}$. This means that if a symbol a_{in_j} is present, then the object a will be moved to the inner membrane labeled j.
- P systems with promoters/inhibitors. In the case of promoters, the rules (reactions) are possible only in the presence of certain objects which can evolve at the same time as objects whose evolutions they support. Inhibitors forbid certain rules (reactions). We denote these features by considering rules of the forms $u \to v|_a$ and $u \to v|_{\neg a}$, respectively.
- P systems with finitely stable catalysts for which the evolution rules containing catalysts are of the form $c_i a \to c_j v$, where $a \in O - C$, $c \in C$ and $i, j \in M_c$, with M_c being a given set of integers associated with c, and v is a string over $\{a_{here}, a_{out}, a_{in} \mid a \in O - C\}$. The catalysts are neither created nor destroyed during the computation, but they can change their "state" as specified by M_c.
- P systems with mobile catalysts. This model is an extension of the original model, permitting catalysts to move between regions. In this case a rule involving catalysts is written in the form $ca \to c_\alpha v$, where $a \in O - C$, $c \in C$, $\alpha \in \{here, out, in\}$, and v is a string over $\{a_{here}, a_{out}, a_{in} \mid a \in O - C\}$.
- P systems with symport/antiport rules. In this case the rules provide synchronized movement of objects. In this way, particular sets of objects may pass together through a membrane in the same or opposite direction (these objects need each other for the transport). Formally, in the case of symport this means that the rules are of the form (ab, in) or (ab, out), indicating that the objects a and b pass together through a membrane. In the antiport case, the rules are of type $(a, out; b, in)$, indicating that while object a goes out, object b enters in the region.
- P systems with membrane dissolution. Rules may act directly on the membrane structure by dissolving membranes. Formally we represent this by rules of type $u \to v\delta$, where δ means that, after finishing the actual step of computation, the current membrane will be dissolved and all the inner objects will pass to the upper region.
- P systems with membrane division. In this case the membranes are allowed to divide and to participate as active components in the process of computation.
- Evolution-communication P systems. In this variant, evolution rules that are simple, i.e., do not move the objects, and communication rules that are pure, i.e., do not change the objects, coexist. See [1, 4] for exact definitions, results, and details. Generally, symport rules of weight 1, antiport rules of weight 1, and non-cooperative simple evolution rules provide enough computational power.
- P systems with string objects. Unlike all other variants considered here, the regions contain strings rather than symbols. In this model, the string objects evolve by rewriting, and the rewriting is sequential at the level of symbols in the strings. It is trivial to perform any computation using

cooperative rules. In our solutions we will employ only non-cooperative rules with priorities to solve the sorting problem.

A P system with the features mentioned starts to evolve from the *initial configuration* by performing all operations in a parallel way for all applicable rules, for all regions at the same time and according to a universal clock. A computation is *successful* if and only if it halts, meaning that no rule is applicable to the objects present in the final configuration. We will assume that the system ejects objects in the environment (external output) and, to obtain strings, we concatenate these symbols according to the order in which they come to the environment. When more objects arrive at the same time in the environment, we consider the strings formed by all permutations of these objects. The language generated by a P system Π is the language $L(\Pi)$, containing all the strings produced by all successful computations of Π according to the output mode.

Another way to define the result of the halting computation is to consider the number of objects in the given output region. In this case, a P system Π generates a set $N(\Pi)$ of numbers, corresponding to all its halting computations. Many types of P systems, using various combinations of ingredients as described before, are known to be universal – they generate the same languages/sets of numbers as Turing machines. We refer the reader to [9], [8], [6], and [7] for details.

These features added to the original model are very useful in designing different ways of performing comparisons, and so the idea of solving the sorting problem in this framework is completely natural.

We also remark that proving universality by using matrix or programmed grammars, register machines, or other constructions (as they are known in the literature) usually involves nondeterministic approaches as well as the use of a "trap symbol" which guarantees that the system works forever in the "wrong" cases. In contrast, when solving a practical problem we have to find an algorithm working in a deterministic or confluent way and this is because we need to receive the answer of the problem in a specified time. In addition, the system has to stop or reach an equilibrium state after finishing the computation so that the output can be read. These issues, combined with the fact that in a multiset we do not have an ordering of the elements, give a hint of the difficulty of addressing the sorting problems in terms of P systems.

2.2 Sorting Networks

One parallel model for studying integer sorting problems is based on the comparison network, where more comparison operations can be executed at the same time. This feature offers the possibility to construct such networks, which sort k numbers in sublinear time.

A comparison network is built of wires and comparators. A comparator is a device that has two inputs X and Y, and computes the function

$$(X,Y) \longrightarrow \Big(\min(X,Y), \max(X,Y)\Big).$$

The sorting network consists of input wires, output wires, and comparators. In a comparison network, wires are responsible for passing information from one comparator to another. Essentially, such a network contains only comparators, linked together with wires, because we can consider that input wires and output wires are also nodes (like comparators), but they do not compute anything, and just store a value.

Formally, a comparison network is a directed acyclic graph where the nodes are comparators, input nodes, or output nodes, and the directed arcs are wires.

A sorting network is a comparison network that produces as output a monotone sequence for any input sequence. The running time of a comparison network is the time elapsed from the initialization of the input nodes to the time when the values reach the output nodes. A typical sorting network is presented in Figure 1.

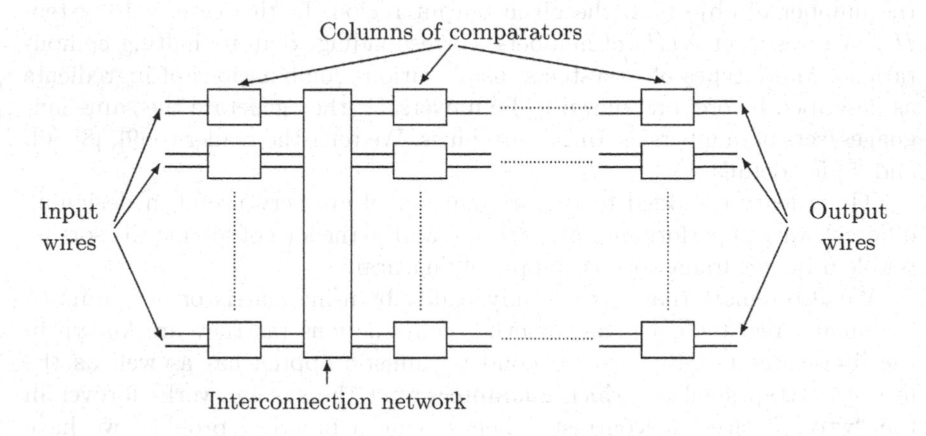

Fig. 1. A general sorting network scheme.

Particular examples are the sorting network Bubblesort and the sorting network Odd-even (see Figure 2).

The sorting network Bubblesort consists of a first diagonal of $n-1$ comparators that move the greatest element to the last position. The remaining $n-1$ elements are sorted recursively by applying the same procedure. Bubblesort consists of $n(n-1)/2$ comparators, arranged in $2n-3$ stages.

The sorting network Odd-even transposition sort for n input elements consists of n comparator stages. Each stage consists of comparators $[i : i+1]$ where i is odd and even in alternating order. The number of comparators is $n(n-1)/2$.

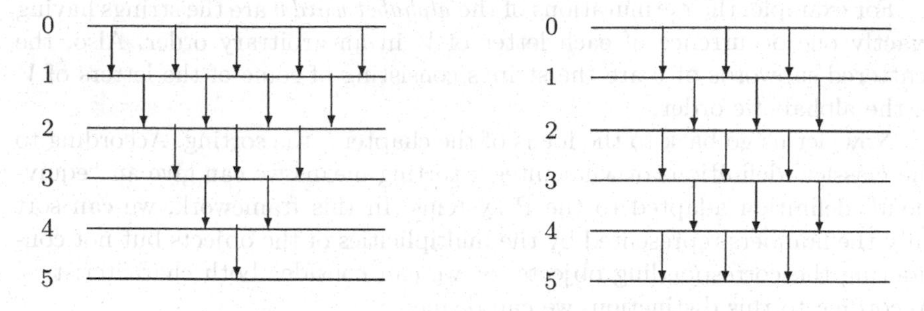

Fig. 2. Bubblesort and Odd-even transposition sort.

3 Sorting Definitions and Notations

The alphabets we consider here are composed of natural numbers. The alphabet symbols are denoted by underlined numbers, so that in this way we can distinguish the symbols and also have the implicit order associated with natural numbers.

Let $V = \{\underline{i} \mid 1 \leq i \leq k\}$ be an alphabet. A word over V is denoted by

$$w = \prod_{j=1}^{m} a_j = a_1 a_2 \ldots a_m,$$

$m \in \mathbb{N}$, where $a_j \in V$ for each $1 \leq j \leq m$. Here, the product symbol $\prod$ represents concatenation.

Example 1. $w = \underline{23}\,\underline{9}\,\underline{157}$.

Let $ord : V \to \{1, \ldots, k\}$ be a bijective function such that $i = ord(\underline{i})$, $1 \leq i \leq k$. Then $i_j = ord(a_j)$ is an ordinal number of the jth letter of w and $a_j = \underline{i_j}$.

Let $v = \prod_{j=1}^{k} \underline{j}$ be the "alphabet word," made by concatenating the elements of the alphabet V in the natural order.

Since in multiset processing we represent multisets by strings, we will need a few formal language definitions and notations.

Let $w = \prod_{j=1}^{m} a_j$ be a string. We denote by

$$Perm(w) = \{\prod_{j=1}^{m} a_{i_j} \mid 1 \leq i_j \leq m, 1 \leq j \leq m, \text{ with } i_j \neq i_l, 1 \leq j, l \leq m\}$$

the set of *permutations* of string w, i.e., the set of all strings that can be obtained from w by changing the order of symbols. We denote by

$$SSub(w) = \{\prod_{j=1}^{l} a_{i_j} \mid 1 \leq i_{j-1} < i_j \leq m, 2 \leq j \leq l, 0 \leq l \leq m\}$$

the set of *scattered subwords* of w, i.e., the set of all strings that can be obtained from w by deleting some (0 or more, possibly all) of its symbols, and concatenating the remaining ones, preserving the order.

For example, the permutations of the *alphabet word* v are the strings having exactly one occurrence of each letter of V in an arbitrary order. Also, the scattered subwords of v are the strings consisting of some of the letters of V in the alphabetic order.

Now, let us go back to the focus of the chapter – the sorting. According to the classical definitions of what integer sorting means we can give an "equivalent" definition adapted to the P systems. In this framework, we can sort only the numbers represented by the multiplicities of the objects but not considering the corresponding objects, or we can consider both characteristics. According to this distinction, we can define:

Definition 1. *Let* $v = \prod_{j=1}^{k} \underline{j}$ *be the alphabet word. The word*

$$w = \prod_{j=1}^{k} a_j \in Perm(v), \ k = card(V) \in \mathbb{N}^+,$$

where $a_j \in V$ *such that* $M(a_j) \leq M(a_{j+1})$, *for each* $1 \leq j \leq k-1$, *is called the* ranking string *of the multiset* M.

Definition 2. *The word* $w = \prod_{j=1}^{k} \underline{j}^{M(a_j)}$ *is called the* weak sorting string *of the multiset* M *if* $\prod_{j=1}^{k} a_j$ *is the ranking string of* M. *Also,* $M' : V \to \mathbb{N}$ *defined as* $M'(\underline{j}) = M(a_j)$ *is the* weak sorting multiset *of* M.

Remark 1. This definition stands for the case when we are interested in sorting the multiplicities of the objects and not in having to look at the corresponding objects. In other words, we sort only "properties" and not "objects and properties." Practically, the symbols from the initial multiset are considered to be in a complete relation order, and, after performing the computation we obtain as the result these objects sorted according to the relation order and having multiplicities sorted.

Definition 3. *The word*

$$w = \prod_{j=1}^{k} a_j^{M(a_j)}$$

is called the strong sorting string *of* M *if* $\prod_{j=1}^{k} a_j$ *is the ranking string of* M.

Remark 2. In this case, we are interested in having as the output of computation the objects with the same associated multiplicities, present in a string in increasing order of their multiplicities.

Example 2. For the alphabet $V = \{\underline{1}, \underline{2}, \underline{3}\}$ and the multiset $M = \{(\underline{1}, 20), (\underline{2}, 10), (\underline{3}, 30)\}$, we have:

- ranking string: $\underline{2}\ \underline{1}\ \underline{3}$
- weak sorting string: $\underline{1}^{10}\ \underline{2}^{20}\ \underline{3}^{30}$
- strong sorting string: $\underline{2}^{10}\ \underline{1}^{20}\ \underline{3}^{30}$

We will typically consider the starting configuration of the sorting P system depending only on the number $k = Card(V)$ of components, taking as the input the multiset

$$\{(\underline{j}, n_j) \mid 1 \leq j \leq k\}$$

over $V \subset O$ placed in a specific region, where O is the system's alphabet.

4 Strong Sorting

Strong sorting is an algorithm which processes a multiset M and gives as a result a word $w = \prod_{j=1}^{m} a_j^{M(a_j)}$ if $\prod_{j=1}^{k} a_j$ is the ranking string of M. A weaker definition could be given: as the result of processing M, a string w is produced such that for each $1 \leq i \leq k$ we have $|w|_{a_i} = M(a_i)$ and for any representation $w = \alpha a_i \beta a_j \gamma$ we have $M(a_i) \leq M(a_j)$. In the case of equality of some numbers either a decision is made whether "=" is "≤" or "≥" (see the below technique of unequalizer) or the weak definition is assumed.

The strong sorting is the most difficult static sorting problem. It is usually implemented by obtaining the ranking, represented one way or the other, and then using it to control the output process.

4.1 Using Promoters

We present an algorithm for the integer sorting problem using promoters and cooperating rules. One can note that despite the fact that the "degree" of cooperation is high and we use other powerful ingredients (promoters), the problem is not trivial because we want to extract order from disorder.

Moreover, the procedure must, in a certain sense (as we will see later, there are some algorithms for which the nondeterminism exists but will not affect the computation) be deterministic. First we will give the general algorithm which will work for a number $k > 1$ of integers.

Algorithm 1. *Let $v = \prod_{i=1}^{k} \underline{i}$. Consider the P system*

$$\Pi = (O, \emptyset, [_1\ ]_1, p_k, R_1, \{1\}, \{0\}) \text{ with input } \prod_{j=1}^{k} \underline{j}^{n_j},$$

where:

$$O = \{\underline{i}, X_i \mid 1 \leq i \leq k\} \cup \{y_i^{(j)} \mid 1 \leq i, j \leq k\} \cup \{p_i \mid 0 \leq i \leq k\}$$
$$\cup \{\langle w \rangle \mid w \in SSub(v) - \{v\}\},$$
$$R = \{w \to \prod_{j=1}^{|w|} x_j \langle \prod_{j=1, j \notin alph(w)}^{k} \underline{j} \rangle |_{p_{|w|}} \mid w \in SSub(v) - \{\varepsilon\}\}$$
$$\cup \{p_i \to p_{i-1} \mid 1 \leq i \leq k\} \cup \{x_j \to \underline{j}_{out} \mid 1 \leq j \leq k\}$$
$$\cup \{y_j^l \to y_j^{l-1} \mid 2 \leq l \leq k\}$$
$$\cup \{\prod_{j=1}^{k-1} \langle \prod_{l=1}^{k} \underline{i_j} \rangle \to \prod_{j=1}^{k} y_i^{(j)} \mid \prod_{j=1}^{k} \underline{i_j} \in Perm(v)\}.$$

For a better understanding, we present an example illustrating the behavior of the system. Let us consider the case of sorting three numbers and also, for simplicity, let us rename variables present in the general algorithm as follows: $\underline{1} = a$, $X_1 = a'$, $y_1^{(j)} = A_j$, $\langle \underline{1} \rangle = \langle a \rangle$, $p_i = p_i$, etc. A similar renaming will be used in all examples below.

Example 3. Initial data: $a^{n_1} b^{n_2} c^{n_3} p_3$
The rules are:

$$abc \to a'b'c'\langle\varepsilon\rangle|_{p_3},$$

$$\begin{array}{lll} p_3 \to p_2, & p_2 \to p_1, & p_1 \to p_0, \\ ab \to a'b'\langle c\rangle|_{p_2}, & ac \to a'c'\langle b\rangle|_{p_2}, & bc \to b'c'\langle a\rangle|_{p_2}, \\ a \to a'\langle bc\rangle|_{p_1}, & b \to b'\langle ac\rangle|_{p_1}, & c \to c'\langle ab\rangle|_{p_1}, \\ \langle a\rangle\langle ab\rangle \to A_1B_2C_3, & \langle a\rangle\langle ac\rangle \to A_1C_2B_3, & \langle b\rangle\langle bc\rangle \to B_1C_2A_3, \\ \langle b\rangle\langle ab\rangle \to B_1A_2C_3, & \langle c\rangle\langle ac\rangle \to C_1A_2B_3, & \langle c\rangle\langle bc\rangle \to C_1B_2A_3, \\ A_3 \to A_2, A_2 \to A_1, & B_3 \to B_2, B_2 \to B_1, & C_3 \to C_2, C_2 \to C_1, \\ a' \to a_{out}|_{A_1}, & b' \to b_{out}|_{B_1}, & c' \to c_{out}|_{C_1}. \end{array}$$

Description of the model. The system starts with objects $a^{n_1}, b^{n_2}, c^{n_3}$, p_3; let us suppose, without loss of generality, that we have $n_1 \geq n_2 \geq n_3$. First, in the presence of the promoter p_3 the rule $abc \to a'b'c'\langle\varepsilon\rangle|_{p_3}$ will be used. We apply this rule (in one step) a number of times equal to the smallest multiplicity of the objects (n_3) and add to the membrane the objects a'^{n_3}, b'^{n_3}, c'^{n_3}, $\langle\varepsilon\rangle^{n_3}$. In the same step, the promoter p_3 will be used by the rule $p_3 \to p_2$. In the second step of computation, the promoter p_2, being present in the membrane, will let the rule $ab \to a'b'\langle c\rangle|_{p_2}$ be applied $n_2 - n_3$ times. As above, at the same time, the promoter p_2 will change to p_1 and so the configuration of the system for the region will contain the objects

$$a'^{(n_3+(n_2-n_3))},\ b'^{(n_3+(n_2-n_3))},\ c'^{(n_3)},\ \langle c\rangle^{(n_2-n_3)},\ \langle\varepsilon\rangle^{n_3},\ a^{(n_1-n_2)},\ p_1.$$

In the third step, the remaining objects a will be transformed into objects a' and $\langle bc\rangle$. Since the number of objects a which remained after the second step was $a^{(n_1-n_2)}$, we will add to the membrane $n_1 - n_2$ copies of a' and the same number of copies of $\langle bc\rangle$. The objects present in the membrane will be:

$$a'^{n1},\ b'^{(n_2)},\ c'^{(n_3)},\ \langle\varepsilon\rangle^{n_3},\ \langle c\rangle^{(n_2-n_3)},\ \langle bc\rangle^{(n_1-n_2)},\ p_0.$$

Now, there are only the objects $\langle c\rangle$ and $\langle bc\rangle$ that can react, and so the symbols $C_1B_2A_3$ will be produced. These symbols, with a corresponding "delay" (implemented through some renaming of the symbols), will be used as promoters to throw out the symbols in the correct order.

Of course, the model, being symmetrical in what concerns the rules, has a behavior that does not depend on the order between the initial multiplicities, and will produce as output the objects in the correct order.

Theorem 1. *The time complexity for the strong sorting algorithm with P systems with promoters is $2k+1$ where k is the number of elements to be sorted, and it is constant with respect to the values of the elements. However, the number $Card(O)$ of different objects used is exponential with respect to k.*

4.2 Unequalizer Technique

In the previous section, the sorting was successful only if all the elements were different. This technique allows us to solve the general problem by reducing it to the sorting of different numbers at the price of adding one step in the beginning and changing the last step.

Algorithm 2. *(considered for one membrane for simplicity) Let* $v = \prod_{i=1}^{k} \underline{i}$. *Given* $\Pi = (O, \emptyset, [_1\ ]_1, \beta, R, \{1\}, \{0\})$ *with input* $\prod_{j=1}^{k} \underline{j}^{n_j}$*, consider* $\Pi' = (O', \mu, p, R', \{1\}, \{0\})$ *with input* $\prod_{j=1}^{k} z_j^{n_j}$*, where* $O' = O \cup \{p\} \cup \{z_j \mid 1 \leq j \leq k\}$ *and* R' *is obtained from* R *by excluding all output rules and adding*

$$\{p \to \beta \prod_{i=1}^{k} \underline{i}^{i-1}\} \cup \{z_i \to \underline{i}^k\} \cup \{\alpha X^k \to \alpha \underline{i}_{out}|_C \mid (\alpha X \to \alpha z_{i\,out}\,|_C) \in R\}.$$

Example 4. Initial data: $a_0^{n_1} b_0^{n_2} c_0^{n_3} p_4$
Rules at the beginning of the computation: add

$$p_4 \to p_3 bcc,\ a_0 \to aaa,\ b_0 \to bbb,\ c_0 \to ccc.$$

Rules for the output: change $X \to Y_{out}$ to $XXX \to Y_{out}$; in this case we have

$$a'a'a' \to a_{0\,out}|_{A_1},\ b'b'b' \to b_{0\,out}|_{B_1},\ c'c'c' \to c_{0\,out}|_{C_1}.$$

Given numbers $(n_i)_{1<i\leq k}$, the original system is used to sort numbers

$$(kn_i + i - 1)_{1 \leq i \leq j},$$

and the result is then obtained by dividing the numbers by k. Constructions with the same idea apply to other algorithms as well.

4.3 Using Inhibitors

Using inhibitors to solve the sorting problem is also a challenging subject that can be discussed. By using this feature, we can specify when the execution of some rules should not happen, and so we can drive the production of objects in the right order. This feature of forbidding the execution of certain rules seems to decrease the "cooperativeness" of the rules compared to the algorithm using promoters. The inhibitors are important only after we have done the computation and have the solution of the corresponding ranking problem, but have not actually sorted. Then, by using the inhibitors we can drive the elimination process in the correct way. The algorithm proposed for the general case is as follows.

Algorithm 3. *Let* $v = \prod_{i=1}^{k} \underline{i}$*. Consider the system*

$$\Pi = (O, \emptyset, [_1\ ]_1, \prod_{i=1}^{k-1} \prod_{j=2}^{k} \langle i, 2, j \rangle \prod_{j=1}^{k} (y_j^j I_j), R_1, \{1\}, \{0\})$$

with input $\prod_{j=1}^{k} \underline{j}^{n_j}$*, where:*

$$\begin{aligned} O &= \{\underline{i}, I_i, y_i \mid 1 \leq i \leq k\} \cup \{x_i^{(j)} \mid 1 \leq i, j \leq k\} \\ &\cup \{\langle i, l, j \rangle \mid 1 \leq i < j \leq k, 0 \leq l \leq 2\}, \\ R_1 &= \{\underline{i} \to \prod_{j=1}^{k} x_i^{(j)} \mid 1 \leq i \leq k\} \\ &\cup \{x_i^{(j)} x_j^{(i)} \to \varepsilon,\ x_i^j \langle i, 0, j \rangle \to y_i,\ x_j^i \langle i, 0, j \rangle \to y_i \mid 1 \leq i < j \leq k\} \\ &\cup \{\langle i, l, j \rangle \to \langle i, l-1, j \rangle \mid 1 \leq i < j \leq k, l \in \{1, 2\}\} \\ &\cup \{I_j y_j \to I_j,\ x_i^{(i)} \to \underline{i}_{out}|_{\neg y_j} \mid 1 \leq j \leq k\}. \end{aligned}$$

As in the previous cases, for an easier understanding we will consider the 3-integer sorting problem.

Example 5. Initial data: $a^{n_1}b^{n_2}c^{n_3}\langle ab\rangle_2\langle ac\rangle_2\langle bc\rangle_2\overline{A}AAA\overline{B}BBB\overline{C}CCC$
The rules are as follows:

$$\begin{array}{lll} a \to a_b a_c a_0, & b \to b_a b_c b_0, & c \to c_a c_b c_0, \\ a_b b_a \to \varepsilon, & a_c c_a \to \varepsilon, & b_c c_b \to \varepsilon, \\ \langle ab\rangle_2 \to \langle ab\rangle_1, & \langle ac\rangle_2 \to \langle ac\rangle_1, & \langle bc\rangle_2 \to \langle bc\rangle_1, \\ \langle ab\rangle_1 \to \langle ab\rangle, & \langle ac\rangle_1 \to \langle ac\rangle, & \langle bc\rangle_1 \to \langle bc\rangle, \\ a_b\langle ab\rangle \to A, & a_c\langle ac\rangle \to A, & b_a\langle ab\rangle \to B, \\ b_c\langle bc\rangle \to B, & c_b\langle ac\rangle \to C, & c_b\langle bc\rangle \to C, \\ \overline{A}A \to \overline{A}, & \overline{B}B \to \overline{B}, & \overline{C}C \to \overline{C}, \\ a_0 \to a_{out}|_{\neg A}, & b_0 \to b_{out}|_{\neg B}, & c_0 \to c_{out}|_{\neg C}. \end{array}$$

It is presented in the example how the sorting algorithm works in the case of three numbers represented as multiplicities of the three objects a, b, and c. The system, which has only one membrane, starts with the multiset

$$a^{n_1}, b^{n_2}, c^{n_3}, \langle ab\rangle_2, \langle ac\rangle_2, \langle bc\rangle_2, \overline{A}, A^3, \overline{B}, B^3, \overline{C}, C^3$$

representing the objects for which we want to sort their multiplicities, plus some other objects which will be used during the computation for delaying the execution of certain rules or for eliminating the objects in the right order. As before, we assume that $n_1 \geq n_2 \geq n_3$. First the device starts by executing in the maximum parallel manner, in one step, the rules $a \to a_b a_c a_0$, $b \to b_a b_c b_0$, and $c \to c_a c_b c_0$. Also, at the same time, the rules $\overline{A}A \to \overline{A}$, $\overline{B}B \to \overline{B}$, $\overline{C}C \to \overline{C}$ are executed, as well as $\langle ab\rangle_2 \to \langle ab\rangle_1$, $\langle ac\rangle_2 \to \langle ac\rangle_1$, $\langle bc\rangle_2 \to \langle bc\rangle_1$. The first group of rules mentioned produces the objects

$$a_b^{n_1},\ a_c^{n_1},\ a_0^{n_1},\ b_a^{n_2},\ b_c^{n_2},\ b_0^{n_2},\ c_a^{n_3},\ c_b^{n_3},\ c_0^{n_3}$$

(all the objects a, b, c being transformed) while the other groups are used only for delaying activities.

The transition to the next configuration is made by the rules $a_b b_a \to \varepsilon$, $a_c c_a \to \varepsilon$, $b_c c_b \to \varepsilon$, which will delete from the membrane n_2 copies of a_b and b_a, n_3 copies of a_c and c_a, and, finally, n_3 copies of b_c and c_b. Simultaneously the rules $\overline{A}A \to \overline{A}$,$\overline{B}B \to \overline{B}$, $\overline{C}C \to \overline{C}$, as well as $\langle ab\rangle_1 \to \langle ab\rangle_0$, $\langle ac\rangle_1 \to \langle ac\rangle_0$, $\langle bc\rangle_1 \to \langle bc\rangle_0$, are applied. This means that the configuration of the P system in terms of the objects will be the following:

$$a_b^{n_1-n_2},\ a_c^{n_1-n_3},\ a_0^{n_1},\ b_c^{n_2-n_3},\ b_0^{n_2},\ c_0^{n_3},\ \overline{A},\ A,\ \overline{B},\ B,\ \overline{C},\ C.$$

Now is the time when the rules $a_b\langle ab\rangle \to A$, $a_c\langle ac\rangle \to A$, $b_a\langle ab\rangle \to B$, $b_c\langle bc\rangle \to B$, $c_b\langle ac\rangle \to C$, $c_b\langle bc\rangle \to C$ can be applied. Their goal is to produce objects A, B, or C (in our case A^2, B) that will be used as inhibitors in the rules and so will forbid, in one step, the execution of corresponding rules. This finishes the execution of the algorithm, because the objects will be eliminated in the correct order.

Theorem 2. *The time complexity for the strong sorting algorithm with P systems with inhibitors is 2k, where k is the number of elements to be sorted.*

4.4 Weak Priorities Feature

Another biologically inspired model is represented by P systems with weak priorities. In nature, systems evolve according to rules that usually are in some order relation. Moreover, systems evolve in time according to sequences of rules based on the priorities, in parallel, up to some moment when they reach what we call an equilibrium state. Below, we present a system that uses weak priorities for solving the sorting problem. One can notice that with very few modifications the system will compute and solve the ranking problem.

Algorithm 4. *Let* $v = \prod_{i=1}^{k} \underline{i}$. *(To define the priority relation among the rules, we will need to refer to them. In order to be able to do that, we will give* `labels` *to some of the rules while defining them, and then write down the priorities between the rules associated with these labels.) Consider*

$$\Pi = (O, \emptyset, [_1\ ]_1, \varepsilon, (R_1, \rho), \{1\}, \{0\}) \text{ with } input \prod_{j=1}^{k} \underline{j}^{n_j},$$

where:

$$\begin{aligned}
O &= \{\underline{i} \mid 1 \le i \le k\} \cup \{x_i^{(j)} \mid 1 \le i, j \le k\},\\
R &= \{\mathtt{r(w)} : w \to \prod_{j=1}^{|w|} x_j \langle \prod_{j=1, j \notin alph(w)}^{k} \underline{j} \rangle \mid w \in SSub(v) - \{\varepsilon\}\}\\
&\cup \{\prod_{j=1}^{k} \langle \prod_{l=1}^{k} \underline{i_j} \rangle \to \langle \prod_{j=1}^{k} x_j \rangle\}\\
&\cup \{\mathtt{r(x_l \prod_{j=1}^{k} x_{i_j})} : x_l \langle \prod_{j=1}^{k} x_{i_j} \rangle \to \underline{l}_{out} \langle \prod_{j=1}^{k} x_{i_j} \rangle \mid \prod_{j=1}^{k} \underline{i_j} \in Perm(v),\\
\rho &= \{r(w_1) > r(w_2) \mid |w_1| > |w_2|, w_1, w_2 \in SSub(v) - \{\varepsilon\}\}\\
&\cup \{r(x_{i_p} \prod_{j=1}^{k} x_{i_j}) > r(x_{i_q} \prod_{j=1}^{k} x_{i_j}) \mid 1 \le p < q \le k, \prod_{j=1}^{k} \underline{i_j} \in Perm(v)\}.
\end{aligned}$$

The example below shows how the system behaves when considering the 3-integer sorting problem:

Example 6. Initial data: $a^{n_1} b^{n_2} c^{n_3}$

The following rules are used:

$$\begin{gathered}
abc \to a'b'c'\langle\varepsilon\rangle >\\
\{ab \to a'b'\langle c\rangle, \quad ac \to a'c'\langle b\rangle, \quad bc \to b'c'\langle a\rangle\} >\\
\{a \to a'\langle bc\rangle, \quad b \to b'\langle ac\rangle, \quad c \to c'\langle ab\rangle\},\\
\langle a\rangle\langle ab\rangle \to \langle ABC\rangle, \quad \langle a\rangle\langle ac\rangle \to \langle ACB\rangle, \quad \langle b\rangle\langle bc\rangle \to \langle BCA\rangle,\\
\langle b\rangle\langle ab\rangle \to \langle BAC\rangle, \quad \langle c\rangle\langle ac\rangle \to \langle CAB\rangle, \quad \langle c\rangle\langle bc\rangle \to \langle CBA\rangle,\\
\langle ABC\rangle a' \to \langle ABC\rangle a_{out} > \langle ABC\rangle b' \to \langle ABC\rangle b_{out} > \langle ABC\rangle c' \to \langle ABC\rangle c_{out},\\
\langle ACB\rangle a' \to \langle ACB\rangle a_{out} > \langle ACB\rangle c' \to \langle ACB\rangle c_{out} > \langle ACB\rangle b' \to \langle ACB\rangle b_{out},\\
\langle BAC\rangle b' \to \langle BAC\rangle b_{out} > \langle BAC\rangle a' \to \langle BAC\rangle a_{out} > \langle BAC\rangle c' \to \langle BAC\rangle c_{out},
\end{gathered}$$

$\langle BCA\rangle b' \to \langle BCA\rangle b_{out} > \langle BCA\rangle c' \to \langle BCA\rangle c_{out} > \langle BCA\rangle a' \to \langle BCA\rangle a_{out}$,
$\langle CAB\rangle c' \to \langle CAB\rangle c_{out} > \langle CAB\rangle a' \to \langle CAB\rangle a_{out} > \langle CAB\rangle b' \to \langle CAB\rangle b_{out}$,
$\langle CBA\rangle c' \to \langle CBA\rangle c_{out} > \langle CBA\rangle b' \to \langle CBA\rangle b_{out} > \langle CBA\rangle a' \to \langle CBA\rangle a_{out}$.

Weak priorities means that in one step of computation the rules are applied in a sequence according to the priority relation, as much as possible for a specific rule and in the maximum parallel manner for those rules for which the relation is not defined or rules that have the same priority (with competition for objects if it is so the case). For instance, in our example, in the first step of computation (considering as usual $n_1 \geq n_2 \geq n_3$), the following rules are executed: $abc \to a'b'c'\langle\varepsilon\rangle$, then $ab \to a'b'\langle c\rangle$, and finally $a \to a'\langle bc\rangle$. After this step, in the membrane we have the objects

$$a'^{n_1},\ b'^{n_2},\ c'^{n_3},\ \langle c\rangle, \langle bc\rangle,\ \langle\varepsilon\rangle.$$

The elements $\langle c\rangle$ and $\langle bc\rangle$ will be used for controlling the sorting process. Practically, in the second step of derivation, these elements will react and will produce the object $\langle CBA\rangle$ which contains all the information needed for eliminating objects in the right order. Therefore, in the last step of the computation, according to the priority relation, one applies the rules: $\langle CBA\rangle c' \to \langle CBA\rangle c_{out} > \langle CBA\rangle b' \to \langle CBA\rangle b_{out} > \langle CBA\rangle a' \to \langle CBA\rangle a_{out}$.

Theorem 3. *The time complexity for the strong sorting algorithm with P systems with weak priorities is $2 + \sum_{i=1}^{k} n_i$, i.e., 2 plus the sum of all the elements to be sorted.*

4.5 Strong Priorities and Finite-State Catalysts

We use now the notion of s-stable catalysts. Such a catalyst has s states $c_1, \ldots, c_s$ and the rules that use this catalyst may switch between states. This means that rules like $c_i a \to c_j v$, with $1 \leq i, j \leq s$, are allowed. Because in our sorting problem we deal with a finite number of objects, the finitely stable catalysts are useful in synchronizing different tasks of the process and "more appropriate" for understanding the algorithm used.

On the other hand, strong priority means that in one step of computation only the rules with the highest priority are applied in the maximally parallel manner, regardless of whether rules with a lower priority can be used for the remaining objects.

Algorithm 5. *Let $v = \prod_{i=1}^{k} \underline{i}$. (As in the weak priority case, to define the priority relation among the rules we will give* `labels` *to some of the rules while defining them, and then write down the priorities between the rules associated with these labels.) Consider the system*

$$\Pi = (O, C, [_1\ ]_1, \prod_{j=1}^{k} y_j^{(0)} \prod_{j=1}^{k} l_j, (R_1, \rho), \{1\}, \{0\}) \text{ with input } \prod_{j=1}^{k} \underline{j}^{n_j},$$

where:

$$
\begin{aligned}
O &= \{\underline{i}, x_i \mid 1 \le i \le k\} \cup \{y_i^{(j)} \mid 1 \le i \le k, 0 \le j \le k\},\\
C &= \{y_i^{(j)} \mid 1 \le i \le k, 0 \le j \le k\},\\
R &= \{\mathtt{r}(\mathtt{y}_\mathtt{i}^{(0)}) : y_i^{(0)}\underline{i} \to y_i^{(0)} x_i \mid 1 \le i \le k\}\\
&\cup \{\mathtt{r}(\mathtt{y}_\mathtt{i}^{(0)}) : y_i^{(0)}\underline{i} \to y_i^{(j)} \mid 1 \le i, j \le k\}\\
&\cup \{\mathtt{r}(\mathtt{y}_\mathtt{i}^{(j)}\mathtt{x}_\mathtt{j}) : y_i^{(j)} x_j \to y_i^{(j)} \underline{j}_{out} \mid 1 \le i, j \le k\},\\
\rho &= \{r(y_i^{(j)}) > r(y_i^{(l)}) \mid 0 \le j < l \le k\}\\
&\cup \{r(y_i^{(p)} x_j) > r(y_j^{(q)} x_j) \mid 1 \le i, j \le k, 1 \le p < q \le k\}.
\end{aligned}
$$

We present an example that illustrates the use of this system on a 3-integer sorting problem. One can notice that the use of the strong priority controls more strictly the computation process and, as a result, the degree of sensitivity is smaller than in the model where we use weak priorities.

Example 7. Initial data: $a^{n_1}b^{n_2}c^{n_3}ABCl_1l_2l_3$.
The rules we use are:

$$
\begin{array}{lllllll}
Aa \to Aa' & > & Al_1 \to A_1 & > & Al_2 \to A_2 & > & Al_3 \to A_3,\\
Bb \to Bb' & > & Bl_1 \to B_1 & > & Bl_2 \to B_2 & > & Bl_3 \to B_3,\\
Cc \to Cc' & > & Cl_1 \to C_1 & > & Cl_2 \to C_2 & > & Cl_3 \to C_3,
\end{array}
$$

$$
\begin{array}{l}
\{A_1a' \to A_1a_{out}\ ,\ B_1b' \to B_1b_{out}\ ,\ C_1c' \to C_1c_{out}\} >\\
\{A_2a' \to A_2a_{out}\ ,\ B_2b' \to B_2b_{out}\ ,\ C_2c' \to C_2c_{out}\} >\\
\{A_3a' \to A_3a_{out}\ ,\ B_3b' \to B_3b_{out}\ ,\ C_3c' \to C_3c_{out}\}.
\end{array}
$$

As in the previous cases we assume that $n_1 \ge n_2 \ge n_3$. The system starts by executing the catalytic rules $Aa \to Aa'$, $Bb \to Bb'$, $Cc \to Cc'$ only once in each step of the computation, during the first n_3 steps. When these steps are completed, all the objects c are transformed in c'. This means that for the next $n_2 - n_3$ steps the rules $Aa \to Aa'$, $Bb \to Bb'$, and $Cl_1 \to C_1$ will be executed. Practically, in the case when $n_3 \ne n_2$, these rules will produce at least one object C_1 which will be used by the rule $C_1c' \to C_1c_{out}$ for eliminating the objects with the smallest initial multiplicity.

If $n_3 = n_2$, then, instead of rule $Bb \to Bb'$, the rule $Bl_1 \to B_1$ will be executed which will further permit the execution of the rule $B_1b' \to B_1b_{out}$. In this way, in the case of objects with the same multiplicities, the output will be with mixed objects.

Coming back to the case when multiplicities of objects are distinct, one can notice that with the system being symmetrical, the order among the multiplicities of the objects makes no difference. In our case, the system will evolve in the following way: while objects c are sent out (in our example by the rule $C_1c' \to C_1c_{out}$), the rules $Aa \to Aa'$ and $Bb \to Bb'$ are executed in the next $n_2 - n_3$ steps because the objects a and b are still present in the membrane. When all the objects b are transformed into b' one can notice that, since the object l_1 was "consumed" by the rule $Cl_1 \to C_1$, the applicable rule of the highest priority for the object B is $Bl_2 \to B_2$. But this will trigger the expelling of objects b in the environment only after all the objects c are sent out (this is due to the priority relation among the sets of rules).

Finally, in a similar way, n_1 copies of object a will be eliminated and this will complete the goal of this P system. As a remark we can observe that in the first part of the algorithm the priority rules act in the same way as if they were weak. Still, the strong priorities are needed in the last part of computation to control the output process.

Theorem 4. *The time complexity for the strong sorting algorithm with P systems with strong priorities and s-stable ($s = k+1$) catalysts is* $\max_{1 \leq i \leq k}(n_k) + \sum_{i=1}^{k} n_k$, *i.e., the sum of the elements to be sorted plus their maximum.*

4.6 Dissolution Model

We now consider a sorting algorithm implemented by a P system with membrane dissolution. If some rule $x \to y\delta$, is executed, then the objects of x are consumed, the objects of y are produced, and the corresponding membrane (the exterior one of the current region) is dissolved. In other words, the contents of the parent region is enriched by the contents of the current region (to which that rule belonged), the current region disappears, and so do its rules. All child membranes of the current region become child membranes of its parent. The skin membrane is never dissolved.

Algorithm 6. *Let* $v = \prod_{i=1}^{k} \underline{i}$. *Consider*

$$\Pi = (O, \emptyset, [_1 \dots [_{2k+1}\ ]_{2k+1} \dots]_1, w_1, \dots, w_{2k+1}, R_1, \dots, R_{2k+1}, \{k+1\}, \{0\})$$

with input $\prod_{j=1}^{k} \underline{j}^{n_j}$, *where:*

$$\begin{aligned}
O &= \{\underline{i}, X_i, X'_i, Y_i \mid 1 \leq i \leq k\} \cup \{z_i^{(j)} \mid 1 \leq i \leq k, 0 \leq i \leq 2k(k-1)\} \\
&\quad \cup \{\langle w \rangle \mid w \in SSub(v) - \{v\}\},\ w_1 = g,\ w_{k+1} = z', \\
w_i &= \varepsilon,\ \mathit{for\ all}\ i \notin \{1, k+1\}, \\
R_1 &= \{x_j \to \underline{j}_{out} \mid 1 \leq j \leq k\} \\
&\quad \cup \{g \textstyle\prod_{j=1}^{k-1} \langle \prod_{l=1}^{k} \underline{i_j} \rangle \to \prod_{j=1}^{k} y_i^{(2k(j-1))} \mid \prod_{j=1}^{k} \underline{i_j} \in Perm(v)\} \\
&\quad \cup \{z_i^j \to z_i^{(j-1)} \mid 1 \leq i \leq k,\ 1 \leq j \leq 2k(k-1)\}, \\
R_m &= R'_m + R''_m,\ 2 \leq m \leq k+1,\ R''_m = \emptyset,\ 2 \leq m \leq k, \\
R'_{m+1} &= \{\textstyle\prod_{j=1}^{m} X_{i_j} \to \prod_{j=1}^{m} \langle \prod_{j=1, \underline{j} \notin alph(w)}^{k} \underline{j} \rangle \mid \\
&\qquad w = \textstyle\prod_{j=1}^{m} \underline{i_j} \in SSub(v) - \{\varepsilon\}\} \\
&\quad \cup \{z \to z\delta\},\ 2 \leq m \leq k+1, \\
R''_{k+1} &= \{\underline{j} \to X'_{j_{in}} y_i\} \cup \{z' \to z\},
\end{aligned}$$

$$R''_{k+1+m} = \{X'_m \to X_m, z_m^{(0)} \to \delta\}$$
$$\cup \{X'_j \to X_{j\,in},\ z_j^{(0)} \to z_j^{(0)}{}_{in},\ X_j \to X_{j\,out} \mid j > m\},$$
$$\textit{for all } 2 \leq m \leq k+1.$$

Example 8. Initial configuration: $[_1 g [_2 [_3 [_4 a^{n_1} b^{n_2} c^{n_3} z' [_5 [_6 [_7\ ]_7]_6]_5]_4]_3]_2]_1$.
Rules:

$$R_1 = \{A \to a_{out},\ B \to b_{out},\ C \to c_{out},$$
$$g\langle a\rangle\langle ab\rangle \to a_0 b_6 c_{12},\ g\langle a\rangle\langle ac\rangle \to a_0 c_6 b_{12},\ g\langle b\rangle\langle bc\rangle \to b_0 c_6 a_{12},$$
$$g\langle b\rangle\langle ab\rangle \to b_0 a_6 c_{12},\ g\langle c\rangle\langle ac\rangle \to c_0 a_6 b_{12},\ g\langle c\rangle\langle bc\rangle \to c_0 b_6 a_{12},$$
$$a_i \to a_{i-1},\ b_i \to b_{i-1},\ c_i \to c_{i-1},\ a_0 \to a_{0in},$$
$$b_0 \to b_{0in},\ c_0 \to c_{0in}\},$$
$$R_2 = \{a' \to \langle bc\rangle,\ b' \to \langle ac\rangle,\ c' \to \langle ab\rangle,\ z \to z\delta\},$$
$$R_3 = \{a'b' \to \langle c\rangle,\ a'c' \to \langle b\rangle,\ b'c' \to \langle a\rangle,\ z \to z\delta\},$$
$$R_4 = \{a \to A'_{in}a',\ b \to B'_{in}b',\ c \to C'_{in}c',\ z' \to z,\ a'b'c' \to \langle\varepsilon\rangle,\ z \to z\delta\},$$
$$R_5 = \{A' \to A,\ B' \to B'_{in},\ C' \to C'_{in},$$
$$a_0 \to \delta,\ b_0 \to b_{0in},\ c_0 \to c_{0in},\ B \to B_{out},\ C \to C_{out}\},$$
$$R_6 = \{B' \to B,\ C' \to C'_{in},\ b_0 \to \delta,\ c_0 \to c_{0in},\ C \to C_{out}\},$$
$$R_7 = \{C' \to C,\ c_0 \to \delta\}.$$

As explained below, the dissolution is used in the first part of the computation to solve the ranking subproblem, and in the second part of the computation to release the output.

In this example, z' changes to z and the rules $a \to A'_{in}a'$, $b \to B'_{in}b'$, and $c \to C'_{in}c'$ of membrane 4 are first executed. This process will send the objects A', B', and C' to the inner membrane (number 5). In one step, objects A' are changed to objects A. In two steps, objects B' will travel from region 5 to region 6 and change to Bs. In three steps, objects C' will come to region 7 and change to Cs.

In the second step of computation, assuming as before that $n_1 \geq n_2 \geq n_3$, we will remain in membrane 4 with $a'^{n_1-n_3} b'^{n_2-n_3} \langle\varepsilon\rangle^{n_3}$, the other copies of a, b, and c being deleted by the rule $a'b'c' \to \langle\varepsilon\rangle$. In the same step, z dissolves membrane 4 and comes to region 3.

Then, in the third step, $a'^{n_1-n_2}\langle c\rangle^{n_2-n_3}\langle\varepsilon\rangle^{n_3}$ will be present in membrane 3 after the rule $a'b' \to \langle c\rangle$ is applied. Again, z dissolves membrane 3 and exits to region 2. In the fourth step, $\langle bc\rangle^{n_1-n_2}\langle c\rangle^{n_2-n_3}\langle\varepsilon\rangle^{n_3}$ will be obtained in membrane 2, also being dissolved. In region 1, $g\langle c\rangle\langle bc\rangle$ evolves to $c_0 b_6 a_{12}$; thus c_0 has no delay and will be used to output objects c, b_6 will wait six turns, and a_{12} will wait even longer, to output objects a in the last turn.

Output works in the following way: a symbol a_0, b_0, or c_0 comes to the region 2, 3, or 4, respectively, where it dissolves the corresponding membrane. Symbols A, B, and C, being in a region with an index smaller than 2, 3, and

4, respectively, travel to the skin region, where they lead to a_{out}, b_{out}, c_{out}, respectively. Thus, the initial multiplicities are preserved, and the order of output corresponds to the order of increasing multiplicities of the corresponding objects.

Theorem 5. *The time complexity for the weak sorting algorithm with P systems with object rewriting rules is* $k+2+2k(k+1)+2 = 2k^2+3k+4$, *where* k *is the number of elements to be sorted. However, the number of different objects used is exponential with respect to* k.

4.7 Ranksort

This is a strong sorting algorithm based on a ranking algorithm discussed in the ranking section.

We will design a P system that will be responsible for the elimination of symbols in the right order. Let us start with a 3-integer sorting example. First, let us denote by *Rank* the module presented above. In order that the elimination of a certain object not interfere with the elimination of another object, we have to distribute each catalyst which represents a signal in different, consecutive membranes in such a way that the first signaling catalyst arrives at the outermost membrane, the second signaling catalyst at the second outermost membrane, and so on. This can be achieved by a construct such as the following one (let us consider for simplicity the case of three signaling catalysts that leave the rank module; moreover, for the sake of simplicity we will show only the rules for the signal catalysts A; all the other rules can be constructed in a very similar way). Figure 3 illustrates the procedure.

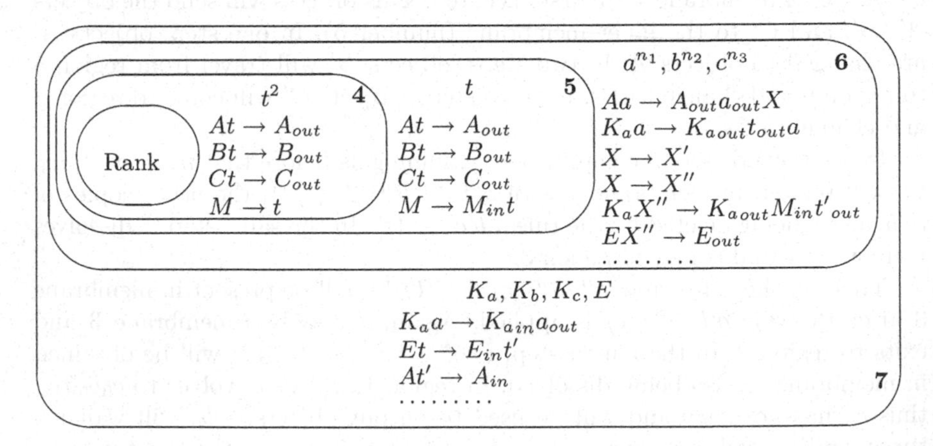

Fig. 3. The P system solving the 3-integer strong sorting problem.

Recall that in the *Rank* example $n_1 < n_2 < n_3$ and so the signal catalysts were eliminated in the order A, B, and finally C. This means that the catalyst A will leave the *Rank* module first and will arrive in region 4. There, by the rule $At \to A_{out}$, it will consume an object t and will go to region 5. From there, by using a rule of the same type as before, it will finally arrive in membrane 6. One can see that if another signal catalyst (say B) leaves the *Rank* module, it will arrive in region 5 and will not go further because all the objects t have been already consumed. In this way, signal catalysts are separated in different membranes according to the multiplicities they represent.

Now, we have the catalyst A in region 6 and also the multiset $a^{n_1}, b^{n_2}, c^{n_3}$ ($a^{n_1}, b^{n_2}, c^{n_3}$ can be transported here from the input membrane using only context-free rules). As we said before, for the sake of simplicity we will consider only the rules involving object a (the other rules are very similar, obtained, for instance, by replacing A with B, a with b, and K_a with K_b in membranes 6 and 7).

In region 6 we want to eliminate objects corresponding to the signal that we receive. Once we finish this task we also would like for the next signal catalyst (which stays in a queue) to enter in region 6 and start the elimination process again. Also, all other signal catalysts must "advance" up one region.

Since in our case the catalyst A enters in region 6, the rule that can be applied is $Aa \to A_{out}a_{out}X$, which will decrease by one the number of objects a. As an effect, an object X will be produced and both the catalyst A and the corresponding object a will arrive into region 7. There, by the rule $K_a a \to K_{a\,in} a_{out}$ the object a will be sent to the environment, while the catalyst K_a (a "checker" – this catalyst will check if there are still objects a present in region 6) will arrive into region 6; if there are still objects a present, then the catalyst K_a will go to its initial position (region 7). Also, the catalyst A will come back to region 6 and the process will start again. If the catalyst K_a does not find any object a in membrane 6, an object M ("message") will be sent to the inner membranes. This object will be responsible for the "advancing" in the "queue" of the signaling catalysts. The catalyst E present initially in region 7 is responsible for "cleaning" operations of undesired symbols, while the objects X, X', and X'' are used for synchronizing different tasks.

For a better understanding we present configurations in Tables 1 and 2. We use the question mark (?) to denote that the indicated objects might be present in the corresponding region.

For the k-integer sorting problem we will have a number of membranes proportional to k. The mechanism used in the example for advancing in the queue of signal catalysts will be preserved. Practically, we will have a system with 3 (ranking problem) + k (the queue device) +1 membranes, and $2k + 2$ catalysts.

The time complexity for the strong sorting algorithm with P systems with mobile catalysts is linear with respect to the maximum number of elements to be sorted. It also depends on the number of components to be sorted and on the sum of the elements.

Table 1. The case when $n_1 \geq 2$:

time	region 5	region 6	region 7	environment
1	$B?$	$A, a^{n_1}, b^{n_2}, c^{n_3}$ $Aa \to A_{out}a_{out}X$	K_a, K_b, K_c	
2	$B?$	$a^{n_1-1}, b^{n_2}, c^{n_3}X$ $X \to X'$	K_a, K_b, K_c, E, A, a $K_a a \to K_{a\,in}a_{out}$	
3	$B?$	$K_a, a^{n_1-1}, b^{n_2}, c^{n_3}, X'$ $K_a a \to K_{a\,out}at_{out}$ $X' \to X''$	K_b, K_c, E, A	a
4	$B?$	$a^{n_1-1}, b^{n_2}, c^{n_3}, X''$	K_a, K_b, K_c, E, A, t $Et \to e_{in}t'$	
5	$B?$	$a^{n_1-1}, b^{n_2}, c^{n_3}, X'', E$ $EX'' \to E_{out}$	K_a, K_b, K_c, A, t' $At' \to A_{in}$	
6=1'	$B?$	$A, a^{n_1-1}, b^{n_2}, c^{n_3}$ ...	K_a, K_b, K_c, E ...	

Table 2. The case when only one object a remains in region 6:

time	region 5	region 6	region 7	environment
1	$B?$	A, a, b^{n_2}, c^{n_3} $Aa \to A_{out}a_{out}X$	K_a, K_b, K_c	
2	$B?$	$b^{n_2}, c^{n_3}X$ $X \to X'$	K_a, K_b, K_c, E, A, a $K_a a \to K_{a\,in}a_{out}$	
3	$B?$	$K_a, b^{n_2}, c^{n_3}, X'$ $X' \to X''$	K_b, K_c, E, A	a
4	$B?$	$K_a, b^{n_2}, c^{n_3}, X''$ $K_a X'' \to K_{a\,out}M_{in}t'_{out}$	K_b, K_c, E, A	
5	$B?, M$ $M \to M_{in}t$	b^{n_2}, c^{n_3}	K_a, K_b, K_c, E, A, t' $At' \to A_{in}$	
6	$B?, t$ $Bt \to B_{out}$	A, b^{n_2}, c^{n_3}	K_a, K_b, K_c, E	
7	$C?$	$A, B?, b^{n_2}, c^{n_3}$ $Bb \to B_{out}b_{out}X$?	K_a, K_b, K_c, E	
8	... ...	... ...	... ...	... ...

The result is a consequence of the facts presented in the previous sections. One can see that the ranking solution of the problem is given in linear time; the transporting of catalysts to their correct positions depends on the number of membranes (and so depends on the number of components). The elimination process depends linearly on the sum of elements.

5 Weak Sorting

Weak sorting is an algorithm which processes a multiset M and gives as a result a word $w = \prod_{j=1}^{k} j^{M(a_j)}$ such that $\prod_{j=1}^{k} a_j$ is the ranking string of M. A weaker definition could be given: the multiset M' corresponding to w can be considered as the result ($M'(j) = M(a_j)$, that is, the objects' multiplicities are ordered) because the order of numbers is represented by the order of the corresponding objects in the alphabet. In the case of equality of some numbers there is no need to make a decision whether "=" is "≤" or "≥" since the result is the same. A typical strategy for weak sorting would be a *sorting network*, i.e., a parallel or sequential usage of compare-swap-if-needed operator

$$(n, n') \to (\min(n, n'), \max(n, n')).$$

Weak sorting is typically easier than strong sorting.

5.1 Bead Sort

One of the sorting algorithms is Bead Sort. This subsection is devoted to the construction from [3] of a P system implementing this algorithm. It actually uses a tissue-like P system[5] (see [9]) with $k \times m$ membranes, where $m = max\{n_1, \dots, n_k\}$ (the number of "rods"). The positive integers from the set to be sorted are represented by a set of "beads"; they slide along the "rods" to their appropriate places.

For a graphical representation of how bead sort works, Figure 4 shows the initial and the final configuration of such a system when it sorts a particular set, namely $\{3, 2, 4, 2\}$. Numbers from the set to be sorted are represented horizontally (each level of the rods contains a number).

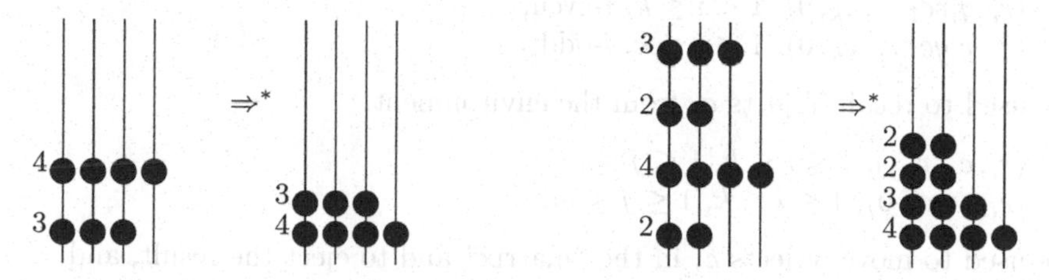

Fig. 4. Sorting the sets $\{4, 3\}$ and $\{3, 2, 4, 2\}$ using beads that slide on rods.

When we encode this simple idea within the P systems framework we will use a tissue-like P system to represent the rods (membranes that can communicate); also, beads will be represented by objects x placed inside membranes

[5] In such a system an antiport rule of the form $(i, x|y, j)$ means the simultaneous exchange of objects of multiset x from cell i with objects of multiset y from cell j. Similarly, a symport rule (i, x, j) means moving the multiset x from cell i to cell j.

(the object # represents the absence of the bead). The membrane structure is of degree $m \cdot k + k$ where $m \cdot k$ represents the m rods with k levels; the extra k membranes are counters useful in solving the sorting problem. Objects x represent the beads, objects # represent the absence of beads, objects p, q, and r control the process, and objects $c_1, \ldots, c_m$ (present in the environment in unbounded quantities) are used for the output.

There is an $m \cdot k$ bidimensional array of cells; let us denote them by $r_{i,j}$, $1 \leq i \leq k$, $1 \leq j \leq m$. The antiport rules

$$(r_{i+1,j}, x|\#, r_{i,j})$$

are used (they exchange objects x and # between membranes that are "above" the others), $1 \leq i < k$, $1 \leq j \leq m$, as the fall-down mechanism of the beads.

In order to read the output, symport rules of weight at most 2 and antiport rules of weight at most $m + 1$ are used. Label 0 represents the environment. Apart from the $m \cdot k$ matrix, there are k membranes, associated with the rows. Let us denote them by $r_1, \cdots, r_k$. The object p, initially present in membrane r_k, "falls down" by the rules

$$(r_{i+1}, p, r_i),\ 1 \leq i < k.$$

Notice that when p enters membrane r_1 the "matrix" is guaranteed to be sorted (no objects x are "above" objects #).

The answer is sent to the environment "row by row": $x^{n_{\sigma(1)}}$, then $x^{n_{\sigma(2)}}$, and so on. For this, the following rules are used (other objects are also moved from and to the environment):

$$(r_1, p|qc_1 \ldots c_k, 0),$$
$$(r_i, q|rc_1 \ldots c_k, 0),\ 1 < i \leq k,\ i\text{-even},$$
$$(r_i, r|qc_1 \ldots c_k, 0),\ 1 < i \leq k,\ i\text{-odd},$$

are used to recall objects c_j from the environment,

$$(r_i, c_j, r_{i,j}),\ 1 \leq i \leq k,\ 1 \leq j \leq m,$$
$$(r_{i,j}, xc_j, 0),\ 1 \leq i \leq k,\ 1 \leq j \leq m,$$

are used to move objects c_j in the "matrix" and to eject the result, and

$$(r_i, q, r_{i+1}),\ 1 \leq i < k,\ i\text{-odd},$$
$$(r_i, r, r_{i+1}),\ 1 \leq i < k,\ i\text{-even},$$

are used to propagate the output process "up."

Thus, the problem is solved in a purely communicative way (just by moving objects). The time complexity of this solution is linear, but the features used are quite powerful (antiport rules of unbounded weight), the descriptional complexity of the solution is rather high ($m \cdot k + k$ membranes), the initial data has to be a priori distributed among the $m \cdot k$ cells, and the result is obtained in the environment.

5.2 Communicative Sorting

Here we will present the static version of sorting with P systems where the dimension of the input data set is given at the beginning of computation and does not change. The construction we discuss is from [5].

As usual, we want to sort k natural integers $X = \{n_1, \ldots, n_k\}$. In the P system we construct we will represent the integer n_j, $1 \leq j \leq k$, as a string a^{n_j} (integer x is represented in a region as x occurrences of symbol $a \in V$).

Having this representation, it is obvious that each region can contain at most one number, because we have used only one object to represent the whole set to be sorted. It follows that to sort k natural integers we will need at least k membranes.

Let us consider a nested structure of membranes. This way, we have total order among membranes; we can therefore use this order to have a codification of the sorting result: for a given input sequence $X = \{n_1, n_2, \ldots, n_k\}$, the solution to the sorting problem is the permutation of k elements, σ, i.e., we have $n_{\sigma(1)} \leq n_{\sigma(2)} \leq \ldots \leq n_{\sigma(k)}$; then, at the end of the process of sorting, the nested membrane structure should hold the integers in ascending order, from the outermost to the innermost membrane.

Such a P system and its associated tree structure are depicted in Figure 5.

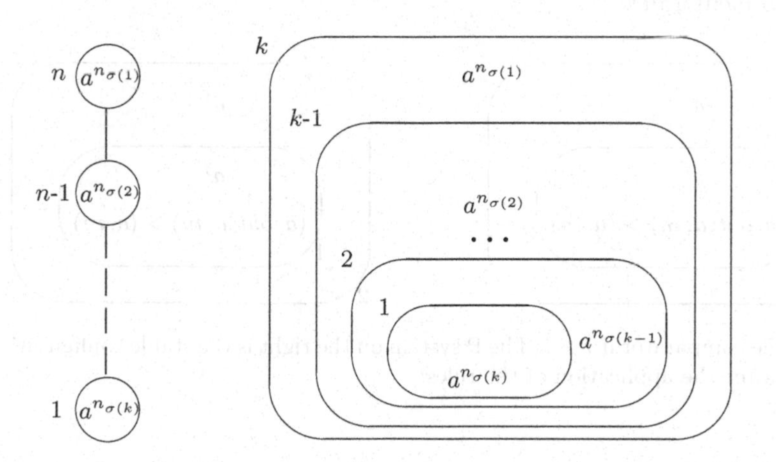

Fig. 5. The result of sorting.

Customarily, in P systems framework a successful computation is defined as a halting one. A weaker definition was given in [5] by introducing the notion of a *stable configuration* (even if some rules are still applicable, their application does not change the multisets present in the system compartments or the membrane structure). Based on this notion, we can define a successful computation as being one that ends in a stable configuration.

We can construct the following P system with symport/antiport rules (with priorities among rules) which compares the multiplicities of objects a:

$$\begin{aligned}\Pi &= (\{a\}, [_2\, [_1\,]_1\,]_2, w_2, w_1, R_2, R_1, \{1,2\}, \{1,2\}),\\ R_2 &= \emptyset,\\ R_1 &= \{(a, out; a, in) > (a, in)\},\end{aligned}$$

with input

- $w_1 = a^i$ in region 1,
- $w_2 = a^j$ in region 2, where i, j are numbers to be compared.

We know that the system works in the maximally parallel way and that objects are associated with rules according to the priorities among rules. This means that exactly $\min(i, j)$ copies of a will be acted upon by the antiport rule $(a, out; a, in)$, and only afterward may the rule (a, in) start its execution. If executed (i.e., $j > i$), rule (a, in) has the role of moving exactly $j - i$ copies of object a to the inner region. In case $i > j$, only rule $(a, out; a, in)$ is executed.

One can remark that after just one computation step we have reached a stable configuration. In addition, in this stable configuration membranes hold objects a with multiplicities in ascending order, from the outermost to the innermost membrane.

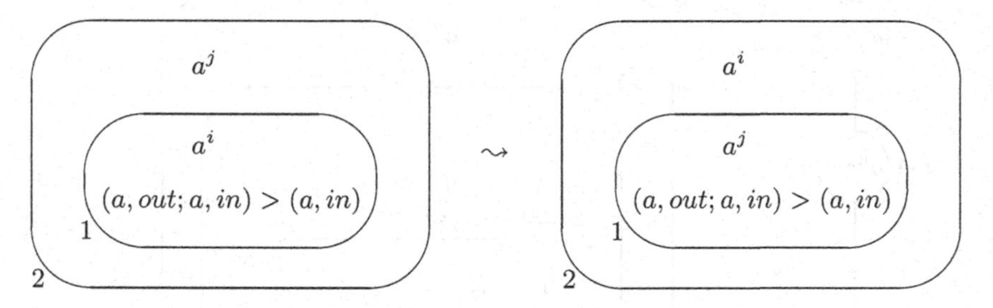

Fig. 6. The comparator if $j > i$. The P system on the right is the stable configuration obtained after the application of the rules.

The developed comparator (a simple P system with two membranes) represents a module that can be further "integrated" in a larger construction. We can construct in this way a P system that simulates the odd-even transposition networks. In order to do this we need to have a mechanism to control the execution time of different comparators. This will be done by considering promoted rules of type $(a, out; a, in)|_p > (a, in)|_p$ (i.e., rules of the comparator presented above are active in the presence of promoter p).

Let us consider the P system

$$\Pi^{o-e} = (O, \mu, w_1, \dots, w_k, R_1, \dots, R_k, \{1, \dots, k\}, \{1, \dots, k\})$$

with input a^{n_i} in membrane i, $1 \leq i \leq k$, where

$$O = \{a, p\},$$
$$\mu = [_k [_{k-1} \cdots [_2 [_1]_1]_2 \cdots]_{k-1}]_k,$$
$$w_i = p, \; 1 \leq i < k, i \text{ odd},$$
$$w_i = \varepsilon, \; 1 \leq i < k, i \text{ even},$$
$$R_i = \{(a, out; a, in)|_p > (a, in)|_p\}, \; 1 \leq i < k, \; i \text{ odd},$$
$$R_i = \{(p, in) > (a, out; a, in)|_p > (a, in)|_p > (p, out)\}, \; 1 \leq i < k, \; i \text{ even},$$
$$R_n = \emptyset.$$

This is how the system works. At the beginning of the computation, each region i contains objects a^{n_i}. In addition, all the odd regions (with odd numbers as labels) contain the promoter p. The computation starts by executing rules in the odd regions since all rules (no matter if they are acting in odd or even regions) are promoted by object p. The first rule that the system tries to execute is (p, in), but since there is no object p in the external region the rule mentioned will not act. Therefore, because of the priority relation among rules and of massive parallelism, rules $(a, out; a, in)|_p > (a, in)|_p$ (the rules of the comparator) are executed in the same computational step. As a result, objects a with smaller multiplicity will be placed in the locally outer region, while objects a with larger multiplicity will be placed in the locally inner region. The configuration reached will be stable. The rule (p, out) will be enabled and promoter p will enter the outer even region.

In this way the system has locally ordered the multiplicities of objects a and, moreover, the initial configuration has been reestablished. Therefore, the process can start again. Figure 7 illustrates the regions of a P system as above, with six membranes (the regions are depicted as lines in a network; the promoter p triggers the action of a comparator).

Based on this construction, we have the following result (see [5]):

Theorem 6. *The P system Π^{o-e} presented above will have in its membranes the integers $n_1, \ldots, n_k$, sorted in increasing order from the outermost to the innermost membrane, after k transitions, in a configuration which is stable over a.*

5.3 Using Weak Priorities

The following algorithm implements a weak sorting using a (rewriting) P system with symbol objects with weak priorities; at most two objects are present on the left hand side of any rule. The idea behind this algorithm is parallel simulation of an *odd-even sorting network*; see also [5].

Algorithm 7. *Let $v = \prod_{i=1}^{k} \underline{i}$. Consider the following P system*

$$\Pi = (O, [_1]_1, \varepsilon, (R, \rho), \{1\}, \{0\}) \textit{ with input } \prod_{j=1}^{k} \underline{j}^{n_j},$$

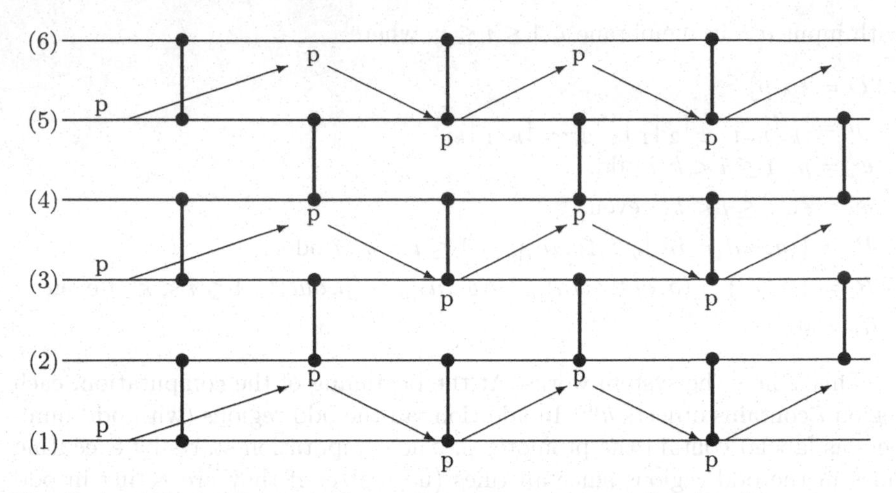

Fig. 7. An odd-even P system for sorting: promoters p travel and activate comparators between regions.

with weak priority *(again defined with the help of* `labels`*), where:*

$$
\begin{aligned}
O &= \{\underline{i} \mid 1 \le i \le k\} \cup \{y_i^{(j)} \mid 1 \le i \le k, 1 \le j \le 2k\},\\
R &= \{\underline{j} \to y_j^{(0)} \mid 1 \le j \le k\}\\
&\cup \{\mathtt{r(j,l,0)} : y_j^{(l)} y_{j+1}^{(l)} \to y_j^{(l+1)} y_{j+1}^{(l+1)},\\
&\quad \mathtt{r(j,l,1)} : y_j^{(l)} \to y_{j+1}^{(l+1)}),\\
&\quad \mathtt{r(j,l,2)} : y_{j+1}^{(l)} \to y_{j+1}^{(l+1)} \mid 1 \le j, l \le k; j \equiv l(mod\ 2)\}\\
&\cup \{y_1^{(l)} \to y_1^{(l+1)} \mid l \equiv 0(mod\ 2)\}\\
&\cup \{y_k^{(l)} \to y_k^{(l+1)} \mid l \equiv k(mod\ 2)\}\\
&\cup \{y_j^{(k-1+j)} \to \underline{j}_{out} \mid 1 \le j \le k\} \cup \{y_j^l \to y_j^{l+1} \mid k \le l \le k+j-2\},\\
\rho &= \{r(j,l,0) > r(j,l,1),\ r(j,l,0) > r(j,l,2) \mid\\
&\quad 1 \le j,\ l \le k,\ j \equiv l(mod\ 2)\}.
\end{aligned}
$$

Example 9. In the case of an even number k of components, $k-1$ comparisons suffice rather than k.
Initial data: $a^{n_1} b^{n_2} c^{n_3} d^{n_4}$.
Rules:

$$
\begin{array}{llll}
a \to a_1, & b \to b_1, & c \to c_1, & d \to d_1, \\
a_i b_i \to a_{i+1} b_{i+1} > \{a_i \to b_{i+1},\, b_i \to b_{i+1}\},\ i \in \{1,3\}, \\
c_i d_i \to c_{i+1} d_{i+1} > \{c_i \to d_{i+1},\, d_i \to d_{i+1}\},\ i \in \{1,3\}, \\
b_i c_i \to b_{i+1} c_{i+1} > \{b_i \to c_{i+1},\, c_i \to c_{i+1}\},\ i \in \{2\}, \\
a_i \to a_{i+1}, & d_i \to d_{i+1}, & i \in \{2\}, \\
a_4 \to a_{out}, & b_5 \to b_{out}, & c_6 \to c_{out}, & d_7 \to d_{out}, \\
b_4 \to b_5, & c_4 \to c_5, & c_5 \to c_6, \\
d_4 \to d_5, & d_5 \to d_6, & d_6 \to d_7.
\end{array}
$$

Here, the rules of type $AB \to A'B' > \{A \to B', B \to B'\}$ perform the compare-swap-if-needed operator:

$$A^n B^{n'} \Rightarrow A'^{\min(n,n')} B'^{\max(n,n')}.$$

This happens because the higher priority rule $AB \to A'B'$ is applied $\min(n, n')$ times, and one of the lower-priority rules is applied $|n_1 - n_2|$ times, due to the maximally parallel nature of the application. This operator is computed in one step, and at least $\lfloor (k-2)/2 \rfloor$ operators are computed in parallel. The needed number of steps is $k - 1 + (k \; mod \; 2)$ where *mod* denotes the remainder. Finally, the objects exit the system in alphabetical order.

Theorem 7. *The time complexity for the weak sorting algorithm with P systems with rewriting is* $2k + 1$, *where* k *is the number of elements to be sorted.*

5.4 Evolution-Communication Systems

The idea of the next algorithm is very similar to that from the previous section (a parallel simulation of an *odd-even sorting network*). However, the type of P system we use is quite different. We now have no target indications in the evolution rules (the result stays in the same regions), but we use communication rules for moving objects (they do not change objects, just move them). These two types of rules are executed in a maximally parallel manner. In the following construction, we will only have non-cooperative evolution rules, symport rules of weight 1, and antiport rules of weight 1.

The *weak* priority can exist between the evolution rules of the same region, between the communication rules of the same membrane, and between the rules of the two classes, where the membrane is the external boundary of the corresponding region.

Example 10. Let $B_4 = b_4$, $D_4 = d_4$, that is b_4 and d_4 are used as synonyms of B_4 and D_4. The initial configuration is $[_1 [_2 a^{n_1} b^{n_2} c^{n_3} d^{n_4}]_2]_1$.
The rules are defined in the tables below.

Region 1	**Region 2**	$i \in$	Step
	$a \to a_0,\ b \to b',\ c \to c_0,\ d \to d'$		1
	$(a_0, out),\ b' \to b_0,\ (c_0, out),\ d' \to d_0$		2
$a_0 \to a_1,\ c_0 \to c_1$	$b_0 \to b_1,\ d_0 \to d_1$		3
	$(b_i, out; a_i, in) > \{(a_i, in), b_i \to a_i\}$ $(d_i, out; c_i, in) > \{(c_i, in), d_i \to c_i\}$	$\{1,3\}$ $\{1,3\}$	4,12
$b_i \to A_{i+1},\ d_i \to C_{i+1}$	$a_i \to B_{i+1},\ c_i \to D_{i+1}$	$\{1,3\}$	5,13
	$(A_i, in), (B_i, out), (C_i, in), (D_i, out)$	$\{2\}$	6
$B_i \to b_i,\ D_i \to d_i$	$A_i \to a_i,\ C_i \to c_i$	$\{2\}$	7
$d_i \to d'_i$	$(c_i, out; b_i, in) > \{(b_i, in), c_i \to b_i\}$ $a_i \to a'_i$	$\{2\}$ $\{2\}$	8
$a'_i \to A_{i+1},\ b_i \to C_{i+1}$	$c_i \to B_{i+1},\ d'_i \to D_{i+1}$	$\{2\}$	9
	$(A_i, out), (B_i, in), (C_i, out), (D_i, in)$	$\{1\}$	10
$A_i \to a_i,\ C_i \to c_{i+1}$	$B_i \to b_i,\ D_i \to d_i$	$\{1\}$	11
$A_4 \to a_4,\ C_4 \to c_4$	$(B_4, out), (D_4, out)$		14

Region 1	Step
$a_4 \to a,\ b_4 \to b_5,\ c_4 \to c_5,\ d_4 \to d_5$	15
$(a, out),\ b_5 \to b,\ c_5 \to c_6,\ d_5 \to d_6$	16
$(b, out),\ c_6 \to c,\ d_6 \to d_7$	17
$(c, out),\ d_7 \to d$	18
(d, out)	19

Here, the rules written in the same row of the table are executed in the same step, and the table also indicates the actual step in which these rules can be applied. The computation proceeds as follows:

$$
\begin{aligned}
&[_1[_2 a^{n_1} b^{n_2} c^{n_3} d^{n_4}]_2]_1 \Rightarrow && [_1[_2 a_0^{n_1} b'^{n_2} c_0^{n_3} d'^{n_4}]_2]_1 \Rightarrow \\
&[_1 a_0^{n_1} c_0^{n_3} [_2 b_0^{n_2} d_0^{n_4}]_2]_1 \Rightarrow && [_1 a_1^{n_1} c_1^{n_3} [_2 b_1^{n_2} d_1^{n_4}]_2]_1 \Rightarrow \\
&[_1 b_1^{n'_1} d_1^{n'_3} [_2 a_1^{n'_2} c_1^{n'_4}]_2]_1 \Rightarrow && [_1 A_2^{n'_1} C_2^{n'_3} [_2 B_2^{n'_2} D_2^{n'_4}]_2]_1 \Rightarrow \\
&[_1 B_2^{n'_2} D_2^{n'_4} [_2 A_2^{n'_1} C_2^{n'_3}]_2]_1 \Rightarrow && [_1 b_2^{n'_2} d_2^{n'_4} [_2 a_2^{n'_1} c_2^{n'_3}]_2]_1 \Rightarrow \\
&[_1 c_2^{n''_2} {d'_2}^{n''_4} [_2 {a'_2}^{n''_1} b_2^{n''_3}]_2]_1 \Rightarrow && [_1 B_3^{n''_2} D_3^{n''_4} [_2 A_3^{n''_1} C_3^{n''_3}]_2]_1 \Rightarrow \\
&[_1 A_3^{n''_1} C_3^{n''_3} [_2 B_3^{n''_2} D_3^{n''_4}]_2]_1 \Rightarrow && [_1 a_3^{n''_1} c_3^{n''_3} [_2 b_3^{n''_2} d_3^{n''_4}]_2]_1 \Rightarrow \\
&[_1 b_3^{n'''_1} d_3^{n'''_3} [_2 a_3^{n'''_2} c_3^{n'''_4}]_2]_1 \Rightarrow && [_1 A_4^{n'''_1} C_4^{n'''_3} [_2 B_4^{n'''_2} D_4^{n'''_4}]_2]_1 \Rightarrow \\
&[_1 a_4^{n'''_1} B_4^{n'''_2} c_4^{n'''_3} D_4^{n'''_4} [_2]_2]_1 \Rightarrow && [_1 b_5^{n'''_2} c_5^{n'''_3} d_5^{n'''_4} [_2]_2]_1 a^{n'''_1} \Rightarrow \\
&[_1 c_6^{n'''_3} d_6^{n'''_4} [_2]_2]_1 a_4^{n'''_1} b^{n'''_2} \Rightarrow && [_1 d_7^{n'''_4} [_2]_2]_1 a^{n'''_1} b^{n'''_2} c^{n'''_3} \Rightarrow \\
&[_1[_2]_2]_1 a^{n'''_1} b^{n'''_2} c^{n'''_3} d^{n'''_4}.
\end{aligned}
$$

The objects are divided into two classes: $\{\underline{j} \in V \mid j \equiv 0(mod\ 2)\}$ and $\{\underline{j} \in V \mid j \equiv 1(mod\ 2)\}$. These classes are stored in different membranes, so the compare-swap-if-needed operator sorting (A, B) to (A', B') is made of an antiport rule $(A, out; B, in)$, having weak priority over a symport rule (A, out) and over a rewriting rule $B \rightarrow A$; after using this rule, A is renamed to $\overline{B}$, B is renamed to $\overline{A}$, both types of objects cross the membrane, and we then use $\overline{B} \rightarrow B'$, $\overline{A} \rightarrow A'$. A comparison operator is executed in four steps.

In this example, the following sorting network $(n_i)_{1 \leq i \leq 4} \Rightarrow (n_i''')_{1 \leq i \leq 4}$ was simulated,

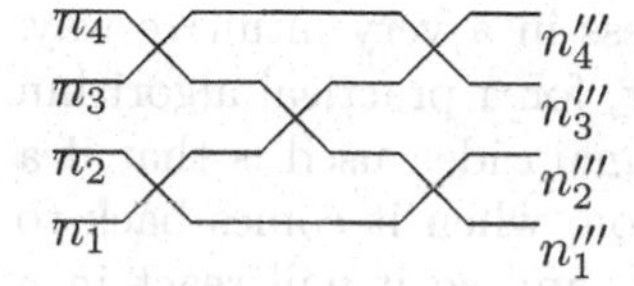

$$(n_1, n_2) \rightarrow (n_1', n_2'),\ (n_3, n_4) \rightarrow (n_3', n_4'),$$
$$n_1'' = n_1',\ (n_2', n_3') \rightarrow (n_2'', n_3''),\ n_4'' = n_4',$$
$$(n_1'', n_2'') \rightarrow (n_1''', n_2'''),\ (n_3'', n_4'') \rightarrow (n_3''', n_4'''),$$

represented by the picture on the left.

The general case of sorting k numbers is addressed in a similar way. The depth of the sorting network should be $k - ((k + 1) mod\ 2)$.

5.5 Strings

Let us define a string weak sorting in a weak sense: the order of the symbols in the resulting string is not important, and the numbers are represented by the number of occurrences of corresponding objects in the string.

Consider the following string rewriting P system with priorities, implementing the compare-swap-if-needed operator:

Algorithm 8. $\Pi = (O, \emptyset, [_1\ [_2\ [_3\ ]_3\ ]_2\ ]_1, \emptyset, \emptyset, \emptyset, R_1, R_2, R_3, \{3\}, \{0\})$ *with input* $a^n b^{n'}$, *where:*

$$O = \{a, b, a', b', \overline{a}\},$$
$$R_1 = \{a \rightarrow b', b \rightarrow b'\} > \{b' \rightarrow b'_{out}\},$$
$$R_2 = \{\overline{a} \rightarrow a'_{out}\} > \{b \rightarrow b'_{in}\} > \{b' \rightarrow b'_{out}\},$$
$$R_3 = \{a \rightarrow a'_{out}\} > \{a' \rightarrow \overline{a}_{out}\}.$$

Here, for each comparison, the string goes back and forth between two regions, and at every step one symbol is marked; when the minimum of the two numbers is reached, the string exits to the third region and the unmarked symbols are rewritten by marked symbols, corresponding to the object which has a higher alphabetic position. The result of the computation will be $w \in Perm(a'^{\min(n,n')} b'^{\max(n,n')})$.

By nesting these constructions and varying the rules according to the objects to compare, a sorting network can be easily implemented. An alternative approach is to use the outer membrane to rewrite all the objects, return the string to region 3, and proceed with the next comparison using a different subset of rules, all subsets being like those in the construction above, with the necessary renames and moves. This would lead to a construction with only three membranes, but a lot of objects and rules.

5.6 Mobile Catalysts and I/O Communication

Catalysts are the "weakest" components that, joined with non-cooperative rules, prove to be enough for reaching universality in the case of P systems with symbol objects. The mobility of catalysts proves to be important because it allows the construction of a deterministic process in a very intuitive way. As we commented at the beginning of the chapter, for a practical algorithm we are interested in deterministic processes. The main idea used is that if a catalyst changes the region during the computation, when it comes back to the original region it will find a different context, and so it will react in a different way. In order to implement this we will need at least two mobile catalysts. First, let us see how to build a comparator – a component that for classical algorithm theory is very important.

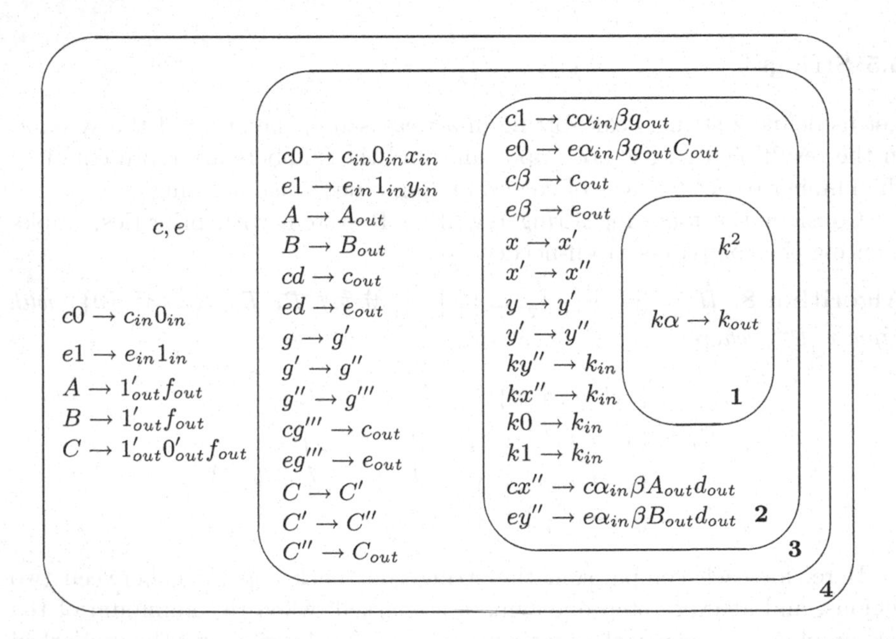

Fig. 8. The comparator. The system computes the maximum and the minimum between multiplicities of the objects 0 and 1 that enter membrane 4.

The module shown in Figure 8 compares the multiplicity of two objects and returns an object having as multiplicity the maximum of the initial two multiplicities and an object having as multiplicity their minimum. The construction will work and the tables given in Figures 9 and 10 explain both cases that can occur:

- the objects have different multiplicities,
- the objects have the same multiplicities.

Another important aspect that must be handled is that we want the process to give the answer at the same time for both cases that could appear. This is why we have to introduce some delays (by applying some renaming rules) in order to control the computation.

Time	Region 4	Region 3	Region 2	Region 1
1	$c, e, 0, 1$ $c0 \to c_{in}0_{in}$ $e1 \to e_{in}1_{in}$	–	–	k^2
2	$0?, 1?$	$0, 1, c, e$ $c0 \to c_{in}0_{in}x_{in}$ $e1 \to e_{in}1_{in}y_{in}$	–	k^2
3	$0?, 1?$	–	$0, 1, c, e, x, y$ $c1 \to c\alpha_{in}\beta g_{out}$ $e0 \to e\alpha_{in}\beta g_{out}C_{out}$	k^2
4	$0?, 1?$	g^2, C $C \to C'$ $g \to g'$	c, e, x, y, β^2 $x \to x'$ $y \to y'$ $c\beta \to c_{out}$ $e\beta \to e_{out}$	k^2, α^2 $k\alpha \to k_{out}$
5	$0?, 1?$	g'^2, C', c, e $C' \to C''$ $g' \to g''$	x', y', k^2 $x' \to x''$ $y' \to y''$	–
6	$0?, 1?$	g''^2, C'', c, e $C'' \to C_{out}$ $g'' \to g'''$	x'', y'', k^2 $kx'' \to k_{in}$ $ky'' \to k_{in}$	–
7	$0?, 1?, C$ $C \to 0'1'f_{out}$	g'''^2, c, e $cg''' \to c_{out}$ $eg''' \to e_{out}$	–	k^2
$8(0'1'f)$	$c, e, 0?, 1?$	–	–	k^2

Fig. 9. The case when objects 0 and 1 enter simultaneously region 4.

We proceed as follows. The idea is to detect when two objects simultaneously enter the comparator and when only one enters. Based on the answer to this decision problem we can trigger a signal stating the largest number. There are two distinct cases that can occur: in membrane 4, the symbols 0 and 1 enter simultaneously or there is only one symbol (0 or 1) that enters at a time. Now, the rules $c0 \to c_{in}0_{in}$ and $e0 \to e_{in}1_{in}$ in membrane 4 are used only to carry the corresponding objects into the comparator. There, by the same type of rules ($c0 \rightarrow c_{in}0_{in}x_{in}$ and $e0 \rightarrow e_{in}1_{in}y_{in}$), the objects will be passed from membrane 3 to membrane 4.

According to these cases, the system will react in different ways. Suppose we have the first case, where 0 and 1 enter membrane 4; see Figure 9. Then, as we said above, by using catalytic rules (we associate the catalyst c with 0 and the catalyst e with 1) the pairs of objects $(c, 0)$ and $(e, 1)$ go up to region 2. There, we switch the associations made before in order to check if 0 and 1 enter simultaneously. If this is so, the objects α and β will be produced and two copies of g will be sent out into region 3. The objects α and β will be used for triggering the "cleaning" process of unneeded symbols. For instance, α is used only to take out the catalyst k from membrane 1, and this catalyst is responsible (in the rules $k0 \rightarrow k_{in}$ and $k1 \rightarrow k_{in}$) for deleting from membrane 2 the unused symbols 0 and 1. In the other case, β will be used in the process of sending the catalysts c and/or e from membrane 2 to membrane 1.

Time	Region 4	Region 3	Region 2	Region 1
1	$c, e, 1$ $e1 \rightarrow e_{in}1_{in}$	–	–	k^2
2	$c, 1?$	$e, 1$ $e1 \rightarrow e_{in}1_{in}y_{in}$	–	k^2
3	$c, 1?$	–	$e, 1, y$ $y \rightarrow y'$	k^2
4	$c, 1?$	–	$e, 1, y'$ $y' \rightarrow y''$	k^2
5	$c, 1?$	–	$e, 1, y''$ $ey'' \rightarrow e\alpha_{in}\beta B_{out}d_{out}$	k^2
6	$c, 1?$	B, d $B \rightarrow B_{out}$	$e, \beta, 1$ $e\beta \rightarrow e_{out}$	$k^2\alpha$ $k\alpha \rightarrow k_{out}$
7	$c, 1?, B$ $B \rightarrow 1'_{out}$	d, e $ed \rightarrow e_{out}$	$k, 1$ $k1 \rightarrow k_{in}$	k
$8(1'f)$	$c, e, 1?$	–	–	k^2

Fig. 10. The case when only object 1 enters region 4.

The second case can occur if only one symbol, 0 or 1, enters membrane 4; see Figure 10. In this case, the rules $c0 \rightarrow c_{in}0_{in}$ and $e1 \rightarrow e_{in}1_{in}$ cannot be used simultaneously. Because of this, only one rule will be executed in membrane 3 ($c0 \rightarrow c_{in}0_{in}$ or $e1 \rightarrow e_{in}1_{in}$). Suppose that 1 enters region 2; then the rule $c1 \rightarrow c\alpha_{in}\beta g_{out}$ will not be applied because the correct catalyst c is missing. So, for instance, the sequence of rules that will be applied in the case of a unique symbol 1 being present is the following.

First the rules $y \rightarrow y'$ and $y' \rightarrow y''$ are used for delaying the process, preventing the entrance of a new symbol 1 (if any) into region 2. Also, one can see that if the catalyst e and the symbol p are at the same time in region

2, then it means that only one symbol 1 entered the region, and so we can notify this to the outer regions. In all cases, the object k is responsible for "deleting" unused symbols that otherwise could cause a conflict between some rules in the next cycle of computation.

Figures 9 and 10 show the timetables for the cases discussed, considering in each case one significant cycle.

The case where only one symbol 0 enters region 4 can be realized in a straightforward way, being similar to the previous case (where object 1 enters membrane 4).

One can note that the delays used during the whole process were important for two reasons: some of them were used for synchronization tasks, while others were meant only to have the output in the same computation time (not depending on the input). For instance, the sequence of rules $C \to C'$, $C' \to C''$, and $C'' \to C_{out}$ will delay the output in the case where 0 and 1 enter simultaneously.

Now, suppose that in membrane 1 we have objects 0^i and 1^j with $i > j$. In this case, after $7 \cdot (|i - j|) + 1$ steps, region 4 will receive the first signal symbol A, indicating the fact that $i > j$. In a similar way, if $j > i$ or $i = j$, at the same time as before, region 4 will receive the signal symbol B or C, pointing to the corresponding cases.

What is interesting is that the same P system with few modifications can serve different tasks, such as obtaining the minimum (maximum) between the multiplicities of two objects, obtaining the difference of multiplicities, obtaining the sum of multiplicities, and so on.

5.7 Using a Comparator

Now, once we have the comparator, in order to solve the sorting problem we have only to construct the comparison network. In this way, we will be able to simulate practically any classical comparison-exchange-based algorithm.

We can use the comparator to make all the comparisons present in the sorting network, but only one comparison at a time due to the number of catalysts present in the comparison module (one can extend the number of catalysts for a more parallel approach, but then we will have synchronization problems). The time complexity for this case will be proportional to the number of comparisons made in a sorting network. The number of membranes in this model is fixed and does not depend on the number of components to be sorted.

The order of how elements enter in the comparison module will be controlled also by mobile catalysts, one associated with each comparison, as well as a fixed number of auxiliary catalysts. The comparison module sends to the outer region signals representing the fact that the work of the module is still in progress. When a comparison is finished, no signal will be sent, and so the system will proceed to the next comparison.

6 Ranking

The ranking is an algorithm which processes a multiset M and gives as a result a word $w = \prod_{j=1}^{k} a_j \in Perm(V)$ such that $M(a_j) \leq M(a_{j+1})$. In the strong sorting by P systems with promoters, another method was used to calculate the ranking: $rank(\underline{j}) = card\{\underline{i} \mid M(\underline{i}) < M(\underline{j})\}$, and $w = \prod_{j=1}^{k} a_j \in Perm(V)$ such that $rank(a_j) \leq rank(a_{j+1})$. In fact, only $\{(j, rank(\underline{j})) \mid 1 \leq j \leq k\}$ was computed. In the case of equality of some numbers either a decision should be made as to whether "=" is "≤" or "≥" (see the technique of unequalizer) or the arbitrary order is produced. In the strong sorting algorithms presented above, also the representation of ranking in some form was computed, so ranking subroutines are hidden in them.

6.1 Employing Mobile Catalysts

If in the previous case the construction of the comparator was intended to prove that the comparison of the multiplicities of two objects could be achieved with only three catalysts, here we will show a generalization of the comparator. Practically, with more catalysts (k+1 catalysts, where the first k are associated with the objects whose multiplicities we want to sort, plus one which will be used to "clean" unneeded symbols) we can reach our goal in an unexpectedly good time with respect to the "weakness" of the ingredients.

Nevertheless, this will give the answer to the ranking problem only, and one can remark that the passing from the ranking problem to the sorting problem (using only catalytic rules) is not a trivial task. This is due to the fact that the rules must be independent of the initial multiplicities of the objects representing the initial vector (no rules of type $ca \to ca_{out}^{n_1}$ or $a \to a_{out}^{n_1}$ are used).

Algorithm 9. *Let* $v = \prod_{i=1}^{k} \underline{i}$. *Consider*

$$\begin{aligned}
\Pi &= (O, C, [_1\ [_2\ [_3\ ]_3\ ]_2\ ]_1, s^k, \textstyle\prod_{j=1}^{k} X_j, K^k, R_1, R_2, R_3, \{1\}, \{0\}), \\
&\quad \textit{with input } \textstyle\prod_{j=1}^{k} \underline{j}^{n_j}, \\
C &= \{X_i \mid 1 \leq i \leq k\} \cup \{K\}, \\
O &= C \cup \{\underline{i}, y_i \mid 1 \leq i \leq k\} \cup \{s, f, t\}, \\
R_1 &= \{\underline{j} \to \underline{j}_{in},\ X_j y_j \to X_j \underline{j}_{out} \mid 1 \leq j \leq k\} \cup \{s \to f,\ f \to s_{in}\}, \\
R_2 &= \{X_j \underline{j} \to X_{j\,in} \underline{j}_{in},\ X_j s \to X_{j\,out} y_{j\,out} \mid 1 \leq j \leq k\} \\
&\quad \cup \{Ks \to K_{in} f_{out}\}, \\
R_3 &= \{K\underline{j} \to K_{out} t,\ X_j t \to X_{j\,out} \mid 1 \leq j \leq k\}.
\end{aligned}$$

In Figure 11 are presented the significant steps of the computation made by the Π when solving the k-integer ranking problem.

Step	Env	Region 1	Region 2	Region 3
0		$s^k\prod_{j=1}^k \underline{j}^{n_j}$	$\prod_{j=1}^k X_j$	K^k
1		f^k	$\prod_{j=1}^k \underline{j}^{n_j}\prod_{j=1}^k X_j$	K^k
1		f^k	$\prod_{j=1}^k \underline{j}^{n_j}\prod_{j=1}^k X_j$	K^k
2			$s^k\prod_{j=1}^k \underline{j}^{n_j-1}$	$K^k Pi_{j=1}^k X_j \prod_{j=1}^k \underline{j}$
3			$s^k K^k \prod_{j=1}^k \underline{j}^{n_j-1}$	$t^k Pi_{j=1}^k X_j$
4		f^k	$\prod_{j=1}^k \underline{j}^{n_j-1}\prod_{j=1}^k X_j$	K^k
$1'$		f^k	$\prod_{j=1}^{k-1} \underline{j}^{n_j}\prod_{j=1}^k X_j$	K^k
$2'$			$s^k X_k \prod_{j=1}^{k-1} \underline{j}^{n_j-1}$	$K^k Pi_{j=1}^{k-1} X_j \prod_{j=1}^{k-1} \underline{j}$
$3'$		$X_k y_k$	$s^{k-1} K^{k-1} \prod_{j=1}^{k-1} \underline{j}^{n_j-1}$	$K t^{k-1} Pi_{j=1}^{k-1} X_j$
$4'$	$\underline{k}$	$f^{k-1} X_k$	$\prod_{j=1}^{k-1} \underline{j}^{n_j-1}\prod_{j=1}^{k-1} X_j$	K^k

Fig. 11. The computation for the k-integer ranking problem.

Example 11. Consider the case of ranking 3 numbers. For simplicity, let us rename the variables as follows: $\underline{1} = a$, $X_1 = A$, $y_1 = a'$, etc.
Initial data: $[_1 a^{n_1} b^{n_2} c^{n_3} s^3 [_2 ABC [_3 K^3]_3]_2]_1$
Rules of region 1: $a \to a_{in}$, $b \to b_{in}$, $c \to c_{in}$, $s \to f$, $f \to s_{in}$,
$Aa' \to Aa_{out}$, $Bb' \to Bb_{out}$, $Cc' \to Cc_{out}$;
Rules of region 2: $Aa \to A_{in}a_{in}$, $Bb \to B_{in}b_{in}$, $Cc \to C_{in}c_{in}$,
$As \to A_{out}a'_{out}$, $Bs \to B_{out}b'_{out}$, $Cs \to C_{out}c'_{out}$, $Ks \to K_{in}f_{out}$;
Rules of region 3: $Ka \to K_{out}t$, $Kb \to K_{out}t$, $Kc \to K_{out}t$,
$At \to A_{out}$, $Bt \to B_{out}$, $Ct \to C_{out}$.

In Figure 12 we present the computation for the case of three numbers.

Step	Env	Region 1	Region 2	Region 3
0		$s^3 a^{n_1} b^{n_2} c^{n_3}$	ABC	K^3
1		f^3	$a^{n_1} b^{n_2} c^{n_3} ABC$	K^3
1		f^3	$a^{n_1} b^{n_2} c^{n_3} ABC$	K^3
2			$s^3 a^{n_1-1} b^{n_2-1} c^{n_3-1}$	$K^3 ABCabc$
3			$s^3 K^3 a^{n_1-1} b^{n_2-1} c^{n_3-1}$	$tttABC$
4		f^3	$a^{n_1-1} b^{n_2-1} c^{n_3-1} ABC$	K^3
$1'$		f^3	$a^{n_1} b^{n_2} ABC$	K^3
$2'$			$s^3 C a^{n_1-1} b^{n_2-1}$	$K^3 ABCabc$
$3'$		Cc'	$ssKKa^{n_1-1} b^{n_2-1}$	$KttAB$
$4'$	c	ffC	$a^{n_1-1} b^{n_2-1} AB$	K^3

Fig. 12. The computation for the 3-integer ranking problem.

Now let us see what the idea is for this system. First we have the symbol f (a checker) which goes back and forth from membrane 1 to membrane 2. It will react in different ways according to what it finds in membrane

2. In this membrane, by applying the rules of type $X_j\underline{j} \rightarrow X_{j\,in}\underline{j}_{in}$ in a step of computation, we "erase" one symbol from each component (where by component we mean the multiset $\underline{j}^{n_j}$, $1 \leq j \leq k$). Up to the time when one component is completely erased, the rules of type $Ks \rightarrow K_{in}f_{out}$ will be applied. The catalyst K is also executing a back and forth oscillation between membranes 2 and 3, but with a different timing than f. This is done because we would like that if there are still elements to delete, the rules of type $X_j\underline{j} \rightarrow X_{j\,in}\underline{j}_{in}$ be executed before the rules of type $Ks \rightarrow K_{in}f_{out}$. When one component is "removed" from the initial multiset, the corresponding catalyst will not leave membrane 2, and so the rule $X_js \rightarrow X_{j\,out}y_{j\,out}$ can be applied. This means that we can send a signal representing the fact that the object with the smallest multiplicity was reached. At the same time, the corresponding catalyst will enter membrane 1 and will not participate in any more tasks. The process will continue until all components have been exhausted. In this way we obtain a ranking algorithm which will have time complexity proportional to the maximum multiplicity of the objects from the initial multiset.

Now, once we have the ranking problem solved, we can go further to solve the sorting. First, one can remark that a signal is a catalyst or a common object (the rule to send signals in membrane 1 is of type $X_js \rightarrow X_{j_{out}}y_{j_{out}}$). Of course, one can send out a finite set of objects but this will not affect essentially the process. Another possibility is that before the process begins we can make a copy of the initial multiset (for example, we replace the rule $\underline{j} \rightarrow \underline{j}_{in}$ with the rule $\underline{j} \rightarrow \underline{j}_{in}\,\underline{j}'$ in membrane 1) for using later when we would like to eliminate objects in the right order. A rule of type $X_jy_j \rightarrow X_jy_j'^{n_1}$ is not good, because we would like to have a system that dissociates the input from the implementation; hence we have to construct the output based on the signals that we receive. Moreover, the elements cannot be eliminated all at once because this implies a non-cooperative rule; however, having such kinds of rules, we cannot control the process.

Therefore, the only solution is to again use catalysts. No matter how this is done, it will be a difficult task because the signals representing the order are not sent out in the "right" interval of time (the interval from which we can easily deduce the multiplicity of a certain object). For instance, consider the case where the multiplicities of the objects are $n_1 > n_2 > n_3 \ldots$; then, the first signal will appear after a time proportional to n_3, the second after a time proportional to $n_2 - n_3$, and so on. As it can be seen, if in our example we have $n_3 > (n_2 - n_3)$, then we will have "overlapping" tasks, meaning that at a certain moment we will have more than one signal present in the same membrane. To overcome this, one can use the same techniques as before to allow only one catalytic rule to work, up to the time when all elements from a specified component will be exhausted.

7 Conclusion

We have studied the possibility of solving the sorting problem in the membrane computing framework by considering the main variants of P systems. An interesting result concerning this topic is that starting with objects that do not have any order and are mixed together in what we formally call a multiset, we constructed the order by computing. We studied in this way many membrane system models that behave in slightly different ways when they address the same problem. The common feature shared by many algorithms presented is that we sort by "carving" (consuming objects iteratively, one symbol from each of the components at once) and signaling when a modification occurs in the system (usually triggering a signal when a certain component has been eliminated). Other ideas included applying the classical approaches of sorting by using a comparator. This comparator was implemented using only input/output operations and catalytic rules. The device presented can be adapted to work in many algorithms that are comparison-based.

Sorting problems are among the most important problems in computer science theory. Besides them, many other problems are waiting to be solved in the framework of P systems. The reason is that we can obtain better results in time complexity than by using classical algorithms because of the massive parallelism of P systems. We believe that some techniques (the comparison method in the case of movable catalysts and non-cooperative rules, the synchronization methods used in the comparison-based algorithms) developed in this chapter can be also used to construct systems that solve such kinds of problems. The improvements of current algorithms (by reducing the number of membranes when this is the case, reducing the number of catalysts, and so on) are also open areas of research. Yet another interesting research topic is designing *dynamic* sorting (universal for an arbitrary number of components) P systems.

Acknowledgments. The first author is supported by the project TIC2002-04220-C03-02 of the Research Group on Mathematical Linguistics, Tarragona. The first author acknowledges IST-2001-32008 project "MolCoNet" and the Moldovan Research and Development Association (MRDA) and the U.S. Civilian Research and Development Foundation (CRDF), Award No. MM2-3034. The second author's work was supported by a doctoral fellowship from Agencia Española de Cooperación Internacional, Spanish Ministry of Foreign Affairs.

References

1. A. Alhazov: Minimizing Evolution-Communication P Systems and Automata. *New Generation Computing*, 22, 4, (2004), 299–310.
2. A. Alhazov, D. Sburlan: Static Sorting Algorithms for P Systems. In *Pre-Proceedings of the Workshop on Membrane Computing, Tarragona, 2003* (A. Alhazov, C. Martín-Vide, Gh. Păun, eds.), GRLMC Report 28/03, Tarragona, 2003, 17–40.
3. J.J. Arulanandham: Implementing Bead-Sort with P Systems. In *Unconventional Models of Computation 2002* (C.S. Calude, M.J. Dinneen, F. Peper, eds.), LNCS 2509, Springer, Berlin, 2002, 115–125.
4. M. Cavaliere: Evolution-Communication P Systems. In *Membrane Computing, International Workshop, WMC-CdeA 2002, Curtea de Argeş, Romania, August 2002, Revised Papers* (Gh. Păun, G. Rozenberg, A. Salomaa, C. Zandron, eds.), LNCS 2597, Springer, Berlin, 2003, 134–145.
5. R. Ceterchi, C. Martín-Vide: P Systems with Communication for Static Sorting. In *Pre-Proceedings of Brainstorming Week on Membrane Computing, Tarragona, February 2003* (M. Cavaliere, C. Martín-Vide, Gh. Păun, eds.), Tech. Rep. No. 26, Rovira i Virgili Univ., Tarragona, 2003.
6. R. Freund, L. Kari, M. Oswald, P. Sosik: Computationally Universal P Systems Without Priorities: Two Catalysts Are Sufficient. *Theoretical Computer Science*, 330, 2 (2005), 251–266.
7. M. Ionescu, D. Sburlan: On P Systems with Promoters/Inhibitors. *Journal of Universal Computer Science*, 10, 5, (2004), 581–599.
8. S.N. Krishna, A. Păun: Three Universality Results on P Systems. In *Pre-Proceedings of Brainstorming Week on Membrane Computing, Tarragona, February 2003* (M. Cavaliere, C. Martín-Vide, Gh. Păun, eds.), Tech. Rep. No. 26/03, Rovira i Virgili Univ., Tarragona, 2003, 198–206.
9. Gh. Păun: *Membrane Computing. An Introduction.* Springer, Berlin, 2002.
10. D. Sburlan: A Static Sorting Algorithm for P Systems with Mobile Catalysts. *Analele Ştiintifice Univ. Ovidius Constanţa*, seria Matematică, 11, 1, (2003), 195–205.

Chapter 9
Membrane-Based Devices Used in Computer Graphics

Alexandros Georgiou, Marian Gheorghe, Francesco Bernardini

Department of Computer Science
The University of Sheffield
Regent Court, Portobello Street, Sheffield, S1 4DP, UK
Alex.Georgiou@hushmail.com, M.Gheorghe/F.Bernardini@dcs.shef.ac.uk

Summary. A model of plant growing based on some variants of P systems is considered. The model (very close to L systems-based approaches) represents a further step toward a more modular way to specify biological systems. A specification language and a tool supporting the model are also presented together with some simple examples. A key operation introduced in this context is rule rewriting – there are rules which rewrite the right hand sides of other rules. Because of the importance of this operation for our approach, we also briefly investigate it from a theoretical point of view in the Annex.

1 Introduction to Plant Modeling

A variety of computational models have been employed for the task of simulating the growth and development of living plants. These models can give us a variety of information regarding the behavior of the modeled plants as well as graphical representations of the simulation. The graphical models are designed according to two driving principles: (i) they should be simple enough to be designed by humans and computed by machines, and (ii) they must be complex enough to mimic the relevant biological processes, a requirement for producing realistic graphics.

Among the most commonly used models to graphically represent plants are those based on grammars. L systems, which were introduced in 1968 by Aristid Lindenmayer, have been successfully used as the rewriting mechanism of this strategy [13], [7]. In short, a string of commands is rewritten repetitively using a set of production rules. The commands will usually be targeted to what is known as a turtle interpreter, a "cursor" able to move, rotate, and draw in two or three dimensions [9]. First, the L systems production rules are applied in parallel, iteratively, starting from an initial string called an axiom, and the resulting string is then passed to the turtle interpreter, which generates the graphics. Other computational models have been applied to the problem of plant modeling, some of which make use of a turtle interpreter, and some not. None, however, have been shown to be as flexible and expressive as L systems. On the other hand, there have been attempts to combine the L systems approach with other similar computational models in order to achieve simpler and more powerful models [23].

P systems are a much newer biologically inspired computational model, introduced in 1998, by Gheorghe Păun [17]. The model is based on the concept of membranes as they appear in living cells. In such a system the membranes separate various compartments called regions. Since membranes can be nested, regions form a hierarchical structure. Adjacent regions may communicate symbols in some circumstances. In some variants of P systems the type of computations performed inside the regions can be string rewriting. It follows that by allowing these strings to be processed by the turtle interpreter, one may use P systems to generate plant graphics. Care must be taken in defining the model; thus, the suitability of string manipulating P systems for the same task may be investigated. The novel features would be the concept of the membrane, which might add to the success of other grammar-based systems and the way symbols are passed around from one region to another.

1.1 L Systems and the Turtle Interpreter

As it is often the case in science, models that work best are the ones that have a strong analogy to the real world systems they mimic.

Perhaps the *L systems* model, which originated from Aristid Lindenmayer in 1968, is the one most widely studied for plant representation. This is a string-rewriting model, which was inspired by the developmental processes occurring in simple algae. A short and informal review of L systems follows. The interested reader may refer to the book *The Algorithmic Beauty of Plants* by Lindenmayer and Prusinkiewicz [13], which is an excellent introduction to L systems and their applications.

String rewriting in L systems proceeds in parallel (unlike Chomsky grammars, where rewriting is sequential); this creates an analogy with cellular growth and division, which also proceeds in parallel in multicellular organisms. An L system consists of an axiom string and a set of production rules. The system changes its state in a sequence of discrete steps. At any step, any symbol in the string that may be rewritten by a rule is rewritten (by exactly one rule). Thus the following axiom ω and rules

$$\omega : B$$
$$r_1 : A \to BA$$
$$r_2 : B \to AB$$

describe a system that progressively produces these strings:

B
AB
$BAAB$
$ABBABAAB$
$BAABABBAABBABAAB$
...

Just like in nature, where RNA links the one-dimensional genes with the three-dimensional proteins, so in this model there must be something that will eventually connect the two worlds; that is, convert the 1D string to a final 3D structure. This is typically accomplished with a turtle interpreter. Here a "turtle" is a cursor that moves and rotates in space, drawing lines in its path. The symbols in the L system string are interpreted as drawing commands by the turtle, thus producing plant-like shapes. These commands instruct the turtle to move forward, rotate, and draw lines, while many other commands exists for fine-tuning the drawing process.

Many extensions of L systems have been suggested, and each increases the power of the model in a different direction.

Parametric L systems are systems where symbols may have zero or more associated arithmetical parameters. Thus, a symbol A with two parameters, 4 and 2, would be denoted as $A(4,2)$. Symbols that are also drawing commands may have a special meaning associated with their parameters. For example, F is a command that makes the turtle proceed by one unit and draw a line in its path; but if the same command has a parameter, the turtle walks and draws a line of length equal to the parameter.

In this way, the parameters facilitate the specification of non-integral lengths, lines of different widths and colours, etc. The turtle interpreter must be programmed to recognize these. Parameters may also have arbitrary (user-defined) meanings in a model, such as counters, signals (hormones and nutrients), etc. Rules can be conditional, so that they rewrite a symbol only if a certain expression involving its parameters is true. Furthermore rules may perform arithmetical operations. For example, the rule

$$A(2,x) : x > 3 \rightarrow BC(x-1)$$

would rewrite any symbol A with its first parameter equal to 2 and its second parameter greater than 3, with the symbols B and C, as specified on the right hand side. The parameter of C is an arithmetical expression involving a parameter from A.

Several other extensions have been used with L systems, the most notable of which are perhaps the *context-sensitive rules* [13] and the *query modules* [15]. For the purposes of this approach we may concentrate on L systems with parameters and conditional rules, i.e., systems with rules as in the example given above. Other methods of generating plant graphics that follow the same general pattern are conceivable. The turtle interpreter could be replaced by some other graphical interpreter, like a vector interpreter [1]. Alternatively, the turtle interpreter could be left intact and the L system model could be replaced with any other suitable developmental model that works on strings. It is this second possibility that will be investigated here.

1.2 Existing L Systems Software

The most widely used plant simulation package is *L-Studio/cpfg* and it uses the combination of L systems and turtle interpreter. It consists of two parts: *L-Studio* is the integrated development environment where a designer creates and edits models, while *cpfg* is the software that runs simulations of the models and produces graphical output using a turtle interpreter. The software package is accompanied by many examples that illustrate its power. For more information one may browse the relevant Website [11].

2 A Hybrid Model

An advantage of L systems is that they are developmental mechanisms. Not only do they construct a plant structure as it would exist at one particular moment in time, but they describe its growth and development. This feature must be preserved in a new P systems-based model.

2.1 Design Decisions

It is desirable to be able to combine models, thus constructing composite developmental models. Let us consider the requirement to create a forest of many types of trees, or to combine a leaf model and a branch model in order to create a tree model in two parts. Modular design is an old and familiar concept. It is highly desirable that the composite complex system remain developmental.

The modular design is normally facilitated with the use of sub-L systems. Multiple systems are labelled with integers and the first one is the main system. One may use special symbols to "call" one system from another (Hanan [8] constructs a loose analogy with subroutine calls in programming). A substring from one system is thus processed by a subsystem as an axiom. The resulting structure of the subsystem may be scaled geometrically, but it may also continue to grow as part of the supersystem. The details of this mechanism are given in the *CPFG User's Manual* that accompanies the software.

In a P system it is reasonable to encapsulate multiple systems within membrane regions. This makes it easier to reason about the hierarchy of subsystem calls. As will be later discussed, this hierarchical structuring may assist in debugging systems, with some relevant software support.

Membranes may also assist in modeling *switching events*, as defined in [13]. A switching event is an event that triggers some sudden change in the growth of a plant. In reality, such an event may be either some change in environmental conditions (weather, availability of water and nutrients, etc.), or an internal delay mechanism. In L systems, this mechanism is facilitated by *table L systems* and conditional rules ([13], pp. 66, 67).

In P systems where strings cross membranes, a set of regions may have a different meaning. It may be interpreted as a set of alternative sets of rules that may be applied to a growing string. Thus a string that travels through different regions in the membrane structure is grown under varying conditions.

Toward this purpose, a somewhat non-standard idea has also been explored: that of making rules mobile, rather than strings. Rules may be associated with lists of *conditional targets.* A conditional target is a region label, associated with a Boolean expression, so that if the expression becomes true, the rule migrates to that region. The simplest strategy for selecting among multiple true expressions in a list would be to pick the first one.

Encapsulating parts of a system into membranes also makes it easier to localise *variables.* In L systems, numerical data can be stored either as parameters associated with individual symbols, or as global variables. In membrane-based systems, variables may be associated with regions. Then the whole problem of switching events does not any more require rule or string migration. In L systems with C-like programming extensions, as defined in the *CPFG User's Manual*, rule preconditions may involve variables declared globally. In membrane systems, preconditions may involve variables local to the region of the rules. Like string-rewriting rules that modify strings, there can be arithmetical rules that modify numerical variables.

The issue of how to collect the resulting strings arises with membrane-based systems. In L systems this is trivial, since one string has always been derived from the axiom after any number of steps. In P systems there can be multiple strings.

In the case of membrane systems with migrating strings, a solution is to use conditional targets to move strings into a special target compartment, called *result.* This produces the complication of how to select among multiple strings sent to the result during one simulation step.

In the more promising approach of membrane systems with nonmigrating strings and rules, a solution with a more "natural" feel exists: strings may be collected and concatenated by visiting regions recursively. For any region there is exactly one string. This string may involve special symbols that reference other regions. When collecting a string, any special symbol is to be replaced by the string of the subregion it points to. This approach constructs an analogy between the branching structure of plants and the hierarchical topology of membrane regions.

In such systems, where strings and rewriting rules are not migrating, communication through membranes is achieved with migrating variables and arithmetical rules. These may move in the usual directions, that is either out one level, or in a submembrane of their current region. Such migrating components may again be given lists of conditional targets.

Once a subsystem ceases computation, its result may or may not need to be available to the containing membrane. In the first case, the submembrane may be dissolved, thus flattening the string. The flattening process would involve inserting the substring into the positions indicated by the special symbols, which would then disappear. The issue of how dissolution may be triggered arises.

Finally the possibility of *rule rewriting* was investigated. This is the possibility of rewriting the right hand side of a string-rewriting rule. These rule-rewriting rules could be used to create an entire set of rules, each of which would have a similar right hand side, but evolved at a different level. For example, X symbols on the string representation of a tree could indicate the positions where various branches are to be appended. The branches might need to be of different ages. One may prepare a set of rules that will each rewrite X with a branch of a different age. For example, the rule-rewriting rule

$$A \to^{l} b\,A\,b$$

could operate on a rewriting rule such as

$$X \to A\,B\,C$$

to give the following set of rules:

$$\begin{aligned} X &\to A\,B\,C \\ X &\to b\,A\,b\,B\,C \\ X &\to b\,b\,A\,b\,b\,B\,C \\ X &\to b\,b\,b\,A\,b\,b\,b\,B\,C \\ &\ldots \end{aligned}$$

This set of rules could be applied nondeterministically on X symbols, to place branches of various ages at random X positions. More usefully, a mechanism may be devised so that the sequence of generated rules rewrites symbols of increasing parameter values, e.g., $X[1], X[2], X[3], \ldots$. However, the specification of how this counting should proceed becomes complex, since many parameters need to be specified or otherwise implied. These would include the number from which counting starts, the amount of every increment, the parameter or set of parameters that is to be modified, the maximal amount of rules generated, etc.

The problem of when a rule may rewrite other rules and when it may be rewritten also arises. This may be resolved by means of a label shown next to the arrow of the production, here denoted by l. In the region r a rule with a label attachment may be called *active* if $l = r$, and *inactive* otherwise. Active rules may rewrite other rules, while inactive rules may have their right hand side rewritten by active rules. Thus, in a setting where rule migration is allowed, rules can switch between these two states as they migrate in the membrane structure. In models where rule migration is not allowed, rules may be characterized as active or inactive according to another custom Boolean condition.

While the idea of rule rewriting seems exciting at first, and indeed might deserve further research, it is obvious that it requires several supporting mechanisms in its most general form. To summarize, the two major problems identified above involve

- resolving the issue of which rules rewrite which ones, and
- differentiating between generated rules.

If these issues are addressed in the context of a membrane-based plant model, then the specification of metarules may become more complex than the problems such rules might address.

2.2 The Sub-LP Systems Model and Its Specification Language

Considering the reasoning presented above, the sub-LP systems model is developed. It is based on recursive strings, and features parametric symbols, conditional string rewriting rules, migrating numerical variables and migrating arithmetical rules; rule rewriting and dissolution have been also considered, though not yet implemented by the software tool supporting the model.

The sub-LP systems model is based on the following four components:

- *strings* of symbols;
- *rewriting rules* that include communication and dissolution; these may be either conditional or non-conditional;
- *numerical variables* that contain real numbers;
- *arithmetical rules* that act on numerical variables.

Before we formally define such a system we introduce for an alphabet V the notation V_n^A as denoting the set of elements $a(r_1, \ldots, r_p)$ where $a \in V$ and r_i is an arithmetical expression constructed with variables from A, $1 \leq i \leq p$, $0 \leq p \leq n$. The set V_n^A identifies all symbols of V with at most n parameters from A.

A sub-LP system may be formally defined as a construct

$$\Pi = (V, A, \mu, w_1, \ldots, w_m, A_1, \ldots, A_m, R_1, \ldots, R_m, T_1, \ldots, T_m),$$

where:

- V is an alphabet;
- A is a set of variables with real values;
- μ is a membrane structure of degree m; the membranes are labelled with integers in the range 1 to m;

- w_i, $1 \leq i \leq m$, represents a multiset of strings over $V_{n_i}^{A_i}$, for a given n_i and a set of variables from A_i, possibly with an initial value;
- A_i, $1 \leq i \leq m$, represents the set of variables from A used by membrane i;
- R_i, $1 \leq i \leq m$, represents the set of rules associated with membrane i; the rules are of the form
  - $u : c \rightarrow^l v$ where c is a Boolean expression built with variables from A_i, l is a value in the range 1 to m, $u \in (V_{n_i}^{A_i})^*$, and $v \in (V_{n_i}^{A_i})^* \times Tar$, where $Tar = \{out, in_j, \lambda\}$ defines the target of the rule;
  - v can also contain the special notation *?name*; the occurrence of this special symbol will import a sub-LP system called *name*, which will be added to the contents of the current membrane; the special symbol is then written in the string in the form *?n*, where n is the region label of the top-level membrane of the newly imported subsystem;
  - v can also contain the symbol δ which means dissolution of the membrane where the rule is applied;
- T_i, $1 \leq i \leq m$, represents the set of arithmetical rules present in the membrane i of the form $n \leftarrow e$, where $n \in A_i$ and e is an arithmetical expression built with variables from A_i.

The rules from R_i and A_i are applied in parallel and it is assumed that no conflicts occur between rule targets. The role of A_i rules is to compute some parameters requested by different variables. When a membrane is dissolved the rules are also delivered to the new context.

Systems are conveyed by means of a dedicated language, which allows for constructs representing the various elements of the model. A detailed description of the model follows.

Membranes are used to separate the structure of the system into regions, each of which performs computation. When regions do not communicate, each performs an isolated computation. Membranes are labelled with distinct positive integers. A membrane's region may contain any number (or none) of variables, arithmetical rules, string rewriting rules, and other membranes. Every membrane contains one and only one symbol string. Thus sub-LP systems are hierarchically structured just like P systems. The top-level membrane is known as the skin membrane. The skin membrane and its symbol string are the only required components of a sub-LP system.

Membranes are specified using curly brackets. The text between the two curly brackets is made up of other membranes, and components that end in a semicolon. The first statement of the membrane is required, and labels the membrane's region with a distinct positive integer. The second statement is always the string associated with the region. For example:

$$\{\ region\ 3;\ \ A\ BC;\}$$

is a membrane with label 3. The membrane contains the two symbol string

$A\ BC$. The first symbol in the string is A and the second one is BC, since symbols are separated by spaces and may have names of multiple characters. An example of nested membranes is

$$\begin{array}{l} \{\ \textit{region}\ 10;\ A; \\ \quad \{\ \textit{region}\ 5;\ B;\ \} \\ \quad \{\ \textit{region}\ 20;\ C\ d\ eF;\ \} \\ \} \end{array}$$

where membrane 10 contains two membranes, with labels 5 and 20.

If a region is labelled with 0, then the simulator assigns a fresh label to it when it is constructed. This is useful for defining subsystems that are to be dynamically imported.

Variables can hold floating-point numbers. Every variable is associated with a region. Defining variables is done using the *define* keyword, which may be abbreviated to *def*. The definition statement may specify a value to be placed in the variable. Otherwise, the variable is taken to be equal to zero. Initial values may be provided in the form of arithmetical expressions, but cannot depend on other variables.

For example, the statement

$$\begin{array}{l} \{\ \textit{region}\ 1; \\ \quad \textit{string}; \\ \quad \textit{def}\ i; \\ \quad \textit{def}\ J = 5; \\ \quad \textit{define}\ z\ =\ 3 * (9+1); \\ \quad \textit{def}\ pi\ =\ 3.14159; \\ \} \end{array}$$

constructs a membrane which contains four variables, named i, J, z, and pi, initialized to $0, 5, 30$, and 3.14159, respectively, while the statement

$$\begin{array}{l} \{\ \textit{region}\ 2; \\ \quad \textit{string}; \\ \quad \textit{def}\ foo = 1; \\ \quad \textit{def}\ bar = foo + 1; \\ \} \end{array}$$

will set both variables to 1 (variables are constructed in parallel and *foo* is initialized to 0).

Arithmetical rules allow for the contents of a variable to be set to the result of an arithmetical expression. The expression may involve decimal numbers, parentheses, other variables, functions, a few special symbols, and arithmetical $(+, -, *, /)$, logical $(==, !=, <=, >=, <, >)$, and boolean (&&, ||) operators. The meaning of the operators is as in the C programming language. As floating-point is the only acceptable data type, comparison operators return either -1 or 0. The boolean operators take precedence over the comparison operators, which take precedence over addition and subtraction, which take precedence over multiplication and division. Functions are discussed later in the text.

Three special values are always available: *true* is equal to -1, *false* is equal to 0, and *region* is the label of the current region. If variables have one of these names, these variables are hidden. It is thus not recommended to name variables with one of these three names. If a variable does not exist in the region it is referenced, its value is assumed to be 0.

The syntax of an arithmetical rule requires a left arrow written as $\leftarrow$. The target variable is on the left of the arrow, where the result is stored after evaluation. On the right of the arrow the expression to evaluate is specified. For example, the system

$$\begin{array}{l} \{\ region\ 1; \\ \quad string; \\ \quad def\ x;\ def\ y = 1; \\ \quad x\ \leftarrow\ y;\ y\ \leftarrow\ x + y; \\ \} \end{array}$$

will compute numbers of the Fibonacci sequence. For every variable definition in the region, one arithmetical rule is applied at maximum. All the arithmetical rules to be applied are applied in parallel. Thus the system

$$\begin{array}{l} \{\ region\ 1; \\ \quad string; \\ \quad def\ x;\ def\ y = 1; \\ \quad y\ \leftarrow\ x + y;\ x\ \leftarrow\ y; \\ \} \end{array}$$

is equivalent to the previous one, and also computes the Fibonacci sequence, although the two arithmetical rules are given in reverse order.

The system

$$\begin{array}{l} \{\ region\ 0; \\ \quad def\ x = 3 + 2; \\ \quad define\ y = 8; \\ \quad def\ z; \\ \quad z\ \leftarrow\ x + 1 <=\ y\ ||\ region\ != 4; \\ \} \end{array}$$

will set variable z to -1 since the condition is true. An equivalent expression with priorities made explicit would be

$z \leftarrow ((x+1) <= y) \,||\, (region \,!= 4);$

Strings are series of symbols. Any number of these symbols may be parametric. As mentioned above, a string is defined as a series of symbols separated by spaces (and terminated with a semicolon, like any other construct).

Symbols have a name and optionally one or more parameters, as in parametric L systems. The name is composed of one or more consecutive letters. Parameters are specified in square brackets and are separated by commas. Each parameter may be specified as an arithmetical expression. For example, the following system

$$
\begin{array}{l}
\{\ region\ 1;\\
\quad \{A\ B\ Foo\ bar\ baz[3, 4.2]\ X[region+1]\ x[x];\\
\quad def\ x = 3;\\
\}
\end{array}
$$

will construct a membrane whose region contains a variable x, equal to 3, and a string with the symbols A, B, Foo, bar, and baz (the last has two parameters, 3 and 4.2), followed by the symbol X with parameter 2 and another symbol x with parameter 0. Note that the variable x is unknown at the time of constructing the symbol $x[x]$ because all contents of the membrane are initialized in parallel. That is why the parameter is set to 0.

Rewriting rules are used to rewrite symbols in the same region with (sequences of) other symbols. Rewriting rules in sub-LP systems are based on the rewriting rules of L systems. Unlike some rewriting P system variants, these rules do not replicate strings. Rewriting rules are specified with the right arrow, $\rightarrow$. On the left hand side of the arrow is the symbol to be rewritten. On the right hand side is a list of zero or more symbols with which to replace the left hand side symbol. An example follows:

$$
\begin{array}{l}
\{\ region\ 1;\\
\quad a\ abc\ ab[1,2]\ ab[1,3]\ a;\\
\quad define\ x = 7;\\
\quad a\ \rightarrow\ c[4]\ a\ ab[1,2]\ c[4];\\
\quad ab[1,2]\ \rightarrow\ cba[x+1];\\
\}
\end{array}
$$

The system shown above contains a string, a variable, and two rewriting rules. After one simulation step, the string will become:

$$c[4.0]\ a\ ab[1.0, 2.0]\ \ c[4.0]\ abc\ cba[8.0]\ ab[1.0, 3.0]\ c[4.0]\ a\ ab[1.0, 2.0]\ c[4.0]$$

From this example several points should be observed:

- All the instances of the symbol a are rewritten in parallel.
- The symbol ab with parameters 1 and 2 is rewritten with the second rule.
- The symbol ab with parameters 1 and 3 is not rewritten.
- The newly constructed symbol ab with parameters 1 and 2 is not rewritten; only the old one is.
- Arithmetic is allowed in new parameters, and the region's variables are available.

Rewriting rules may be conditional and are specified as in parametric L systems. In the system

$$
\begin{array}{l}
\{ \ region\ 1; \\
\quad A[1,2]\ A[2,3]\ A[3,2]; \\
\quad define\ y = 8; \\
\quad A[x,2]\ :\ x > 2\ \rightarrow\ B[2, x+10]\ C; \\
\quad A[x,y]\ :\ x == 1\ \rightarrow\ A[x+1, y+1]; \\
\}
\end{array}
$$

the string becomes

$$A[2.0, 3.0]\ A[2.0, 3.0]\ B[2.0, 13.0]\ C\ .$$

This is because the first rule applies to $A[3,2]$ only and the second rule applies to $A[1,2]$. The formal parameter y hides the region variable y. Note also that in the first rule the two conditions are specified in two ways: as a literal value of 2 in formal parameters, and as an explicit condition that the formal parameter x is greater than 2.

The special notation *?name* may appear on the right hand side of a rule. The occurrence of this special symbol will import a system archived under *name* in a system library. *name* may not be a number. The system will be added to the contents of the current membrane. The special symbol is then written in the string in the form *?n*, where n is the region label of the top level membrane of the newly imported subsystem. For example,

$$
\begin{array}{l}
\{ \ region\ 1; \\
\quad A\ X\ Y; \\
\quad X\ \rightarrow\ B\ ?test\ Z; \\
\}
\end{array}
$$

may become

$$
\begin{array}{l}
\{ \ region\ 1; \\
\\
\quad A\ B\ ?2\ Z\ Y\ ; \\
\quad \{ \ region\ 2;
\end{array}
$$

```
    f;
    f → f f;
  };
  X → B ?test Z;
}
```

if the appropriate system named *test* exists. The result of this entire system would now be $A\ B\ f\ Z\ Y$. In the next step it would be $A\ B\ f\ f\ Z\ Y$. If a rule uses a subsystem multiple times in one rewrite operation, the subsystem is loaded once and multiple special symbols point to it. Variables and arithmetical rules can be specified so that they migrate within the membrane structure according to some conditions (this is more general than the definition of a sub-LP systems). A list of target-condition pairs can be appended at the end of any of the two components mentioned above, as follows:

```
component //sendto target1 ifcondition1
    //sendto target2 if condition2
      ...
```

The conditions are evaluated left to right and the component migrates to the region associated with the first condition found to be true. If no condition is true, then the component does not migrate. Conditions are again logical/arithmetical expressions.

The target can be one of:

- *out*, meaning that the component is to be moved to the next upper level region;
- *inx*, where x is the label of a region immediately enclosed by the current region; if x is not directly enclosed by the current region, the component does not migrate and subsequent target-condition pairs are not evaluated.

In the following example the variable migrates:

```
{ region 1;
  a;
  X ← X + 1;
  def X = 1 //sendto in2 if X > 3
          // sendto out if X < 1;
  { region 2; b;
    X ← X − 1; };
}
```

Here the variable will oscillate between regions 1 and 2. As mentioned before, arithmetical rules may migrate in a similar fashion. Anything sent out of the skin membrane is discarded.

2.3 Sub-LP Systems Software Support

A software system has been prepared for testing the sub-LP systems models. The SubLP-Studio application [6] parses code of the form discussed in the previous section to produce models and simulate their behavior. As it produces symbol strings at each simulation step, it passes them to *cpfg* for rendering. Reusing cpfg makes the software simpler. Also, the comparison of L systems and sub-LP systems is more objective, since the interpreting mechanism is kept constant.

SubLP-Studio is a Java application. It includes a simple text editor for preparing models and a window for viewing the progress of the simulation schematically. This view may be updated after each simulation step, helping the user observe errors and debug the system.

Systems that are referred to and loaded dynamically using the question mark ("?") symbol are actually loaded from files with .lps extensions.

Wherever numerical values are used in the systems, the software allows the use of external functions. These external functions take the form of Java *.class* files which may be compiled separately. A small set of examples accompanies the software, and more may be defined by the user. These extend the flexibility of the arithmetic-based control mechanism (i.e., the variables and arithmetical rules).

3 Examples

We examine first a simple example using conditional and parametric L systems. The L system structure is

$\#define\ BODYSTEPS\ 3$
$\#define\ TERMINALSTEPS\ 2$
$Lsystem:\ 1$
$derivation\ length:\ BODYSTEPS + TERMINALSTEPS$
$Axiom:\ A(1)$
$A(i) : i < BODYSTEPS\ \rightarrow\ I[+++A(i+1)][---A(i+1)]$
$A(i) : i == BODYSTEPS\ \rightarrow\ I[--A(i)][-A(i)][+A(i)][++A(i)]$
$homomorphism$
$A\ \rightarrow\ F$
$I\ \rightarrow\ ,F;$
$end\ Lsystem$

This system produces the graph structure shown in Figure 1. It may be observed that the first rule applied iteratively (*BODYSTEPS* times) produces the tree body whereas the last one (applied *TERMINALSTEPS* times) is responsible for the terminal structure. We will model this problem using sub-LP systems without parameters, but with active and dissolution rules. Using the specification language presented previously, the corresponding sub-LP system is

```
{ region 1;
    { region 2;
        { region 3;
            A; 0;
            n → 3 n + 1; -- n = 0, 1, 2
            3 → 3 <>; -- dissolution
            A → 3 I PUSH L L L A POP PUSH R R R A POP;
        };
        { region 4;
            0;
            n → 4 n + 1; -- n = 0, 1
            2 → 4 <>; -- dissolution
            A → 2 A;
            A → 4 I PUSH R R A POP PUSH R A POP
                PUSH L A POP PUSH L L A POP;
        };
        2 → 2 <>; -- dissolution
    };
    A → 1 F;
    I → 1 F;
}
```

This specification requires some explanations. It is obvious that this specification is much longer than the previous one, but it is more modular and has no parameters. Membrane 3 computes the string for the body structure of the tree. It involves symbols A which represent the terminal structures. Membrane 4 computes terminal structures in the form of a rule that rewrites A (the right hand side of $A \rightarrow^2 A$). When membranes 3 and 4 are dissolved, membrane 2 combines the results and is dissolved. The skin membrane contains the homomorphism transformations. The rules $n \rightarrow x\ n+1$, $x = 3, 4$, might be replaced by arithmetical rules $n \leftarrow n+1$ which will act as incrementing rules for values stored in n in membranes 3 and 4.

Even more modularization is presented by the next example, that produces the structure shown in Figure 2.

The L systems-based specification is

```
#define STEPS 4
Lsystem : 2
derivation length : STEPS
Axiom : A
A → I[+A][-A]IA
I → II
```

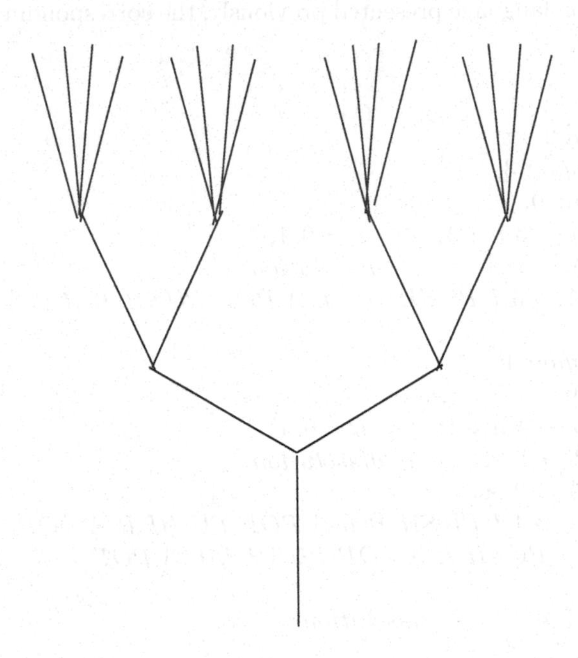

Fig. 1.

homomorphism
$A \rightarrow F$
$I \rightarrow ,F;$
endlsystem

The L system is admittedly represented in a compact way. Stems are represented with A symbols if they are about to branch, or with I symbols if they are about to double in size. The special homomorphism section maps them all to the F drawing symbol before each rendering. The same structure would be produced by the following equivalent sub-LP system, which is split into two files:

example2.lps:

```
{ region 0;
    F;
    F  → ?example2b
        PUSH L ?example2 POP
        PUSH R ?example2 POP
        ?example2b ?example2 ;
}
```

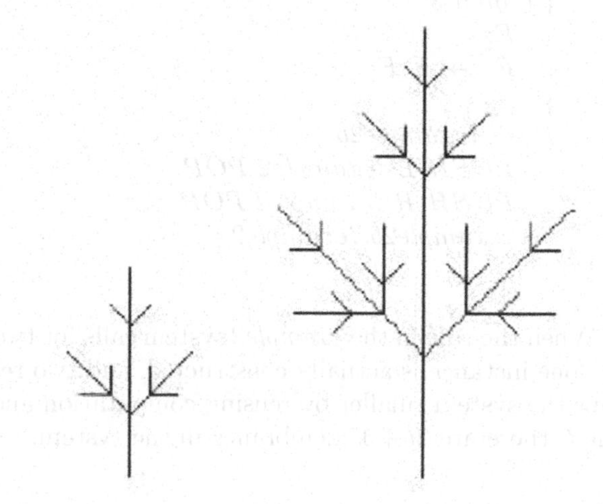

Fig. 2.

example2b.lps:

```
{ region 0;
    F;
    F → F F;
}
```

example2b is a subsystem that is used in place of the $I \rightarrow II$ rule, thus avoiding the need for a homomorphism. *example2* refers to itself, thus again avoiding the need to distinguish between A and I stems. The sub-LP system has a longer textual representation, partly because of the multicharacter symbols, partly because of the extra subsystem. However, symbols that are meant to draw stems are always F; membranes are used to control which stems grow and which ones branch.

The initial state of the system is a system loaded from the *example2.lps* file. The system after the first transition constructs two other membranes.

```
{ region 0;
    ?3 PUSH L ?2 POP PUSH R ?2 POP ?3 ?2;
    { region 2;
        F;
        F → ?example2b
            PUSH L ?example2 POP
            PUSH R ?example2 POP
            ?example2b ?example2 ;
```

```
    }
    { region 3;
        F;
        F → F F;
    }
    F → ?example2b
        PUSH L ?example2 POP
        PUSH R ?example2 POP
        ?example2b ?example2 ;
}
```

When the rule in the *example2* system calls for two new copies of *example2*, only one instance is actually constructed, and two references point to it. This keeps the system smaller by reusing computation and memory. At simulation time t, there are $2t + 1$ membranes in the system.

4 Discussion

4.1 Advantages of Sub-LP Systems

P systems are systems of nested membranes. This nestedness implies a tree-like hierarchy of regions, which has an analogy with branching in plant structures. This analogy is only loosely used in sub-LP systems, as branching is modeled by both grammatical rules and membrane nesting. Allowing this option gives the designer freedom to either exploit the representational compactness of L system-like grammatical rules, or to use membranes to localise computation.

Using membranes of subsystems gives an opportunity for additional control mechanisms to be localized. In the particular model presented here, the control mechanism consists of arithmetical rules and numerical variables. In the sub-LP systems language, these are defined close to the parts of the plant structure they affect.

The analogy with programming would be that of local procedures and variables. The designer is encouraged to solve a modeling problem in a bottom-up fashion, since subsystems can also be self-sustained entities. The designer is also encouraged to reuse computation by exploiting self-similarity in plants.

While such features as the ones discussed above are technically available in the sub-LP systems model, membranes model locality in a more intuitive way. An increase in designers' productivity is expected of the proposed model; this could be investigated with rigorous experimentation of the kind typically applied to software engineering to evaluate programming languages.

To continue the analogy of plant modeling with software engineering, the simulator software includes some primitive form of *debugging*. In between simulation steps the designer can view the state of the system; variable contents and strings are readily inspected, and membrane nesting is represented graphically.

This possibility arises due to the fact that membranes "factor" the system into small and easy to understand units. The single string of L systems on the other hand quickly becomes long and unintelligible. The hierarchical structure is difficult for humans to follow in the seemingly linear sequence of symbols. A more tight loop between design and simulation should allow for more productive interaction with the modeling software.

4.2 Language Legibility

There are some technical differences between the language employed for L systems and the one employed for sub-LP systems. In particular, the newly proposed language has the following features:

- Curly brackets ({}) to denote the concept of subsystems.
- Multicharacter symbols, an idea proposed in [23].
- Easy to remember "mnemonics" for the turtle commands.
- Rules that cross line boundaries.

The purpose of these design decisions is to increase readability. Complex L systems can be compactly represented, but will tend to appear like "line noise" to the untrained eye. Again, a relevant programming analogy is to compare compact Perl code to Java code: the latter is lengthier but requires less training to understand. If enough emphasis is put on the readability of model representations, then eventually plant modeling will become accessible to non-scientists, such as artists and graphics designers.

5 Conclusions

In this chapter we studied the opportunity to model plants growing and their graphical representation by using some variants of P systems. The approach develops a model and a specification language based on L systems models and their use in graphics. The approach presents a more modular view on modeling in general and plant development in particular. It shows some advantages in this respect illustrated by two very simple examples, but it requires more thorough investigation for us to prove its entire capabilities. We also studied the power of a class of P sysems (with rule rewriting) that is used in modeling plant development for graphical purposes.

ANNEX: Some Theoretical Results Regarding Rule Rewriting

In the previous discussion were presented a number of operations associated with rewriting rules that require further research in order for us to establish their properties and computational power. Rewriting rules were defined containing on the right hand side references to other membrane structures that yield a new structure containing at the rule level (a) an instance of the membrane structure referenced by the rule and (b) a reference in the strings rewritten by that rule to the newly instantiated membrane structure. This powerful mechanism deserves some further investigation in order for us to see to what extent the computation capabilities of membrane systems with this feature increases. Another operation suggested in the same subsection deals with rewriting the right hand side of some rules when they are inactive (rule rewriting).

Here, we present some results concerning the power of rule rewriting in P systems with string objects by formalizing some of the ideas mentioned before. Specifically, we introduce a model of rewriting P systems that includes both the feature of rule rewriting and the feature of rule moving. This means allowing both strings and rules to be moved from one region to another with the right hand side of the rules possibly modified during this process. Formally, we give the following definition.

Definition 1. *A rule rewriting P system is a construct*

$$\Pi = (V, T, \mu, M_1, \ldots, M_n, R_1, \ldots, R_n),$$

where:

1. V *is a finite alphabet of abstract symbols;*
2. $T \subset V$ *is the terminal alphabet;*
3. μ *is a membrane structure with n membranes injectively labeled with numbers in* $\{1, \ldots, n\}$*;*
4. M_i*, for each* $1 \leq i \leq n$*, is a finite language over* V *that contains the strings initially present in region* i*;*
5. R_i*, for each* $1 \leq i \leq n$*, is a finite set of rules that are initially assigned to the region* i*; they can be of the following forms:*
   a) $a \to (x, t)$*, with* $a \in V$*,* $x \in V^*$*, and* $t \in \{here, in, out\}$ *(rewriting rules),*
   b) $a \to (x, t_1; y, t_2)$*, with* $a \in V$*,* $x, y \in V^*$*, and* $t_1, t_2 \in \{here, in, out\}$ *(rule-rewriting rules).*

A rule rewriting P system is characterized by a hierarchical arrangement of $n \geq 1$ membranes injectively labeled $\{1, \ldots, n\}$, which identify n distinct regions inside the system. Each region i, with $i \in \{1, \ldots, n\}$, gets assigned a finite set of strings M_i and a finite set of rules R_i. More precisely, the structure of the system is given by μ, a string of pairs of matching square brackets injectively labeled $\{1, \ldots, n\}$, such as

$$[_1 [_2]_2 [_3 [_4]_4 [_5]_5]_3]_1, \qquad (1)$$

where each pair of matching square brackets corresponds in a one-to-one manner to a membrane in the system. This representation makes it possible to point out the relationships of inclusions among the regions delimited by the membranes. For instance, for the membrane structure (1), we can say that region 3 (i.e., the region delimited by membrane 3) directly includes or contains region 4 and region 5. Moreover, membrane 1 that delimits the outmost region is called the skin membrane, and this defines the boundaries of the system. Finally, regions that do not contain any other regions, like regions 2, 4, and 5, are called elementary regions.

A rule rewriting P system Π evolves by applying inside each region i the set of rules currently assigned to region i to the strings currently assigned to region i. Specifically, if a region i contains a string $x_1\, a\, x_2$ and a rewriting rule $a \to (x, t)$, with $a \in V$, $x, x_1, x_2 \in V^*$, and $t \in \{here, in, out\}$, then the string $x_1\, a\, x_2$ can be replaced by a string $x_1\, x\, x_2$, which is immediately moved form region i to another region according to the target t. This target can be either *here*, that is the string will remain in region i, or *out*, that is the string will reach the region that directly contains region i, or *in*, that is the string will reach one of the regions, nondeterministically chosen, directly contained by region i. In a similar way, if a region i contains a string $x_1\, a\, x_2$ and a rule rewriting rule $a \to (x, t_1; y, t_2)$, with $a \in V$, $x, x_1, x_2 \in V^*$, and $t_1, t_2 \in \{here, in, out\}$, then the string $x_1\, a\, x_2$ can be replaced by a string $x_1\, x\, x_2$, which is immediately moved form region i to another region according to the target t_1. Moreover, as a consequence of this operation, all rules of the form $b \to (y_1\, a\, y_2, t)$ or $b \to (y_3\, a\, y_4, t_1'; y', t_2')$, with $a, b \in V$, $y_1, y_2, y_3, y_4, y' \in V^*$ and $t, t_1', t_2' \in \{here, in, out\}$, can be replaced by rules of the form $b \to (y_1\, y\, y_2, t)$ or $b \to (y_3\, a\, y_4, t_1'; y', t_2')$, respectively, which are immediately moved from region i to another region according to the target t_2. As usual, all these rules are applied in a nondeterministic maximally parallel manner. In each step, in each region, all the strings that can be rewritten by some rewriting rules or rule-rewriting rules must be rewritten in a sequential way by nondeterministically choosing a single rule for each string. Then, all the rules that can be rewritten by some of the rule-rewriting rules currently used must be rewritten in a sequential way by nondeterministically choosing a single rule-rewriting rule for each rule.

Therefore, by operating in this way, a computation in a rule rewriting P system Π is obtained by starting from the initial configuration and varying the contents of the various regions of the system in terms of sets of rules and sets of strings they get assigned. A computation in Π is said to be successful if it reaches a configuration where no more rules can be applied to the strings and the rules contained in the system. A successful computation in Π produces as output the language obtained by collecting the strings of symbols in T that are moved out from the region delimited by the skin membrane during the computation. The language $L(\Pi)$ generated by the P system Π is then defined by the union of all these languages generated as outputs of all the successful computations in Π. Finally, we introduce $ELrRP_n$, with $n \geq 1$, as the families of languages generated by rule rewriting P systems with at most n membranes.

In the context of P systems, rewriting alone has been shown not to be enough to obtain the whole class of recursively enumerable languages; the largest family of languages generated by rewriting P systems coincides in fact with the family of languages generated by matrix grammars without appearance checking [18]. For this reason other special features have been considered for rewriting P systems, such as a priority relationship among the rules, permitting and forbidding conditions associated with the rules, replicated rewriting, and control of membrane permeability [18], which increase the power of rewriting P systems and lead to various characterizations of recursively enumerable languages.

Here we want to prove a similar result for rule rewriting P systems by showing that they are able to generate the whole family of recursive enumerable languages. For this, we need to formally introduce the notion of *matrix grammars with appearance checking* from [3]. A matrix grammar with appearance checking is a construct $G = (N, T, S, M, F)$ where N, T are disjoint alphabets, $S \in N$, M is a finite set of sequences of the form $(A_1 \to x_1, \ldots, A_n \to x_n)$, $n \geq 1$, with $A_i \to x_i$ a context-free rule over $N \cup T$, for all $1 \leq i \leq n$, and F a set of occurrences of rules in M. Given $w, z \in (N \cup T)^*$, we write $w \Longrightarrow z$ if and only if there is a matrix $(A_1 \to x_1, \ldots, A_n \to x_n) \in M$ such that $w = w_1$, $z = w_{n+1}$, and, for all $1 \leq i \leq n$, either $w_i = w_i' A_i w_i''$, $w_{i+1} = w_i' x_i w_i''$, for some $w_i', w_i'' \in (N \cup T)^*$, or $w_i = w_{i+1}$, A_i does not appear in w_i, and the rule $A_i \to x_i$ appears in F (the rules of a matrix are applied in sequence, possibly skipping the rules in F if they cannot be applied; we say that the rules in F are applied in *appearance checking mode*).

The language generated by a matrix grammar with appearance checking G is defined by $L(G) = \{ w \in T^* \mid S \Longrightarrow^* w \}$. The family of languages of this form is denoted by MAT_{ac}; it is known that $MAT_{ac} = RE$ (see [3]), where RE denotes the family of recursively enumerable languages. We refer to [22] for further details about formal language theory.

A matrix grammar with appearance checking $G = (N, T, S, M, F)$ is said to be in *binary normal form* if $N = N_1 \cup N_2 \cup \{S, \#\}$, with the three sets mutually disjoint, and the matrices in M assume one of the following forms:

1. $(S \to X A)$, with $X \in N_1$, $A \in N_2$,
2. $(X \to Y, A \to x)$ with $X, Y \in N_1$, $A \in N_2$, $x \in (N_2 \cup T)^*, |x| \leq 2$,
3. $(X \to Y, A \to \#)$, with $X, Y \in N_1$, $A \in N_2$,
4. $(X \to \lambda, A \to x)$, with $X \in N_1$, $A \in N_2$, $x \in T^*$.

There is only one matrix of type 1, and F consists exactly of all rules $A \to \#$ appearing in matrices of type 3; # is a trap symbol, and once introduced, it is never removed. A matrix of type 4 is used only once, in the last step of the derivation. Note that, given a matrix grammar G, in every step of a derivation, the strings produced by applying the matrices in M assume the form αw, with either $\alpha \in N_1$, $w \in (N_2 \cup T)^*$ or $\alpha = \lambda$, $w \in T^*$. That is, in every step we have only one symbol in N_1 which is used to control the rewriting of symbols in N_2; the unique symbol in N_1 is deleted in the last step of the derivation by applying a matrix of type 4. It is known that for each matrix grammar there exists an equivalent matrix grammar in the binary normal form (see Lemma 1.3.7 in [3]). Moreover, it has been proved in [4] that each recursively enumerable language can be generated by a matrix grammar with appearance checking in binary normal form, G, with $ac(G) \leq 2$, where $ac(G)$ is the cardinality of the set $\{A \in N_2 \mid A \rightarrow \# \in F\}$. This means all the matrices of type 3 are of the form $(X \rightarrow Y, A \rightarrow \#)$, with $X, Y \in N_1$, $A \in \{B_1, B_2\} \subseteq N_2$.

Now, we can prove the following result.

Theorem 1. $ELrRP_7 = RE$.

Proof. Consider a matrix grammar with appearance checking $G = (N, T, S, M, F)$ where

- $F = \{B_1, B_2\}$,
- matrices of type 2 and 4 are injectively labeled with values in $\{1, \ldots, k\}$,
- matrices of type 3 of the form $(X \rightarrow Y, B_1 \rightarrow \#)$ are injectively labeled with values in $\{k+1, \ldots, h\}$,
- matrices of type 3 of the form $(X \rightarrow Y, B_2 \rightarrow \#)$ are injectively labeled with values in $\{h+1, \ldots, n\}$.

It is proved in [14] that each matrix grammar without appearance checking can be simulated by a rewriting P system with three membranes. This means we can construct a P system $\Pi = (V, T, [_1\, [_2\, [_3\,]_3]_2]_1, M_1, \emptyset, \emptyset, R_1, R_2, R_3)$ that is able to simulate all the matrices of type 2 and 4 in G in exactly the same way as in the proof of Theorem 4.1 in [14]. Then, in order to simulate matrices applied in appearance checking mode, we define a rule rewriting P system Π' such that

$$\Pi' = (V', T, \mu', M'_1, M'_2, \emptyset, \emptyset, M'_4, \emptyset, M'_6, \emptyset, R'_1, R'_2, R'_3, R'_4, R'_5, R'_6, R'_7)$$

where:

$$V' = V \cup \{X_i, X'_i, X'', \bar{X}, \$_i \mid X \in N_1, k+1 \leq i \leq n\} \cup \{\$_1, \$_2\},$$
$$\mu' = [_1\, [_2\, [_3\,]_3]_2\, [_4\, [_5\,]_5]_4\, [_6\, [_7\,]_7]_6]_1,$$
$$M'_1 = \{XA \mid (S \rightarrow XA) \text{ is the unique matrix of type 1 in } M\},$$
$$M'_2 = \{\$_1\$_2\},$$
$$M'_4 = \{\$_1, \$_2\},$$
$$M'_6 = \{\$_1, \$_2\},$$
$$\begin{aligned} R'_1 = R_1 &\cup \{X'' \rightarrow (X, here) \mid X \in N_1, k+1 \leq i \leq n\} \\ &\cup \{X \rightarrow (X_i, here), X_i \rightarrow (X'_i, in; \lambda, in) \mid X \in N_1, k+1 \leq i \leq n\} \\ &\cup \{\# \rightarrow (\#, here)\} \\ &\cup \{\$_1 \rightarrow (\$_i X_i, here) \mid X \in N_1, k+1 \leq i \leq h\} \\ &\cup \{\$_2 \rightarrow (\$_i X_i, here) \mid X \in N_1, h+1 \leq i \leq n\}, \end{aligned}$$

$$R'_2 = R_2 \cup \{X'_i \to (\#, out), \$_i \to (\#, out) \mid X \in N_1, k+1 \le i \le n\},$$
$$R'_3 = R_3,$$
$$R'_4 = \{X'_i \to (\#, out), \$_i \to (\#, out) \mid X \in N_1, h+1 \le i \le n\}$$
$$\cup \{\$_i \to (\bar{Y}, in; \$_i X_i, out) \mid i : (X \to Y, B_1 \to \#) \in M\}$$
$$\cup \{X'' \to (\$_1, here; \bar{X}, in) \mid X \in N_1\}$$
$$\cup \{B_1 \to (\#, out)\},$$
$$R'_5 \cup \{X'_i \to (\bar{Y}, out) \mid i : (X \to Y, B_1 \to \#) \in M\}$$
$$\cup \{\bar{Y} \to (Y'', out; Y'', out) \mid Y \in N_1\},$$
$$R'_6 = \{X'_i \to (\#, out), \$_i \to (\#, out) \mid X \in N_1, k+1 \le i \le h\}$$
$$\cup \{\$_i \to (\bar{Y}, in; \$_i X_i, out) \mid i : (X \to Y, B_2 \to \#) \in M\}$$
$$\cup \{X'' \to (\$_2, here; \bar{X}, in) \mid X \in N_1\}$$
$$\cup \{B_2 \to (\#, out)\},$$
$$R'_7 \cup \{X'_i \to (\bar{Y}, out) \mid i : (X \to Y, B_2 \to \#) \in M\},$$
$$\cup \{\bar{Y} \to (Y'', out; Y'', out) \mid Y \in N_1\}.$$

The P system Π' simulates the matrices of type 2 and 4 in M in exactly the same way as the P system Π by using the rules in R_1, R_2, and R_3. The simulation of matrices of type 3 is done in the following way.

At a given time in any computation in Π', we can assume to have produced inside region 1 a string of the form Xw, with $X \in N_1$, $w \in (N_2 \cup T)^*$. This string represents a sentential form of the grammar G produced after some applications of the matrices in M. In order to simulate a matrix $i : (X \to Y, B_1 \to \#) \in M$, with $k+1 \le i \le h$, we apply first the rewriting rule $X \to (X_i, here)$ from R'_1 and then the rule-rewriting rule $X_i \to (X'_i, in; \lambda, in)$ from R'_1. In this way, we can move both the string $X'_i w$ and the rule $\$_1 \to (\$_i, here)$ inside region 2, region 4, or region 6; these two objects can obviously finish in two different regions.

If the string $X'_i w$ reaches region 2, then an infinite computation is generated because of the rules $X'_i \to (\#, out)$ in R'_2 and $\# \to (\#, here)$ in R'_1. In a similar way, if the string $X'_i w$ reaches region 6, an infinite computation is generated because the unique rules from R'_6 that can be used are $X'_i \to (\#, out)$ and $B_2 \to (\#, out)$. This means the unique chance to complete the simulation of the matrix i is by string $X'_i w$ being moved inside region 4. In this latter case, we generate an infinite computation if (and only if) the string $X'_i w$ contains a symbol B_1 (first the rule $B_1 \to (\#, out)$ from R'_4 is applied and then the rule $\# \to (\#, here)$ from R'_1 is used). Therefore, if the string $X'_i w$ does not contain any symbol B_1, then we can complete the simulation of the matrix i by replacing the symbol X'_i with the nonterminal Y. For this we need the rule $\$_1 \to (\$_i, here)$ to also be in region 4. Notice that:

- if the rule $\$_1 \to (\$_i, here)$, with $k+1 \le i \le h$, reaches region 2, then an infinite computation is generated because of the string $\$_1\$_2$ in M'_2, the rule $\$_i \to (\#, out)$ in R'_2, and the rule $\# \to (\#, here)$ from R'_1;

- if the rule $\$_1 \rightarrow (\$_i, here)$, with $k+1 \leq i \leq h$, reaches region 6, then an infinite computation is generated because of the string $\$_1$ in M'_6, the rule $\$_i \rightarrow (\#, out)$ in R'_6, and the rule $\# \rightarrow (\#, here)$ from R'_1.

If the rule $\$_1 \rightarrow (\$_i, here)$, with $k+1 \leq i \leq h$, reaches region 4, then the string $\$_1$ can be replaced by $\$_i$ and, in the next step of computation, we can use the rule $\$_i \rightarrow (\bar{Y}, in; \$_i X_i, out)$ from R'_4. This rule returns the rule $\$_1 \rightarrow (\$_i X_i, here)$ into region 1 and moves the string $\bar{Y}$ into region 5. At this point we can apply the rule $\bar{Y} \rightarrow (Y'', out; Y'', out)$ from R'_5, which moves back the string Y'' into region 4 together with the rule $X'_i \rightarrow (Y'', out)$. Next, we apply this rule to the string $X'_i w$ in region 4 by moving in this way the string $Y''w$ into reach region 1. At the same time, by using the rule $Y'' \rightarrow (\$_1, here; \bar{Y}, in)$ from R'_4, the string Y''_4 in region 4 is replaced by a string $\$_1$ and the rule $X'_i \rightarrow (\bar{Y}, out)$ is returned to region 5. Finally, when the string $Y''w$ reaches region 1, the symbol Y'' is replaced by the nonterminal Y and the simulation of the matrix i ends.

A matrix $i : (X \rightarrow Y, B_2 \rightarrow \#) \in M$, with $h+1 \leq i \leq n$, is simulated in a similar way by using regions 6 and 7 instead of regions 4 and 5.

Therefore, we can say the rule rewriting P system Π' correctly simulates the matrix grammar G and $L(\Pi') = L(G)$. □

Theorem 1 shows that the feature of rule rewriting effectively increases the power of rewriting P systems by providing a characterization of recursively enumerable languages. Moreover, Theorem 1 shows that the hierarchy on the number of membranes collapses at level 7; the optimality of this result is not known though, and it is very likely that this result will be further improved.

However, the model of rule rewriting P systems considered here is based on sequential rewriting: in each region each string is rewritten by at most a rule at a time. In this respect, parallel rewriting represents an obvious and interesting alternative, especially in the context of LP systems models. Parallel rewriting is in fact one of the defining characteristic of every L system. In the case of P systems, parallel rewriting can be introduced in various forms [18], with the further difficulty of defining how to move a string from one region to another once this has been rewritten by many different rules at the same time. An elegant solution to this problem that can be found in [2] is based on parallel rewriting P systems without target conflicts. In these systems, the rules to rewrite a certain string might be chosen according to a specific parallel rewriting method, but from the set of rules that are applicable at a certain moment, only those that agree on the target specification can be applied. If there are rules in the applicable set of rules with different targets, then only a proper subset of these rules will be actually used. The main result obtained in [2] shows that the largest family of languages generated by parallel rewriting P systems coincides with the family of languages generated by extended tabled 0L systems, denoted by $ET0L$.

Here, we want to consider a variant of parallel rewriting P systems without target conflicts augmented with the feature of rule rewriting as specified in

Definition 1. Specifically, we choose to adopt the parallel rewriting method called maximal parallelism: at any time, all the symbols in a string that can be rewritten according to some rules must be rewritten; the rule to be applied to a specific symbol is nondeterministically chosen from among all those that can be applied. In the case of P systems with rule rewriting, this means:

- in each step, in each region, each string that can be rewritten by some rules must be rewritten; each of these strings is rewritten according to the maximal parallelism rewriting method by choosing for each string a set of rules without target conflicts;
- in each step, each rule that can be rewritten by some of the rewriting rules used in that step of computation must be rewritten; each of these rules is rewritten according to the maximal parallelism rewriting method by choosing for each string a set of rules without target conflicts.

Then, the notions of computation and language generated by a parallel rewriting P system with rule rewriting is defined in exactly the same way as in the sequential case. The family of languages generated by parallel rewriting P systems with rule rewriting and with at most $n \geq 1$ membranes is denoted by $ELPrP_n$.

Next, parallel rewriting P systems with rule rewriting are proved to be more powerful than parallel rewriting P systems without target conflicts by showing that they can generate languages that are not in $ET0L$. However, a more precise characterization of the families of languages $ELPrP_n$, with $n \geq 1$, remains to be found.

Theorem 2. $ET0L \subset ELPrP_3$.

Proof. The inclusion $ET0L \subseteq ELPrP_3$ is a direct consequence of the results obtained in [2]. In fact, it is proved in [2] that, in the case of maximal parallelism, each language in $ET0L$ can be generated by a parallel rewriting P system without target conflicts with at most two membranes. This means we just need to prove strict inclusion by showing that parallel rewriting P systems with rule rewriting and at most three membranes can generate languages that are not in $ET0L$. For this, consider the P system Π such that

$$\Pi = (V, T, [_1\, [_2\, [_3\,]_3\,]_2\,]_1 \emptyset, \{A\}, \{B\}, R_1, R_2, R_3),$$

where:

$$\begin{aligned}
V &= \{a, b, A, B, C, X, \#\},\\
T &= \{a, b\},\\
R_1 &= \{A \to (\lambda, out), X \to (\lambda, out)\},\\
R_2 &= \{A \to (AX, here), a \to (a, out)\},\\
R_3 &= \{B \to (\lambda, here; \lambda, out), B \to (B, here; bB, here)\}\\
&\quad \cup \{X \to (abB, here)\}.
\end{aligned}$$

It is easy to see that the parallel rewriting P system Π with rule rewriting generates the language $L(\Pi) = \{(ab^n)^m \mid 1 \leq n \leq m\}$, which is known not to be in $ET0L$ (e.g., see [22]). In fact, by starting from the initial configuration, we can apply the rule $B \rightarrow (B, here; bB, here)$ from R_3 and the rule $A \rightarrow (AX, here)$ from R_2 an arbitrary number of times. This means we obtain a string of the form AX^{n-1} in region 2 and a string of the form $b^{n-2}B$, a rule $B \rightarrow (b^{n-1}B, here; bB, here)$, and a rule $X \rightarrow (ab^nB, here)$ in region 3, with $n \geq 1$. Next, we can apply the rule $B \rightarrow (\lambda, here; \lambda, out)$ from R_2, which moves the rule $X \rightarrow (ab^n, here)$ into region 2 together with the rule $B \rightarrow (b^{n-1}, here; bB, here)$. At the same time, the rule $A \rightarrow (AX, here)$ from R_2 is applied once more, and this produces the string AX^n inside region 2. At that point, the rule $X \rightarrow (ab^n, here)$ becomes available in region 2, and each occurrence of the symbol X in the string AX^n can be replaced by an occurrence of the string ab^n. This is done in parallel with the application of the rule $A \rightarrow (AX, here)$. The parallel application of the rules $X \rightarrow (ab^n, here)$ and $A \rightarrow (AX, here)$ can be then repeated an arbitrary number of times in order to produce a string of the form $AX(ab^n)^m$, with $1 \leq n \leq m$, inside region 2. After that, we can finish the computation by applying a rule $a \rightarrow (a, out)$ for each occurrence of the symbol a in the string $AX(ab^n)^m$. In this way, the string $AX(ab^n)^m$ reaches region 1 and the string $(ab^n)^m$ can be produced as output of the computation by applying in parallel the rules $A \rightarrow (\lambda, out)$ and $X \rightarrow (\lambda, out)$ from R_1. □

Rule rewriting has been introduced in P systems with the specific motivation of capturing aspects of modularity typical of some of the existing extensions of L systems. In this respect, rule rewriting can be interpreted as an operation to develop a particular substring inside a region of the system in the form of a rule that can be moved from one region to another and used to insert the substring into another string present in the system. Another motivation for rule rewriting P systems comes from [19] where the research topic is formulated that concerns P systems where the rules are moved instead of the objects. In the model of rule rewriting P systems considered here, both rules and strings can be moved, but rules can be moved only as a consequence of the rules applied to the strings currently associated with the various regions of the system. In a sense, this feature introduces a certain level of interaction among the strings in various regions of the system. This is an interesting feature that is considered for rewriting P systems and it might deserve further and deeper investigation.

Acknowledgement. The research of M. Gheorghe and F. Bernardini was supported by the Engineering and Physical Sciences Research Council (EPSRC) of the United Kingdom, Grant GR/R84221/01.

References

1. M. Alfonseca, A. Ortega: Representation of Some Cellular Automata by Means of Equivalent L Systems. *Complexity International*, 7 (2000), `http://www.complexity.org.au/ci/vol07/alfons01`.
2. D. Besozzi, G. Mauri, G. Vaszil, C. Zandron: Collapsing Hierarchies of Parallel Rewriting P Systems without Target Conflicts. In *Membrane Computing, International Workshop, WMC 2003, Tarragona, July 2003, Selected Papers* (C. Martín-Vide, Gh. Păun, G. Rozenberg, A. Salomaa, eds.), LNCS 2933, Springer, Berlin, 2004, 55–69.
3. J. Dassow, Gh. Păun: *Regulated Rewriting in Formal Language Theory.* Springer, Berlin, 1989.
4. R. Freund, Gh. Păun: On the Number of Non-Terminal Symbols in Graph-controlled, Programmed, Matrix Grammars. In *Machine, Computations, and Universality* (M. Margenstern, Yu. Rogozhin, eds.), LNCS 2055, Springer, Berlin, 2001, 214–225.
5. A. Georgiou: *Sub-LP Systems – A Computational Model for Plant Simulation.* MSc Dissertation, University of Sheffield, 2003.
6. A. Georgiou: *SubLP-Studio Software*, available from the P systems web page at `http://psystems.disco.unimib.it/software.html`, 2003.
7. M. Hammel, R. Mech, P. Prusinkiewicz: The Artificial Life of Plants. In volume 7 of *SIGGRAPH '95 Course Notes*, 1995, 1–38.
8. J.S. Hanan: *Parametric L Systems.* PhD Thesis, University of Regina, Regina, Saskatchewan, Canada, 1992.
9. J. Hanan, R. Mech, P. Prusinkiewicz: *Extensions to the Graphical Interpretation of L systems Based on Turtle Geometry.* Research Report No. 97/599/01, Department of Computer Science, University of Calgary, 1997.
10. J.A. Kaandorp, J.E. Kubler: *The Algorithmic Beauty of Seaweeds, Sponges, and Corals.* Springer, Berlin, 2001, 91–99.
11. R. Karwowski: *L-studio v. 3.1.* Department of Computer Science, University of Calgary, 2001. Available at `http://www.cpsc.ucalgary.ca/Research/bmv/lstudio`.
12. S.N. Krishna, R. Rama: P Systems with Replicated Rewriting. *Journal of Automata, Languages and Combinatorics*, 6, 3 (2001), 345–350.
13. A. Lindenmayer, P. Prusinkiewicz: *The Algorithmic Beauty of Plants.* Springer, New York, 1990.
14. M. Madhu: *Studies of P Systems as a Model of Cellular Computing.* PhD Thesis, Indian Institute of Technology, Madras, India, 2003.
15. R. Mech, P. Prusinkiewicz: *Visual Models of Plants Interacting with Their Environment.* Department of Computer Science, University of Calgary, 1996.
16. M. Mutyam, K. Krithivasan: P Systems with Membrane Creation: Universality and Efficiency. In *Machine, Computations, and Universality* (M. Margenstern, Yu. Rogozhin, eds.), LNCS 2055, Springer, Berlin, 2001, 276–286.
17. Gh. Păun: Computing with Membranes. *Journal of Computer and System Sciences*, 61, 1 (2000), 108–143.
18. Gh. Păun: *Membrane Computing. An Introduction.* Springer, Berlin, 2002.
19. Gh. Păun: Membrane Computing: Some Non-Standard Ideas. In *Aspects of Molecular Computing* (N. Jonoska, Gh. Păun, G. Rozenberg, eds.), LNCS 2950, Springer, Berlin, 2004, 322–337.

20. P. Prusinkiewicz: *Modelling and Visualisation of Biological Structures.* Department of Computer Science, University of Calgary, 1993.
21. P. Prusinkiewicz, M. Hammel, E. Mjolsness: Animation of Plant Development. In *SIGGRAPH '93*, 1993, 351–360.
22. G. Rozenberg, A. Salomaa, eds.: *Handbook of Formal Languages.* 3 volumes, Springer, Berlin, 1997.
23. S. Vanac: *Perfect Plants.* Dissertation, University of Sheffield, 1999.

Chapter 10
An Analysis of a Public Key Protocol with Membranes

Olivier Michel[1], Florent Jacquemard[2]

[1] LaMI CNRS umr 8042 – Université d'Évry
Tour Évry-2, 523 Place des Terrasses de l'Agora, 91000 Évry, France
michel@lami.univ-evry.fr

[2] INRIA FUTURS and LSV, CNRS umr 8643 – ENS de Cachan
61 Avenue du Président Wilson, 94235 Cachan Cedex, France
florent.jacquemard@lsv.ens-cachan.fr

Summary. We develop an analysis of the Needham-Schroeder public key protocol in the framework of membrane computing. This analysis is used to validate the protocol and exhibits, as expected, a well known logical attack. The novelty of our approach is to use multiset rewriting in a nest of membranes. The use of membranes enables us to make airtight the conditions for detecting an attack. The approach has been validated by developing a full implementation for several versions of the analysis.

1 Goal and Motivations

Since the 1994 landmark demonstration by Adleman of the possibilities of DNA to solve a class of combinatorial problems, biocomputing has often been advocated to develop "chemically combinatorial problem solvers." In this chapter, we want to use an approach belonging to the membrane computing [23] area to address a well known combinatorial problem: the analysis of a cryptographic protocol.

Our starting point is the logical analysis of the Needham-Schroeder public key protocol (NSPK). The goal of the logical analysis is to find an interleaving of elementary actions (sending and answering messages) that allows an intruder to obtain confidential information. We have chosen this problem because it is simple to explain, and, at the same time, requires sophisticated data structures for the exploration of its state space, is paradigmatic of this kind of application, and has a well known solution, so that we can validate our result.

The approach taken in this chapter is brute force and consists of the exploration of the state space of the protocol for a systematic search of attacks.

Indeed, we are interested in the study of the representation and generation of states, rather than in designing a new and smart search strategy. This approach is motivated by the opinion that the representation of data is a central problem in biocomputing.

The rest of this chapter is organized as follows. In Section 2 we give some background on the logical analysis of cryptographic protocols. Section 3 describes precisely the Needham-Schroeder public key protocol. Section 4 presents the technical meat of the chapter. We develop a version of the analysis of the NSPK that improves a similar analysis initially proposed within the ELAN rewriting framework [5], with a more accurate representation of states using nesting. In the appendix are given a short presentation of the MGS language, which enables a kind of membrane computing, and the MGS code of the algorithms detailed in Section 4.

2 Formal Verification of Cryptographic Protocol

In this section, we give a brief introduction to the verification problem we shall consider. Cryptographic protocols define the exchange of a few messages between parties in order to distribute some secret data like cryptographic keys or to authenticate themselves. These messages are built with cryptographic primitives, like encryption, signature, or hash functions, and therefore the security of protocols relies on the strength of the cryptographic functions in use. However, it appears that even though these functions are assumed unbreakable, the security of a protocol can be compromised by an unexpected interleaving of messages between honest agents and a malicious intruder which has some limited control over the communication network (e.g., wiretapping some messages or impersonating identities while sending new ones). For instance, the well known problems of the distribution of keys for symmetric cryptosystems like AES (Advanced Encryption Standard) and the authenticity of public keys in PKIs (Public Key Infrastructure) are beyond the scope of the study of encryption functions.

Such *logical* attacks can be realized at almost no computational cost and hence can have disastrous consequences. Various formal methods have been proposed for the automation of the analysis of the vulnerability of cryptographic protocols to logical attacks, both for searching for flaws of this kind or for the formal proof of their absence. Several systems have been implemented for the purpose of searching for flaws, e.g., [18, 17, 13]. But many general purpose languages and tools have also appeared appropriate in this setting, with the advantage of a greater expressive power, efficiency, and maturity. To cite only a few examples, there are model checkers like FDR [16] or murφ [21], first order theorem provers [25, 14] and declarative languages used as model checkers [7, 5].

Our purpose in this chapter is to describe an experiment using membranes for modeling a cryptographic protocol and finding attacks by state exploration. The declarative style supported by the membrane computing framework is strongly advocated by the intruder-centric model which is generally considered in order to apply formal methods to cryptographic protocol verification. In this model, often referred as the "Dolev-Yao model" [8], the agents executing the protocol communicate asynchronously via a unique channel which has been compromised by an intruder. The intruder is able to spy and divert every message on the channel, and to analyze read messages, with the restriction that he must know the appropriate encryption key in order to decipher an encrypted message. It can also build and send new messages, possibly under a fake identity. The global state of the system can hence be represented by a heterogeneous set containing the local states of each agent (with a bounded memory), the messages and submessages known to the intruder, and the messages sent and not yet received by an agent. The actions of the agents (receiving and sending messages) as well as of the intruder can be modeled using rewriting rules on multisets. The search for an interleaving leading to an attack can be coded very simply with an appropriate pattern expression to find sequences of value or arbitrary length.

The problem of finding attacks of protocols is highly undecidable, the state space being infinite for several reasons: the unboundedness of the number of agents present, the ability of agents to generate fresh random data (nonces), the unlimited size of terms generated by the intruder. In order to restrict our exploration to a finite search space, while keeping our procedure reasonably complete, we shall rely on some theoretical results about protocol verification. It is shown in [24] that the problem of protocol security (non-existence of attacks) becomes decidable when the number of agents considered is bounded. Indeed, [24] shows that in this case, whenever there exists an attack, there exists an attack involving messages of a bounded size. We can use this result here to ensure the completeness of our attack search procedure, given a finite number of agents.

3 The Needham-Schroeder Public Key Protocol

The Needham-Schroeder public key protocol [22] (NSPK for short) is the favorite example for the application of formal methods to the verification of cryptographic protocols. This popularity comes from one of the most famous success stories in this domain, which is the discovery in 1994 by G. Lowe [16] of a replay attack in this protocol 16 years after its publication. In [16], G. Lowe models the protocol in the CSP process algebra and uses the model checker FDR to explore the state space. We obtain here the same result with a model based on membrane computing, implemented in the language MGS.

3.1 Description of the Protocol

The Needham-Schroeder public key protocol involves two participants, Alice (A) and Bob (B), willing to authenticate reciprocally with three messages using public keys. The original protocol of [22] also involves a server distributing the public keys to A and B with three additional messages. We omit the server and its three messages here, assuming that A and B both initially know each other's public key, since they are not necessary in Lowe's attack. The messages are described below in the usual notation (see also Figure 1):

$$\begin{array}{l} \texttt{REQ}\ A \rightarrow B : \{A, N_a\}_{K(B)} \\ \texttt{CHAL}\ B \rightarrow A : \{N_a, N_b\}_{K(A)} \\ \texttt{AUTH}\ A \rightarrow B : \{N_b\}_{K(B)} \end{array}$$

In the first message (labelled `REQ`), Alice generates a random number (*nonce*) N_a, appends it to her name A (the append operator is denoted $\_, \_$) encrypts the results with Bob's public key $K(B)$ (public key encryption is denoted with the binary operator $\{\_\}_\_$) and sends the result to the network. When Bob receives a message in the form of `REQ`, he deciphers it and retrieves the identity A of Alice and the nonce N_a. Then he generates a second random number N_b, appends it to N_a and sends back the result encrypted with Alice's public key $K(A)$ (message `CHAL` for challenge). Alice, receiving message `CHAL`, can decipher it and check whether the first component corresponds to the nonce she sent in message `REQ`. Then, she resends Bob's nonce N_b encrypted with Bob's public key (message `AUTH`). Bob can check that the message `AUTH` contains the nonce N_b he has generated at second step (`CHAL`).

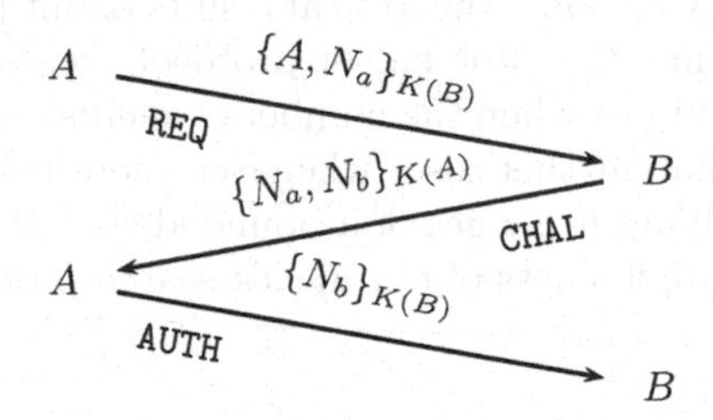

Fig. 1. Description of the NSPK protocol.

3.2 A Replay Attack

Receiving the message `AUTH` assures Bob that Alice has really received the message `CHAL` and answered, because Alice is the only one able to decipher this message. We assume that each agent, as well as the intruder (let us call him Charly, C), knows only its own private key, and that this key is necessary to decipher a message encrypted with the corresponding public key.

Similarly, when receiving the message `CHAL`, Alice is assured that it really comes from B (and is not a fake message from Charly), as proven by the presence of N_a, because the knowledge of Bob's private key is necessary for the extraction of N_a from the message `REQ`. Hence, N_a and N_b are used as *authenticators* in this protocol, and they must remain secret. However, the attack of [22], described in Figure 2, shows that this is not the case, even with the above hypotheses concerning the private keys.

This attack involves two sessions in parallel. In the first session, Alice enters into communication with Charly (without knowing that he is an intruder). Since the message REQ is encrypted with Charly's public key $K(C)$, Charly can retrieve A, N_a and encrypts it with Bob's public key $K(B)$. He then sends this message as the first message REQ′ of a second session between A and B. In this step, Charly impersonates A, denoted by $C(A)$. Bob answers to REQ′ and Charly diverts this message CHAL′ (it is by denoted $C(A)$). Then Charly, with two messages CHAL and AUTH of the first session, uses A as an oracle in order to obtain Bob's nonce N_b.

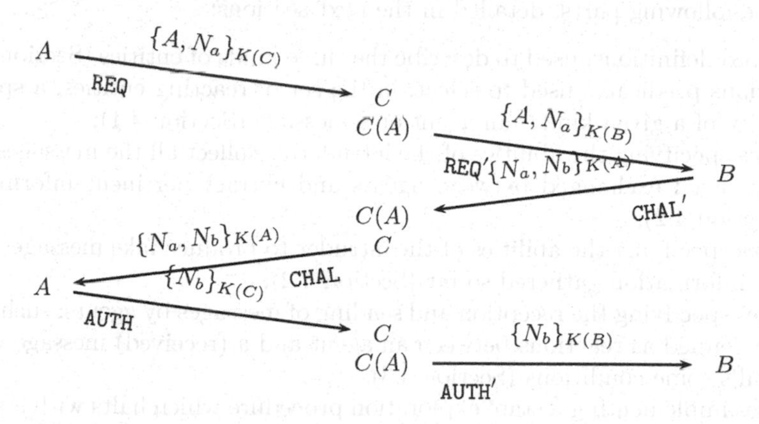

Fig. 2. A Replay attack following G. Lowe.

4 Finding an Attack on the NSPK Using Membranes

We shall describe here the specification of the Needham-Schroeder publi-key protocol and the implantation of an attack-search procedure using rules and membranes. For the implementation, we rely on rewriting modulo associativity and commutativity (AC) in terms representing nested multisets (membranes). Rewriting rules can be guarded by arbitrary conditions. This model is similar to the chemical computations presented in [1], and we use the term "chemical solution" to denote the content of a membrane. Examples of systems implementing such models of computations are Gamma, ELAN [2], MAUDE [3], and MGS [9]. The ingredients of this model of computation are rather sophisticated but a translation into the fundamental core mechanisms of P systems is possible, and we give in [19] some elements that support this assertion.

We present in this chapter the principles of a simple version of the attack-search procedure. This version improves a similar analysis initially proposed within the ELAN rewriting framework, because we use a more accurate representation of states using nesting. The functional representation of the interleaving of actions is also a new idea. Note that another version is described and fully detailed in [20]; it goes further by generalizing the approach to the exploration of general state spaces and does not rely on the assumption that attacks involve messages of bounded size.

The description of the protocol involves two different kinds of components: *entities* and *evolution rules*. The entities are records and evolution rules are given by rewrite rules. A *system state*, for which we shall also write *solution*, is a finite collection of entities which are of three kinds: agents, messages transmitted through the network, and messages components memorized by the intruder. Several entities in a state shall react, firing an evolution rule which transforms a system state into a successor state. The model is organized into the following parts, detailed in the next sections:

- record definitions, used to describe the three kinds of entities (Section 4.1);
- various predicates used to select, in the set of reacting entities, a specific entity of a given kind – an agent or a message (Section 4.1);
- rules specifying the abilities of the intruder to collect all the messages that have been exchanged between agents and extract pertinent information (Section 4.2);
- rules specifying the abilities of the intruder to produce fake messages from the information gathered so far (Section 4.2);
- rules specifying the reception and sending of messages by agents: such rules are defined as reactions between an agent and a (received) message which fulfills some conditions (Section 4.3);
- rules implementing a state exploration procedure which halts with a predicate checking whether a bad state is reached, and hence whether the search of an attack is successful (Section 4.3).

4.1 Representing Agents, Messages, and Intruder Knowledge

The three different kinds of entities (unstructured information) found in the system states (solutions) are represented using records (the `MGS` code for this section can be found in Appendix B.1).

Agents. We shall distinguish the *roles*, Alice and Bob in our example, which are programs, from the *agents* executing the programs, characterized by an identifier (agent's name), a role, and a bounded memory. In particular, there can be several agents for one role. An agent consists of:

- an *identity* `id` (its name; several agents may have the same identity),
- two *stores* `ni` and `nr` to memorize the session-specific values of the nonces N_a and N_b,
- a *program counter* `pc`, which can take the value described below.

Every agent with role either Alice or Bob shall create a nonce and receive another one during the execution of the protocol of Section 3.1. The fields `ni` and `nr` store these two values for Alice, `ni` stores N_a and `nr` stores N_b, and reciprocally for Bob (ni stands for *nonce initial*, because we can assume that each agent initially creates the nonces before starting a session of the protocol, and nr stands for *nonce received*).

The program counter pc of an agent can take the following values (these values are arbitrary symbols and prefixed by a backquote), according to the role: 'REQ, 'AUTH, and 'FINISHED for Alice and 'CHAL, 'WAIT, and 'FINISHED for Bob. For Alice, pc = 'REQ means that the agent is about to send the message with the corresponding label in the protocol specified in section 3.1, and similarly for pc = 'AUTH (role Alice) and 'CHAL (role Bob). For an agent playing the role of Bob, pc = 'WAIT means that he is waiting for the answer of Alice to his challenge CHAL, and pc = 'FINISHED means that the agent has completed his session of the protocol.

Messages. Three different kinds of messages are exchanged between Alice and Bob during the protocol. We define a predicate to recognize each kind of message: REQ, CHAL, and AUTH. Messages are also records, and are characterized by the kind of information that they hold. For instance, messages of type REQ contain a field na representing the content of the message and a field kb which is the public key used for encryption. For the sake of simplicity in our program every public key or private key is represented by the identity of the owner.

Intruder Knowledge. The knowledge of the intruder is also represented by records with fields name, nonce, pub, and priv. We define several predicates (info_name, info_nonce, info_pub, and info_priv) for each kind of information that the intruder will be able to reveal from the whole history of exchanged messages: name, nonce, public key, and private key. These predicates are used to determine the presence of a message of a given kind with given information in the solution.

4.2 The Intruder Transformation Rules

The network is common to all agents and the intruder; hence the latter is able to read and produce new messages. This behavior is implemented by the rules presented in the two following sections (the MGS code for this section can be found in Appendix B.2).

Reading and Analyzing Messages. In our approach, the existing messages are read by the intruder from the current state and are put back unchanged. Moreover, the encrypted contents of a message are added as new known information to the state if decryption is possible. More precisely, the intruder can learn plaintext encrypted with a public key (for instance the nonce nb encrypted with kb in message AUTH) only if he knows the corresponding private key.

The following three rules define the evolution of the knowledge of the intruder, according to the messages present in the network. There is exactly one rule for each kind of message. They will actually not generate all the information that the intruder can extract from collected message. However, these transformations are sufficient to extract all the information needed to build messages with the forging rules below. For instance, if a message m present in the solution has type REQ, and the intruder knows the private key associated with m.kb, then he learns the components m.na and m.a of m. Theoretically, he also learns the pair (m.na, m.a) but storing such information is useless since we assume that the intruder is able to build pairs arbitrarily.

$$m \longrightarrow m, \{\texttt{nonce} = m.\texttt{na}\}, \{\texttt{name} = m.\texttt{a}\} \quad \texttt{where}\ m \in \texttt{REQ} \wedge \exists k \in \text{self s.t. } k.\texttt{priv} = m.\texttt{kb}$$

$$m \longrightarrow m, \{\texttt{nonce} = m.\texttt{na}\}, \{\texttt{nonce} = m.\texttt{nb}\} \quad \texttt{where}\ m \in \texttt{CHAL} \wedge \exists k \in \text{self s.t. } k.\texttt{priv} = m.\texttt{ka}$$

$$m \longrightarrow m, \{\texttt{nonce} = m.\texttt{nb}\} \quad \texttt{where}\ m \in \texttt{AUTH} \wedge \exists k \in \text{self s.t. } k.\texttt{priv} = m.\texttt{kb}$$

The keyword "self" used in the rules denotes the current multiset (i.e., the multiset from which m is chosen). The existential quantifier in the guard of the rules checks whether some condition is satisfied by an element k in a given multiset: such a kind of predicate is easily computed by the set of reduction rules.

Forging Some New Messages. In the previous section, we have described the `intruder` rules set which reveals information only according to already known messages and keys. The following rule *produces* a new fake REQ message from known information in the solution:

$$k, n, m \longrightarrow \{\texttt{na} = m.\texttt{nonce}, \texttt{a} = n.\texttt{name}, \texttt{kb} = k.\texttt{pub}\} \quad \texttt{where info\_pub}(k) \wedge \texttt{info\_name}(n) \wedge \texttt{info\_nonce}(m)$$

There is one such rule for the two other kinds of messages, CHAL and AUTH. These rules are used to produce by saturation (fixed point computation) all possible fake messages that can be forged from the known facts in a multiset.

An attack consists of revealing all possible information using the above rules of the intruder after having forged all possible fake messages. Actually, we will see in the following sections that a *real* attack always consists of the application of the attack rules of Section 4.3 until a fixed point is reached.

4.3 *Nested* Multiset Rewriting to Explore the State Space

The first idea to implement the logical analysis of NSPK is to aggregate all the entities involved into the protocol in a single multiset acting as a chemical solution containing the agents, the messages, and the revealed information. The agents and the intruder will react with messages to augment the solution with new information. All information is in the solution at the same level. An attack on the NSPK protocol consists here of finding an interleaving of the agents actions described below such that Bob's nonce is revealed (the MGS code corresponding to this section can be found in Appendix B.3).

This approach suffers from the following problem: let S be a solution and a be an agent in a state where it might reply to two different messages m_1 and m_2. The two following scenarios could occur:

1. The agent replies to both messages: to m_1 to give m'_1 and to m_2 to give m'_2. Here, after the agent action, S becomes $S \cup m'_1 \cup m'_2$. In the future evolution of the protocol, another agent may react to *both* m'_1 *and* m'_2 leading to an incorrect situation, even where the intruder may break the protocol and reveal the nonce.
2. The agent replies to only one of the two messages: to m_i to produce m'_i. In that case, an attack might not be found because the case where the reply should have concerned the other message has not been considered. The protocol analysis is therefore too weak.

The consequence is that we have to take into account the different evolutions of the protocol that might occur when an agent receives more than one message. To model such a situation, we make use of several multisets (membranes) to localize the computation and to avoid the (possible) interferences. The initial state consists of a multiset of multisets. Each element in the top multiset (the *skin* in the language of P systems) is a possible state in the protocol and represents some possible evolution, as depicted in Figure 3.

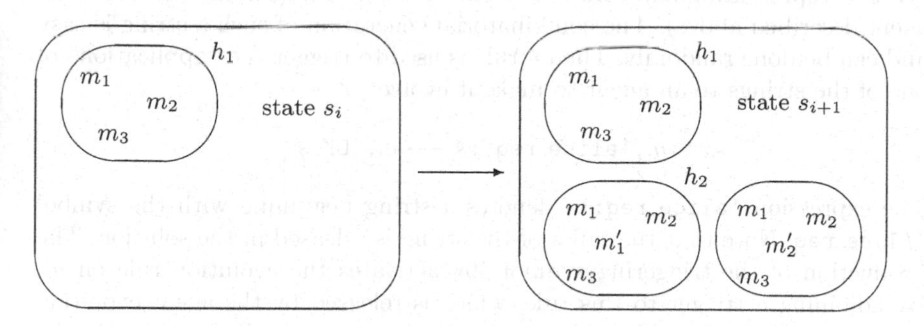

Fig. 3. Creation of membranes.

The Agents. The behavior of each agent, at each possible pc, is described by a set of rules. For example, the behavior of Alice with pc = 'AUTH is to switch to the state 'AUTH and to produce a new message:

$$x, t \longrightarrow (x + \{\texttt{pc} = \texttt{'AUTH}\}), \{\texttt{kb} = x.\texttt{dest}, \texttt{na} = x.\texttt{ni}, \texttt{a} = x.\texttt{id}\}$$
$$\texttt{where } x \in \texttt{REQ} \land x.\texttt{id} = \texttt{alice} \land t = \texttt{'OK}$$

the operator + is the asymmetric merge of records and the results of $x+\{\texttt{pc} = \texttt{'AUTH}\}$ is a record equal to x except for the field pc that takes the value 'AUTH.

The variable t matches a symbol used to inhibit or activate the rule: if the symbol is present (e.g., $t =$ `'OK`) then the rule can be triggered. If the symbol `'OK` is not present in the chemical solution, the rule is inhibited.

There are three additional similar rules to describe the evolution of Alice waiting for the authentication, Bob waiting for a challenge, and Bob in the finishing state.

Note that the messages addressed to an agent must not be removed from the solution and are available for other rule applications.

The Initial State. The initial state for the attack search consists of a multiset (of multisets) with only one element:

- the two agents, Alice and Bob, initialized with their respective identities, the destination of the message for Alice, initial nonces to arbitrary integer values, and program counter;
- intruder knowledge (public keys of all participants and its own private key).

Looking for an Attack. In our definition of the initial state, the number of agents is fixed and remains such. Therefore, the number of execution steps is bounded accordingly. The problem consists of finding the correct interleaving of Alice's and Bob's actions leading to a successful attack.

The basic idea is to generate all strings of bounded length made of four symbols representing an evolution of one of the agents (see the rule set of an agent described above). The combinatorial generation of such a string is easy and can be done randomly. Then a rule is used to trigger the "application" of one of the strings to an agent to make it evolve:

$$m, \texttt{'alice\_req::}s \longrightarrow m, \texttt{'OK}, s$$

The expression `'alice_req::`s denotes a string beginning with the symbol `'Alice_req`. Note that the tail s of the string is released in the solution. The production of the triggering symbol `'OK` activates the evolution rule on m. By adjoining a trigger to this rule, which is released by the agent evolution rule and consumed by the rule application, we can interleave correctly the evolution of an agent until the exhaustion of the string s.

We still look for an interleaving leading to revealing the nonce. A successful attack is to find the nonce of Bob revealed in the chemical solution. This is done by adding a specific rule, e.g., a rule leading to a dissolution of all enclosing membranes.

Validation in the `MGS` Programming Language. To validate our propositions, we have completely implemented and validated several versions of the logical analysis using the `MGS` programming language. A presentation of `MGS` and the commented code can be found in the Appendix.

`MGS` is a research project devoted to the design and the development of a programming language dedicated to the simulation of biological processes [9, 11]. Based on topological notions, MGS supports the notion of transformation: a localized computation specified by rules. One can, for example, define multiset rewriting rules [1] that act on a nest of multisets (i.e., membranes). These rules can be used to move values from one multiset to another, as well as to dissolve, divide, and create new multisets. So, MGS can potentially be used to process membranes. However, we outline that the MGS project focuses on the design of a programming language rather than the development of a well founded computational model.

5 Summary

In this chapter, we have used the membrane computing approach to describe and analyze the NSPK protocol. This application of membrane computing is new to the best of our knowledge. It has been shown that using our approach, the well known security hole of [16] is easily (in less than one second) discovered by our state exploration procedure.

In the proposed version, we are searching for the correct interleaving of the agents' actions leading to a possible attack. Using membranes permits us to handle correctly the fact that an agent may have to react to more than one message, leading to more than one evolution of the state.

Nevertheless, this method is tailored for the search of an interleaving of agents' actions leading to the revelation of the nonce. This is possible because we actually know that such an interleaving *will* lead to a successful attack. We have proposed in [19] a more general approach where a full state space search is done. The complete running code of the two versions has been implemented in MGS and is detailed in [20]. The complete code is particularly simple and readable. Moreover, it is also easy to evolve the initial analysis into a more sophisticated one.

The approach presented here has been developed for this special protocol and heavily relies on the nesting of membranes to localize the computation and to avoid evolution interference, leading to more approximate analysis. We believe that the principles of our modeling are general enough to envision a systematic way to derive a program for searching for attacks from an abstract description of the messages of a protocol given with the notations of Section 3.1, following [14].

Acknowledgments. The authors are grateful to Jean-Louis Giavitto, Julien Cohen, and Antoine Spicher at LaMI for stimulating discussions and thoughtful remarks. This research is supported in part by the CNRS, the GDR ALP, the University of Évry, Genopole©, INRIA and ENS Cachan, the RNTL project PROUVÉ, and the ACI-SI Rossignol.

References

1. J.-P. Banâtre, P. Fradet, D. Le Métayer: Gamma and the Chemical Reaction Model: Fifteen Years After. In *Multiset Procssing. Mathematical, Computer Science, and Molecular Computing Points of View* (C.S. Calude, Gh. Păun, G. Rozenberg, A. Salomaa, eds.), LNCS 2235, Springer, Berlin, 2001, 17–44.
2. P. Borovansky, C. Kirchner, H. Kirchner, P.E. Moreau, M. Vittek: ELAN – A Logical Framework Based on Computational Systems. *Electronic Notes in Theoretical Computer Science*, 4 (1996).
3. M. Clavel, F. Durán, S. Eker, P. Lincoln, N. Martí-Oliet, J. Meseguer, J.F. Quesada: The Maude System. LNCS 1631, Springer, Berlin, 1999, 240–243
4. I. Cervesato, N. Durgin, P.D. Lincoln, J.C. Mitchell, A. Scedrov: A Meta-Notation for Protocol Analysis. In *Proc. 12th IEEE Computer Security Foundations Workshop (CSFW1999)*, Mordano, Italy, 55–69.
5. H. Cirstea: Specifying Authentication Protocols Using ELAN. In *Workshop on Modeling and Verification*, 1999.
6. D.L. Dill, A.J. Drexler, A.J. Hu, C.H. Yang: Protocol Verification as a Hardware Design Aid. In *International Conference on Computer Design, VLSI in Computers and Processors (ICCD1992)*, 522–525, Los Alamitos, Ca., USA, 522–525.
7. G. Denker, J. Meseguer, C. Talcott: Protocol Specification and Analysis in Maude. In *Workshop on Formal Methods and Security Protocols*, 1998.
8. D. Dolev, A. Yao: On the Security of Public Key Protocols. *IEEE Transactions on Information Theory*, IT-29, 2 (1983), 198–208.
9. J.-L. Giavitto: Topological Collections, Transformations and Their Application to the Modeling and the Simulation of Dynamical Systems. In *Rewriting Technics and Applications (RTA'03)*, LNCS 2706, Springer, Berlin, 2003, 208–233.
10. J.-L. Giavitto, O. Michel: The Topological Structures of Membrane Computing. *Fundamenta Informaticae*, 49 (2002), 107–129.
11. J.-L. Giavitto, G. Malcolm, O. Michel: Rewriting Systems and the Modeling of Biological Systems. *Comparative and Functional Genomics*, 5 (2004), 95–99.
12. S. Peyton Jones, C. Hall, K. Hammond, W. Partain, P. Wadler: The Glasgow Haskell Compiler: A Technical Overview. In *Joint Framework for Information Technology Technical Conference*, 1993.
13. A. Huima: Efficient Infinite-State Analysis of Security Protocols. In *Proceedings of FLOC'99 Workshop on Formal Methods and Security Protocols*, 1999.
14. F. Jacquemard, M. Rusinowitch, L. Vigneron: Compiling and Verifying Security Protocols. In *Logic for Programming and Automated Reasoning (LPAR'00)*, LNCS 1955, Springer, Berlin, 2000.
15. X. Leroy: The Objective CAML System, Release 3.07. Documentation and User's Manual. Technical report, INRIA, 2004.
16. G. Lowe: An Attack on the Needham-Schroeder Public Key Authentication Protocol. *Information Processing Letters*, 56, 3 (1995).
17. C.A. Meadows: The NRL Protocol Analyzer: An Overview. *Journal of Logic Programming*, 26, 2 (1995), 113–131.
18. J.K. Millen, S.C. Clark, S.B. Freedman: The Interrogator: Protocol Security Analysis. *IEEE Transactions on Software Engineering*, SE-13, 2 (1987).
19. O. Michel, F. Jacquemard: An Analysis of the Needham-Schroeder Public Key Protocol with MGS. In *Pre-Proceedings of Fifth Workshop on Membrane Computing (WMC5)*, Milano, 2004, 295–315.

20. O. Michel, F. Jacquemard, J.-L. Giavitto: Three Variations on the Analysis of the Needham-Schroeder Public Key Protocol with MGS. Technical Report LaMI-98-2004, Univ. d'Évry - CNRS, 2004, 25 pages.
21. J. Mitchell, M. Mitchell, U. Stern: Automated Analysis of Cryptographic Protocols Using Murphi. In *Proceedings of the IEEE Symposium on Security and Privacy*, 1997, 141–151.
22. R.M. Needham, M.D. Schroeder: Using Encryption for Authentication in Large Networks of Computers. *Communications of the ACM*, 21, 12 (1978), 993–999.
23. Gh. Păun: *Membrane Computing. An Introduction.* Springer, Berlin, 2002.
24. M. Rusinowitch, M. Turuani: Protocol Insecurity with Finite Number of Sessions is NP-Complete. In *Proceedings of the 14th Computer Security Foundations Workshop (CSFW2001)*, 174–190.
25. C. Weidenbach: Towards an Automatic Analysis of Security Protocols in First-Order Logic. LNCS 1632, Springer, Berlin, 1999, 378–382.

Appendix A. A Brief Introduction to the `MGS` Language

We briefly present the `MGS` language in this section. We do not detail all the features of the language but, rather, focus on the notions required to understand the next section.

A.1 `MGS` as a Functional Language

`MGS` embeds a complete, impure, dynamically typed, strict, functional language. We describe here only the major differences between the constructions available in `MGS` with respect to functional languages like `OCAML` [15] or `HASKELL` [12].

Values. Atomic values (such as integers, floats, booleans, and strings), with their usual functions, are available. Constants are denoted with a backquote, as in `‘REQ` (reminiscent of `LISP` symbols). The only operations allowed on a constant are to store it and to compare it for equality with another value.

Records (cartesian products with labels) are defined using braces: `{x=0, y=1}` creates a pair with label `x` and `y` (an `MGS` record is similar to Pascal's `record` or C's `struct`). The fields are accessible using the dot notation: `let v = {x=0, y=1} in v.x` has value `0`. Since records are used in `MGS` to define a particular state of an entity, `MGS` allows the definition of predicates based on the fields found in a record. The keyword `record` is used to define such predicates:

```
record agent = \{id, ni, nr, pc\}
```

defines the predicate `agent` that holds only if applied to a record value that has at least the fields `id`, `ni`, `nr`, and `pc`. Record `alice`, defined as `record alice = {dest} + agent`, extends predicate `agent` with the additionally required field `dest`. So far, the record predicates required only have the fields to hold. The predicate `req` defined as `record req = {pc = ‘REQ}` holds only if its argument has a field `pc` with a value equal to the constant `‘REQ`.

Imperative Variables and Sequencing. Variables in a functional languages are not true variables: they refer to values and cannot be updated. `MGS` has a notion of *imperative* variable (also called *mutables*) that can be updated. The `:=` operator allows us to define such variables. For example `imp := 0` defines `imp` with value `0` that can be later updated with the same construction.

The semicolon operator `;` is used to express the sequencing of expressions: the value of `f();g()` is the value returned by `g()` where `f()` has been computed before.

Functions. Since `MGS` is a functional language, it has functions as first-class values. Functions are defined either using the construction `fun`, as in `fun max(x, y) = if (x > y) then x else y fi`, or using the classical lambda notation, as in `\x.\y.if (x > y) then x else y fi`

Computations by fixpoints are heavily used in applications like simulations or state space explorations. `MGS` provides an operator to compute iterations and fixpoints of functions. Let `f` be a function, then `f[iter = n](x)` computes `fⁿ(x)` and `f[*](x)` denotes the fixpoint of `f` starting from `x`.

Functions together with mutables and iterations allow us to define functions that pass information between calls. For example, function `f` defined as `fun f[acc=0](x)=(acc := acc+1; x+acc)` allows to define an accumulator `acc` which stores a value that is incremented between each call. The value of `f['iter = 10, acc = 0](1)` is 56.

A.2 Topological Collections and Their Transformations

A distinctive feature of the `MGS` language is its handling of entities structured by *abstract topologies* using *transformations* [10]. A set of entities organized by an abstract topology is called a *topological collection*. Topological means here that each collection type defines a neighborhood relation inducing a notion of *subcollection*. A subcollection B of a collection A is a subset of connected elements of A, inheriting its organization from A.

Collection Types. Many different predefined and user-defined collection types are available in `MGS`. We won't describe them here since sets, multisets, and sequences are the only collection type used in this chapter.

For any collection type `T`, the corresponding empty collection is written `():T`. The name of a collection type is also a predicate used to test if a value is of the type: `T`(v) holds only if v is of type `T`. Each collection type can be subtyped. The type declaration `collection U = T` introduces a new collection type `U` which is a subtype of `T`. The new type `U` shares the same topology as `T`. However, a value of type `U` can be distinguished from a value of type `T` using the `U` predicate (i.e., the subtyping relation implies that `U`$(u) \Rightarrow$ `T`(u) for any value u, but not the reverse). Elements in a collection can be of any type, including collections.

Operations on Collections. The join of two collections C_1 and C_2 (written with a comma: C_1`,`C_2) is the main operation on collections. The comma operator is overloaded in MGS and can be used to build any collection (the type of the arguments disambiguates the collection built). So, the expression `1, 1+1, 2+1, ():set` builds the set with the three elements 1, 2, and 3, while the expression `1, 1+1, 2+1, ():bag` makes a multiset with the same three elements.

Transformations. The *global transformation* of a topological collection C consists of the *parallel application* of a set of *local transformations.* A local transformation is specified by a rewriting rule r that specifies the replacement of a subcollection by another one. The application of a rewriting rule $\beta \Rightarrow f(\beta, \ldots)$ to a collection A:

1. selects a subcollection B of A whose elements match the *pattern* β,
2. computes a new collection C as a function f of B and its neighbors,
3. and specifies the insertion of C in place of B into A.

One should pay attention to the fact that, due to the parallel application strategy of rules, *all distinct instances* B_i *of the subcollections matched by the* β *pattern are "simultaneously replaced" by the* $f(B_i)$. This is very different from the evaluation strategies followed by classical rewriting tools like MAUDE [3], ELAN [2], Murφ [6], MSR [4], etc.

The MGS experimental programming language implements the idea of transformations of a topological collection into the framework of a simple dynamically typed functional language. Collections are new kinds of values and transformations are functions acting on collections and defined by a specific syntax using rules. Transformations (like functions) are first-class values and can be passed as arguments or returned as the result of an application.

Subcollection Patterns. A transformation is defined by a set of rules (listed between braces). A pattern β that appears on the left hand side of a rule is an expression used to select a subcollection to be replaced. Several operators are available; we will review here only a few of them:

- **Literal**: a literal value matches an element with the same value. For example, `123` matches an element with the integer value 123.
- **Variable**: a pattern variable a matches exactly one element. The variable a can then occur elsewhere in the rule and denotes the value of the matched element. The identifier of a pattern variable can be used only once in a pattern. To match an element without giving it a name, an underscore, _, can be used.
- **Alias**: the pattern p as X associates the variable X with the value matched by the pattern p. X is a regular variable that can be used as previously described.
- **Neighbor**: the pattern b , p matches a subcollection composed of an element matched by b neighbor of a subcollection matched by p.
- **Guard**: p/exp matches a subcollection matched by p such that the predicate exp hold. For instance, x , y / $y > x$ matches two neighbor elements such that the second one is greater than the first.

- **Repetition:** $p*$ matches a subcollection made of a (possibly empty) repetition of subcollections matched by p. If p is a pattern variable, then its value refers to the sequence of matched elements, and not to one of the individual values. For example, 3+ matches a nonempty subcollection made of only 3s.

MGS and Membrane Computing. The MGS language enables a kind of membrane computing. It embeds the rewriting of multisets (or sets) in the following way: in a multiset, an element is susceptible to interacting with any other element, so the abstract topology of a multiset is the topology of a complete connected graph, and the neighbors of an element are all the other elements in the multiset. Then, a pattern β can select an arbitrary sub-multiset, and a multiset rewriting rule is simply a local transformation in this topology.

A.3 Example: Computing All n-Tuples in a Set

Let S be a set of values. To compute all the n-tuples one can use the transformation:

```
trans n_tuple[acc, n] = {
   (_* as c) / size(c) == n / (acc := c::acc; false) => !!(0);
                                                       => return(acc)
   -
}
```

In transformation `n_tuple`, parameters `acc` and `n` are mutables whose definitions are local to the transformation. They are set at the first call of the transformation. Applied to a collection C, the pattern of the first rule `((_* as X) / size(X) == n)` matches a subcollection c of C of size n such that all elements of c are neighbors (with respect to the topology induced by C). Once c is found, the predicate `(acc := c::acc; false)` is calculated: collection c is added to the accumulator (`::` is the concatenation of a value to a collection) and the value `false` is returned. Since the predicate does not hold, the right hand side of the rule is not evaluated (the expression `!!(0)` aborts the program) and the rule is tried against another instance, storing each time the solution of the matching into the accumulator. Once all the possibilities have been tried and failed, the second rule is tried. That rule succeeds in matching anything and returns the value of the accumulator. Transformation `n_tuple[acc=set:(), n=2]((3,4,5,6,set:()));;` computes all the pairs

```
((3, 4):'seq, (3, 5):'seq, (3, 6):'seq, (4, 3):'seq, (4, 5):'seq,
(4, 6):'seq, (5, 3):'seq, (5, 4):'seq, (5, 6):'seq, (6, 3):'seq,
(6, 4):'seq, (6, 5):'seq):'set
```

where `(3, 4):'seq` is a pair holding the two integer values.

Appendix B. MGS Code for the Description of the Attack

We give in the following sections the MGS code that implements the search for an attack that is described in Section 4.

B.1 Representing Agents, Messages and Intruder Knowledge

The code presented in this section implements the data structures defined in Section 4.1.

Agents. One set of records is used to define *agents*, defined as:

```
record agent    = { id, ni, nr, pc};;
record alice    = { dest } + agent;;
record bob      = agent;;
```

Some records for the possible agent pc are defined as follows:

```
record req      = { pc = ‘REQ };;
record chal     = { pc = ‘CHAL };;
record auth     = { pc = ‘AUTH };;
record wait     = { pc = ‘WAIT };;
record finished = { pc = ‘FINISHED };;
```

Messages. A predicate is defined for each kind of message:

```
record messageReq  = { na, a, kb  };;
record messageChal = { na, nb, ka };;
record messageAuth = { nb, kb     };;
```

Intruder Knowledge. Finally, we define a predicate for each kind of information that the intruder will be able to reveal from the whole history of exchanged messages:

```
record info_name  = { name  };;
record info_nonce = { nonce };;
record info_pub   = { pub   };;
record info_priv  = { priv  };;
```

Predicates are defined for each kind of message to determine the presence of a message of a given kind in the solution:

```
fun messageReqCond(a, m)  = messageReq(m) & (m.kb == a.id);;
fun messageChalCond(a, m) = messageChal(m) & (m.ka == a.id)
                            & (m.na == a.ni);;
fun messageAuthCond(a, m) = messageAuth(m) & (m.kb == a.id)
                            & (m.nb == a.ni);;
fun PmessageReq(b, all)   = exists(messageReqCond(b), all);;
fun PmessageChal(a, all)  = exists(messageChalCond(a), all);;
fun PmessageAuth(a, all)  = exists(messageChalCond(a), all);;
```

B.2 The Intruder Transformation Rules

The intruder's behavior described in Section 4.2 is defined here in terms of MGS transformations.

Reading and Analyzing Messages. The following transformation rules define the evolution of the knowledge of the intruder, according to the messages present in the network:

```
trans intruder = {
  m / messageReq(m)  & exists((\k.(info_priv(k)
                       & (k.priv == m.kb))), neighbors(m))
   => m, {nonce = m.na}, {name = m.a};
  m / messageChal(m) & exists((\k.(info_priv(k)
                       & (k.priv == m.ka))), neighbors(m))
   => m, {nonce = m.na}, {nonce = m.nb};
  m / messageAuth(m) & exists((\k.(info_priv(k)
                       & (k.priv == m.kb))), neighbors(m))
   => m, {nonce = m.nb}
};;
```

The function neighbors used in the transformation is a special form that returns all the neighbors of the element denoted by a pattern variable.

Forging Some New Messages. In the previous section, we have described the intruder transformation which reveals only information according to already known messages and keys. The following transformation *produces* new fake messages from information known in the solution. There is one transformation for each kind of message:

```
trans forge_req[acc = set:()] =
{
     ((k:info_pub), (n:info_name), (m:info_nonce)) as X
     / acc := {na = m.nonce, a = n.name, kb = k.pub},acc; false
     => !!(0);
     _   => return(acc)
};;
trans forge_chal[acc = set:()] =
{
     ((k:info_pub), (n:info_nonce), (m:info_nonce)) as X
     / acc := {na=m.nonce, nb=n.nonce, ka=k.pub},
              {nb=m.nonce, na=n.nonce, ka=k.pub},acc; false
     => !!(0);
     _ => return(acc)
};;
trans forge_auth[acc = set:()] =
{
     ((k:info_pub), (m:info_nonce)) as X
     / acc := {nb=m.nonce, kb=k.pub}, acc; false
     => !!(0);
     _ => return(acc)
};;
fun forge(s)  =
     s, forge_req[acc=set:()](s), forge_chal[acc=set:()](s),
     forge_auth[acc=set:()](s);;
fun attack(s) = intruder(forge(s));;
```

Consider the first transformation; one should remark that, since the record made of info_pub, info_name, and info_nonce might not be unique, we have to use the same kind of procedure as that described in Section A.3 to produce *all* matching triples. This way, we produce all possible fake messages knowing public keys, names of agents involved in the sessions, and revealed nonces.

Function `forge` applied to the solution `s` adds to the original solution the result of the application of the three `forge` transformations.

An attack, represented by `attack`, consists in the revealing of all possible information by the `intruder` after having forged all possible fake messages.

B.3 Nested Multiset Rewriting

A new collection type, `membrane`, is defined which derives from the collection type `seq` (`membrane` is then a sequence with a different name). An empty collection of that type is `():membrane`.

```
collection membrane = seq;;
```

The Agents. The transformations describing the behavior of each agent are described below:

```
trans alice_req = {
   x / (req(x) & alice(x))  => (x + {pc = 'AUTH}),
                               {kb = x.dest, na = x.ni, a = x.id}
};;
trans bob_chal = {
   y / bob(y) & chal(y) & PmessageReq(y, neighbors(y))
   => let all_messages = filter(messageReqCond(y), neighbors(y))
      in return(map((\m.((y + {pc = 'WAIT, nr = m.na}),
                            {ka = m.a, na = m.na, nb = y.ni},
                            setify(neighbors(y)))), all_messages))
};;
trans alice_auth = {
  x / auth(x) & alice(x) & PmessageChal(x, neighbors(x))
  => let all_messages = filter(messageChalCond(x), neighbors(x))
     in return(map((\m.((x + {pc = 'FINISHED}),
                            {kb = x.dest, nb = m.nb},
                            setify(neighbors(x)))), all_messages))
};;
trans bob_finish = {
  y / bob(y) & wait(y) & PmessageAuth(y, neighbors(y))
  => let all_messages = filter(messageAuthCond(y), neighbors(y))
     in return(map((\m.((y + {pc = 'FINISHED}),
                   setify(neighbors(y)))),
                   all_messages))
};;
```

Notice that the messages addressed to Alice are not removed from the solution. Since they do not appear in the pattern part of the rule, they are not matched and therefore not "consumed" from the solution.

Care has been taken in the previous transformations to generate the correct membrane structure (`setify(neighbors(y))` for the argument in the `map` of the right hand side of each transformation; `setify` computes the set of elements of its collection argument and the function `neighbors` returns all the neighbors of the element denoted by a pattern variable).

Revealing a Successful Attack. A successful attack is to find in the chemical solution the nonce of Bob revealed. Since we have a membrane of sets, revealing a successful attack consists of looking in each set for the nonce:

```
fun isbroken(x) = member({nonce = 1}, x);;
fun broken(x)   = exists(isbroken, x);;
```

The Initial State. The initial state is a membrane of sets with only one set:

```
initial := ({id = "alice", ni = 0, nr, pc = 'REQ, dest = "charly",},
            {id = "bob",   ni = 1, nr, pc = 'CHAL},
            {priv = "charly"}, {pub = "charly"}, {pub = "alice"},
            {pub = "bob"}, set:()
           ):: membrane:();;
```

Note that the `nr` field is not set in the definitions: in this case, it is defined with an unspecified value (and will later be set to a relevant value once a message is received).

Looking for an Attack. As stated in Section 4.3, the problem consists of finding the correct interleaving of Alice's and Bob's actions leading to a successful attack. Transformation `breaks` succeeds if such an interleaving exists. It is applied on `functions`, which is the set of the transformations describing the agents' behavior. The `MGS` pattern expression `(_*) as F` will match all possible permutations of the elements of `functions`. For the sake of explanation, let `F` be the sequence $[f_1, \ldots, f_n]$ of *one possible permutation.* The guard checks whether `broken` holds for an attack on the state `attack`$^* \circ f_1 \circ \cdots \circ$ `attack`$^* \circ f_n$`(initial)`.

As for the search of an attack, we still look for an interleaving leading to the revealing of the nonce. We now have to `map` and `flatten` the attack that follows an action of one of the agents:

```
fun fmap(f, e) = flatten(map(f, e));;
trans break = {
   (_*) as F / broken(fold((\fn.\s.(fmap(attack[*], fmap(fn,s)))),
                           initial, F))
     => return(true)
};;
functions := alice_req, alice_auth, bob_chal, bob_finish, set:();;
successful := break(functions);;
```

The search for an attack succeeds in less than a second on a `AMD-1.4Ghz Linux Debian/Woody` computer, and reveals that the correct interleaving of functions is, as expected, `bob_finish` $\circ$ `alice_auth` $\circ$ `bob_chal` $\circ$ `alice_req`. The implemented code and the `MGS` interpreter is available from `mgs.lami.univ-evry.fr`.

Chapter 11
Membrane Algorithms: Approximate Algorithms for NP-Complete Optimization Problems

Taishin Yasunobu Nishida

Faculty of Engineering, Toyama Prefectural University
Kosugi-machi, Toyama 939-0398, Japan
nishida@pu-toyama.ac.jp

Summary. A new type of approximate algorithm for optimization problems, called the membrane algorithm, is proposed. A membrane algorithm consists of several regions separated by means of membranes; in each region we place a few tentative solutions of the optimization problem and a subalgorithm. The subalgorithms improve the tentative solutions simultaneously. The best and worst solutions in a region are sent to adjacent inner and outer regions, respectively. By repeating this process, a good solution will appear in the innermost region. The algorithm terminates if a termination condition is satisfied. A simple condition is the number of iterations, while a little more sophisticated condition becomes true if the good solution is not changed during a predetermined number of iterations. Computer experiments show that the membrane algorithms solve the traveling salesman problem better than the simulated annealing algorithm.

1 Introduction

Studies on approximate algorithms for **NP**-complete problems [1, 2, 9] are a very important issue in computer science, because:

- There are thousands of **NP**-complete problems.
- Almost all **NP**-complete problems correspond to practical problems.
- There are very few (I think no) expectations for $\mathbf{P} = \mathbf{NP}$, that is, for strictly solving **NP**-complete problems in deterministic polynomial time.

Recently, we have suggested a new type of approximate algorithms for solving **NP**-complete optimization problems [4, 5]. The algorithms use the P system paradigm [7]; that is why they are called *membrane algorithms*. A membrane algorithm borrows nested membrane structures, rules in membrane separated regions, transporting mechanisms through membranes, and dynamic structures of rules and membranes from P systems, and uses all these membrane computing ingredients to solve **NP**-complete optimization problems approximately.

In the next section, the outline of membrane algorithms is explained. Details about a membrane algorithm constructed in order to solve the traveling salesman problem approximately are given in Section 3. The section also describes results of computer experiments. Improved membrane algorithms are mentioned in Section 4.

2 Outline of Membrane Algorithms

Here we explain the new type of algorithm, called *membrane algorithm*.

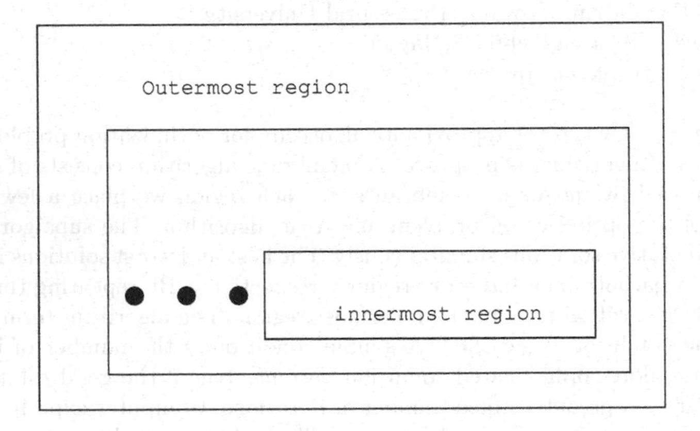

Fig. 1. Membrane structure of the suggested algorithm.

A membrane algorithm consists of three different kinds of components:

1. A number of regions which are separated by nested membranes – see Figure 1.
2. For every region, a subalgorithm and a few tentative solutions of the optimization problem to be solved.
3. Solution transporting mechanisms between adjacent regions.

After initial settings, the membrane algorithm works as follows:

1. In every region, the solutions are updated by the subalgorithm placed in the region, simultaneously.
2. In every region, the best and worst solutions, with respect to the optimization criterion, are sent to the adjacent inner and outer regions, respectively.

3. The membrane algorithm repeats updating and transporting solutions until a termination condition is satisfied. A simple termination condition is the number of iterations, while a little more sophisticated condition becomes true if the good solution is not changed during a predetermined number of steps.

The best solution in the innermost region is the output of the algorithm.

A membrane algorithm can have a number of subalgorithms which can be any approximate algorithm for optimization problems, for example, a genetic algorithm, tabu search, simulated annealing, local search, etc. The membrane algorithm is expected to be able to escape from local minima by using a subalgorithm which enhances a random search in the outer regions. On the other hand, the membrane algorithm can improve good solutions in the inner regions by a subalgorithm which enhances the local search. So, assigning appropriate subalgorithms for a given problem, the performance of membrane algorithms will be excellent.

Because the subalgorithms are separated by membranes and communications occur only between adjacent regions, a membrane algorithm will be easily implemented in parallel, distributed, or grid computing systems. This is the second important advantage of this approach.

3 First Experiment: Solving the Traveling Salesman Problem

In this section we fix the components of a membrane algorithm to solve the traveling salesman problem (TSP for short). Then, we implement and experiment with the algorithm on a computer.

3.1 Details of the Algorithm

Let m be the number of membranes, let regions 0 and $m-1$ be the innermost and outermost regions, respectively.

An instance of the TSP with n nodes contains n pairs of real numbers (x_i, y_i), for $i = 0, 1, \ldots, n-1$, which correspond to points in the two dimensional space. The distance $d(v_i, v_j)$ between two nodes $v_i = (x_i, y_i)$ and $v_j = (x_j, y_j)$ is the Euclidean distance $d(v_i, v_j) = \sqrt{(x_i - x_j)^2 + (y_i - y_j)^2}$. A solution of the instance is a list of nodes $(v_0, v_1, \ldots, v_{n-1})$ in which $v_i \neq v_j$ for every $i \neq j$. The *value* of a solution $v = (v_0, v_1, \ldots, v_{n-1})$ denoted by $W(v)$ is given by

$$W(v) = \sum_{i=0}^{n-2} d(v_i, v_{i+1}) + d(v_{n-1}, v_0).$$

For two solutions u and v, v is better than u if $W(v) < W(u)$. The solution which has the minimum value among all possible solutions is said to be the *strict solution* of the instance. A solution which has a value close to the strict solution is called an approximate solution.

The algorithm we propose has one tentative solution in region 0 and two solutions in all regions from 1 to $m-1$.

We use a tabu search as the subalgorithm in the innermost region, region 0. The tabu search searches a neighbor of the tentative solution by exchanging two nodes in the solution. In order to prevent a node from appearing in the same solution twice, the tabu search has a tabulist which consists of nodes already exchanged. Nodes in the tabulist are not exchanged again. The tabu search resets the tentative solution and the tabulist if one of the following three conditions occurs:

1. The value of the neighboring solution is less than that of the tentative solution. The neighboring solution becomes the new tentative solution.
2. The value of the best solution in region 1 is less than that of the tentative solution. The best solution in region 1 becomes the new tentative solution.
3. The neighbor search exceeds a predetermined turns (in this case $\frac{n}{5}$). The tentative solution remains unchanged, and only the tabulist is reset.

In case 3, no improvement occurs; however, the tabu search tries to search other neighbors, since there are many unsearched neighbors.

The tentative solutions in regions 1 to $m-1$ (there are two solutions in each region) are improved by a subalgorithm briefly described as follows:

1. If the two solutions have the same value, then a part of one solution (which is selected probabilisticaly) is reversed.
2. Recombinate the two solutions and produce two new solutions.
3. Modify the two new solutions by point mutations. In the ith region, a mutation occurs with probability $\frac{i}{m}$.

Obviously the subalgorithm described above resembles genetic algorithms. But the subalgorithm always recombinates the two solutions in a region while genetic algorithms randomly select solutions to be recombinated (from a large population of candidate solutions). If the two solutions in a region are identical, then the recombination produces no new solutions. Step 1 avoids this case and introduces a new solution using the reverse operation, which is a kind of mutation.

The overall algorithm looks as follows:

1. Consider an instance of the TSP.
2. Randomly construct one tentative solution for region 0 and two tentative solutions for every region from 1 to $m-1$.
3. Repeat 3.1 to 3.3 d times (d is given as a parameter).
   3.1 Modify tentative solutions simultaneously in each region using the subalgorithm associated with the region.
   3.2 For each region i $(1 \leq i \leq m-2)$, send the best solution in the region (old solutions and modified solutions) to region $i-1$ and the worst

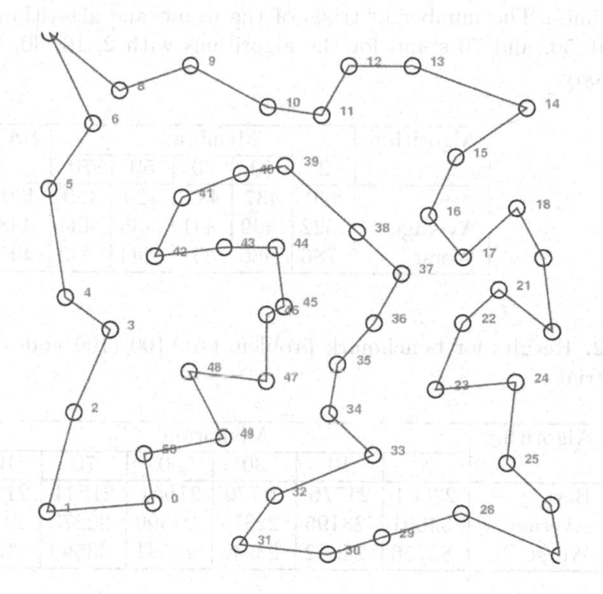

Fig. 2. An example obtained by a membrane algorithm. The instance is a benchmark problem of TSP, called eil51. The value of this solution is 429, while the strict value is 426.

solution to region $i+1$. (In region 0, send the worst solution to region 1, and in region $m-1$, send the best solution to region $m-2$.)

3.3 For each region 1 to $m-1$ remove all solutions but the best two.

4. Output the tentative solution in region 0 as the output of the algorithm.

In the above algorithm, steps 3.2 and 3.3 correspond to solution transporting mechanisms between adjacent regions.

3.2 Computer Experiments

We have implemented the algorithm using the Java programming language. By using Java, modifications of the algorithm have been easily tested on a computer. For example, we have implemented several recombination methods and have found that edge exchange recombination (EXX) [3] exhibits the best performance when used as subalgorithm.

Figure 2 shows an example of a solution provided by the computer experiment.

Table 1. Results of the membrane algorithm and simulated annealing (SA) for the benchmark problem eil51 (51 nodes). The membrane algorithm repeats step 3 40,000 times. The number of trials of the membrane algorithm is 20. Membranes 2, 10, 30, 50, and 70 stand for the algorithms with 2, 10, 30, 50, and 70 regions, respectively.

Algorithm	Membrane					SA
	2	10	30	50	70	
Best	440	437	432	429	429	430
Average	522	449	441	435	434	438
Worst	786	466	451	444	443	445

Table 2. Results for benchmark problem kroA100 (100 nodes). 100000 iterations and 20 trials.

Algorithm	Membrane						SA
	2	10	30	50	70	100	
Best	23564	21776	21770	21651	21544	21299	21369
Average	34601	23195	22878	22590	22275	21941	21763
Worst	82756	24862	23940	24531	23569	22954	22564

Tables 1 and 2 indicate results of the program for the TSP benchmark problem eil51[1] and kroA100[2] from TSPLIB [8]. Results of simulated annealing from [10] are also shown in the tables. From Tables 1 and 2, we can observe that the membrane algorithm gets slightly better results for eil51 and slightly worse results for kroA100 than the simulated annealing. Since the differences are very small, we may conclude that the membrane algorithm is as good as the simulated annealing. The more membranes a membrane algorithm has, the better the results obtained. Of course, the computation time is proportinal to the number of membranes. But it seems that the "number of membrane effect" will be saturated with 50 to 70 membranes for the eil51 problem. It is an issue for future research to find how many membranes give the best approximation for the kroA100 problem.

Figure 3 shows the changes of the average value of solutions for the kroA100 problem solved by a membrane algorithm with 50 membranes. One can see that the algorithm converges to remarkably good solutions in a few steps, about 2,000 to 3,000 steps.

[1] The value of the strict solution is 426.
[2] The value of the strict solution is 21,282.

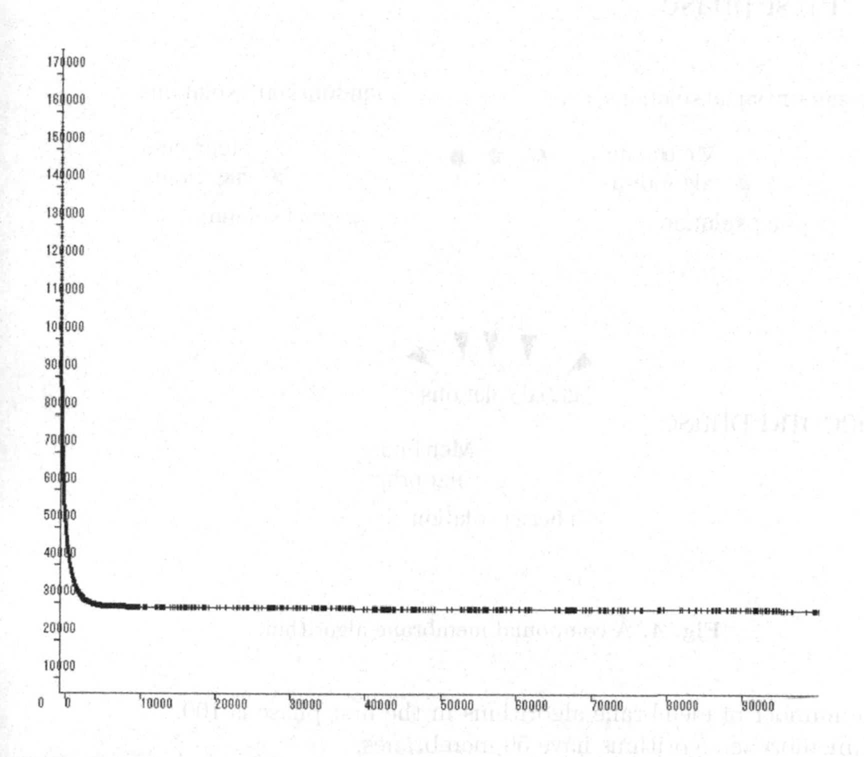

Fig. 3. Changes of the average value of solutions for the kroA100 problem solved by a membrane algorithm with 50 membranes.

4 Improved Membrane Algorithms

In this section we discuss some improved membrane algorithms. These improvements are performed by incorporating the concepts of tissue P systems and of P systems with a dynamic membrane structure.

4.1 Compound Membrane Algorithms

First we introduce *compound membrane algorithms*, which have two phases (Figure 4). In the first phase, a number of membrane algorithms produce good solutions from randomly generated initial solutions. The good solutions, in turn, become the initial solutions of the second phase, and in this way a better solution is obtained.

We examine a compound membrane algorithm with the following parameters:

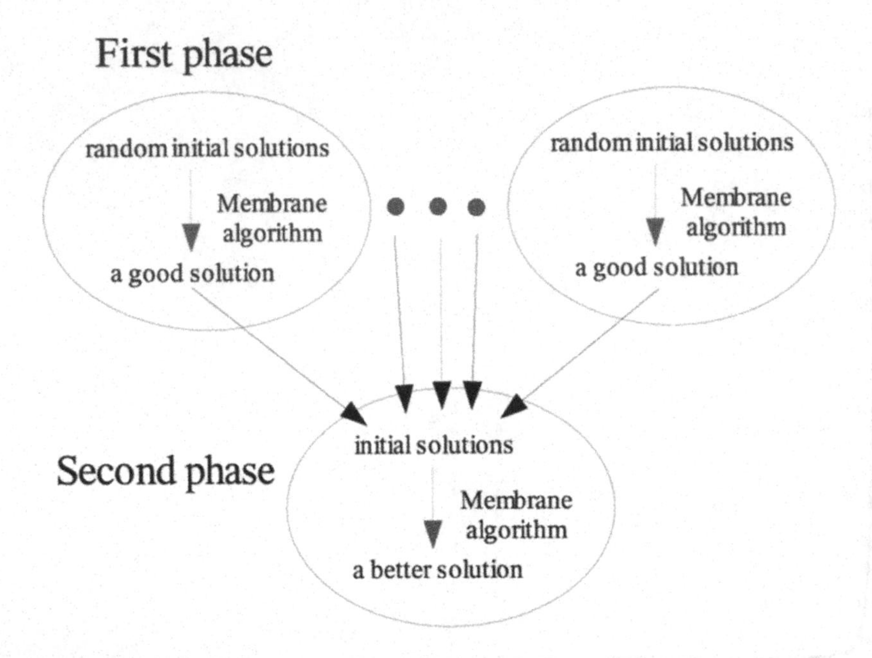

Fig. 4. A compound membrane algorithm.

- The number of membrane algorithms in the first phase is 100.
- All membrane algorithms have 50 membranes.
- Each membrane algorithm in the first phase terminates if the best solution does not improve during 500 iterations[3].
- The membrane algorithm in the second phase terminates if the best solution does not improve in 5,000 iterations[3].

Table 3. Results of the compound membrane algorithm. The number of trials of the compound and simple membrane algorithms is 20.

	eil51			kroA100		
	compound	membrane 70	SA	compound	membrane 100	SA
best	429	429	430	21316	21299	21369
average	432	434	438	21607	21941	21763
worst	438	443	445	21816	22954	22564

[3] These numbers are selected according to the observation that a membrane algorithm converges fast (Figure 3).

The results of computer experiments with the compound membrane algorithm are shown in Table 3. We can see that the compound membrane algorithm always outputs almost strict solutions.

On a single processor, the computation time of the compound membrane algorithm is, as expected, much longer than that of the simple membrane algorithm. However, because the membrane algorithms in the first phase work completely independent, the compound membrane algorithm will easily be implemented on a distributed computing system, and the computation time will then be only twice as long as that of a simple membrane algorithm.

4.2 Shrink Membrane Algorithms

We consider that the good results of the compound membrane algorithm come from:

1. The large number of random initial solutions that are used. Indeed, there are
$$99 \times 100 = 9,900$$
initial solutions.
2. The first phase selects "good seeds" for good solutions from the initial solutions.
3. The second phase generates very good solutions by recombining the "good seeds" obtained in the first phase.

However, it may be ineffective to use the same membrane algorithm in the first and second phases. That is why we propose a *shrink membrane algorithm* by incorporating dynamic membrane structures, as customary in the P systems area.

A shrink membrane algorithm also consists of two phases. Its first phase starts with five membranes and GA type subalgorithms in all regions. If the best solution in region 0 does not change during $100n$ iterations (where n is the size of the instance, i.e., the number of nodes), then the number of membranes becomes two with tabu search in region 0 and GA type subalgorithm in region 1. The two regions have the same solution, which is the best solution obtained so far. Then the algorithm improves solutions until the solution in region 0 does not change during $300n$ iterations.

The second phase of the shrink membrane algorithm is identical to that of the compound membrane algorithm, but the computation terminates if the best solution does not change during $100n$ iterations.

Figure 5 and Table 4 illustrate the flow and parameters of the shrink membrane algorithm.

Table 5 shows results of the shrink membrane algorithm. The experiments are done on a Sun Fire V210 computer with one 1 GHz UltraSPARC IIIi processor, 1.5 GB main memory, Solaris 9 OS, and Version 1.4.1 Java virtual machine.

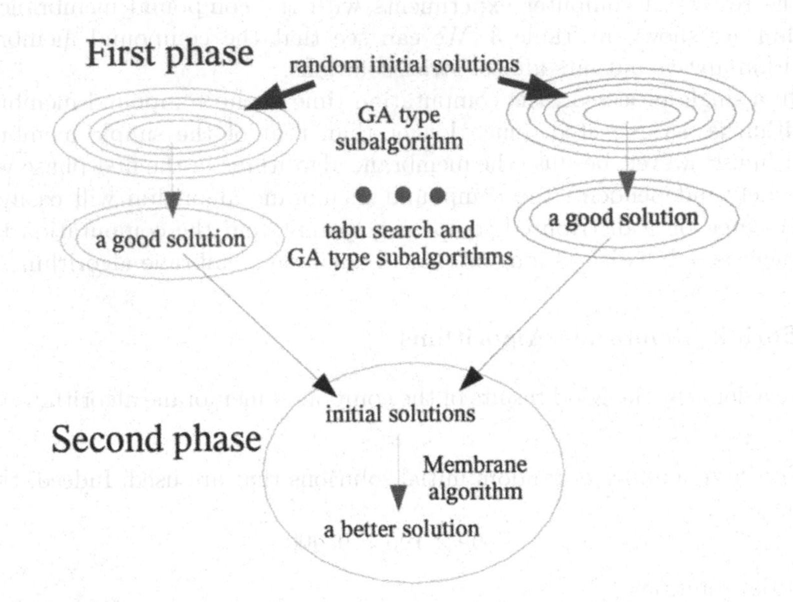

Fig. 5. A shrink membrane algorithm.

Table 4. Parameters of the shrink membrane algorithm used in the computer experiment.

Phase	number of		subalgorithms	terminate conditions
	algorithms	membranes		unchange during
1–1	100	5	GA type only	$100n$
1–2	100	2	GA type and tabu search	$300n$
2	1	50	GA type and tabu search	$100n$

We can see that the shrink membrane algorithm obtains better approximate solutions than the compound membrane algorithm. Because the differences are very small and the number of trials is not very large, there might be no meaningful differences between the shrink and compound strategies. We, however, stress that computation times of shrink algorithms are much shorter than those of compound algorithms. The speedup of computations should be a consequence of using dynamic membrane structures in the shrink membrane algorithm; it starts with a relatively small number of regions and GA type subalgorithms, then "shrinks" to only two membranes and does tabu search.

Table 5. Results of the shrink membrane algorithm. We perform 20 trials.

	eil51		kroA100	
	shrink	compound	shrink	compound
best	429	429	21299	21316
average	430	432	21504	21607
worst	433	438	21750	21816
average computation time (second)	748	1357	4312	7295

These processes can produce sufficiently "good seeds" for the second phase, in an efficient way.

5 Conclusion

We have proposed and implemented a new algorithm, called the membrane algorithm, for solving **NP**-complete optimization problems. Computer experiments have shown that the membrane algorithms get as good approximate solutions for the TSP as the simulated annealing algorithm. The convergence of the membrane algorithm is rather fast. Improved membrane algorithms, in the form of compound membrane algorithms and shrink membrane algorithms, always give almost strict solutions to the TSP.

There are many possibilities for improving membrane algorithms: using different subalgorithms, different variable structures, and different termination conditions, considering further P systems ingredients, and so on. It is also interesting to consider membrane algorithms solving other **NP**-complete optimization problems.

All programs used in the computer experiments are downloadable from the Website [6].

References

1. C.A. Floudas, P.M. Pardalos, eds.: *Encyclopedia of Optimization.* Kluwer, Dordrecht, 2001.
2. M.R. Garey, D.S. Johnson: *Computers and Intractability: A Guide to the Theory of* **NP**-*Completeness.* Freeman, 1979.
3. K. Maekawa et al.: A Solution of Traveling Salesman Problem by Genetic Algorithm (in Japanese). *SICE*, 31 (1995), 598–605.
4. T.Y. Nishida: An Application of P Systems: A New Algorithm for **NP**-Complete Optimization Problems. In *Proceedings of the 8th World Multi-Conference on Systems, Cybernetics and Informatics* (N. Callaos et al., eds.), vol. V, 2004, 109–112.

5. T.Y. Nishida: An Approximate Algorithm for **NP**-Complete Optimization Problems Exploiting P Systems. In *Proceedings of Brainstorming Workshop on Uncertainty in Membrane Computing*, Palma de Mallorca, 2004, 185–192.
6. T.Y. Nishida: URL `http://www.comp.pu-toyama.ac.jp/nishida/`.
7. Gh. Păun: Computing with Membranes. *Journal of Computer and System Sciences*, 61 (2000), 108–143.
8. G. Reinelt: TSPLIB URL `http://www.iwr.uni-heidelberg.de/group/comopt/software/TSPLIB95/`.
9. A. Salomaa: *Computation and Automata*. Cambridge University Press, Cambridge, 1985.
10. M. Yoneda: URL `http://www.mikilab.doshisha.ac.jp/dia/research/person/yoneda/research/2002_7_10/SA/07-sareslut.html`.

Chapter 12
Computationally Hard Problems Addressed Through P Systems

Mario J. Pérez-Jiménez, Alvaro Romero-Jiménez,
Fernando Sancho-Caparrini

Department Computer Science and Artificial Intelligence
E.T.S. Ingeniería Informática, University of Seville
Avda. Reina Mercedes s/n, 41012 Sevilla, Spain
{Mario.Perez,Alvaro.Romero,Fernando.Sancho}@cs.us.es

Summary. In this chapter we present a general framework to provide efficient solutions to decision problems through families of cell-like membrane systems constructed in a semi-uniform way (associating with *each instance* of the problem one P system solving it) or a uniform way (all instances of a decision problem having the same *size* are processed by the same system). We also show a brief compendium of efficient semi-uniform and uniform solutions to hard problems in these systems, and we explicitly describe some of these solutions.

1 Introduction

Many interesting problems of the real world are presumably intractable (unless **P** = **NP**) and hence it is not possible to obtain algorithmic solutions when we address large instances of those problems on an electronic computer.

From this point of view, membrane systems have two attractive features: they are inherently parallel and nondeterministic. Can the parallelism and nondeterminism of P systems be used to solve hard problems in feasible time? The answer is yes, but we must point out two facts. On the one hand, we have to deal with the nondeterminism in such a way that the solutions obtained by using these devices are algorithmic solutions in the classical sense, that is, the answers of the computations of the system must be reliable. On the other hand, the drastic decrease of the execution time from exponential to polynomial is not achieved for free, but with the use of an exponential amount of space, although this space is created in polynomial time.

In this chapter we present the theoretical requirements for a P system to provide an *algorithmic solution* to an abstract decision problem (a precise definition of the latter is given in Section 2). First, all computations of the system must halt, providing a positive or negative answer to the problem (i.e., for a particular instance of it). Second, we impose that the systems be confluent. This is a generalization of the notion of determinism for which we require all possible computations to provide the same answer. This way we do not obtain a contradictory result. In Section 3, P systems verifying these two properties are called recognizer systems.

It is important to note that all the feasible solutions to hard problems obtained by means of these biologically inspired devices so far presented do not use a single P system, but a family of systems. However, there are significative differences between those solutions, dividing them in two groups: the *semi-uniform* solutions, which associate with *each instance* of the problem one P system solving it, and the *uniform* solutions, which associate with *each possible size* of the instances of the problem one P system that can solve all instances of that size. A formal definition of these two concepts can be found in Sections 4 and 5.

Another possible classification can be considered with respect to the existence or not in the system of a membrane where the input data is introduced before the computation starts. Usually, the semi-uniform solutions are performed by P systems without input, whereas the uniform solutions are performed by P systems with input. In Section 4 we present a compendium of known semi-uniform solutions to hard problems by P systems without input, and a detailed description of two of these solutions: one to the *Satisfiability Problem* and the other to the *Hamiltonian Path Problem.* Finally, in Section 5 we do the same for known uniform solutions to hard problems by P systems with input, detailing the ones corresponding to the *Decision Knapsack Problem (0/1)* and to the *Common Algorithmic Decision Problem.*

2 Abstract Problems

Membrane systems provide devices with the ability to solve problems. But these machines are equipped only with tools able to handle inputs and outputs that are multisets of symbol objects. Hence, we are forced to treat these problems as collections of multisets.

In general we define an *abstract problem* X to be a pair $X = (I_X, S_X)$ where I_X is a language over a finite alphabet, whose elements are called *instances* of the problem, S_X is a function whose domain is I_X, and for any instance $i \in I_X$, the set $S_X(i)$ is finite (the elements of this set are called *candidate solutions*). Observe that the set of instances of an abstract problem is a finite or enumerable set.

As an example, consider the problem MAX-CLIQUE of finding a largest clique in a given unidirected graph. Recall that a *clique* in a graph is a set of vertices such that their each pair is connected by an edge. An instance of MAX-CLIQUE is an undirected graph, and a candidate solution is a set of vertices. An *exact solution* will be a set of vertices which is a clique with the maximal number of vertices. The problem MAX-CLIQUE itself is the relation that associates each undirected graph with each largest cliques in the graph (an exact solution). Since largest cliques in an uniderected graph are not necessarily unique, a given problem instance may have more than one exact solution; that is, the binary relation associated with the problem is not necessarily univocal.

In this chapter we will work only with *decision problems*, that is, abstracts problems that require either a *yes* or a *no* answer. Formally, a *decision problem* X is an abstract problem (I_X, S_X) such that S_X is a total boolean function (that is, a predicate) over I_X.

For example, the following is a decision version of the problem CLIQUE: given an undirected graph G and a natural number k, determine whether or not G has a clique of size at least k. An instance for CLIQUE is a pair (G, k) where G is an undirected graph and k is a natural number, and a candidate solution is 1 (*yes*) or 0 (*no*). If $i = (G, k)$ is an instance of this problem, then $S_{CLIQUE}(i) = 1$ (*yes*) if there exists a clique in G of size at least k, and $S_{CLIQUE}(i) = 0$ (*no*) otherwise.

There exists a natural correspondence between languages and decision problems in the following way: each language L over an alphabet Σ has a decision problem, $X_L = (I_{X_L}, S_{X_L})$, associated with it, where $I_{X_L} = \Sigma$ and $S_{X_L} = \{(x, 1) \mid x \in L\} \cup \{(x, 0) \mid x \notin L\}$; reciprocally, given a decision problem $X = (I_X, S_X)$, the language L_X over I_X corresponding to it is defined as $L_X = \{a \in I_X \mid S_X(a) = 1\}$.

Let M be a Turing machine such that the result of any halting computation is *yes* or *no*. Let L be a language over an alphabet Σ. If M is a deterministic device then we say that M *recognizes* or *decides* L whenever, for any string a over Σ, if $a \in L$ then the answer of M on input a is *yes*, and *no* otherwise. If M is a nondeterministic Turing machine, then we say that M *recognizes* or *decides* L if the following is true: for any string a over Σ, $a \in L$ if and only if there exists a computation of M with input a such that the answer is *yes*. That is, an input string a is accepted if there is *some* accepting computation of M on input a.

Notice the difference in the definition of acceptance by deterministic and nondeterministic Turing machines. An input string a is accepted by a deterministic Turing machine M if *the* computation of M on input a halts and answers *yes*. A nondeterministic Turing machine M accepts a string a if there exists *some* computation of M on input a answering *yes*; that is, there exists a sequence of nondeterministic choices that results in *yes*. In this case, it is possible that we accept a string but that there exists another computation with the same input that either halts and answers *no*, or does not halt.

In some sense, we can state that nondeterministic devices do not properly capture the intuitive idea underlying the concept of algorithm, because the result of such a machine on an input is not reliable, since the answer of the device is not always the same.

In the context of computability theory, we consider a problem X to be solved when we have a *general method* (described in a model of computation) that works for any instance of the problem. From a practical point of view, such methods run only over a finite set of instances whose cardinality depends on the available resources.

We say that a Turing machine M solves a decision problem X if M *recognizes* the language associated with X; that is, for any instance a of the problem: (a) in the deterministic case, the machine (with input a) outputs *yes* if the answer of the problem is *yes*, and outputs *no* otherwise, and (b) in the nondeterministic case, some computation of the machine (with input a) outputs *yes* if the answer of the problem is *yes*.

As for when the instances of abstract problems are strings, we can consider their size in a natural manner: the size of an instance is the length of the string. Then, how do the resources required to execute a method increase according to the size of the instance? This is a fundamental question in computational complexity theory.

Let M be a deterministic Turing machine. Let R be a resource used by the device (for example, the number of steps executed before the machine halts, the *time* of execution). We consider a function f_R mapping nonnegative integers to nonnegative integers defined as follows: $f_R(n)$ is the maximum, over all instances a of size n, of the amount of resource R used when the device M is executed with input a. For example, if R is the time of execution, then we say that M operates in time $f_R(n)$; if f_R turns to be a polynomial function, then we say that M works in polynomial time.

3 Recognizer P Systems

In this chapter we use membrane computing as a framework to address the resolution of decision problems; that is why we consider P systems as *recognizer* devices. Since P systems work in a nondeterministic manner, we need to adapt the usual definition of the processes of acceptance in nondeterministic Turing machines.

In order to accept or reject an input string/multiset it should be enough to read the answer of *any* computation of the system. Hence it is necessary to require a condition of *confluence* in the following sense: every computation of the system is a halting computation, and (on the same input, if any) all computations have the same output.

Definition 1. *A* recognizer P system *is a P system with external output such that:*

1. *The working alphabet contains two distinguished elements* yes *and* no.
2. *All computations halt.*
3. *If $\mathcal{C}$ is a computation of the system, then either some object* yes *or some object* no *(but not both) must have been released into the environment, and only in the last step of the computation.*

For recognizer P systems, we say that a computation $\mathcal{C}$ is an *accepting computation* (or *rejecting computation*) if the object *yes* (or *no*) appears in the external environment associated with the corresponding halting configuration of $\mathcal{C}$. Hence, these devices send to the environment an accepting or rejecting answer at the end of their computations. Therefore, if we want these kinds of systems to properly solve decision problems, we have to require all responses to be consistent, in the sense that the system must always give the same answer.

4 Recognizer P Systems Without Input

The first results about *solvability* of **NP**-complete problems in polynomial (even linear) time by membrane systems were given by Gh. Păun [24], C. Zandron et al. [40], S.N. Krishna et al. [10], and A. Obtulowicz [15] in the framework of P systems that lack an input membrane. Thus, the constructive proofs of such results need to design *one* system for *each* instance of the problem.

This method for solving problems provides a *specific* algorithmic solution, in the following sense: if we wanted to apply such a method to some decision problem then a system should be constructed for every instance of the problem. However, the construction is done in polynomial time, which prevents the solving of the problem during the "programming" phase.

Now, we formalize these ideas in the following definition.

Definition 2. *Let $X = (I_X, \theta_X)$ be a decision problem. We say that X is solvable in polynomial time by a family of recognizer membrane systems* without input $\mathbf{\Pi} = (\Pi(w))_{w \in I_X}$ *if the following are true:*

- *The family $\mathbf{\Pi}$ is "Turing polynomially uniform"; that is, there exists a deterministic Turing machine which, in polynomial time, constructs the system $\Pi(w)$ starting from the instance $w \in I_X$.*
- *The family $\mathbf{\Pi}$ is polynomially bounded; that is, there exists a polynomial function $p(n)$ such that for each $w \in I_X$, every computation of $\Pi(w)$ halts in at most $p(|w|)$ steps.*
- *The family $\mathbf{\Pi}$ is sound; that is, for each instance of the problem $w \in I_X$, if there exists an accepting computation of $\Pi(w)$, then $\theta_X(w) = 1$ (the corresponding answer of the problem is* yes*).*
- *The family $\mathbf{\Pi}$ is complete; that is, for each instance of the problem $w \in I_X$, if $\theta_X(w) = 1$ (the answer to the problem is* yes*), then every computation of $\Pi(w)$ is an accepting computation (the system also responds* yes*).*

The soundness property means that if we obtain an *acceptance response* of the system (associated with an instance) through some computation, then the answer to the problem (for that instance) has to be *yes*. The completeness property means that if we obtain an *affirmative* response to the problem, then any computation of the system must be an accepting one.

Notice that in the above definition we consider two different tasks. The first one is the construction of the family solving the problem, which we require to be done in polynomial time; that is, there exists a deterministic Turing machine M, and a polynomial function $q(n)$ such that for each $w \in I_X$, $M(w)$ provides a complete description of the system $\Pi(w)$ in at most $q(|w|)$ steps. This is precomputation time, expressed in the number of *sequential steps*.

The second task is the execution of the systems $\Pi(w)$ of the family, and for this task we impose that the total number of steps performed by the computations of system $\Pi(w)$ is bounded by the polynomial function $p(n)$. This is the real computation time, and it is described by the number of *parallel steps*.

In this section we use recognizer membrane systems *without input membrane* in order to solve decision problems. In this context, we need to associate with each instance of the problem such a system, accepting or rejecting it. Bearing in mind the nondeterminism of the system, we require the confluence condition; that is, all branches of a computation associated with the instance eventually reach a unique configuration.

Unless **P**=**NP**, to solve an **NP**-complete problem in an efficient way using P systems we have to create an exponential workspace in polynomial time. This is possible in various types of membrane systems, for example, through *membrane division* [24] (we can repeatedly divide membranes in order to obtain 2^n membranes in n steps), *membrane creation* [9], [13] (new membranes are produced under the influence of the existing objects in a membrane), *string replication* [4] (in a rewriting membrane system using string objects we can generate exponentially many strings in linear time), or by using *precomputed resources* [25], [6] (we start from an arbitrarily large initial membrane structure, without objects placed in its regions, and we trigger a computation by introducing both objects and rules related to a given problem in a specified membrane).

In [23] Gh. Păun explains the convenience of solving many **NP**-complete problems in a *uniform manner*, that is, with P systems which are very similar to each other. This idea about the uniformity of the P systems able to solve a decision problem was formulated for the first time by A. Obtulowicz [17] in relation to a proposed solution to *SAT*: the family of P systems (without input membrane) described in [24] is generated in a logarithmic space by a multitape deterministic Turing machine. We say that this construction is *semi-uniform*: the systems of the family solving the decision problem are constructed starting not from the *size* of an instance, but from an instance only; that is, in a recursive manner, for each instance of the problem a P system associated with it is constructed.

In contrast, in Section 5 we will define the concept of *uniform* construction of families of P systems solving a decision problem.

Now, we briefly comment the different efficient solutions of **NP**-complete problems in the framework of P systems without input, described in a semi-uniform way, proposed until now.

The first efficient solution to SAT is given by Gh. Păun in [24], using division for non-elementary membranes. This result was improved by Gh. Păun, Y. Suzuki, H. Tanaka, and T. Yokomori in [29] using only division for elementary membranes (in that paper a solution to HPP using membrane creation is also presented).

In [2], A. Alhazov and T.O. Ishdorj present a linear time solution to SAT through P systems with active membranes without polarizations but using some membrane rules (merging, separation, and release).

The first efficient solution to HPP by P systems is due to S.N. Krishna and R. Rama [10], but using rules for d-division, with an arbitrary d (a solution to Vertex Cover is also provided in this paper). The result of S.N. Krishna and R. Rama about HPP was improved by A. Păun in [21] by using only rules for 2-division.

Different efficient solutions to several variants of SAT (3-SAT, *Not-all-equal* 3 SAT, *One-in-three* 3SAT, and *Minimum 2-satisfiability*) are decribed by Gh. Păun, Y. Suzuki, H. Tanaka, and T. Yokomori in [29]. Also, in that paper solutions to several problems of graph theory (*Vertex Cover*, *k-closure*, *Independent set*, *Graph 3-colourability*, and *Monochromatic triangle*) are given by P systems with active membranes using cooperative rules.

Other efficient solutions to SAT and HPP are given by J. Castellanos et al. [4], through P systems using string replications (similar solutions are given by S.N. Krishna and R. Rama in [11]), and by E. Czeizler [6], through P systems using precomputed resources.

A polynomial solution to *Integer Factorization Problem* is presented by A. Obtulowicz in [16] through deterministic P systems with active membranes but using cooperative rules.

P. Sosik in [39] provides a semi-uniform efficient solution to QBF (satisfiability of quantified propositional formulas), a well known **PSPACE**-complete problem in the framework of P systems with active membranes but using 2-division for nonelementary membranes.

S.N. Krishna and R. Rama [12] show how P systems with membrane division can theoretically break the most widely used cryptosystem, DES. That is, given an arbitrary (plain text, cipher text) pair, one can recover the DES key in linear time with respect to the length of the key.

In this section we present a semi-uniform solution to *SAT* problem due to C. Zandron, G. Mauri, and C. Ferretti [41] (with slight modifications the above solution can be adapted to a linear time solution by recognizer P systems, as we point out at the end of subsection 4.1). We also describe a quadratic time semi-uniform solution to the Hamiltonian Path Problem, a variant of a solution presented by M.J. Pérez-Jiménez, A. Romero-Jiménez, and F. Sancho-Caparrini in [35].

4.1 A Linear Time Solution to SAT

The Satisfiability Problem (SAT) is the following decision problem:

> *Given a propositional formula in conjunctive normal form, determine if there exists a truth assignment of its variables which makes the formula true.*

In what follows we present a family of P systems with active membranes using 2-division that solves SAT in linear time.

Given an instance ϕ of SAT we construct a P system $\Pi(\phi)$ whose functioning can be divided into the following stages:

(a) Generate all the possible truth assignments for the variables of ϕ. (This will be done in a nondeterministic way.)
(b) Apply the assignments generated in the previous stage to the formula.
(c) Check if all the clauses have value *true*.
(d) Answer *yes* or *no* depending on the results from (c).

Let us suppose that ϕ is a formula with m clauses and n variables. That is, $\phi = C_1 \wedge \cdots \wedge C_m$ for some $m \geq 1$, with $C_i = y_{i,1} \vee \cdots \vee y_{i,p_i}$ for some $p_i \geq 1$ and $y_{i,j} \in \{x_k, \neg x_k \mid 1 \leq k \leq n\}$, for each $1 \leq i \leq m, 1 \leq j \leq p_i$.

The P system $\Pi(\phi)$ is defined as follows.

- The working alphabet is

$$\begin{aligned}\Gamma(\phi) = &\{a_i, t_i, f_i \mid 1 \leq i \leq n\} \cup \{r_i \mid 0 \leq i \leq m\} \\ &\cup \{c_i \mid 1 \leq i \leq m-1\} \cup \{d_i \mid 0 \leq i \leq n\} \cup \{yes\}.\end{aligned}$$

- The set of labels is $\{1, 2\}$.
- The initial membrane structure is $[_1 [_2\ ]_2^0]_1^0$ (each membrane with label 2 is said to be an internal membrane).
- The initial multisets associated with the membranes are $\mathcal{M}_1 = \lambda$ and $\mathcal{M}_2 = a_1 a_2 \ldots a_n d_0$.
- The rules are:

(a.1) $[_2 a_i]_2^0 \to [_2 t_i]_2^0 [_2 f_i]_2^0$, for $1 \leq i \leq n$,

(a.2) $[_2 d_k \to d_{k+1}]_2^0$, for $0 \leq k \leq n-2$,

(a.3) $[_2 d_{n-1} \to d_n c]_2^0$,

(a.4) $[_2 d_n]_2^0 \to [_2]_2^+ d_n$,

(b.1) $[_2 t_i \to r_{h_{i,1}} \ldots r_{h_{i,j_i}}]_2^+$, for $1 \leq i \leq n$, and the clauses $C_{h_{i,1}}, \ldots, C_{h_{i,j_i}}$ contain the literal x_i,

(b.2) $[_2 f_i \to r_{h_{i,1}} \ldots r_{h_{i,j_i}}]_2^+$, for $1 \leq i \leq n$, and the clauses $C_{h_{i,1}}, \ldots, C_{h_{i,j_i}}$ contain the literal $\neg x_i$,

(c.1) $[_2 r_1]_2^+ \to [_2]_2^- r_1$,

(c.2) $[_2 c_i \to c_{i+1}]_2^-$, for $1 \le i \le m$,
(c.3) $[_2 r_k \to r_{k-1}]_2^-$, for $2 \le k \le m$,
(c.4) $r_1 [_2]_2^- \to [_2 r_0]_2^+$,
(c.5) $[_2 c_{m+1}]_2^+ \to [_2]_2^+ yes$,

(d.1) $[_1 yes]_1^0 \to [_1]_1^0 yes$.

Let us see now that this P system indeed solves SAT for the formula ϕ.

The objects a_i contained in the initial internal membrane of the system correspond to the variables x_i. By using a rule from (a.1), for i nondeterministically chosen, we produce the truth values *true* and *false* assigned to variable x_i. The truth values are represented by the objects t_i and f_i, respectively, and are placed in two separate copies of the internal membrane. Since the charge remains neutral for both membranes, the process can continue.

In this way, in n steps we assign truth values to all variables. Hence, we get all the 2^n different truth assignments for the formula ϕ, placed in 2^n separate internal membranes, which, in turn, are placed in the skin membrane. Note that, in spite of the fact that in each step the object a_i is nondeterministically chosen, after n steps we get the same result, regardless of the objects used in each step.

On the other hand, the objects d_i are used as counters, to control when the process described above has finished. The initial internal membrane starts with the object d_0 and at each division step we pass from d_k to d_{k+1} using the rules from (a.2). Also, each new internal membrane created by division gets copies of these objects, keeping their own counter this way. At the step before the last division, the rule from (a.3) introduces both d_n and c_1. The former object will exit the membrane (at step $n+1$ using the rule from (a.4)), changing its polarization to positive, and this finishes the first stage and starts the second one in that membrane. The latter object will be used in the third stage.

In the second stage we look for the clauses satisfied by the truth assignments associated with each internal membrane. We do this in one step in parallel for all the 2^n membranes and in parallel for all the truth values present in them, using the rules from (b.1) and (b.2). These rules introduce an object r_i in the membrane for each clause satisfied by the truth value being considered.

If, after completing this step, there is at least one internal membrane which contains all symbols $r_1, \dots, r_m$, this means that the truth assignment associated with that membrane satisfies all the clauses, and therefore satisfies the formula ϕ. Otherwise, if no internal membrane gets all the objects $r_1, \dots, r_m$, the formula ϕ is not satisfiable. Note that the satisfiability problem has already been solved, in $n+2$ steps, making an essential use of the parallelism, but we still have to "read" the answer and send a suitable signal out of the system.

We use the rules from (c.1–4) in the third stage to check if some internal membrane has an associated truth assignment that makes true all the clauses of ϕ. To do this we carry out a loop in which we check in each step if the object r_1 is present, eliminate it, and perform a rotation of the objects $r_2, \dots, r_m$, decreasing their subscripts by 1. It is clear that an internal membrane contains all the objects $r_1, \dots, r_m$ (that is, the truth assignment associated with it satisfies all the clauses and, hence, the formula) if and only if we can run m steps of the loop.

Let us take a closer look at how one step of the loop is performed. First the rule from (c.1) checks whether or not the object r_1 is present in the membrane. If this is the case, r_1 is sent out of the membrane, and the polarization of the membrane is changed to negative. The membranes which do not contain the object r_1 remain positively charged and will no longer evolve; no further rule can be applied to them. Next, for all the internal membranes negatively charged (that is, those that had contained an object r_1) the rules from (c.3) decrease the subscripts of the objects $r_2, \dots, r_m$ in that membrane, and the rule from (c.4) gives back the object r_1 from the skin membrane, changed to a dummy object r_0 which will never evolve again, and changing the polarization of the membrane to positive so that the process can be started again. Note the important fact that in the skin membrane we have a number of copies of the object r_1 equal to the number of membranes with a negative charge. Because a rule from (c.4) can only involve one membrane and because these rules are applied in the maximally parallel way, each membrane which contained before an object r_1 will now contain an object r_0 and will be able to continue running the loop.

Simultaneously, in an internal membrane negatively charged, the rule (c.2) increases the subscripts of the objects c_i. These objects are used to count the number of steps of the previous loop that have been carried out. Thus, if an object c_{m+1} appears in the membrane, it means the membrane contained all the objects $r_1, \dots, r_m$ since all the m steps of the loop have been able to run. Also, the membrane is positively charged because the rule from (c.4) has just been applied. Therefore, we can apply the rule from (c.5), which sends out the object c_{m+1} to the skin as an object *yes*. Finally, this object is sent out to the environment by means of the rule from (d.1).

The family of P systems $\Pi(\phi)$ constructed as above for each propositional formula ϕ in conjunctive normal form is *Turing polynomially uniform*. Indeed, the evolution rules are computable in a uniform way; the size of the working alphabet is $4n + 2m + 2$; the number of membranes in the initial membrane structure is 2; the maximum cardinality of the initial multisets is $n + 1$; the total number of rules is $4n + 2m + 4$ and the maximum length of a rule is $\max(7, m + 3)$.

However, this family should be adjusted in several respects in order to fulfill the conditions requested by Definition 2 from Section 4 for a linear time solution to SAT.

1. First, the P systems constructed above, although confluent, are nondeterministic. This is not really a problem, but if we want deterministic P

systems, we have to change the rules for stage 1 so that they cover all the objects a_i sequentially, instead of choosing them nondeterministically.
2. Then, the P systems may not halt in linear time, because the rule from (d.1) can only be applied once at a given time. If, for example, the formula ϕ is valid, then we may have 2^n objects *yes* in the skin membrane at the last step of the computation. These objects are then sent out to the environment one by one, so the total time used will be exponential. The solution is easy: we change the polarization of the skin when the first object *yes* is sent out, so that the rule from (c.2) cannot be applied any more and the system halts.
3. Third, the P systems above are not recognizer systems. If the formula is satisfiable, we obtain the answer *yes* in the environment, but if the formula is not satisfiable the system halts without answering anything. To solve this problem we can add objects to the skin membrane that count the number of computation steps that have been carried out. With the aid of these counters we can check if the system should have already sent an object *yes* in case the formula is satisfiable. Otherwise, we can be sure that the formula is not satisfiable and send an object *no* to the environment.

4.2 A Quadratic Time Solution to HPP

A Hamiltonian path in a graph G is a path that visits each of the nodes of G exactly once. The Hamiltonian Path Problem (HPP) is the following decision problem:

> *Given a graph G, determine if there exists a Hamiltonian path in G.*

In the following discussion we will construct, in a way similar to that of Section 7.2.2 of [25], a family of recognizer P systems with active membranes using 2-division that solves HPP in quadratic time. Specifically, given an instance G of HPP of size n (that is, with n nodes), we construct a P system $\Pi(G)$ whose functioning can be divided into the following stages:

(a) Generate all the possible paths in G of length n.
(b) For each of the previous paths, determine whether it visits all the nodes of G.
(c) Answer *yes* or *no* depending on the results from (b).

Consequently, this P system solves HPP for G, because a path in G is Hamiltonian if and only if it has length n and visits all the nodes of G.

Before describing $\Pi(G)$ in full detail, we have to fix several notations. Let G be a graph of size n. We denote by $V = \{v_1, \ldots, v_n\}$ the set of nodes and by E the set of edges of G. Given a node v_i, we assume that the set of nodes $adj(v_i) = \{v_{j_1^i}, \ldots, v_{j_{k(i)}^i}\}$ adjacent to v is ordered such that $j_1^i < \cdots < j_{k(i)}^i$ (and, since we suppose G to be connected, it is nonempty).

For example, let us consider the graph in Figure 1. For this graph, $V = \{v_1, v_2, v_3\}$ and $E = \{\{v_1, v_2\}, \{v_1, v_3\}\}$. Also, $adj(v_1) = \{v_2, v_3\}$, so $k(1) = 2$ and $j_1^1 = 2, j_2^1 = 3$; $adj(v_2) = \{v_1\}$, so $k(2) = 1$ and $j_1^2 = 1$; $adj(v_3) = \{v_1\}$, so $k(3) = 1$ and $j_1^3 = 1$.

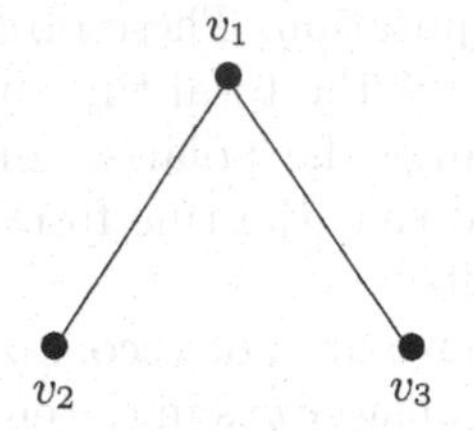

Fig. 1. Example graph.

The P system $\Pi(G)$ is defined as follows.

- The working alphabet is

$$\begin{aligned}\Gamma(G) = &\{v_i \mid 1 \leq i \leq n\} \cup \{v_{i,j} \mid 1 \leq i, j \leq n\} \cup \{a_{i \to j,l} \mid 1 \leq i, j, l \leq n\} \\ &\cup \{r'_i \mid 1 \leq i \leq n\} \cup \{r_i \mid 0 \leq i \leq n\} \cup \{c_i \mid 1 \leq i \leq n+1\} \\ &\cup \{d_i \mid 0 \leq i \leq n^2 + 2n + 4\} \cup \{c, a_{n+1}, yes, no\}.\end{aligned}$$

- The set of labels is $\{1, 2\}$.
- The initial membrane structure is $[_1\ [_2\ ]_2^0\ ]_1^0$ (each membrane with label 2 is said to be internal).
- The initial multisets associated with the membranes are $\mathcal{M}_1 = d_0$ and $\mathcal{M}_2 = v_1$.
- The rules are:
  (a.1) $[_2 v_i]_2^0 \to [_2 v_{i,1}]_2^- [_2 v_{i+1}]_2^0$, for $1 \leq i \leq n-2$,
  (a.2) $[_2 v_{n-1}]_2^0 \to [_2 v_{n-1,1}]_2^- [_2 v_{n,1}]_2^-$,
  (a.3) $[_2 v_{i,l} \to r'_i a_{i \to j_1^i, l}]_2^-$, for $1 \leq i \leq n, 1 \leq l \leq n-1$,
  (a.4) $[_2 v_{i,n} \to r'_i c a_{n+1}]_2^-$, for $1 \leq i \leq n$,
  (a.5) $[_2 a_{i \to j_k^i, l}]_2^- \to [_2 v_{j_k^i, l+1}]_2^- [_2 a_{i \to j_{k+1}^i, l}]_2^-$, for $1 \leq i \leq n, 1 \leq k \leq k(i) - 1, 1 \leq l \leq n-1$,
  (a.6) $[_2 a_{i \to j_{k(i)}^i, l} \to v_{j_{k(i)}^i, l+1}]_2^-$, for $1 \leq i \leq n, 1 \leq l \leq n-1$,
  (a.7) $[_2 a_{n+1}]_2^- \to [_2]_2^+ a_{n+1}$,
  (b.1) $[_2 r'_i \to r_i]_2^+$, for $1 \leq i \leq n$,
  (b.2) $[_2 c \to c_1]_2^+$,
  (b.3) $[_2 r_1]_2^+ \to [_2]_2^- r_1$,
  (b.4) $[_2 r_i \to r_{i-1}]_2^-$, for $1 \leq i \leq n$,

(b.5) $[_2 c_i \to c_{i+1}]_2^-$, for $1 \le i \le n$,
(b.6) $r_1 [_2]_2^- \to [_2 r_0]_2^+$,
(b.7) $[_2 c_{n+1}]_2^+ \to [_2]_2^+ c_{n+1}$,

(c.1) $[_1 c_{n+1}]_1^0 \to [_1]_1^+ yes$,
(c.2) $[_1 d_i \to d_{i+1}]_1^0$, for $0 \le i \le n^2 + 2n + 3$,
(c.3) $[_1 d_{n^2+2n+4}]_1^0 \to [_1]_1^- no$.

Let us see now that this P system indeed solves HPP for G.

The rules from (a.1) to (a.7) perform the first stage; that is, they generate all the possible paths in G of length n. In this stage, the objects v_i denote the starting node of the path to be generated, whereas the objects $v_{i,l}$ means that we are considering the node v_i as the lth element of the path.

We first create by succesive divisions, using the rules from (a.1) and (a.2), n internal membranes in which the paths starting from each of the nodes of G will be generated. Each of these membranes contains a different object $v_{i,1}$; that is, the first node in the path being considered in the corresponding membrane is the node v_i. Also, these membranes are negatively charged, and they remain this way through the first stage.

Suppose now that an internal membrane contains an object $v_{i,l}$, meaning that the current node being considered as the lth element of the path is the node v_i. The rule (a.3) keeps a record of this fact by means of the object r_i' and starts the process of choosing the next node in the path. This process is performed by the rules from (a.5) and (a.6) which generate a new membrane for each of the nodes adjacent to v_i. This is done by successive divisions of the internal membrane using the objects of type $a_{i \to j,l}$ to cover all the nodes in $adj(v_i)$. These objects transform themselves into an object $v_{j,l+1}$ in one of the new membranes created, to indicate that the $(l+1)$th node to include in the path is v_j, and into another object $a_{i \to j',l}$ in the other new membrane, to consider the next node adjacent to v_i.

Observe that the generation of the paths is done in parallel but not simultaneously, because we continue generating the path as soon as we choose the node adjacent to the current one. In other words, we do not wait until all the membranes considering a node adjacent to the current one have been generated to continue with the next element of the path.

Finally, when the last node of the path is reached, the rules from (a.4) are applied, keeping a record of this last node and introducing an object c to be used in the next stage and an object a_{n+1} to start the stage.

For the rules corresponding to the second stage not to be fired at the same time as the rules corresponding to the first stage, the sets of objects used in each of them are disjoint. At the beginning of the second stage the objects r_i' and c produced in the first stage are transformed into objects r_i and c_1, respectively. Now we can check for the existence of a membrane containing all the objects $r_1, \dots, r_n$ the same way as is done in the third stage of the solution to SAT presented in subsection 4.1.

Finally, suppose that there is a Hamiltonian path in G. Then, at least one of the internal membranes created along the first stage contains all the objects $r_1, \ldots, r_n$. An object c_{n+1} is then sent out of this membrane at the end of the second stage. Using the rule (c.1) this object is sent to the environment as an object *yes* and the charge of the skin is changed to positive. Hence, the rule (c.3) cannot be applied and, since the P system eventually halts, the answer to the HPP problem for G is positive.

On the other hand, if there is no Hamiltonian path in G, no internal membrane contains all the objects $r_1, \ldots, r_n$. Therefore, all these membranes stall at some moment of the second stage without sending out an object c_{n+1}, which, in turn, cannot be sent out to the environment to produce a positive answer.

We focus our attention on the rule (c.2); this rule is applied at each step of the computation and uses objects d_i to count how many steps have been performed. It is easy to see that stage 1 lasts at most $n^2 + 1$ steps and stage 2 lasts at most $2n + 2$ steps. Hence, if at step $n^2 + 2n + 4$ the charge of the skin has not changed to positive by means of the rule (c.1), then the answer must be negative; this is shown using the rule (c.3) to send an object *no* to the environment. The P system then halts.

Now, we are going to justify that the family $\mathbf{\Pi} = (\Pi(G))_{G \in \mathbf{HPP}}$ solves the problem HPP in quadratic time.

First, the above description of the evolution rules is computable in a uniform way. It is also a polynomial description, since the size of the working alphabet is $n^3 + 2n^2 + 6n + 10$; the number of membranes in the initial membrane structure is 2; the maximum cardinal of the initial multisets is 1; the total number of evolution rules is at most $n^3 + 7n + 9$; and the maximum length of a rule is 7. Hence, the family $\mathbf{\Pi}$ is *Turing polynomially uniform.*

Second, the family $\mathbf{\Pi}$ is *polynomially bounded*, since $\Pi(G)$ is deterministic for every G and the total number of steps performed by the computation of $\Pi(G)$ is at most $n^2 + 2n + 5$, which is quadratic in the size of G.

Third, from the description of the functioning of the P system $\Pi(G)$ it can be seen that the family $\mathbf{\Pi}$ is sound and complete.

5 Recognizer P Systems with Input

Recall that a P system with input is a tuple (Π, Σ, i_Π), where (a) Π is a P system with working alphabet Γ and p membranes labeled $1, \ldots, p$, with initial multisets $\mathcal{M}_1, \ldots, \mathcal{M}_p$ associated with them; (b) Σ is an (input) alphabet strictly contained in Γ, and the initial multisets are over $\Gamma - \Sigma$; and (c) i_Π is the label of a distinguished (input) membrane. If m is a multiset over Σ, then the *initial configuration of* (Π, Σ, i_Π) *with input* m is $(\mu, \mathcal{M}_1, \ldots, \mathcal{M}_{i_\Pi} \cup m, \ldots, \mathcal{M}_p)$.

In this section we deal with recognizer P systems *with input membrane* and we propose to solve hard problems in a *uniform* way in the following sense: all instances of a decision problem that have the same *size* (according to a given polynomial time computable criterion) are processed by the same system, to which an apropriate input, that depends on the concrete instance, is supplied.

This method for solving problems provides a *general purpose* algorithmic solution in the following sense: if we want to implement such a solution, a system constructed to solve an instance of the problem can also to be used when trying to solve another instance of the same *size*.

Now, we formalize these ideas in the following definition.

Definition 3. *Let* $X = (I_X, \theta_X)$ *be a decision problem. We say that* X *is solvable in polynomial time by a family of recognizer membrane systems* with input $\mathbf{\Pi} = (\Pi(n))_{n \in \mathbf{N}}$ *if the following is true:*

- *The family* $\mathbf{\Pi}$ *is Turing polynomially uniform; that is, there exists a deterministic Turing machine that constructs in polynomial timc the system* $\Pi(n)$ *starting from* $n \in \mathbf{N}$.
- *There exist two polynomial time computable functions, cod and s over the set* I_X *of instances of* X, *such that:*
  - *For every* $w \in I_X$, $s(w)$ *is a natural number and* $cod(w)$ *is an input multiset of the system* $\Pi(s(w))$.
  - *The family* $\mathbf{\Pi}$ *is polynomially bounded; that is, there exists a polynomial function* $p(n)$ *such that for each* $w \in I_X$ *every computation of the system* $\Pi(s(w))$ *with input* $cod(w)$ *is halting and, moreover, performs at most* $p(|w|)$ *steps.*
  - *The family* $\mathbf{\Pi}$ *is sound; that is, for each* $w \in I_X$ *if therc exists an accepting computation of the system* $\Pi(s(w))$ *with input* $cod(w)$, *then* $\theta_X(w) = 1$.
  - *The family* $\mathbf{\Pi}$ *is complete; that is, for each* $w \in I_X$, *if* $\theta_X(w) = 1$, *then every computation of the system* $\Pi(s(w))$ *with input* $cod(w)$ *is an accepting computation.*

The soundness property means that if given an instance we obtain an *acceptance responsc* of the system associated with it (and individualized by the apropriate input multiset) through some computation, then the answer to the problem (for that instance) has to be *yes*. The completeness property means that if we obtain an *affirmative* response to the problem, then any computation of the system associated with it (and individualized by the apropriate input multiset) must be an accepting one.

Note that in the above definition we consider three different tasks. The first is the construction of the family solving the problem, which we require to be done in polynomial time. The second is the task performed by the polynomial time computable functions *cod* and *s*. The third is the execution of the systems $\Pi(n)$ of the family (with the appropriate input multiset), for which we impose the total number of steps performed by their computations to be bounded by a polynomial function.

Hence, to solve an instance w, we first of all need to compute the natural number $s(w)$, obtain the input multiset $cod(w)$, and construct the system

$\Pi(s(w))$. This is a *precomputation stage*, running in polynomial time represented by the number of *sequential steps*. Next, we execute the system $\Pi(s(w))$ with input $cod(w)$. This is the proper *computation stage*, also running in polynomial time, but now the time is represented by the number of *parallel steps*.

Consequently, in order to solve a decision problem in this context, we need two polynomial time computable functions over the set of instances of the problem, in such a way that the first function assigns the P system that will process the instance when we give a suitable input provided by the second function. Then the system will accept or reject the instance. Bearing in mind the nondeterminism of the system, we require the confluence condition; that is, all computations of the system associated with the instance must always have the same answer.

Now, we briefly comment on the different efficient solutions of **NP**-complete problems obtained in the framework of P systems with input, described in a uniform way, so far proposed. Unless stated otherwise, the solutions cited are described in the usual framework of P systems with active membranes using 2-division, with three electrical charges, without change of membrane labels, without cooperation, and without priority.

M.J. Pérez-Jiménez, A. Romero-Jiménez, and F. Sancho-Caparrini present in [34] a linear time solution to *SAT* in a semi-uniform manner but in such a way that we can easily decribe that solution in an uniform manner (see [35]). Other interesting uniform linear time solutions to *SAT* are given by:

- A. Alhazov [1], through P systems with active membranes and using only two electrical charges.
- L. Pan, A. Alhazov, and T.O. Ishdorj [18], in the framework of P systems with active membranes, and without division, without polarizations, but using three types of membrane rules (separation, merging, and release).
- L. Pan and T.O. Ishdorj [19], through P systems with separation rules instead of division rules, in two different cases: in the first, using polarizations and separation rules; and in the second without polarizations and without change of membrane labels, but using separation rules with change of membrane labels.
- Gh. Păun, M.J. Pérez-Jiménez, and A. Riscos-Núñez [26], using P systems with tables of rules (each membrane is associated with several sets of rules, one of which is nondeterministically chosen in each computation step); in particular, they consider tables with obligatory rules, which are distinguished rules which must be applied at least once when the table is applied.

In [36], M.J. Pérez-Jiménez, A. Romero-Jiménez, and F. Sancho-Caparrini present the first linear time uniform solution to the *VALIDITY* problem, for formulas in conjuctive normal form.

Different efficient solutions to graph problems (*Vertex Cover*, *Clique*) in a uniform way are presented by A. Alhazov, C. Martín-Vide, and L. Pan in [3].

The first efficient and uniform solutions to numerical **NP**-complete problems were given by M.J. Pérez-Jiménez and A. Riscos-Núñez in [30] where a solution to the *Knapsack* problem was presented. A uniform solution to *Multiset 0–1 Knapsack* problem is given in the same framework by L. Pan and C. Martín-Vide in [20].

Other uniform solutions to numerical problems are the following:

- M.J. Pérez-Jiménez and A. Riscos-Núñez provide in [31] a solution to *Subset Sum*, and in [7] give a solution to the *Partition* problem.
- M.J. Pérez-Jiménez and F.J. Romero-Campero present in [32] a solution to the *Bin Packing* problem, and in [33] give a solution to *Common Algorithmic Problem*, a problem with a property of local universality in the sense that many other interesting **NP**-complete problems can be reduced to it in linear time.

In this section we present a linear time uniform solution to the *Knapsack* problem, and a quadratic time uniform solution to the decision version of the *Common Algorithmic Problem.*

5.1 A Linear Time Solution to the Decision Knapsack Problem

The decision *Knapsack* problem (0/1) is as follows:

> *Given a knapsack of capacity $k \in \mathbf{N}$, a set A of n elements, where each element has a "weight" $w_i \in \mathbf{N}$ and a "value" $v_i \in \mathbf{N}$, and given a constant $c \in \mathbf{N}$, decide whether or not there exists a subset of A such that its weight does not exceed k and its value is greater than or equal to c.*

The instances of the problem will be represented by tuples of the form $(n, (w_1, \ldots, w_n), (v_1, \ldots, v_n), k, c)$, where n is the size of the set A, $(w_1, \ldots, w_n)$ and $(v_1, \ldots, v_n)$ are the weights and the values, respectively, of the elements from A, and k and c are the constants mentioned above. The funtions w and v can be extended to every subset of A in a natural way from the data in the instance. Also, we consider the size of the instance to be $\langle n, k, c \rangle$ (where $\langle\rangle$ is a polynomial encoding of the tuple, for example, using the Cantor pair function).

In the discussion that follows we will construct a family of recognizer P systems with active membranes using 2-division and with input that solves the decision *Knapsack* problem (0/1) in linear time. Specifically, given the size n of the set A and two constants k and c, we construct a P system $\Pi(n, k, c)$ that solves all the instances of size $\langle n, k, c \rangle$, given an appropriate encoding for the input membrane of the weights and values of the elements from A. The functioning of this P system can be divided in the following stages:

(a) Generate all the subsets of A, computing simultaneously the weights and the values of the subsets.

(b) For all the subsets check if the condition $w(B) \leq k$ holds.
(c) For all the subsets of A that satisfy the first condition, check if the condition $v(B) \geq c$ holds.
(d) Answer *yes* or *no* according to the results of the two checking stages.

The P system $\Pi(n,k,c)$ is defined as follows.

- The input alphabet is $\Sigma(n,k,c) = \{x_1, \dots, x_n, y_1, \dots, y_n\}$.
- The working alphabet is

$$\begin{aligned}\Gamma(n,k,c) = \{&a_0, a, \bar{a}_0, \bar{a}, b_0, b, \bar{b}_0, \bar{b}, \hat{b}_0, \hat{b}, d_+, d_-, e_0, \dots, e_n,\\ &q_0, \dots, q_{2k+1}, q, \bar{q}, \bar{q}_0, \dots, \bar{q}_{2c+1}, x_0, \dots, x_n, y_0, \dots, y_n,\\ &yes, no, z_0, \dots, z_{2n+2k+2c+6}, \#\}.\end{aligned}$$

- The set of labels is $\{s, e\}$.
- The initial membrane structure is $[_s [_e\ ]_e^0]_s^0$ (each membrane with label e is said to be internal).
- The input membrane is the one with label e.
- The initial multisets associated with the membranes are $\mathcal{M}_s = z_0$ and $\mathcal{M}_e = e_0 \bar{a}^k \bar{b}^c$.
- The rules are:

(a.1) $[_e e_i]_e^0 \to [_e q]_e^- [_e e_i]_e^+$, for $0 \leq i \leq n$,
(a.2) $[_e e_i]_e^+ \to [_e e_{i+1}]_e^0 [_e e_{i+1}]_e^+$, for $0 \leq i \leq n-1$,
(a.3) $[_e x_0 \to \lambda]_e^+$, $\quad [_e x_i \to x_{i-1}]_e^+$, for $1 \leq i \leq n$,
(a.4) $[_e x_0 \to \bar{a}_0]_e^0$,
(a.5) $[_e y_0 \to \lambda]_e^+$, $\quad [_e y_i \to y_{i-1}]_e^+$, for $1 \leq i \leq n$,
(a.6) $[_e y_0 \to \bar{b}_0]_e^0$,
(a.7) $[_e q \to \bar{q} q_0]_e^-$, $\quad [_e \bar{a}_0 \to a_0]_e^-$, $\quad [_e \bar{a} \to a]_e^-$, $\quad [_e \bar{b}_0 \to \hat{b}_0]_e^-$, $\quad [_e \bar{b} \to \hat{b}]_e^-$,

(b.1) $[_e a_0]_e^- \to [_e\ ]_e^0 \#$, $\quad [_e a]_e^0 \to [_e\ ]_e^- \#$,
(b.2) $[_e q_{2j} \to q_{2j+1}]_e^-$, for $0 \leq j \leq k$,
(b.3) $[_e q_{2j+1} \to q_{2j+2}]_e^0$, for $0 \leq j \leq k-1$,
(b.4) $[_e q_{2j+1}]_e^- \to [_e\ ]_e^+ \#$, for $0 \leq j \leq k$,
(b.5) $[_e \bar{q} \to \bar{q}_0]_e^+$, $\quad [_e \hat{b}_0 \to b_0]_e^+$, $\quad [_e \hat{b} \to b]_e^+$, $\quad [_e a \to \lambda]_e^+$,

(c.1) $[_e b_0]_e^+ \to [_e\ ]_e^0 \#$, $\quad [_e b]_e^0 \to [_e\ ]_e^+ \#$,
(c.2) $[_e \bar{q}_{2j} \to \bar{q}_{2j+1}]_e^+$, for $0 \leq j \leq c$,
(c.3) $[_e \bar{q}_{2j+1} \to \bar{q}_{2j+2}]_e^0$, for $0 \leq j \leq c-1$,
(c.4) $[_e \bar{q}_{2c+1}]_e^+ \to [_e\ ]_e^0 yes$, $\quad [_e \bar{q}_{2c+1}]_e^0 \to [_e\ ]_e^0 yes$,

(d.1) $[_s z_i \to z_{i+1}]_s^0$, for $0 \leq i \leq 2n+2k+2c+5$, (d.2) $[_s z_{2n+2k+2c+6} \rightarrow d_+ d_-]_s^0$,

(d.3) $[_s d_+]_s^0 \rightarrow [_s]_s^+ d_+, \quad [_s d_- \rightarrow no]_s^+$,

(d.4) $[_s yes]_s^+ \rightarrow [_s]_s^0 yes, \quad [_s no]_s^+ \rightarrow [_s]_s^0 no.$

Let us see if this P system solves the *Knapsack* problem for every instance of size $\langle n, k, c \rangle$. First of all we must define a polynomial encoding of the problem into the family $\mathbf{\Pi}$ in order to give a suitable input to the system. Given an instance $u = (n, (w_1, \ldots, w_n), (v_1, \ldots v_n), k, c)$ of the *Knapsack* problem, we define $cod(u) = x_1^{w_1} \ldots x_n^{w_n} y_1^{v_1} \ldots y_n^{v_n}$. Now we will informally describe how the system $\Pi(n, k, c)$ with input $cod(u)$ works.

As we have just shown, the objects x_i and y_i will represent the weights and values of the elements of A. On the other hand, the objects $\bar{a}$ and $\bar{b}$ (the first will change to a for the second stage, while the second will change to $\hat{b}$ for the second stage and to b for the third stage), included from the beginning by definition of the system, represent the constants k and c, respectively.

In the first stage of the computation the initial internal membrane is continuosly divided by means of the rules from (a.1) and (a.2). These membrane divisions are controlled by the objects e_i, which represent the elements of the set A being considered. The charges of the newly created membranes indicate whether or not the element has been included in the subsets of A that are being generated. We show in Figure 2 an example for $n = 4$ of the membrane generation tree that is obtained.

Let us introduce the concept of subset *associated* with an internal membrane through the following recursive definition:

- The subset associated with the initial internal membrane is the empty one.
- When an object e_j appears in a neutrally charged internal membrane (with $j < n$), the jth element of A is selected and added to the previously associated subset. Once the stage is over, the associated subset will not be modified any more.
- When a division rule is applied, the two newborn internal membranes inherit the associated subset from the original one.

What we intend to get is a single internal membrane for each subset of A but, as we will later see, the membranes are not generated simultaneously (it can be shown that the membrane corresponding to the subset $\{a_{i_1}, \ldots, a_{i_r}\}$ is generated at the $(i_r + r + 2)$th computation step).

After a division rule from (a.1) is applied, the two new membranes will behave in quite different ways. On the one hand, in the negatively charged membrane (we have marked such membranes in Figure 2 with a circle) the first stage ends, and in the next step the rules from (a.5) will be applied, renaming the objects to prepare the third stage. This is a significant step, so we will designate as *relevant* those membranes that have negative charge and contain an object q_0. A relevant membrane will not further divide during the computation, and its associated subset will remain unchanged. On the other hand, the positively charged membranes will continue the generation of

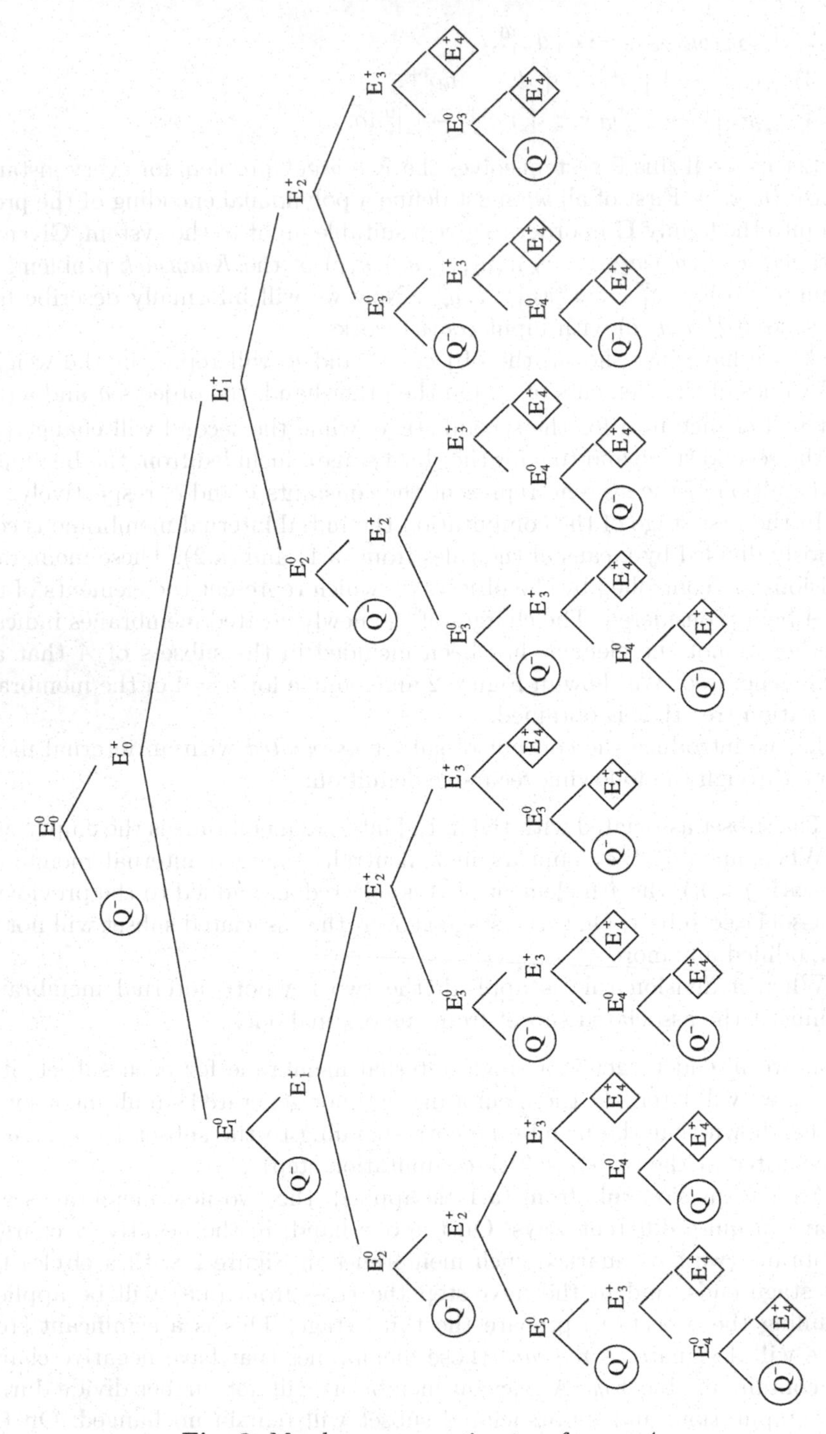

Fig. 2. Membrane generation tree for $n = 4$.

subsets of A; they will give rise to membranes associated with subsets that are obtained by adding elements of A of index $i+1$ or greater to the current subset. Note that if $i = n$, then the membrane cannot continue generating subsets, since it is not possible to add elements of indices greater than n. This has been taking into account because, as there is no rule from (a.1) or (a.2) that can be applied to an object e_n in a positively charged membrane (see the membranes surrounded by a diamond in Figure 2), the membrane halts.

Thus, since the indices of objects e_i never decrease, the relevant membranes are generated in a kind of lexicographic order, in the following sense: if the jth element of A has already been added to the associated subset, then no element with index lower than j will be added later to the subset associated with that membrane, or to the subsets associated with its descendants.

The generation of subsets and the computation of their weights and values are carried out in parallel. In fact, there is only a gap of one step of computation between the time an element is added to the associated subset and the time the new weight and value of the subset are updated. The rules involved here are those from (a.3–6). Recall that the objects x_i represent the weights of the elements of A and the objects y_i represent their values. The rules from (a.3) and (a.5) performed a rotation of these objects so that the objects x_0 and y_0 correspond to the weight and the value of the current element of A; that is, if a positively charged internal membrane contains an object e_i, meaning that the element of A being considered is the ith one, then the objects x_0 represent the weight of the element and the objects y_0 represent the value. On the other hand, if an internal membrane contains an object e_i and is neutrally charged, it means that the ith element of A is going to be added to the subset associated with the membrane. Then the weight and value of the subset have to be updated. This is done by means of the rules from (a.4) and (a.6), which transform the objects x_0 and y_0 into objects $\bar{a}_0$ and $\bar{b}_0$ representing the partial weight and value of the subset (note that the objects $\bar{a}_0$ will change to a_0 for the second stage, and the objects $\bar{b}_0$ will change to $\hat{b}_0$ for the second stage and to b_0 for the third stage).

The purpose of the rules from (a.5) is to prevent the rules for the second stage from being applied to an internal membrane that has not yet reached the end of the first stage. For that, we rename the objects obtained at the end of the latter, $q, \bar{a}_0, \bar{a}, \bar{b}_0$, and $\bar{b}$, to objects $\bar{q}, q_0, a_0, a, \hat{b}_0$, and $\hat{b}$ that are not used in it. The objects $\bar{q}, \hat{b}_0$, and $\hat{b}$ are not used until the third stage, so they remain unchanged through the second stage. The objects a_0 and a are used in a loop that checks if the weight of the subset of A being considered satisfies the condition of being less than or equal to k. The rules from (b.1) eliminate alternatively (the alternation is controlled by the negative and neutral charges of the membrane), at each step of the loop, one by one, until we run out of one (or both) of the objects. It is easy to see that the required condition is verified if and only if the loop can make an even number of steps. The objects q_i are then used to count the number of steps performed by the loop.

Let us see in a more detail how this loop works. Let B be a subset of a certain weight w_B. The evolution of the relevant membrane associated with it along the second stage is described in Table 1.

Multiset	Charge	Parity of q_i
$q_0 a_0^{w_B} a^k\ \bar{q}\hat{b}_0^{v_B} \hat{b}^c$	$-$	EVEN
$q_1 a_0^{w_B-1} a^k\ \bar{q}\hat{b}_0^{v_B} \hat{b}^c$	0	ODD
$q_2 a_0^{w_B-1} a^{k-1}\ \bar{q}\hat{b}_0^{v_B} \hat{b}^c$	$-$	EVEN
$\vdots$	$\vdots$	$\vdots$
$q_{2j}\ a_0^{w_B-j} a^{k-j}\ \bar{q}\hat{b}_0^{v_B} \hat{b}^c$	$-$	EVEN
$q_{2j+1}\ a_0^{w_B-(j+1)} a^{k-j}\ \bar{q}\hat{b}_0^{v_B} \hat{b}^c$	0	ODD
$\vdots$	$\vdots$	$\vdots$

Table 1. Comparison of weight with k.

Note 1. Observe that the index of q_i coincides with the total number of copies of a and a_0 that have already been erased during the comparison.

Note 2. If $B = \{a_{i_1}, \ldots, a_{i_r}\}$ with $i_r \neq n$, then there will be in the multiset some objects x_j and y_j, for $1 \leq j \leq n - i_r$, but they are irrelevant for this stage and therefore they will be omitted.

If the number w_B of objects a_0 is less than or equal to the number k of objects a, then the result of this stage is successful and we can proceed with the next stage. This situation is described in Table 2.

Multiset	Charge	Parity of q_i
$\vdots$	$\vdots$	$\vdots$
$q_{2w_B-1}\ a^{k-w_B+1}\ \bar{q}\hat{b}_0^{v_B} \hat{b}^c$	0	ODD
$q_{2w_B}\ a^{k-w_B}\ \bar{q}\hat{b}_0^{v_B} \hat{b}^c$	$-$	EVEN
$q_{2w_B+1}\ a^{k-w_B}\ \bar{q}\hat{b}_0^{v_B} \hat{b}^c$	$-$	ODD
$a^{k-w_B}\ \bar{q}\hat{b}_0^{v_B} \hat{b}^c$	$+$	ODD

Table 2. Weight less than or equal to k.

If the number w_B of objects a_0 is greater than the number k of objects a, then every time the rules from (b.2) can be applied (that is, for $j = 0, \ldots, k$), the first rule from (b.1) will also be applied. Thus, we can never get to a situation where the index of the counter q_i is an odd number and the charge of the internal membrane is negative. This means that the rule (b.4) can never be applied, and moreover, the membrane gets *blocked* (it will not evolve anymore during the computation). This situation is described in Table 3.

Multiset	Charge	Parity of q_i
$\vdots$	$\vdots$	$\vdots$
$q_{2k-1}\ a_0^{w_B-k}\ a\ \bar{q}\hat{b}_0^{v_B}\hat{b}^c$	0	ODD
$q_{2k}\ a_0^{w_B-k}\ \bar{q}\hat{b}_0^{v_B}\hat{b}^c$	−	EVEN
$q_{2k+1}\ a_0^{w_B-(k+1)}\ \bar{q}\hat{b}_0^{v_B}\hat{b}^c$	0	ODD

Table 3. Weight greater than k.

Let us suppose that the second stage has successfully finished in an internal membrane. That means that this membrane encodes a subset $B \subseteq A$ such that $w(B) \leq k$. Then, after applying the rule (b.4), this membrane gets a positive charge. For the objects used in this stage, so as not to fire the rules corresponding to the next stage, they are renamed by means of the rules from (b.5). In this way, the objects $\hat{b}_0$ are transformed to objects b_0 and the objects $\hat{b}$ to objects b. The counter $\bar{q}_i$ is initialized to $\bar{q}_0$.

The third stage works in a similar way as the second one, using the rules from (c.1), (c.2), and (c.2) corresponding to the rules from (b.1), (b.2), and (b.3), respectively. The end of the stage, however, is different. In this stage, we have to check if the number of objects b_0, corresponding to the value of the subset of A associated with the membrane, is greater than or equal to the number of objects b, corresponding to the constant c. Therefore, to pass to the final stage the two rules from (c.1) must have been applied c times each. Consequently, the rules from (c.2) and (c.3) take the counter of the loop to q_{c+1}, when the rules from (c.4) send it out to the skin as a *yes* object. Table 4 summarizes all the process described above.

Finally, rules from (d.1–3) are associated with the skin membrane and take care of the output stage. The counter z_i, used by the rules from (d.1) and (d.2), waits through $2n + 2k + 2c + 7$ steps ($2n + 3$ steps for the first stage, $2k + 2$ steps for the second stage and $2c + 2$ for the third stage). After all these steps are performed, we are sure that all the inner membranes have already finished their checking stages (or have already got blocked), and thus, the output process is activated.

Then, the skin will be neutrally charged and will contain the objects d_+ and d_-. Furthermore, some objects *yes* will be present in the skin if and only if both checking stages have been successful in at least one internal membrane.

Multiset	Charge	Parity of q_i
$\bar{q}_0 b_0^{v_B} b^c$	+	EVEN
$\bar{q}_1 b_0^{v_B - 1} b^c$	0	ODD
$\vdots$	$\vdots$	$\vdots$
$\bar{q}_{2c-1} b_0^{v_B - c} b$	0	ODD
$\bar{q}_{2c} b_0^{v_B - c}$	+	EVEN
$\bar{q}_{2c+1} b_0^{v_B - (c+1)}$ (if $v_B > c$)	0	ODD
or $\bar{q}_{2c+1}$ (if $v_B = c$)	+	ODD
$b_0^{v_B - (c+1)}$ or $\emptyset$	+	

Table 4. Comparison of value with c.

The output process then begins. First the object d_+ is sent out to the environment, giving a positive charge to the skin. Then the object d_- evolves to *no* inside the skin and, simultaneously, if there exists any object *yes* present in the membrane, it is sent out of the system, giving a neutral charge to the skin and making the system halt (in particular, further evolution of the object *no* is avoided).

Otherwise, if none of the membranes has successfully passed both checking stages, then there will be no object *yes* present in the skin when the output stage begins. Thus, after the object *no* is generated, the skin will still have a positive charge, so the object will be sent out to the environment and the system will halt.

Now we are going to justify that the family $\mathbf{\Pi} = (\Pi(n,k,c))_{n,k,c \in \mathbf{N}}$ solves the decision *Knapsack* problem (0/1) in linear time.

First, the above description of the system is computable in a uniform way. It is also a polynomial description, since the size of the input alphabet is $2n$; the size of the working alphabet is $5n + 4k + 4c + 31$; the size of the set of labels is 2; the number of membranes in the initial membrane structure is 2; the maximum cardinal of the initial multisets is 3; the total number of evolution rules is $6n + 5k + 4c + 34$; and the maximum length of a rule is 7. Hence, the family $\mathbf{\Pi}$ is *Turing polynomially uniform.*

Second, the family $\mathbf{\Pi}$ is linearly bounded, since given an instance $u = (n, (w_1, \ldots, w_n), (v_1, \ldots, v_n), k, c)$ of the problem, the total number of steps performed by the unique computation of the system $\Pi(n,k,c)$ given the input $cod(u)$ is at most $2n + 2k + 2c + 10$.

Third, from the description of the functioning of the P system $\Pi(n,k,c)$ it can be seen that the family $\mathbf{\Pi}$ is sound and complete.

5.2 A Quadratic Time Solution to CADP

The Common Algorithmic Problem (`CAP`) [8] is the following optimization problem:

> *Let S be a finite set and F be a family of subsets of S, called forbidden sets. Find the cardinality of a maximal subset of S which does not include any set belonging to F.*

The Common Algorithmic Problem can be transformed into a roughly equivalent decision problem by supplying a target value to the quantity to be optimized, and asking the question as to whether or not this value can be attained.

The Common Algorithmic Decision Problem (`CADP`) is the following decision problem:

> *Let S be a finite set, F be a family of subsets of S, and $k \in \mathbf{N}$. Determine if there exists a subset A of S such that $|A| \geq k$, and which does not include any set belonging to F.*

We will say that a problem X is a subproblem of another problem Y if there exists a linear time reduction from X to Y (using logarithmic bounded space). That is, X is a subproblem of Y if we can pass from the former to the latter by a simple rewriting process.

Next, we present some **NP**-complete decision problems that are subproblems of `CADP`.

- The Independent Set Decision Problem (`ISD`): *Given an undirected graph G, and $k \in \mathbf{N}$, determine whether or not G has an independent set of size at least k.*
- The Vertex Cover Decision Problem (`VCD`): *Given an undirected graph G, and $k \in \mathbf{N}$, determine whether or not G has a vertex cover of size at most k.*
- The Clique Decision Problem (`CDP`): *Given an undirected graph G, and $k \in \mathbf{N}$, determine whether or not G has a clique of size at least k.*
- The Hamiltonian Path Problem (`HPP`).
- The Satisfiability Problem (`SAT`).
- The Tripartite Matching Problem: *Given three sets B, G, and H, each containing n elements, and a ternary relation $T \subseteq B \times G \times H$, determine whether or not there exists a subset T' of T such that $|T'| = n$ and no two triples belonging to T' have a component in common.*

In what follows we will construct a family of recognizer P systems with active membranes using 2-division and with input that solves `CADP` in polynomial time. Specifically, given the size n of the set S, the size m of the set F, and a constant k, we construct a P system $\Pi(n, m, k)$ that solves the problem for all the instances of size $\langle n, m, k\rangle$, given as input an appropriate encoding of the subsets of S belonging to F. The functioning of this P system can be divided into the following stages:

(a) Generate maximal subsets A of S not including any element of F. For this, we start from the complete set S and eliminate one element from each of the forbidden sets.
(b) For all the previous subsets, compute their cardinality.
(c) Check if any of the subsets has cardinality greater than or equal to k (in fact, we check if the cardinality is greater than $k-1$).
(d) Answer *yes* or *no* according to the results from the previous stage.

The P system $\Pi(n, m, k)$ is defined as follows.

- The input alphabet is $\Sigma(n, m, k) = \{s_{i,j} \mid 1 \le i \le m, 1 \le j \le n\}$.
- The working alphabet is

$$\begin{aligned}\Gamma(n, m, k) = \Sigma(n, m, k) &\cup \{a_i \mid 1 \le i \le m\} \cup \{c_i \mid 0 \le i \le 2n+1\} \\ &\cup \{ch_i \mid 0 \le i \le 2k-1\} \cup \{f_j \mid 1 \le j \le n+1\} \\ &\cup \{e_{i,j,l} \mid 1 \le i \le m, 1 \le j \le n, -1 \le l \le j+1\} \\ &\cup \{g_j \mid 0 \le j \le nm+m+1\} \\ &\cup \{z, s_+, s_-, S_+, S_-, S, o, \tilde{O}, O, t, neg, \#, yes, preno, no\}.\end{aligned}$$

- The set of labels is $\{1, 2\}$.
- The initial membrane structure is $[_1 [_2 \ ]_2^0]_1^0$ (each membrane with label 2 is said to be internal).
- The input membrane is the one with label 2.
- The initial multisets associated with the membranes are $\mathcal{M}_1 = \lambda$ and $\mathcal{M}_2 = g_0 z^m s_+^n o^{k-1}$.
- The rules are:

(a.1) $[_2 s_{1,j} \to f_j]_2^0$, for $1 \le j \le n$,
(a.2) $[_2 s_{i,j} \to e_{i,j,j}]_2^0$, for $2 \le i \le m, 1 \le j \le n$,
(a.3) $[_2 f_1]_2^0 \to [_2 \#]_2^0 [_2 s_-]_2^+$,
(a.4) $[_2 f_j \to f_{j-1}]_2^0$, for $2 \le j \le n+1$,
(a.5) $[_2 f_j \to \lambda]_2^+$, for $1 \le j \le n+1$,
(a.6) $[_2 e_{i,j,l} \to e_{i,j,l-1}]_2^0$, for $2 \le i \le m, 1 \le j \le n, 0 \le l \le j+1$,
(a.7) $[_2 e_{2,j,l} \to f_{j+1}]_2^+$, for $1 \le j \le n, -1 \le l \le j+1, l \ne 0$,
(a.8) $[_2 e_{i,j,l} \to e_{i-1,j,j+1}]_2^+$, for $3 \le i \le m, 1 \le j \le n, -1 \le l \le j+1, l \ne 0$,
(a.9) $[_2 e_{i,j,0} \to a_{i-1}]_2^+$, for $2 \le i \le m, 1 \le j \le n$,
(a.10) $[_2 z]_2^+ \to [_2]_2^0 \#$,
(a.11) $[_2 a_1]_2^0 \to [_2]_2^+ \#$,
(a.12) $[_2 a_1 \to \lambda]_2^+$, $[_2 a_i \to a_{i-1}]_2^+$, for $2 \le i \le m$,
(a.13) $[_2 g_j \to g_{j+1}]_2^0$, $[_2 g_j \to g_{j+1}]_2^+$, for $0 \le j \le nm+m$,
(a.14) $[_2 g_{nm+m+1} \to c_0 neg]_2^0$,
(a.15) $[_2 neg]_2^0 \to [_2]_2^- \#$,

(a.16) $[_2 s_+ \to S_+]_2^-$, $[_2 s_- \to S_-]_2^-$, $[_2 o \to \tilde{O}]_2^-$,
(a.17) $[_2 z]_2^- \to \#$,

(b.1) $[_2 S_-]_2^- \to [_2]_2^+ \#$, $[_2 S_+]_2^+ \to [_2]_2^- \#$,
(b.2) $[_2 c_i \to c_{i+1}]_2^-$, $[_2 c_i \to c_{i+1}]_2^+$, for $0 \leq i \leq 2n$,
(b.3) $[_2 c_{2n+1} \to ch_0 t]_2^-$,
(b.4) $[_2 t]_2^- \to [_2]_2^0 \#$,
(b.5) $[_2 S_+ \to S]_2^0$, $[_2 \tilde{O} \to O]_2^0$,

(c.1) $[_2 S]_2^0 \to [_2]_2^+ \#$, $[_2 O]_2^+ \to [_2]_2^0 \#$,
(c.2) $[_2 ch_i \to ch_{i+1}]_2^0$, $[_2 ch_i \to ch_{i+1}]_2^+$, for $0 \leq i \leq 2k-2$,
(c.3) $[_2 ch_{2k-1}]_2^+ \to [_2]_2^+ yes$, $[_2 ch_{2k-1}]_2^0 \to [_2]_2^0 preno$,

(d.1) $[_1 yes]_1^0 \to [_1]_1^+ yes$,
(d.2) $[_1 preno \to no]_1^0$,
(d.3) $[_1 no]_1^0 \to [_1]_1^- no$.

Let us see if this P system solves CADP for every instance of size $\langle n, m, k\rangle$. First of all we must define a polynomial encoding of the problem into the family $\mathbf{\Pi}$ in order to give a suitable input to the system. Given an instance $u = (S, F, k)$ of the problem, where $S = \{a_1, \ldots, a_n\}, F = \{B_1, \ldots, B_m\}$, and $B_i = \{a_{j_1^i}, \ldots, a_{j_{k(i)}^i}\}$, we define $cod(u) = s_{1,j_1^1} \ldots s_{1,j_{k(1)}^1} \ldots s_{m,j_1^m} \ldots s_{m,j_{k(m)}^m}$. That is, the object $s_{i,j}$ introduced in the initial membrane will represent the fact that the set B_i contains the element a_j.

Now we informally describe how the system $\Pi(n, m, k)$ with input $cod(u)$ works.

To perform the first stage we start from the complete set S and make a loop to consider sequentially all the sets $B_1, \ldots, B_m$. Inside this loop we make another loop in which we generate a number of diferent subsets of S obtained by eliminating only one element of the current set B_i.

The core of this stage are the rules from groups (a.3–8). For these rules, the objects f_i represent the elements of the current set B_i, while the objects $e_{i,j,l}$ represent the elements of the sets B_j not yet considered. The purpose of the rules from (a.1) and (a.2) is now clear; we have to change the representation of the sets B_i from the objects $s_{i,j}$ to the objects f_i and $e_{i,j,l}$, and we do this in such a way that B_1 is the first forbidden set considered.

In the rule from (a.3) the object f_1 represents the element to eliminate from the current forbidden set. This rule creates two new membranes, one neutrally charged and the other positively charged. The former means that we have decided not to eliminate the element, so with the rule (a.4) we perform a rotation of the subscripts of the objects f_i, for the elements of B_i to be considered for elimination in a sequential way. The latter membrane means that we have eliminated the element and that we can proceed with the next forbidden set – but before that we have to do several things.

First we eliminate, by means of the rules from (a.5), the remaining objects f_i, since to meet the cardinal maximality condition we do not eliminate any other object of the forbidden set. This takes us to the following question: what if the element eliminated also belonged to a forbidden set B_j not considered yet? In that case the condition $B_j \not\subseteq A$ is fulfilled, and we do not have to eliminate any object from B_j. To control this from happening at the same time as for the elements f_i, we make a rotation of the subscripts of the objects $e_{i,j,l}$ (representing the elements of the forbidden sets not considered yet) so that they are always in correspondence with the objects f_i (representing the elements of the forbidden set being considered). The rotation is done to the third subscript of the objects, using the rules from (a.6), while the two first subscripts keep a record of which element it represented and which forbidden set contained it. In this way, before passing to the next step of the main loop we can "restore" the objects, by means of the rules from (a.7) and (a.8), and, using the objects a_i, we "memorize" which additional forbidden sets B_j satisfy $B_j \not\subseteq A$ by means of the rules from (a.9). Also, the rule from (a.10) uses the object z to count how many of these sets satisfy the previous condition.

Note that when restoring the objects, as described above, the third subscript gets a value which exceeds by one what it should be. This is because before continuing with the next step of the main loop we have to check if the next forbidden set to consider is not included in A; that is, we have to check for the existence of an object a_1. If this is the case, the rule from (a.11) skips that step, changing the polarization of the membrane to positive. We can then restore again the objects $e_{i,j,l}$, using the rules from (a.7) and (a.8), and perform a rotation of the subscripts of the objects a_i, using the rules from (a.12).

To synchronize the finalization of this first stage in all the generated internal membranes, we use the objects g_i as counters (rules from (a.13)). Since the worst case lasts at most $nm + m$ computation steps, when the rule from (a.14) is applied introducing the objects c_0 and neg, we can be sure that the first stage has reached the end in all the internal membranes. The object neg is then sent out to change the polarization of the membrane to negative (rule from (a.15)), so that the second stage can start, before which we make a renaming of objects to obtain the ones that will be used in this stage (rules from (a.17)).

A careful look at how the internal membranes are created by the rule from (a.3) shows two things. The first one is that, when the element of the forbidden set is eliminated (that is, in the membrane with positive charge), an element s_- is introduced. This object counts how many elements have been eliminated. It is also possible to obtain an internal membrane where none of the elements of the forbidden set being considered have been eliminated. In this case, when the end of the first stage is reached in the membrane, there will be z objects left. The rule from (a.17) allows us to dissolve these disturbing membranes.

The task of the second stage is to compute the cardinality of the subsets of S that are associated with the internal membranes created in the first stage. We have seen in the previous paragraph that the object s_- (changed to S_- in this stage) represents the number of objects eliminated. On the other hand, the object s_+ (changed to S_+ in this stage) represents the total cardinality of the set S, and this is why this object has multiplicity n in the initial multiset of the initial internal membrane. It is clear then that we have only to do a subtraction. For that, the rules from (b.1) alternatively erase the objects S_- and S_+. The latter can only be erased after the former has been erased. Hence, we have only to wait $2n$ computation steps to get the result of the subtraction. This is the purpose of the rules from (b.2), that use the objects c_i as counters. When the object c_{2n+1} appears, we can pass to the next stage, so the rule from (b.3) introduces the initial counter for that stage, ch_0, and an object t that allows us to change the polarization of the membrane to neutral (rule from (b.4)), which in turn allows us to rename the objects to obtain the ones used in the third stage (rules from (b.5)).

To check if the cardinality of the subset is greater than or equal to k, we check if it is greater than $k - 1$. This is why we keep $k - 1$ objects o (now transformed to O) from the beginning of the computation. The rules from (c.1) erase once and again an object S, changing the polarization to positive, and an object O, changing the polarization back to neutral. If we wait $2k - 2$ computation steps (this is done by the rules from (c.2) using the objects ch_i as counters), the comparison is finished. If the final polarization is positive, then the cardinality is greater than $k - 1$, so the rule from (c.3) sends an object *yes* to the skin. Otherwise, what is sent is an object *preno*.

Finally, the output stage is very simple. We have only to be careful, looking for the positive answers before looking for the negative ones. If there is an object *yes* in the skin, it is sent out to the environment, the charge of the skin membrane is changed to positive (rule from (d.1)), and the system halts. If not, the objects *preno* are changed to objects *no* (rule from (d.2)) and one of them is sent out to the environment, the charge of the skin membrane is changed to negative (rule from (d.3)), and the system halts.

Now we are going to justify that the family $\mathbf{\Pi} = (\Pi(n, m, k))_{n,m,k \in \mathbf{N}}$ solves CADP in polynomial time.

First, the above description of the system is computable in a uniform way. It is also a polynomial description, since the size of the input alphabet is mn; the size of the working alphabet is at most $mn^2 + 4mn - m + 3n + 2k + 18$; the size of the set of labels is 2; the number of membranes in the initial membrane structure is 2; the maximum cardinality of the initial multisets is $m + n + k$; the total number of evolution rules is at most $2mn^2 + 8mn + 3m - 2n^2 + n + 4k + 23$; and the maximum length of a rule is 7. Hence, the family $\mathbf{\Pi}$ is *Turing polynomially uniform.*

Second, the family $\mathbf{\Pi}$ is quadratically bounded, since, given an instance $u = (S, F, k)$ of the problem, the total number of steps performed by the unique computation of the system $\Pi(n, m, k)$, given the input $cod(u)$, lasts at most $mn + m + 2n + 2k + 8$ steps.

Third, from the description of the functioning of the P system $\Pi(n, m, k)$ it can be seen that the family $\mathbf{\Pi}$ is sound and complete.

6 Conclusions

The possibility of finding a systematic and suitable framework to address in an efficient way the resolution of many practical problems that are presumably intractable (unless **P**=**NP**) is studied in this chapter.

We consider P systems as recognizer devices. Solutions to **NP**-complete problems are looked for in this framework by making use of appropriate families of P systems that can be constructed in a semi-uniform or uniform way. In this chapter we have discussed the differences between these constructions, and have presented a survey of the different solutions known in the current literature of membrane systems. Also, we have described in some detail two semi-uniform solutions, to `SAT` and `HPP`, and two uniform solutions, to the *Knapsack* problem and the *Common Algorithmic problem.*

Acknowledgement. The authors wish to acknowledge the support of the project TIC2002-04220-C03-01 of the Ministerio de Ciencia y Tecnología of Spain, cofinanced by FEDER funds.

References

1. A. Alhazov: P Systems with Active Membranes and Two Polarizations. In [27], 20–35.
2. A. Alhazov, T.-O. Ishdorj: Membrane Operations in P Systems with Active Membranes. In [27], 37–52.
3. A. Alhazov, C. Martín-Vide, L. Pan: Solving Graph Problems by P Systems with Restricted Elementary Active Membranes. In *Aspects of Molecular Computing* (N. Jonoska, Gh. Păun, G. Rozenberg, eds.), LNCS 2950, Springer, Berlin, 2004, 1–22.
4. J. Castellanos, Gh. Păun, A. Rodríguez-Patón: P Systems with Worm-Objects. *IEEE 7th International Conference on String Processing and Information Retrieval, SPIRE 2000*, La Coruña, Spain, 64–74.
5. A. Cordón-Franco, M.A. Gutiérrez-Naranjo, M.J. Pérez-Jiménez, F. Sancho-Caparrini: Implementing in Prolog an Effective Cellular Solution for the Knapsack Problem. In [14], 140–152.
6. E. Czeizler: Self-Activating P Systems. In *Membrane Computing. International Workshop WMC-CdeA 2002, Curtea de Argeş, Romania, August 2002, Revised Papers* (Gh. Păun, G. Rozenberg, A, Salomaa, C. Zandron, eds.), LNCS 2597, Springer, Berlin, 2003, 234–246.

7. M.A. Gutiérrez-Naranjo, M.J. Pérez-Jiménez, A. Riscos-Núñez: A Fast P System for Finding a Balanced 2-Partition. *Soft Computing*, 9, 7 (2005).
8. T. Head, M. Yamamura, S. Gal: Aqueous Computing: Writing on Molecules. *Proceedings of the Congress on Evolutionary Computation 1999*, IEEE Service Center, Piscataway, NJ, 1999, 1006–1010.
9. M. Ito, C. Martín-Vide, Gh. Păun: Characterization of Parikh Sets of ET0L Languages in Terms of P Systems. In *Words, Semigroups, and Transducers* (M. Ito, Gh. Păun, S. Yu, eds.), World Scientific, Singapore, 2001, 239–254.
10. S.N. Krishna, R. Rama: A Variant of P Systems with Active Membranes: Solving NP-Complete Problems. *Romanian Journal of Information Science and Technology*, 2, 4 (1999), 357–367.
11. S.N. Krishna, R. Rama: P Systems with Replicated Rewriting. *Journal of Automata, Languages and Combinatorics*, 6, 1 (2001), 345–350.
12. S.N. Krishna, R. Rama: Breaking DES Using P Systems. *Theoretical Computer Science*, 299, 1-3 (2003), 495–508.
13. M. Madhu, K. Kristhivasan: P Systems with Membrane Creation: Universality and Efficiency. In *Proceedings of Third International Conference on Universal, Machines and Computations*, Chişinău, Moldova, 2001 (M. Margenstern, Y. Rogozhin, eds.), LNCS 2055, Springer, Berlin, 2001, 276–287.
14. C. Martín-Vide, Gh. Păun, G. Rozenberg, A. Salomaa, eds.: *Membrane Computing. International Workshop WMC2004, Tarragona, Spain, July 2003, Revised Papers.* LNCS 2933, Springer, Berlin, 2004,
15. A. Obtulowicz: Deterministic P Systems for Solving SAT Problem. *Romanian Journal of Information Science and Technology*, 4, 1-2 (2001), 551–558.
16. A. Obtulowicz: On P Systems with Active Membranes: Solving the Integer Factorization Problem in a Polynomial Time. In *Multiset Processing. Mathematical, Computer Science, and Molecular Computing Points of View* (C.S. Calude, Gh. Păun, G. Rozenberg, A. Salomaa, eds.), LNCS 2235, Springer, Berlin, 2001, 267–285.
17. A. Obtulowicz: Note on Some Recursive Family of P Systems with Active Membranes. Submitted, 2004.
18. L. Pan, A. Alhazov, T.-O. Ishdorj: Further Remarks on P Systems with Active Membranes, Separation, Merging, and Release Rules. In [27], 316–324.
19. L. Pan, T.-O. Ishdorj: P Systems with Active Membranes and Separation Rules. *Journal of Universal Computer Science*, 10, 5 (2004), 630–649.
20. L. Pan, C. Martín-Vide: Solving Multiset 0–1 Knapsack Problem by P Systems with Input and Active Membranes. In [27], 342–353.
21. A. Păun: On P Systems with Membrane Division. In *Unconventional Models of Computation* (I. Antoniou, C.S. Calude, M.J. Dinneen, eds.), Springer, London, 2000, 187–201.
22. Gh. Păun: Computing with Membranes. *Journal of Computer and System Sciences*, 61, 1 (2000), 108–143, and *Turku Center for Computer Science-TUCS Report* Nr. 208, 1998.
23. Gh. Păun: Computing with Membranes: Attacking **NP**-Complete Problems. In *Unconventional Models of Computation* (I. Antoniou, C.S. Calude, M.J. Dinneen, eds.), Springer, London, 2000, 94–115.
24. Gh. Păun: P systems with Active Membranes: Attacking **NP**-Complete Problems. *Journal of Automata, Languages and Combinatorics*, 6, 1 (2001), 75–90.
25. Gh. Păun: *Membrane Computing. An Introduction.* Springer, Berlin, 2002.

26. Gh. Păun, M.J. Pérez-Jiménez, A. Riscos-Núñez: P Systems with Tables of Rules. In *Theory is Forever, Essays Dedicated to Arto Salomaa on the Ocassion of His 70th Birthday* (J. Karhumaki, H. Maurer, Gh. Păun, G. Rozenberg, eds.), LNCS 3113, Springer, Berlin, 2004, 235–249.
27. Gh. Păun, A. Riscos-Núñez, A. Romero-Jiménez, F. Sancho-Caparrini, eds.: *Proceedings of the Second Brainstorming Week on Membrane Computing*, Sevilla, February 2004, Report RGNC 01/04, Univ. of Sevilla, 2004.
28. Gh. Păun, G. Rozenberg: A Guide to Membrane Computing. *Theoretical Computer Science*, 287 (2002), 73–100.
29. Gh. Păun, Y. Suzuki, H. Tanaka, T. Yokomori: On the Power of Membrane Division in P Systems. *Theoretical Computer Science*, 324, 1 (2004), 61–85.
30. M.J. Pérez-Jiménez, A. Riscos-Núñez: A Linear Time Solution to the Knapsack Problem Using Active Membranes. In [14], 250–268.
31. M.J. Pérez-Jiménez, A. Riscos-Núñez: Solving the Subset-Sum Problem by P Systems with Active Membranes. *New Generation Computing*, in press.
32. M.J. Pérez-Jiménez, F.J. Romero-Campero: An Efficient Family of P Systems for Packing Items Into Bins. *Journal of Universal Computer Science*, 10, 5 (2004), 650–670.
33. M.J. Pérez-Jiménez, F.J. Romero-Campero: Attacking the Common Algorithmic Problem by Recognizer P Systems. In *Pre-proceedings of the Machines, Computations and Universality, MCU'2004 (abstracts)*, September 21-26, 2004, Sankt Petesburg, p. 27.
34. M.J. Pérez-Jiménez, A. Romero-Jiménez, F. Sancho-Caparrini: *Teoría de la Complejidad en Modelos de Computación con Membranas*. Ed. Kronos, Sevilla, 2002.
35. M.J. Pérez-Jiménez, A. Romero-Jiménez, F. Sancho-Caparrini: Complexity Classes in Models of Cellular Computing with Membranes. *Natural Computing*, 2, 3 (2003), 265–285.
36. M.J. Pérez-Jiménez, A. Romero-Jiménez, F. Sancho-Caparrini: Solving VALIDITY Problem by Active Membranes with Input. In *Proceedings of the Brainstorming Week on Membrane Computing*, Tarragona, February 2003 (M. Cavaliere, C. Martín-Vide, Gh. Păun, eds.) Report GRLMC 26/03, 2003, 279–290.
37. A. Romero-Jiménez: *Complexity and Universality in Cellular Computing Models*. PhD. Thesis, University of Seville, Spain, 2003.
38. A. Romero-Jiménez, M.J. Pérez-Jiménez: Simulating Turing Machines by P Systems with External Output. *Fundamenta Informaticae*, 49, 1-3 (2002), 273–287.
39. P. Sosik: The Computational Power of Cell Division. *Natural Computing*, 2, 3 (2003), 287–298.
40. C. Zandron, C. Ferreti, G. Mauri: Solving NP-Complete Problems Using P Systems with Active Membranes. In *Unconventional Models of Computation, UMC'2K* (I. Antoniou, C.S. Calude, M.J. Dinneen, eds.), Springer, London, 2000, 289–301.
41. C. Zandron, G. Mauri, C. Ferreti, Universality and Normal Forms on Membrane Systems. In *Proceedings of International Workshop on Grammar Systems, 2000* (R. Freund, A. Kelemenova, eds.), Bad Ischl, Austria, July 2000, 61–74.

Chapter 13
Linguistic Membrane Systems and Applications

Gemma Bel Enguix[1,2], Maria Dolores Jiménez López[1,3]

[1] Research Group on Mathematical Linguistics
Rovira i Virgili University
Pl. Imperial Tàrraco, 1, 43005 Tarragona, Spain
[2] Department of Computer Science, University of Milan-Bicocca
Via Bicocca degli Arcimboldi, 8, 20126 Milan, Italy
gbe@astor.urv.es
[3] Department of Computer Science, University of Pisa
Via F. Buonarroti, 2, 56127 Pisa, Italy
mdjl@astor.urv.es

Summary. We introduce a general model for linguistics based on P systems, called linguistic P systems (LPSs). Several applications of LPSs to linguistic issues are suggested. Two variants of LPSs are developed: conversational P systems (CPSs) and dynamic meaning P systems (DMPSs). The former type of systems are used to provide a membrane computing approach to the analysis of conversational acts, while the latter type offer a bio-inspired framework for dealing with semantics.

1 Introduction

The main goal of this chapter is to introduce a general model for dealing with linguistic issues by means of P systems. In order to reach this goal we start by defining a new variant of P systems, called *linguistic P systems* (LPSs), adapted to the features that characterize the functioning of a natural language.

An early application of P systems to linguistics was proposed in [3]. The most important intuition for translating this natural computing model to natural languages – mainly to semantics and pragmatics – is that membranes can be understood as *contexts.* Contexts may be different words, persons, social groups, historical periods, or languages. They can accept, reject, and produce changes in elements they have inside. At the same time, contexts/membranes and their rules evolve, that is, change, appear, vanish, etc. Therefore, membranes, objects, and rules of the system are constantly interacting.

After introducing formal definitions of LPSs, we suggest several applications to different linguistic topics. Being a general framework with a high degree of flexibility, LPSs can be adapted in order to deal with semantics, language evolution, sociolinguistics, dialogue, anaphora resolution, etc. Space constraints prevent us from dealing with all these areas here. We focus our attention on two of the possibilities, *dialogue* and *lexical semantics.*

Conversation has been tackled from very different points of view. Philosophy, psychology, sociology, linguistics, cognitive science, artificial intelligence, human-computer interaction, and software engineering have examined conversation from a variety of perspectives. One of the goals of this chapter is to show the possibility of describing conversation by means of LPSs. What is important for our purposes is that P systems provide a powerful framework for formalizing any kind of *interaction*, both among agents and between agents and the environment. An important idea of P systems is that computation is done by evolution, when the configuration of membranes undergoes some modifications, given by certain rules. These rules generate elements inside the membranes and may also perform some kind of evolution in the other membranes, promoting, therefore, interaction between membranes. Membranes can explain, simulate, and possibly predict how the elements involved in the communicative process are able to modify the structure and the meaning of the message, and also how the message can create new contexts or transform those which already exist.

In turn, formalizing semantics is one of the most challenging topics in linguistics. By using P systems we try to introduce a new *dynamic* model for semantics. Taking into account that P systems can be understood as evolving systems, we consider that they may be a good framework for developing a dynamic semantic model in the line of dynamic semantics [11] and dynamic interpretation theory [4].

This chapter is the first attempt to construct a complex framework dealing with the simulation of conversation and semantics by means of P systems. Since this is an introductory work, we want only to set the model, giving some general intuitions about how these systems can work when they are applied to conversation and lexical semantics. In Section 2, by introducing linguistic P systems, we show how P systems should be adapted to linguistics. In Section 3, we suggest possible applications of LPS to linguistics. In Section 4, we refer to some special features needed in order to use LPSs to describe conversation, introduce the formal definition of conversational P systems (CPSs), and provide examples of their functioning. In Section 5, by defining dynamic meaning P systems (DMPSs), we apply LPSs to semantics, providing a P systems' redefinition of some of the most important lexical semantic relations and offering examples of them. We close the chapter with suggestions for further investigations.

2 Linguistic P Systems

Differences between formal and natural languages are widely known. Among them, we mention the following: a) the structure of natural languages is not stable, and is always evolving; b) the words have a meaning, and the meaning is also evolving; c) the mechanisms of decodification are different depending on the context, addresser, and addressee, that is, not only semantics but also pragmatics must be taken into account.

As generative devices for formal languages, P systems do not consider the problem of meaning; hence, in order to use P systems as linguistic models it is necessary to supplement them with several features, which will lead us to LPSs. The main distinctive features of these systems are:

- Introduction of domains, understood as sets of words, statements, ideas and other linguistic units, a membrane is able to work with in a given state of the computation. A domain is an active context. During a computation, a membrane receives some elements which are accepted by its domain and some which are not. We will explain below how the latter evolve.
- Introduction of several alphabets, a fact which can help explain the adaptation of one linguistic element from one context to another.
- Introduction of rules for adapting symbols traveling trough the membranes.
- Description of ways of communication between membranes, related to the fact that not everybody is able to communicate; there can exist people or communities which are separated, non-communicating.
- LPSs are understood as completely evolving devices. Every feature of an LPS can be changed during a computation: alphabets, domains, configuration of membranes, rules, and relations between membranes.

2.1 Main Features

We deal in this section with important concepts for linguistic P systems, as definitions of alphabets, domains, and adaptation of elements to different membranes.

An LPS has one or more *alphabets* $V_1, V_2, ..., V_n$, which can change or evolve during the computation. Then, each membrane has a *domain*, which specifies the symbols it accepts. Domains are related to one or more alphabets, or even to a single symbol. For example, given the alphabets $V_1 = \{1, 2, 3\}, V_2 = \{a, b, c\}$, a domain D_r assigned to a membrane M_r can be defined as associated with only one alphabet, for instance, $D_r = V_1$, to both alphabets, $D_r = V_1 \cup V_2$, or as a set of symbols belonging to one or more alphabets, such as $D_r = \{1, b, c\}$.

The domain of the skin membrane is the union of the domains of all its internal membranes. Several membranes of the same system can have the same domain.

During the computation, membranes can receive some elements which are not in their domain. Then, there are two possibilities, to reject or to accept them. The acceptance or rejection of these elements is given by what we call the *function of emigration*, which establishes a correspondence between symbols placed in different membranes or between symbols belonging to different alphabets, in a way that, in different places, the same symbol evolves into another symbol.

If the function of emigration is established at the level of alphabets, then the adaptation is independent of the membrane where the symbol is placed. In this case, the function is given in the form $h(V_i \leftrightarrow V_j)$ or $h(V_i \rightarrow V_j)$. The symbol "$\leftrightarrow$" means that the function has the property of reciprocity. Functions with reciprocity are called *returning emigration rules.* "$\rightarrow$" means that the function does not have the property of reciprocity. Functions without reciprocity are called *non-returning emigration rules.*

For instance, consider two alphabets $V_i = \{a, b, c\}$ and $V_j = \{\alpha, \beta, \gamma\}$, and three membranes, M_1, M_2, and M_3, with the following domains: $D_1 = V_i$, $D_2 = V_j$, and $D_3 = V_j$. Then, the function of emigration may be $h(V_i \leftrightarrow V_j) = \{a \leftrightarrow \alpha, c \leftrightarrow \gamma\}$. Thus, every a will be adapted to α in M_2 and M_3, and every α will be adapted to a in M_1. On the other hand, b and β will be never adapted, so they will be rejected if they are in a membrane whose domain does not contain it.

If the function of emigration is established at the level of membranes, then the adaptation will depend on the membrane where the element is. Such a function is given in the form $h(M_n \leftrightarrow M_m)$ or $h(M_n \rightarrow M_m)$. Continuing the previous example, we can consider $h(M_1 \rightarrow M_2) = \{a \leftrightarrow \alpha, c \leftrightarrow \gamma\}$ or $h(M_1 \rightarrow M_3) = \{b \leftrightarrow \beta\}$. In this way, any a and c going from M_1 to M_2 will be accepted and adapted, but the same symbols going to M_3 will be rejected. On the other hand, b can go just to M_3 if it wants to be accepted; otherwise, the symbol will be rejected. In this way, there is the possibility of playing with the same symbols with different behavior, depending on the contexts.

Symbols which cannot be accepted in a membrane are called *non-adapted symbols.* When arriving in a membrane where a symbol is not accepted, we mark it by adding the subscript i. Elements marked in this way are not taken into account in the output of the system when the evolution stops.

The linguistic meaning of domains and rules of emigration is intuitive as well as versatile. One of the immediate referents which can be given to domains are languages: English, Russian, Chinese, etc. But there are many possibilities for interpreting domains. In general, the way domains are understood will give rise to different applications of LPSs. If domains are related to several states of a language, then it is possible to study language evolution. If domains are different languages, then the interaction between them can be approached. For domains that are contexts, semantics and pragmatics arise. For societies or social groups, it seems sociolinguistics could be studied. Therefore, this is a key point for a theory of linguistic P systems. In our approaches, the use of domains, with the meanings we give to them, is the main point leading to different applications of P systems.

Let us consider an example, understanding domains as languages. For the sake of simplicity, such domains will be reduced. Let us imagine three membranes M_1, M_2, and M_3 with $D_1 = V_1$, $D_2 = V_2$, and $D_3 = V_3$, where $V_1 = \{$*correu, granota, neu*$\}$, $V_2 = \{$*correo, caballo, rana*$\}$, and $V_3 = \{$*chip, mail, tennis*}. We see that M_1 is a context for Catalan, M_2 for Spanish, and M_3 for English.

We consider the following rules of emigration for these membranes:

$h(M_1 \leftrightarrow M_2) = \{correu_{M_1} \leftrightarrow correo_{M_2},\ granota_{M_1} \leftrightarrow rana_{M_2}\}$,
$h(M_1 \leftrightarrow M_3) = \{correu_{M_1} \leftrightarrow mail_{M_3}\}$,
$h(M_2 \leftrightarrow M_3) = \{correo_{M_2} \leftrightarrow mail_{M_3}\}$.

When the word *correu* in M_1 is sent to M_2 or M_3, it is immediately adapted to *correo* or *mail*. The same process is carried out when *granota* is sent form M_1 to M_2, adapting it to *rana*. But the other words travelling through membranes will not be adapted, because they are not part of the domain and there is no rule to do that. They cannot be used in the other membranes.

2.2 Evolution

LPSs have been defined as dynamical systems. They are able to evolve during the computation in the same way as societies or languages are constantly evolving. In the present section, we deal with evolution in different parts of the system.

Evolution of alphabets. Alphabets may evolve during the computation, mainly by means of processes of addition and deletion.

1. *Addition* of some symbols. The rule *add* $\{\alpha\}$ *to* V adds a new symbol, α, to the alphabet V. In our example, *add* $\{\texttt{mail}\}$ *to* V_1 increases the alphabet to $V_1 = \{\texttt{correu}, \texttt{granota}, \texttt{neu}, \texttt{mail}\}$.
2. *Deletion* of some symbols. The rule *del* $\{a\}$ *from* V removes the symbol a from the alphabet. In our example, the rule *del* $\{\texttt{correu}\}$ *from* V_1 decreases the alphabet to $V_1 = \{\texttt{granota}, \texttt{neu}, \texttt{mail}\}$.
3. *Merging* is the process by means of which two alphabets are put together. The rule is written $merg(V_i, V_j)$, and its result is the union of V_i and V_j.
4. *Substitution* is a process of addition and deletion. For substitution to take place, it is necessary to have a rule of emigration between two symbols, $a \in V_i$ and $\alpha \in V_j$, in a given membrane M_n. Thus, if $D_n = V_j$ and there exists the rule $h(V_i \leftrightarrow V_j) = (a_{V_i} \leftrightarrow \alpha_{V_j})$, then, when *add* $\{a\}$ *to* V_j is applied, *del* $\{\alpha\}$ *from* V_j is also performed. For the alphabets and the rules of emigration given in Section 2.1, when the rule *add* $\{mail\}$ *to* V_1 is applied, *correu* is deleted; hence we get $V_1 = \{\texttt{mail}, \texttt{granota}, \texttt{neu}\}$ in just one step.

Evolution of domains. If an alphabet is associated with a domain, when an alphabet evolves, it entails the evolution of the domain. In this way, each of the processes explained above is related to the evolution of a domain. But domains may also evolve without modification in any alphabet. We mention the following processes:

- *Addition* of new alphabets to the domain. For example, for V_2, V_3 above and $D_n = V_2$, the rule *add* V_3 *to* DM_n has as a result $D_n = V_2 \cup V_3$, that is, $D_n = \{\texttt{correo}, \texttt{caballo}, \texttt{rana}, \texttt{chip}, \texttt{mail}, \texttt{tennis}\}$.
- *Deletion* of some alphabets from the domain. *del* V_1 *from* D_n, applied to $D_n = V_1 \cup V_2$, has as a result $D_n = V_2$. This is a process opposite to the one in the previous rule.

By means of the rules given for evolution in alphabets and domains, it is easy to explain language evolution, lexical substitution, creole and pidgin languages, and several sociolinguistic processes. For instance, bilingualism corresponds to merging two alphabets.

Evolution of rules. We deal now with the procedure for achieving *activation* or *deactivation* of rules. So far, we do not know how to create rules. Nevertheless we introduce here some very simple mechanisms which can help us use in given moments certain rules which were inactive during most of the computation. We distinguish two types of inactive rules:

1. Rules belonging to a membrane can use only the objects that the membrane accepts as a domain. If a membrane has any rule with an element that does not belong to its domain, then that rule becomes inactive. If the membrane evolves during the computation and it accepts the necessary element, then the rule is immediately activated.
2. *Sleeping rules* are well formed rules in a given membrane, which are inactive until some element activates them. Sleeping rules are denoted by σ. For example, there can exist a rule such that $\sigma r_n : c \rightarrow \delta$, but this rule cannot be applied before being activated. There are several ways of activating a sleeping rule (the symbol ς is used to denote the activation).

The first type of inactive rules do not need a special rule for activation, because they become active when every element of them is in the domain of the membrane they belong to. Adaptation of elements, evolution of domains, changes in alphabets or creation of new symbols can help them become active. For *sleeping rules*, an activation is necessary. We distinguish three ways to do that:

1. The sleeping rule can be put into another rule which is active. For example, having a sleeping rule σr_5, we can write $r_1 : b \rightarrow ab\varsigma r_5 c$.
2. There can exist conditional rules; for instance, consider *if* $\delta M_3 \rightarrow \varsigma r_5$. In this case, if M_3 is deleted, then the rule σr_5 is activated.
3. Finally, there can be a clock regulating the activation of some sleeping rules. It can be programmed as a counter, so that after some rule or set of rules has been applied several times, another one is activated. For example, a counter c_x can be considered that increases every time a rule is applied in a system. Therefore, we can formulate, for instance, a rule with the form $r_n : c_x = 14 \rightarrow \varsigma r_5$ (rule r_5 becomes active when $c_x = 14$)

The same procedures may be used for deactivating a rule. The only difference is the adjunction of σ (the symbol of sleeping) instead of ς.

New rules for activation or deactivation can be introduced, depending on the field to which LPSs are applied, the elements intervening in the computation, and the mechanisms established for the symbols to travel through the membranes. A good example is given by lexical substitution. In that framework, when a complete domain has been replaced by another one, it is normal that some rules related to the syntactical level of the second domain are activated in order to deal with the new words.

2.3 Structural Relations Between Membranes

The study of the possibilities of relation, interaction, and communication between membranes, as well as the rules that can regulate them is another interesting aspect of LPSs. For linguistics, it is very important what happens with contexts, because most of the time they have a strong influence on the evolution of languages. For instance, we can express the fragmentation of Latin that gave rise to romance languages as a partition of a membrane which generated several different contexts for words and syntax. This is just an example, but many analogies can be made with the connection and the interaction of contexts in order to get a specific result in communication. In the sequel to this discussion, we deal with three important aspects of the theory of LPSs: a) structural relations between membranes (Section 2.3), b) communication among them (Section 2.4), and c) rules for interaction in membranes (Section 2.5).

We start by examining the possible types of structural relation between membranes in a system. The way the membranes are related is important when they have to interact, and also in the configuration of the communication we are going to deal with later. In what follows, two types of relations are of direct interest: *nesting* and *sibling*.

Given two membranes M_1 and M_2, it is said that M_2 is *nested* in M_1 if it is placed inside M_1. The outer membrane M_1 is called *parent membrane* and the inner membrane M_2 is called *nested membrane*. We denote this relation by $M_2 \subset M_1$. The set of all membranes nested in M_1 is denoted by $\subset M_1$.

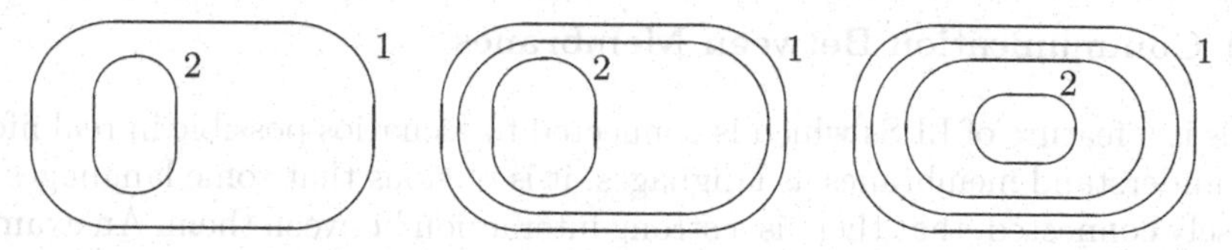

Fig. 1. $M_2 \subset M_1$ with degree 1, 2, and 3.

Nesting is a strict order, as it is non-reflexive, asymmetric, and transitive.

The *degree of nesting* refers to the number of membranes between the nested one and the parent. Figure 1 illustrates this idea.

Two membranes M_n and M_m are related by *sibling* if they are adjacent to each other or nested in adjacent membranes and have the same depth. Sibling is denoted $M_n \approx M_m$. The notion is illustrated in Figure 2. The set of all sibling membranes for M_n is denoted by $\approx M_n$.

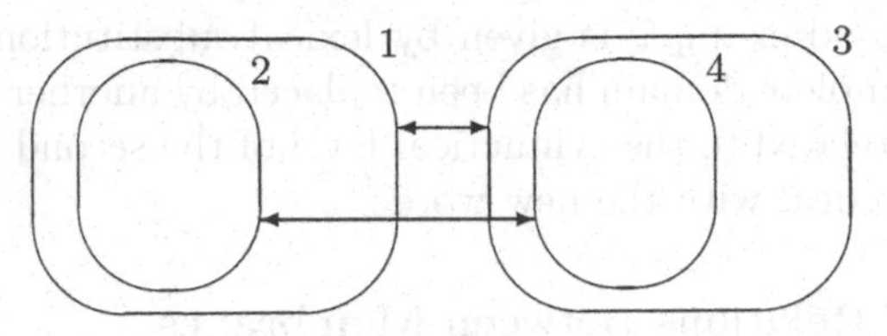

Fig. 2. Sibling.

The sibling is reflexive, symmetric, and transitive; hence it is an equivalence relation.

The *degree of sibling* refers to the proximity of two membranes related by sibling:

- Two membranes are siblings of degree 0 when they have the same direct parent membrane.
- For two sibling membranes $M_n \approx M_m$ which are not of degree 0 we obtain the degree of sibling by subtracting from the depth of M_n and M_m (they have the same depth) the depth of M_i and M_j, where M_i, M_j are two membranes such that $M_i \approx M_j$ with degree 0, and $M_n \subset M_i$, $M_m \subset M_j$.

An illustration is given in Figure 3.

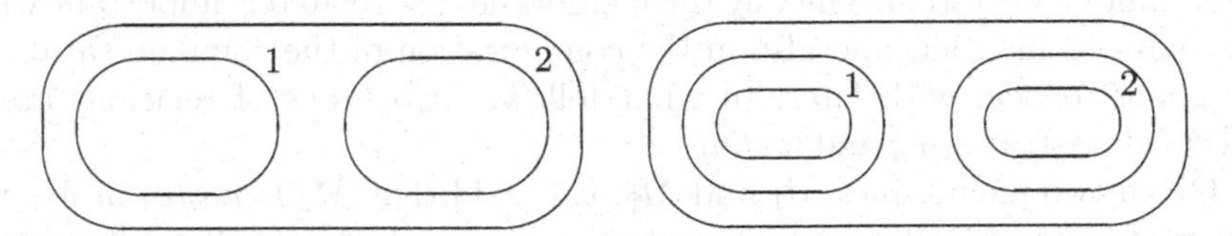

Fig. 3. $M_1 \approx M_2$ with degree 0 and 1.

2.4 Communication Between Membranes

This is a feature of LPSs which is connected to scenarios possible in real life. If we understand membranes as languages, it is obvious that some languages are closely connected, that there is a strong interaction between them. An example may be Spanish and English in the USA. If membranes are understood as social groups, conditions of marginality may be modeled by means of the non-connection (isolation) between a specific group and the others. If membranes refer to different agents in a dialogue, then it is easy to find agents which do not participate, while some others keep the attention all the time. These concepts can be approached by means of the description of the types of communication for membranes. In a preliminary approach, three states in communication are established: *connection*, *isolation*, and *inhibition*.

Connection is the situation in which communication channels are open, that is, membranes can interact and exchange elements.

The main features and notions related to connection are the following:

- Two membranes are connected when the communication channel between them is open (we denote this by $\odot$).
- By definition, the skin membrane is connected with every membrane in the system.
- A membrane M_n is connected when it is connected with every membrane in the system (we denote this by $\odot M_n$).
- A membrane is semi-connected when it has at least one of its channels open and it has at least one of its channels closed.
- A system is called *supra-connected* when every communication channel is open (we denote this by $\odot\mu$).

Clearly, connection is reflexive and symmetric, but not necessarily transitive.

By definition, two membranes related by nesting are connected. Moreover, if $M_n \approx M_m$ and $M_n \odot M_m$, then $M_n\odot\ (\subset M_m)$ and $M_m\odot\ (\subset M_n)$.

When we define a system, if nothing is said about connection, then it is supposed that every membrane is connected.

Isolation refers to the situation where the communication between two membranes is not established, but is possible. Its main features and definitions are as follows:

- Two membranes are isolated when the communication channel between them is closed (we denote this by $\otimes$).
- Closed communication channels can be opened by means of some rules while the system works.
- A membrane M_n is isolated when the communication with every membrane in the system is closed (we denote this by $\otimes M_n$).
- A system is called *supra-isolated* when every communication channel is closed. *Supra-isolation* is denoted by $\otimes\mu$.

The isolation is a non-reflexive and symmetric, but not necessarily a transitive, relation.

For two membranes M_n and M_m, if $Mn \otimes Mm$, then $M_n\otimes\ (\subset M_m)$ and $M_m\otimes\ (\subset M_n)$.

Inhibition is the state of complete and irreversible isolation, and has the following main features:

- Two membranes are inhibited when the communication between them is closed, and there is no rule in the system for opening it (we denote this by $\oslash$).

- A membrane M_n is inhibited when every communication channel is closed to it, and it cannot be opened (we denote this by $\oslash M_n$).
- A system is *supra-inhibited* when every membrane in it is inhibited (we denote this by $\oslash$).
- In inhibited systems the skin cannot communicate with membranes inside the system, and no evolution is possible.

The inhibition is a non-reflexive and symmetric, but not necessarily a transitive, relation.

For two membranes M_n and M_m such that $M_n \approx M_m$, if $M_n \oslash M_m$, then $M_n \oslash (\subset M_m)$ and $M_m \oslash (\subset M_n)$.

2.5 Operations with Membranes

During the evolution, the structure of P systems, which is dynamic by definition, can undergo several variations. For instance, some contexts can disappear or are extended during the progress of the conversation; others can merge, or can be copied many times. We consider here several types of rules for handling the membranes of a P system.

Deletion. By means of deletion, a membrane is dissolved and its elements go to the membrane immediately above. These elements will be accepted or rejected according to the configuration of the parent membrane. The rule for deleting a membrane M_n is written as δM_n, and its effect is $[\ \delta[\ u]_n]_m \rightarrow [\ u]_m$.
Erasing is the operation by which a membrane completely disappears, together with all its elements. The rule is written as ηM_n, and its effect is $[\ \eta[\ u]_n]_m \ \rightarrow [\ ]_m$.
Extraction is the operation by which a membrane nested in another one is extracted, such that they become related by sibling of degree 0 in the resulting configuration. It is denoted by ∇M_n and the effect is $[\ [\ \nabla[u]_n\ ]_m\ ] \rightarrow [\ [\ ]_m[u]_n\ ]$.
Exchange is the operation by which two membranes exchange their elements, but not their domains. It is denoted by $M_n \rightleftharpoons M_m$.

2.6 Formalizing LPSs

A linguistic P system (LPS) is a 5-tuple $\Pi = (\mu, \mathcal{V}, \mathcal{M}, \mathcal{H}, \mathcal{R})$, where:

1. μ is the membrane structure,
2. $\mathcal{V}$ is the set of alphabets associated with each membrane,
3. $\mathcal{M}$ is the initial configuration of each membrane (a configuration is a triple specifying the domain, the elements present in the membrane, and its communication state),
4. $\mathcal{H}$ is the set of emigration functions in the system,

5. $\mathcal{R} = R \cup VR \cup DR \cup AR \cup MR$ is the set of rules, where R is the set of evolution rules for objects, VR is the set of rules for evolving alphabets, DR is the set of rules for domains, AR is the set of rules for activation or deactivation of sleeping rules, and MR is the set of rules for membranes.

So far, our purpose has been to provide a general framework for dealing with linguistics. In the next section, we will present some possible applications, and we will enter into details for some of them. The general definition we have just given should be adapted to each part of linguistics, depending on the requirements of each concrete application.

3 Possible Applications of LPSs to Linguistics

The general definition of LPS that has been given above provides a general framework which allows the treatment of most of linguistic disciplines. However, some aspects must be adjusted in the general description for them to be adequate for each field.

	Domains	Elements	Turn-taking	Output
Semantics	Contexts	Linguemes	No	Membranes
Lang. evolution	Languages	Lgc. units	No	Membranes
Sociolinguistics	Social groups	Lgc. units	No	Membranes
Dialogue	Competence	Speech Acts	Yes	CR
Anaphora resol.	Contexts	Anaphores	No	Output Memb.

Table 1. Features of several applications of LPSs in linguistics.

As we show in Table 1, thanks to the flexibility of their formalization, LPSs can be adapted to deal with different aspects of linguistics. The first differencing feature are the domains. When they are contexts, LPSs are to a large extent suitable for semantics and pragmatics. Within the same framework, anaphora resolution, which is a syntactical and pragmatical problem, can also be approached. When domains are languages, language evolution and languages interaction can be studied. For approaching sociolinguistics, a good idea is to take domains as social groups. Finally, for dialogue we suggest understanding the domain as the personal background (education, context, knowledge of the world) of the agent, which is the membrane.

Among the different suggested interpretations, we see that dialogue is the field able to be formalized without domains and with the use of more than one alphabet. Dialogue is also the only area which does not always need operations with membranes, although it always needs some kind of interaction between them.

In the present chapter, we cannot deal with every one of these disciplines, but we will discuss a very preliminary approach to dialogue and a foundation of semantics from the perspective of P systems. Both aspects are treated here for the first time in such a bio-inspired framework. The development of the other aspects of linguistics we have suggested in Table 1, as well as other syntactical concepts, remains a task for the future.

4 Applying LPSs to Dialogue

Motivation. In the present section we suggest an approach to dialogue by means of P systems. It is widely known that this framework provides a powerful tool for formalizing any kind of *interaction*, both among agents and between agents and the environment. We have already emphasized that one of key ideas of P systems is that computation is done by evolution, when the configuration of membranes undergoes some modifications, given by certain rules. Therefore, most evolving systems can be formalized by means of P systems. In linguistics, we think P systems are quite suitable for dealing with some fields where contexts, and mainly evolving contexts, are a central part of the theory. Such an approach is especially suitable in the study of semantics and pragmatics. Morris [21] defined pragmatics as the science of the relation between signs and interpreters. A different definition, given by Katz and Fodor [9], refers to disambiguation of utterances depending on the contexts in which they are produced. The final results of sentences after the process of disambiguation, taking into account the intention of the speaker, are *speech acts*. The theory of speech acts was introduced by Austin [2] and Searle [26], and is now the central theory of pragmatics and human communication.

In what follows, our purpose is to compute speech acts by means of P systems, that is, to generate the final product of interaction in human communication.

4.1 A Theoretical Framework for Dialogue

In this section we follow three goals: a) to introduce the concept of *act*, b) to classify the agents participating in dialogue, and c) to offer a taxonomy of types of conversation.

A speech act may, by definition, be a communication act whose final meaning is related not only to syntax but also to the illocutionary strength of the speaker.

Speech acts are not only the central topic of pragmatics, but also have an increasing importance in dialogue games [5], which are an attempt to start a formal study of pragmatic situations. Combining both theories, and adapting their main concepts to a computational description, we propose to distinguish the following types of acts in human communication: *Query-yn*, *Query-w*, *Answer-y*, *Answer-n*, *Answer-w*, *Agree*, *Reject*, *Prescription*, *Explain*, *Clarify*, and *Exclamation*. The list includes the most usual acts, and may be modified at any time depending on the convenience and accuracy of the theory.

Acts are usually gathered in topics during the conversation, even if it is not a task-oriented one. From here we infer the existence of structural acts in order to indicate the beginning and the end of the goal, and also the change of topic. Therefore, we add the sequences *Open*, *Close*, and *Changetopic* to the list given above.

Structural acts are in some way different from the others. *Open* is the first act, or at least the first instruction, in every dialogue or human interaction. However, the act *Close* is not always present in the same way that many times topics are not closed in conversations, and new ones arise without an ending to the previous ones. On the other hand, *Changetopic* is a sequence of transitions which cannot be applied to and replied to by every agent. Nevertheless, these concepts have to be accommodated due to the diversity of realistic situations, which may be quite unexpected.

Several *classifications* of dialogue have been introduced which can be useful also in the framework of P systems. The most classical distinction in computational dialogues is the one established between *task-oriented* (TO) and *non-task-oriented* (NTO), based on the cooperation of participants in the consecution of conversational goals. The distinction between TO and NTO can be drawn as follows:

- In task-oriented (TO) dialogues, (1) agents collaborate, (2) there exist conversational goals, and (3) the opening and termination are defined.
- In non task-oriented (NTO) dialogues, (1) agents coact, but they do not necessarily collaborate, (2) there are no conversational goals or, if they exist, they are private, and (3) the opening and termination are not defined.

Besides the above classification, Clark provides in [6] an interesting picture, which includes the following items:

1. *Personal Settings*: Conversations may be devoted to gossip, business transactions, or scientific matters, but they are all characterized by the free exchange of turns among the two or more participants.
2. *Institutional Settings*: The participants engage in speech exchanges that resemble ordinary conversation, but are limited by institutional rules. In these settings, what is said is more or less spontaneous even though turns at speaking are allocated by a leader, or are restricted in other ways (e.g., a lawyer interrogating a witness in court).
3. *Prescriptive Settings*: In contrast with the previous item, there may be exchanges, but the words actually spoken are completely, or largely, fixed beforehand. Prescriptive settings can be viewed as a subset of institutional settings (e.g., church, basketball referee calling foul).

We think that prescriptive settings are included in institutional settings, although they have some special features. Indeed, the main characterization is the same for both types. From here, we finally obtain:

1. *Personal settings.* They define a type of dialogue characterized by:
   - random or not established turn taking,
   - there being no moderator or distributor.
2. *Institutional settings.* They have the following features:
   - well established turn taking, even if it looks like normal conversations,
   - the possibility of existence of a moderator or distributor,
   - the inclusion of *prescriptive settings*, a type of institutional settings where what is to be said is established before starting the dialogue.

Now, taking into account the characterization of institutional and personal settings, besides features of TO and NTO dialogues, it is possible to combine the main traits of both systems as in Table 2.

SETTINGS	TO	TURN-TAKING	DISTRIBUTED
Personal	(N)TO	free or random	no
Institutional	TO	non-free	yes

Table 2. Comparing personal and institutional settings.

The table above allows trying a distinction of three types of dialogues, which can be named *primary*, *oriented*, and *formal*. Their characteristics are shown in Table 3.

TYPE	SETTINGS	TASKS	TURN-TAKING	DISTRIBUTED
PRIMARY	Personal	NTO	free	no
ORIENTED	Personal	TO	free	no
FORMAL	Institutional	TO	non-free	yes

Table 3. Structural types of dialogues.

The degree of institutionalization is increasing from primary to institutional. This means we need more rules and established protocols for constructing the last type of dialogue. On the other hand, the complexity of description and formalization, as well as the computational complexity, is decreasing from primary to institutional dialogues.

An important distinction for our purposes has already been established. The next step is to give another relevant classification for dealing with dialogues, which has been developed in the framework of multi-agent theory and refers mainly to a perlocutive taxonomy of conversation. It is the one introduced by Walton and Krabbe [27] and Reed and Long [24]. According to these authors, it is possible to distinguish five classes of dialogue belonging to the TO group that can hence be included in both groups, oriented and formal. We give the distinctive features of each one of them, gathered in a) initial situation, b) goal of dialogue, and c) participants' aim, according to the description made in [1]:

1. *Persuasion*:
   a. Initial situation: conflicting points of view.
   b. Main goal: resolve conflicts by verbal means.
   c. Participants' aim: persuade the other(s).
2. *Negotiation*:
   a. Initial situation: need for action.
   b. Main goal: reach a decision.
   c. Participants's aim: influence the outcome.
3. *Inquiry*:
   a. Initial situation: general ignorance.
   b. Main goal: grow knowledge and agreement.
   c. Participants's aim: find a proof or refute one.
4. *Deliberation*:
   a. Initial situation: conflicting interests and need for cooperation.
   b. Main goal: make a deal.
   c. Participants's aim: get the best deal for oneself.
5. *Information Seeking*:
   a. Initial situation: personal ignorance.
   b. Main goal: spread knowledge and reveal positions.
   c. Participants' aim: gain, pass on, show, or hide personal knowledge.

From what we have explained throughout this section, we propose a final distribution of dialogues as shown in Figure 4.

The first group, the personal non-tasks-oriented non-distributed dialogues, called *primary*, depend exclusively on the creativity of participants. From gossip to scatology, everything is accepted. They are the common conversations in the street, between friends, at home. In *oriented* dialogues, which are also personal, non-distributed, and task oriented, agents have a goal to reach. Asking an address in the street, giving advice, trying to convince some person, and meetings or scientific discussions belong to this group. Finally, *formal* dialogues correspond to situations with a well established structure, namely, trials, liturgy, interrogations, or several types of formal academic interaction.

Figure 4 graphically shows that, while TO dialogues, namely, oriented and formal ones, can be divided into smaller units, depending on the goals they have, it is not possible to try a similar classification for primary dialogues.

In the next section, we will formalize a conversational P system (CPS), which aims to provide a general framework for dealing with every kind of dialogue.

4.2 From LPSs to CPSs

For adapting P systems to the generation of human interaction so as to obtain a conversational P system (CPS), we need to do some adjusting, concerning mainly the following aspects:

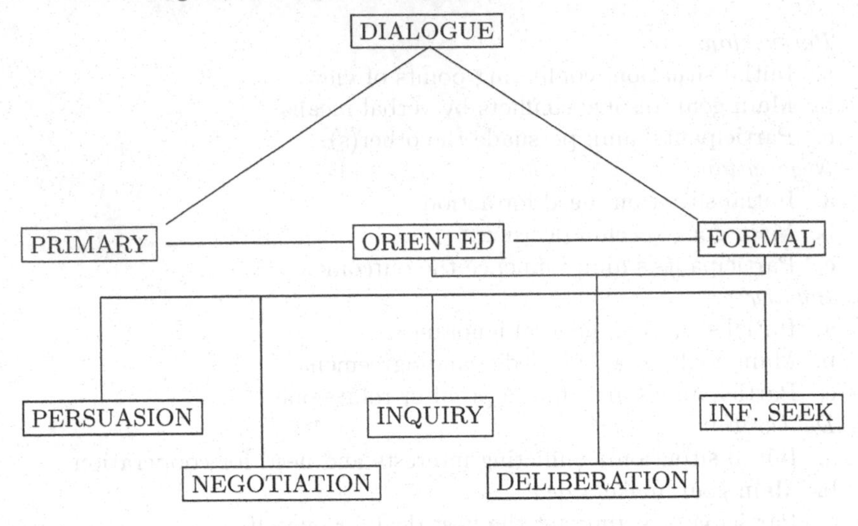

Fig. 4. Types of dialogues.

1. configuration of alphabets;
2. domains;
3. the turn-taking protocol $\mathcal{T}$;
4. halting criteria;
5. the output configuration.

Configuration of alphabets. The main elements of CPS are speech acts which are gathered as several types, following the classification given above. Each of these types is considered to be an alphabet, even if, more than an alphabet, it is a stack of elements. Indeed, each element of the alphabet, that is, each speech act, can be used just once during the computation. When it is used, it is deleted from the alphabet, the way elements are deleted from a stack. We define a maximal set of alphabets $\mathcal{V} = \{\omega, \#, \kappa', \kappa, \alpha^y, \alpha^n, \alpha, \gamma, \varphi, \tau, \varepsilon, \lambda, \xi\}$, where every element is a set of speech acts, as follows:

$\omega = \{o_1, o_2, \ldots, o_k\}$, speech acts of type *open*.
$\# = \{\#_1, \#_2, \ldots, \#_k\}$, speech acts of type *close*.
$\kappa' = \{q'_1, q'_2, \ldots, q'_k\}$, speech acts of type *query-yn*.
$\kappa = \{q_1, q_2, \ldots, q_k\}$, speech acts of type *query*.
$\alpha^y = \{a^y_1, a^y_2, \ldots, a^y_k\}$, speech acts of type *answer-y*.
$\alpha^n = \{a^n_1, a^n_2, \ldots, a^n_k\}$, speech acts of type *answer-n*.
$\alpha = \{a_1, a_2, \ldots, a_k\}$, speech acts of type *answer-q*.
$\gamma = \{g_1, g_2, \ldots, g_k\}$, speech acts of type *agree*.
$\varphi = \{f_1, f_2, \ldots, f_k\}$, speech acts of type *reject*.
$\pi = \{p_1, p_2, \ldots, p_k\}$, speech acts of type *prescription*.
$\varepsilon = \{e_1, e_2, \ldots, e_k\}$, speech acts of type *explain*.

$\lambda = \{l_1, l_2, \ldots, l_k\}$, speech acts of type *clarify*.
$\xi = \{x_1, x_2, \ldots, x_k\}$, speech acts of type *exclamation*.

Domains. In CPS, the *domain* of a membrane is related to the competence of an agent in a dialogue, that is, what the agent knows and can say. It is defined as the set of speech acts from $\mathcal{V}$ that every membrane is able to utter or understand. It can include entire sets of acts defined for the system or just single acts coming from some set.

Turn-taking protocol. We distinguish three types of dialogues: primary, oriented, and formal. We have established for them two types of turn-taking, free and non-free. Free turn-taking is related to primary and oriented dialogue, whereas non-free turn-taking is related to formal dialogue.

In what concerns the *free turn-taking protocol for primary and oriented dialogues*, we assume that it is given as a set of active elements, at the beginning of the computation. Every turn is distributed by the agent that is talking. When someone asks, explains, or clarifies something, he/she does it to someone among the others. Therefore, we assume that the addresser in each turn can choose the next speaker. The way to do that is through the *turn-taking rule* included at the end of each rule of the system. This is denoted by a symbol related to the speech act uttered in the rule. We consider the following turn-taking symbols: O (open), # (close), Q′ (Query-yn), Q (Query-w), A^y (Answer-y), A^n (Answer-n), A (Answer-q), G (Agree), F (Reject), P (Prescription), E (Explain), L (Clarify), X (Exclamation). To these symbols, we add H for "Changetopic," which is not related to any set of speech acts, because any type (except Answer) can follow it.

Every rule in the system has on the left hand side the indication of the turn-taking, and on the right hand side the reply to the explicit invitation to talk, and the agent, if it exists, to whom the speech act is addressed. The turn-taking allows applying just one rule.

In the case when no indication of turn is given in a rule, the turn goes to every agent/membrane able to reply, that is, every membrane containing a rule with the given turn on the left hand side. If there are several membranes able to act, then the turn is indicated by the number of the membrane, that is, M_1 precedes M_2, which precedes M_3, and so on.

As for the *non-free turn-taking protocol for formal dialogues*, we have to say the following:

- The starting membrane is the skin membrane. In every initial configuration of a CPS, the skin membrane has at least one element and the other membranes are empty. This means that the first rule is always applied to this membrane. The initial step is not part of the turn-taking protocol.
- The protocol $\mathcal{T}$ is a sequence of membranes establishing the turn-taking. No restrictions are applied to this sequence, and $\mathcal{T} = \{1_n\}$ is forbidden because it does not generate anything, since the skin membrane is not an output membrane.

- If a membrane of the sequence cannot participate, the next one takes its turn.
- If the skin membrane is in the $\mathcal{T}$ sequence, the system is called *distributed*; otherwise is is called *non-distributed*.
- The turn in the sequence allows a membrane to apply just one rule. If the application of a given rule implies some transformations in the system, they are carried out automatically; but the membrane cannot participate again.

Referring to the *halting criteria*, we establish that the system stops if:

1. No rule can be applied in any membrane.
2. Just one membrane remains in the system.
3. No more acts are available.

Configuration of the output.

- CPSs are not final step systems; therefore, the configuration obtained in every step is stored in a *configuration register* (CR), and the final result gives account of the situation in each configuration.
- The last result obtained and the membrane or membranes still active are stored as a special output of the system in the last register of the CR called FC (final configuration).
- CPSs do not have output membranes. For the *configuration of the output*, we define the generation register (GR), which gives account of the changes in the configuration of the system in every step. The examples we will discuss below will clarify these notions.

4.3 Defining a CPS

If we take into account all the above necessary elements in order to apply LPS to conversation and formalize them, what we obtain is a conversational P systems that can be defined as constructs $\Pi = (\mu, \mathcal{U}, \mathcal{M}, \mathcal{T}, \mathcal{R})$, where:

- μ is the membrane structure,
- $\mathcal{U} \subseteq \mathcal{V}$ is the set of alphabets associated with the types of speech acts,
- $\mathcal{M}$ indicates the initial configuration of each membrane, specifying the set of speech acts associated with the membrane, the domain D of the membrane, the state C of communication channels of the membrane, and an element T of $\mathcal{T}$ for free turn-taking systems,
- $\mathcal{T}$ is the configuration of turn-taking, given by a set of act distributors in primary and oriented dialogues, and by a sequence in institutional dialogues,
- $\mathcal{R} = R \cup VR \cup DR \cup AR \cup MR$ is the set of rules, where R is the set of evolution rules for objects, VR is the set of rules for alphabets, DR is the set of rules for domains, AR is the set of rules for activation or deactivation of sleeping rules, and MR is the set of rules for membranes.

Notice that the concept of a membrane configuration is different in CPSs with respect to LPSs. Besides the set of acts, the membranes of a CPS contain the domain and the state of the communication channels in a given state, and the possible turn-taking configuration for the membrane if the CPS has free turn-taking.

Now, we have a general framework for dealing with dialogue. Usually, for primary dialogue no more adaptations will be necessary, but for TO dialogues most of the situations require special definitions.

In the sequel, we will search for some general laws for oriented dialogues and will give some examples of how primary dialogue develops in real life, in the way theatre can be (or, at least, simulate) real life. Formal dialogues, with preestablished and deterministic structures, will not be considered in this chapter.

4.4 Some Particular Cases of CPSs for Oriented Dialogues

In this section, we adapt CPSs to each of the five types of oriented dialogue considered in [27]: inquiry, information seeking, deliberation, persuasion, and negotiation.

Oriented dialogues are a) personal, b) task oriented, c) free turn-taking structures, and d) non-distributed. Hence, for applying general CPSs to them, we have to take into account that (i) the turn-taking is free, (ii) the acts *Open*, *Close*, and *Changetopic* are present, by definition, in $\mathcal{V}$, and (iii) there is a final goal. If the goal is achieved, then the system stops. If the system stops before reaching the goal, then it is not correctly defined.

Taking into account the traits of *initial situation*, *main goal*, and *participants aim*, explained in [1], we think it is possible to gather the five types of oriented dialogues in two groups: Dialogues devoted to *spread knowledge* (*inquiry* and *information seeking*), and dialogues aimed to *reach agreements* (*deliberation*, *persuasion*, and *negotiation*).

From here, we establish two main characterizations in CPS for oriented dialogues corresponding to *spreading knowledge conversational P systems*, SKCPSs for short, and *reaching agreements conversational P systems*, RACPSs for short.

SKCPSs start from a situation of ignorance and end in a state of increasing knowledge. We think the best way of modeling the process is the extension of membrane domains, with ignorance indicated by $D_n = \emptyset$. Within this general scenario, two main differences should be pointed out between *inquiry* and *information seeking*:

- Inquiry starts from general ignorance, whereas in information seeking the situation affects only one agent of the dialogue.
- In inquiry, every agent collaborates in reaching a solution, whereas in information seeking, the non-ignorant agent(s) may have the personal goal of showing or hiding information.

Taking into account these differences, we propose two different definitions of SKCPSs for inquiry and information seeking.

In the first case, that of *inquiry*, we have to take into account the following facts:

- In the starting configuration, which is general ignorance, every element participating in the "process of inquiry" has an empty domain.
- There is a *final goal*, defined by the situation in which a given symbol is accepted by *every membrane*.
- Communication channels are always open, because every agent collaborates in the search for knowledge. Therefore, the communication state is not specified.

Thus, formalization of inquiry should include a new element, the final goal, while some of the components of a CPS must be reformulated, obtaining a system of the form $\Pi = (\mu, \mathcal{U}, \mathcal{M}, \mathcal{T}, G, \mathcal{R})$, where all components are as in a general CPS, with the following differences: $\mathcal{U}$ includes the alphabet $\{O, C, H\}$, every domain is $\emptyset$, $\mathcal{T}$ is a set of act distributors, and G is the final goal, applied to every agent.

The process for obtaining an information seeking SKCPS is the same as for inquiry. The specific features of the system are now the following:

- In the starting configuration, the domain of the membrane that is seeking information is empty.
- There is a *final goal*, defined by the situation in which a given symbol, already present in the system, but not in the configuration of the "ignorant" membrane, is accepted by *the membrane seeking information*.
- Agents which are providers of information may have several behaviors, more or less collaborative, which can be modeled by rules or by different communication states.

Thus, information seeking not only needs a goal, like inquiry, but also needs an output membrane, because the agent trying to get the information is unique. In this way, a new element has been added, which is not common in LPSs. We also recall that the membranes which are not searching do not necessarily have to collaborate, which means that the state of communication channels must be reported.

The definition of the system may be $\Pi = (\mu, \mathcal{U}, \mathcal{M}, \mathcal{T}, i_O, G, \mathcal{R})$, where the components are the same as in a general CPS, with the following differences: $\{O, C, H\} \in \mathcal{U}$, i_O is the output membrane, and G is the final goal, which the output membrane i_O has to reach.

RACPSs differ from SKCGSs in the initial situation, the main goal, and the participants' goals; hence they are different types of oriented dialogue. However, deliberation, persuasion, and negotiation are very similar types of dialogue due to the fact that they all look for a final agreement where one of the agents takes advantage of the others, with different personal aims, but with the same result. The main difference appears in the initial situation:

while persuasion and deliberation have an initial conflict, such a feature may or may not appear in negotiation.

From the formal point of view, deliberation and persuasion are, thus, very similar. The difference, clearly, is the attitude of agents. Whereas deliberation is a *collaborative* dialogue, persuasion is only a *coaction.* This feature is related to the presence of information about the state of communication channels in the system. As this is the only *formal* distinction between the two types of conversation, we think it is possible to define only one model for both dialogues (we are not saying that persuasion and deliberation are the same, but only that they can be formalized in the same way).

Therefore, we will define two kinds of devices for three different dialogues.

Deliberation and persuasion have two explicit features: a) it is necessary to formalize an initial conflict; the best way to do this seems to be to describe membranes with disjoint domains; and b) the final goal is given by the influence of one of the agents, which is able to impose its arguments. We introduce, then, a winner W, which is the membrane to which the element accepted by the other domains belongs. Because this is an essential element, we introduce it in the definition of the system, but the winner is decided by the configuration register, CR, looking at the final configuration, FC.

The state of communication channels is reported because, although in deliberation it seems every communication channel should be open, it is not clear for persuasion, where some agents can isolate themselves to not be influenced by the others.

Thus, a deliberation and persuasion RACPS has to be written as $\Pi = (\mu, \mathcal{U}, \mathcal{M}, \mathcal{T}, G, W, \mathcal{R})$, where all components are as above, G is the final goal defined by $D_1 \cap D_2 \cap \ldots \cap D_n = \emptyset$, and W is the winner.

In *negotiation*, the initial configuration of membranes is not given, that is, empty domains or disjoint domains could exist, but they are not compulsory for the definition of the system. On the other hand, the main goal is to reach an agreement shared by every agent. This is why it must exist as a goal achieved by every membrane for the system to be successful. From here, the definition of the negotiation RACPS is $\Pi = (\mu, \mathcal{U}, \mathcal{M}, \mathcal{T}, G, \mathcal{R})$, where all components are as above, with G being the final goal of the system, reached by every membrane.

We illustrate below these very general models with examples from the literature.

4.5 An Example of a CPS Generating a Primary Dialogue

In this section we will describe a scene from Shakespeare's play *The Comedy of Errors*, Act 4, Scene 4. The structure is quite complex, because of the number of the agents that participate as well as the functioning of the turn-taking, which is not always well established.

The system we need in order to generate this dialogue is the following:

$$\Pi = (\mu, \mathcal{U}, \mathcal{M}, \mathcal{T}, \mathcal{R}),$$

where:

$\mu = [_0\ [_1\ ]_1\ [_2\ ]_2\ [_3\ ]_3\ [_4\ ]_4\ [_5\ ]_5\ [_6\ ]_6\ ]_0$,
$\mathcal{U} = \{\kappa, \alpha, \varepsilon, \tau, \xi\}$, where
$\kappa = \{q_1$: `wherefore dost thou mad me?`,
q_2 : `Will you be bound for nothing?`,
q_3 : `Say now, whose suit is he arrested at?`,
q_4 : `do you know him?`,
q_5 : `What is the sum he owes?`,
q_6 : `Say, how grows it due?`$\}$,
$\alpha = \{a_1$: `One Angelo, a goldsmith:`,
a_2 : `I know the man.`,
a_3 : `Two hundred ducats.`,
a_4 : `Due for a chain your husband had of him.`$\}$,
$\varepsilon = \{e_1$: `Master, I am here entered in bond for you.`$\}$,
$\pi = \{p_1$: `Out on thee, villain!`,
p_2 : `be mad, good master: cry 'The devil!'`,
p_3 : `Go bear him hence. Sister, go you with me.`$\}$,
$\xi = \{x_1$: `O most unhappy day!`,
x_2 : `O most unhappy strumpet!`,
x_3 : `God help, poor souls, how idly do they talk!`$\}$,
$\mathcal{M} = \{M_1, M_2, M_3, M_4, M_5, M_6\}$, where
$M_1 = (\kappa \cup \alpha \cup \pi \cup \xi, \odot, O)$,
$M_2 = (\kappa \cup \pi \cup \xi, \odot)$,
$M_3 = (\kappa \cup \varepsilon \cup \pi, \odot)$,
$M_4 = (\xi, \odot)$,
$M_5 = (\alpha \cup \{q_4\}, \odot)$,
$M_6 = (\xi, \odot)$.
$\mathcal{T} = \{Q, A, E, P, X, CH\}$.
$\mathcal{R} = R_1 \cup R_2 \cup R_3 \cup R_4 \cup R_5$, where
$R_1 = \{r_1 : O \to x, X,\ \ r_2 : Q \to aq, Q_5,$
$r_3 : X \to p \odot (2,3,6)\ q, Q_5,\ \ r_4 : A \to q, Q_5\}$,
$R_2 = \{r_1 : X \to x, CH,\ \ r_2 : E \to pq, Q_3\}$,
$R_3 = \{r_1 : H \to e, E_2,\ \ r_2 : Q \to qp, P_4,\ \ r_3 : A \to c, C_2\}$,
$R_4 = \{r_1 : P \to x, X_1\}$,
$R_5 = \{r_1 : Q \to aq, Q_1,\ \ r_2 : Q \to a, A_1\}$.

The system starts with every agent connected, even if one of them, M_6, has no possibility of intervening in the dialogue because it has no rule to apply. After the first step, the turn-taking is not addressed to any membrane, so it goes to any one that has a rule with X to the left (M_2 and M_4). Because of the rule of precedence, M_2 is the next one to play. The situation is similar in the next step, where *Changetopic* is not defined. But in this case, the only membrane able to reply is M_3. The computation goes on and, at a given moment, the application of the rule r_3 from R_1 inhibits three of the membranes, which exit from the scene. Figure 5 shows the configurations of the system.

The system stops because there is no act in the stack for replacing A with an act in M_1. The record of the system evolution is given in Table 4.

C_0 :	$[O]_1$					
C_1 :	$[x_1]_1$,	$[X]_2$,		$[X]_4$,		
C_2 :	$[x_1]_1$,	$[x_2]_2$,	$[CH]_3$,			
C_3 :	$[x_1]_1$,	$[x_2E]_2$,	$[e_1]_3$,			
C_4 :	$[x_1]_1$,	$[x_2p_1q_1]_2$,	$[e_1Q]_3$,			
C_5 :	$[x_1]_1$,	$[x_2p_1q_1]_2$,	$[e_1q_2p_2]_3$,	$[P]_4$,		
C_6 :	$[x_1]_1$,	$[x_2p_1q_1]_2$,	$[e_1q_2p_2]_3$,	$[x_3]_4$,		
C_7 :	$[x_1p_3q_3]_1$,	$\oslash[x_2p_1q_1]_2$,	$\oslash[e_1q_2p_2]_3$,	$[x_3]_4$,	$[Q]_5$,	$\oslash[\ ]_6$
C_8 :	$[x_1p_3q_3Q]_1$,	$\oslash[x_2p_1q_1]_2$,	$\oslash[e_1q_2p_2]_3$,	$[x_3]_4$,	$[a_1q_4]_5$,	
C_9 :	$[x_1p_3q_3a_2q_5]_1$,	$\oslash[x_2p_1q_1]_2$,	$\oslash[e_1q_2p_2]_3$,	$[x_3]_4$,	$[a_1q_4Q]_5$,	
C_{10} :	$[x_1p_3q_3a_2q_5A]_1$,	$\oslash[x_2p_1q_1]_2$,	$\oslash[e_1q_2p_2]_3$,	$[x_3]_4$,	$[a_1q_4a]_5$,	
C_{11} :	$[x_1p_3q_3a_2q_5q_6]_1$,	$\oslash[x_2p_1q_1]_2$,	$\oslash[e_1q_2p_2]_3$,	$[x_3]_4$,	$[a_1q_4a_3Q]_5$,	
C_{12} :	$[x_1p_3q_3a_2q_5q_6A]_1$,	$\oslash[x_2p_1q_1]_2$,	$\oslash[e_1q_2p_2]_3$,	$[x_3]_4$,	$[a_1q_4a_3a_4]_5$,	

Table 4. Configuration register.

In order to know the final result of the system it is necessary to go to the generation register, represented in Table 5.

C_0 :	$[O]_1$	$\odot\mu$				
C_1 :	$[x]_1$,	$[X]_2$,		$[X]_4$,		$[X]_6$
C_2 :		$[x]_2$,	$[CH]_3$,	$[CH]_4$,	$[CH]_5$,	$[CH]_6$
C_3 :		$[E]_2$,	$[e_1]_3$,			
C_4 :		$[t_1q_1]_2$,	$[Q]_3$,			
C_5 :			$[q_2t_2]_3$,	$[T]_4$,		
C_6 :	$[X]_1$,			$[x_3]_4$		
C_7 :	$[t_3]_1$,	$\oslash[\ ]_2$,	$\oslash[\ ]_3$,			$\oslash[\ ]_6$
	$[q_3]_1$,				$[Q]_5$,	
C_8 :	$[Q]_1$,				$[a_1q_4]_5$	
C_9 :	$[a_2q_5]_1$,				$[Q]_5$	
C_{10} :	$[A]_1$,				$[a_3]_5$,	
C_{11} :	$[q_6]_1$,				$[Q]_5$	
C_{12} :	$[A]_1$,				$[a_4]_5$,	

Table 5. Generation register.

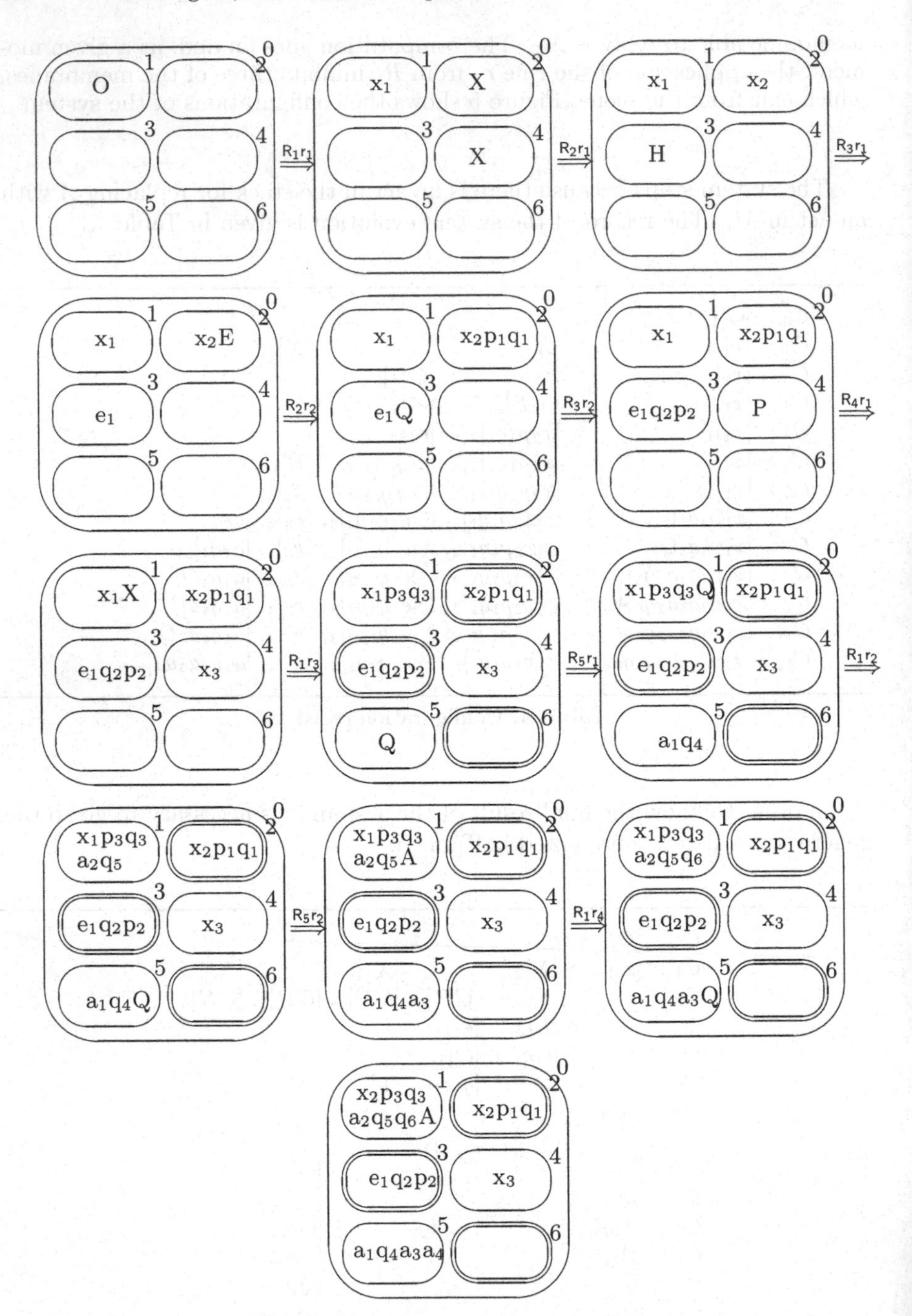

Fig. 5. Development of computation.

Finally, substituting M_1 by Adriana, M_2 by Antipholus, M_3 by Dromio, M_4 by Luciana, M_5 by Officer, and M_6 Courtezan, the result is:

Adriana: O most unhappy day!
Antipholus: O most unhappy strumpet!
Dromio: Master, I am here entered in bond for you.
Antipholus: Out on thee, villain! wherefore dost thou mad me?
Dromio: Will you be bound for nothing? be mad, good master: cry 'The devil!'
Luciana: God help, poor souls, how idly do they talk!
Adriana: Go bear him hence. Sister, go you with me.
Exeunt all but Adriana, Luciana, Officer, and Courtezan.
Say now, whose suit is he arrested at?
Officer: One Angelo, a goldsmith: do you know him?
Adriana: I know the man. What is the sum he owes?
Officer: Two hundred ducats.
Adriana: Say, how grows it due?
Officer: Due for a chain your husband had of him.

4.6 Some Suggestions for Future Research

Even if a lot of work is yet to be done in this direction, we think that to apply P systems to linguistic issues – conversation modeling, for example – has several advantages, among which we stress the generality and the flexibility of the model. Even though the model has not been completely developed yet, it is easy to see that this bio-inspired framework offers good and simple tools to account for main elements and functioning of conversation, such as the following: a) CPSs are able to model interaction, cooperation, and evolution, three vital features in any formal framework that intends to model conversation; b) they are suitable to model turn-taking, one of the basic mechanisms to organize conversation; c) they offer simple tools to account for closings in conversation; d)important elements in conversation, such as contexts, are easily formalized in CPSs.

One of the most interesting projects for the future is to implement these systems for generating a real conversation, testing the aspects which remain to be correctly tuned. P systems are theoretical generative devices quite easily implementable on a computer, and this is true also for CPSs. Another interesting development of CPSs may be the formalization of parallelism in conversation. Such an approach could help, for instance, to model the interaction between agents in a realistic way.

5 Applying LPS to Semantics

Motivation. Formalizing semantics is one of the most challenging topics in linguistics, especially because the systems we have to consider in semantics are very interactive and constantly evolving. In this sense, we remember the sentence of Robert Musil: *"No word means the same thing twice."* Beyond literary references, the fact is that meaning is a continuosly changing concept in linguistic systems; contexts, speakers, cultures, and even education are determinant factors for establishing the meaning of an utterance. Our proposal here tries to combine formal aspects with environmental influence and interaction between agents. Agents can be individual people or groups of people with the same background. The idea of explaining meaning as something dynamic is not new. It can be found in theories like dynamic semantics [11] and dynamic interpretation theory (DIT) [4]. While the former is concerned with *sentence* meaning, the latter is concerned with *utterance* meaning. The main idea of DIT is that the meaning of an utterance in a dialogue should be defined dynamically, in terms of the changes that the agent intends to bring about. Here we propose a new dynamic model for the *lexical* meaning, based on P systems, and we introduce dynamic meaning P systems (DMPSs) by combining P systems with several concepts relating language change and evolution with biological concepts, extracted from Croft [7]. The main elements in the process of communication are the addresser, the addressee, the message, and the context. Membranes are a good model for approaching utterances from the point of view of addresser, addressee, and context, because they can explain, simulate, and possibly predict how the elements involved in the communicative process are able to modify the structure or the meaning of the message, and also how the message can create new contexts or transform those which already exist. We consider our approach here only a first step toward a complex game dealing with the simulation of construction and evolution of meaning. For building the "game," we first give some basic semantic definitions adapted to DMPSs. Then, we introduce definitions for adapting linguistic membrane systems to semantic description. We redefine lexical semantic concepts in terms of DMPSs, and finally provide some examples of the functioning of the DMPS approach to semantic relations.

5.1 A Theoretical Framework for Semantics

Croft points out in [7] that it is possible to give two different definitions of meaning, based on *"the distinction between a language as a population of utterances produced by a speech community, and a grammar as an individual speaker's knowledge about the conventions of the speech community."* The definitions are as follows:

- The **community's meaning** of a linguistic form (a lingueme) is the lineage of replication of its use, in its full encyclopedic, contextual value.
- The **individual's meaning** of a lingueme is a mental structure that emerges from the individual's exposure to (necessarily partial) lineages of the community's meaning, including, of course, the use of the lingueme by the individual.

In order to adapt semantic definitions to DMPSs we take the notion of lingueme referred to in the above definitions. A *lingueme* is a linguistic unit, an utterance, no matter whether it is a word, a sentence, or a discourse. A lingueme, in the first state, has no meaning. From a lingueme we infer another concept, the *semanteme*, which is a semantic unit. The process of meaning assignment is given by the application of a semanteme to a lingueme. Such an application is called a *convention*. The alphabet of linguemes will be denoted below by $V = \{a, \ldots, z\}$, and that of semantemes by $\Sigma = \{\alpha, \ldots, \omega\}$.

5.2 From LPS to DMPS

We will modify now some definitions of linguistic P systems – introduced in Section 2 – in order to obtain a DMPS:

1. The *semantic domain* of a membrane M_n is a subset D_n of Σ which represents the set of semantemes this membrane accepts at any step of a computation.
2. The *linguistic domain* of a membrane M_n is a subset E_n of V which represents the set of linguemes this membrane accepts at any step of a computation.
3. A *membrane* M_n in a DMPS is defined in each state by means of two items, its *semantic domain* D_n and its *linguistic domain* E_n.
4. The correspondence between elements of E_n and D_n associated with a membrane M_n is established by a mapping h_n, which is given by means of *rules*, in the following form: $h_n(D \to E) = \{a \to \alpha, \ldots, m \to \pi\}$. Non-injective functions are allowed.
5. The subscript i attached to an element in a membrane M_n means that there is no transposition rule for the element in h_n. When computation finishes, every lingueme will be converted in the corresponding semanteme, following the transposition rules, but elements marked with i are not taken into account; they simply disappear.
6. The semantic domain of the skin membrane is the union of the semantic domains of all membranes from the system.

Domains associated with membranes can evolve (if they do, they are called *variable domains*) by adding or deleting semantemes or linguemes, through rules of the forms *add* $\{\alpha\}$ *to* D_n, *del* $\{\alpha\}$ *from* D_n, *add* $\{a\}$ *to* E_n, and *del* $\{a\}$ *from* D_n. Here are some rules for handling membranes:

1. If $M_m \subset M_n$ and $D_m = D_n$, then δM_m.
2. If $M_m \subset M_n$ and $D_m = E_n$, then χM_m.
3. If $M_m \subset M_n$ and $E_n = \emptyset$, then ψM_m.
4. If $M_m \odot M_n$ and $D_n = D_m$, then $M_n \nu M_m$.
5. If $M_n \subset M_m$, $D_n = D_m$, and $V_m = V_n$, then ηM_n.
6. If $M_n \subset M_m$ and $V_m = \emptyset$, then ηM_m.

5.3 Definition of Lexical Semantic Concepts from DMPSs

Semantic relations play an essential role in lexical semantics, the area of semantics devoted to word meaning; they intervene at many levels in natural language comprehension and production; and they are a central element in the organization of lexical semantics knowledge bases. Taking into account the relevance of semantic relations, we think that a good way to check the suitability of DMPSs for semantics is to try to describe some of these semantic relations by using our model. To this end, we introduce a simple DMPS account for polysemy, homonymy, synonymy, hyponymy, taxonomy, and meronymy. It is important to recognize that a language can vary along a number of dimensions: historically, geographically, socially, contextually, and stylistically. The recognition of these dimensions will be useful when trying to model the above semantic relations and, of course, for explaining semantic change. We remind you here that membranes in DMPSs are understood as contexts and that contexts may be different historical periods, words, persons, social groups, languages, etc. Before starting with our account of semantic relations, it is important to mention that for each membrane M_r we use in the following examples we should specify the semantic domain D_r, the linguistic domain E_r, the function h_r, and the skin membrane. For simplicity, we will assume that in every example the skin membrane represents a natural language.

Meanings and Readings: Homonymy, Polysemy, and Synonymy

According to [18], when we look at words as meaningful units we also have to deal with the fact that, on the one hand, a single form may be combined with several meanings and, on the other hand, the same meaning may be combined with several word forms. Homonymy is directly related to the fact that many words have more than one meaning and even complete sentences may allow for several readings. The technical term for this phenomenon is *ambiguity.* We say that an expression or an utterance is ambiguous if it can be interpreted in more than one way. The notion of ambiguity can be applied to all levels of meaning: to expression meaning, utterance meaning, and communicative meaning. Here we focus on lexical ambiguity, i.e., the ambiguity of words at the level of expression meaning. We will distinguish between two forms of ambiguity: *homonymy* and *polysemy.* Polysemy is directly related to the phenomenon of *synonymy*, that is, the relation between two (or more) lexemes with the same sense. We will show that DMPSs offer the necessary instruments to account for these important lexical semantic relations. This way we provide evidence of the adequacy of P systems in general, and LPSs in particular, to approach linguistic issues.

The first two lexical semantic concepts we deal with here are *homonymy* and *polysemy.* The problems posed by these two semantic relations are probably at the very heart of semantics. If one consults a monolingual dictionary one will hardly find a word with just one given meaning. If one lexeme has strictly only one meaning, then any variation in meaning would result in two different lexemes. For example, *bank* as in *'The Bank of England'* and *bank* in *'the river bank'* would be regarded as two different words which just happen to have the same sound form and spelling. But *body*, when used to denote the whole physical structure of a human being or an animal, or just the trunk, or a corpse, or a group of people working or acting as a unit, would rather be considered one word with several meanings, because, unlike with *bank*, the meanings of *body* are interrelated. In order to distinguish the two phenomena, the first is called *homonymy*, and the second is called *polysemy*. Roughly speaking, *homonymy* means lexemes with different meanings that happen to have the same sound form or spelling. Ideally, homonyms agree in all points that make up a lexeme, except in meaning. In contrast, *polysemy* deals with one lexeme having several interrelated meanings. In general, different meanings are assigned to different lexemes if they have different historical sources. The idea is that, as long as their meanings remain distinct, different words do not develop into one, even if their sound forms and/or spellings happen to coincide for independent reasons. Hinging on the criterion as to whether or not different meanings are interrelated, the distinction between homonymy and polysemy is vague. It is best taken as characterizing two extremes on a scale. Both phenomena constitute *lexical ambiguity*: the same lexical *form* has different lexical *meanings*. According to [18], what is traditionally described as *homonymy* is illustrated by means of examples of $bank_1$ and $bank_2$, the former meaning 'financial institution' and the latter 'sloping side of a river'. The traditional definition of homonymy is imprecise. Homonyms are traditionally defined as different words with the same form. [18] improves the definition by establishing a notion of *absolute homonymy*. Absolute homonyms will satisfy the following three conditions (in addition to the necessary minimal condition for all kinds of homonymy – an identity of at least one form):

- they will be unrelated in meaning;
- all their forms will be identical;
- the identical forms will be grammatically equivalent.

Absolute homonymy is common enough. Examples as $bank_1$ and $bank_2$, $sole_1$ ('bottom of foot or shoe') and $sole_2$ ('kind of fish') are instances of this type of homonymy. But there are also many different kinds of what [18] calls *partial homonymy*: cases where a) there is an identity of (minimally) one form, and b) one or two, but not all three, of the above conditions are satisfied. For example, the verbs 'find' and 'found' share the form *found*, but not *finds, finding, founds, founding,* etc.; and *found* as a form of 'find' is not grammatically equivalent to *found* as a form of 'found'. In this case, as generally in English, the failure to satisfy (ii) correlates with the failure to satisfy (iii). In our DMPS approach to homonymy we will define just *absolute homonymy*, without providing any definition for what has been called partial homonymy. While homonymy is a rare and accidental phenomenon, polysemy is abundant. It is rather the rule than the exception. A lexeme constitutes a case of *polysemy* if it has two or more interrelated meanings, or better, *meaning variants*. Each of these meaning variants has to be learned separately in order to be understood. The phenomenon of polysemy is independent of homonymy: of two homonyms, each can be polysemous. It results from a natural economic tendency of language. Rather than inventing new expressions for new objects, activities, experiences, etc. to be denoted, language communities usually opt for applying existing terms to new objects, terms hitherto used for similar things. According to [18], while homonymy (whether absolute or partial) is a relation that holds between two or more distinct lexemes, polysemy ('multiple meaning') is a property of single lexemes. This is how the distinction is traditionally drawn. But everyone who draws this distinction also recognizes that the difference between homonymy and polysemy is not always clear-cut in particular instances. What, then, is the difference in theory between homonymy and polysemy? The two criteria that are usually invoked in connection with this are 1) etymology and 2) relatedness of meaning. *Homonymous* is a word that is written and pronounced the same way as another, but which has a different meaning. *Polysemous* is a word that has two or more similar meanings. Homonymy refers to cases in which two words have the same phonological form, whereas polysemy refers to the phenomenon that one and the same word acquires different, though obviously related, meanings often with respect to particular contexts. So, it seems that homonymy and polysemy get resolved in *context*. How do we account for those two semantic relations by using a DMPS? It seems clear that different meanings of a polysemous word are due to different uses, that is, they are directly related to different contexts of use. Taking into account this fact, we can easily describe polysemy in a DMPS by associating each different meaning of a polysemous word with a different membrane. That is, the first meaning of a polysemous word will be possible in membrane 1 while the second will be possible in membrane 2; or, two meanings of the same word, if it is polysemous, cannot be simultaneously in the same membrane as shown in Figure 6, where all elements are identified by the notation conventions mentioned at the beginning of this chapter: $D_2 = \{x\}$, $E_2 = \{\alpha\}$, $D_3 = \{w\}$, and $E_3 = \{\alpha\}$; the mappings h_2 and h_3 are specified in the figure.

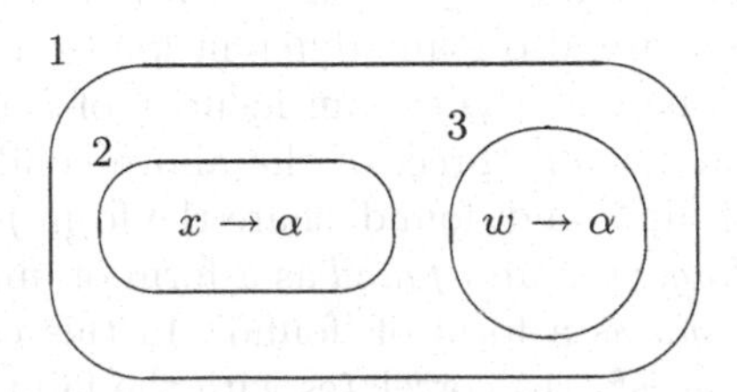

Fig. 6. Polysemy.

So, if what we have is polysemy, then we have the same lingueme α in two different membranes (contexts), and in each of them α is related to a different semanteme (x or w); if what we have is not polysemy but homonymy, that is, two different words that share the same phonological form, then we do not require these two words (linguemes) to be used in different contexts, but they can be used simultaneously in the same membrane, as shown in Figure 7, because they are two different linguemes that happen to have the same form. Note that here we have just one membrane that contains the two linguemes (with the same form) to which h_2 associates different semantemes.

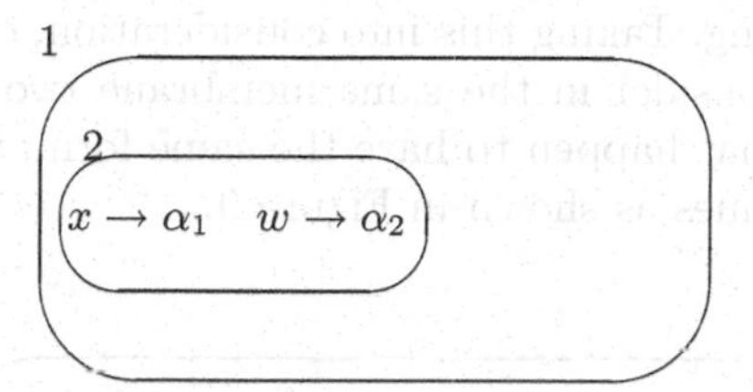

Fig. 7. Homonymy.

Figures 6 and 7 may, though in a very simplistic way, account for the difference between polysemy and homonymy. So, we could say that Figures 6 and 7 offer a definition of those two important semantic relations in terms of DMPSs.

Let us see some examples of polysemy and homonymy. Consider the noun *light*. What we have here is one lingueme, *light*, that can be related to different semantemes, such as 'certain sort of visible radiation', 'electric lamps', 'traffic lights', or 'illuminated areas', in different membranes. So, different interrelated meanings of a polysemous word are found in different membranes/contexts as shown in Figure 8.

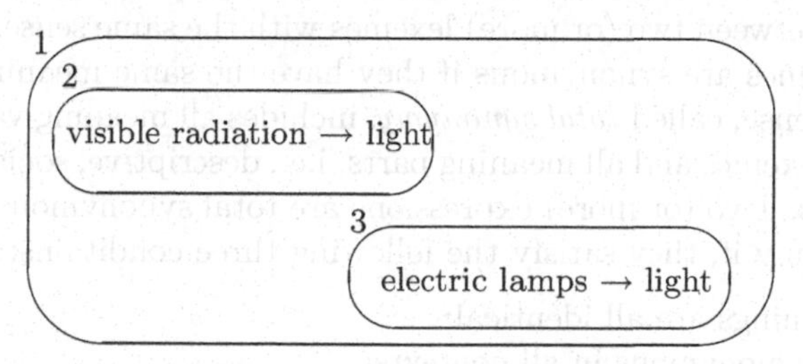

Fig. 8. Example of polysemy: *light*.

A similar DMPS representation can be given for the polysemy of the word *body*. This word can be used to denote the 'whole physical structure of a human being or an animal', or just the 'trunk', or a 'corpse', or a 'group of people working or acting as a unit'. *Body* can be considered as one lingueme that can be associated with different semantemes in different membranes. Again, the meanings of *body* are interrelated, and they get solved in context.

The above are examples of polysemy. Now, if we turn to homonymy, what we have are different words that happen to have the same sound form and spelling. Cases of *absolute* homonymy are, for example, *bank* and *sole*. *Bank*, as in the examples

(1) 'My salary is paid directly in my *bank*'
(2) 'Several people were fishing from the river *bank*'

would be regarded as two different words which just happen to have the same sound form and spelling. Taking this into consideration, our DMPS's account for homonymy will consider in the same membrane two different linguemes ($bank_1$ and $bank_2$), that happen to have the same form, which are related to two different semantemes as shown in Figure 9.

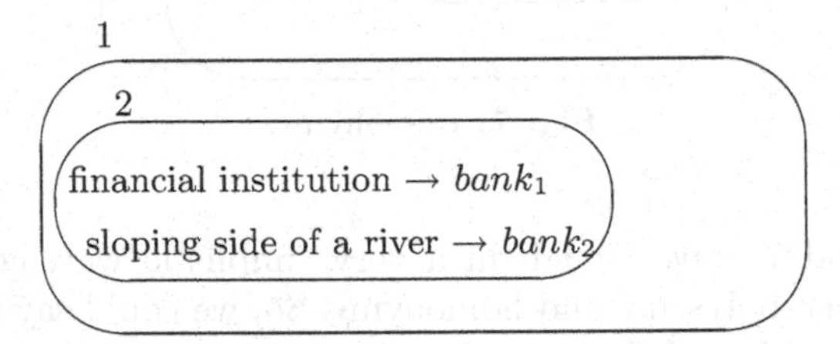

Fig. 9. Example of homonymy: *bank*.

The same can be said for *sole*.

Summing up, examples we have provided in this section show, in a DMPS's way, the difference between polysemy and homonymy: whereas in the first case we have the *same lingueme* associated with different semantemes in *different membranes*, in the second case we have *different linguemes* associated with different semantemes in the *same membrane*.

Another important semantic relation is *synonymy*. Synonymy is defined as the relation between two (or more) lexemes with the same sense. According to [17], two lexemes are synonymous if they have the same meaning. Synonymy in the strict sense, called *total synonymy*, includes all meaning variants for two polysemous lexemes and all meaning parts, i.e., descriptive, social, and expressive meanings. Two (or more) expressions are total synonymous, according to [18], if, and only if, they satisfy the following three conditions:

1. their meanings are all identical;
2. they are synonymous in all contexts;
3. they are semantically equivalent (i.e., their meaning or meanings are identical) on all dimensions of meaning, descriptive and nondescriptive.

Total synonymy is extremely rare. It is limited mostly to technical terms and groups of words that differ only in collocational properties, or the like. More frequently, words that are close in meaning are not fully interchangeable, but varying in their shades of denotation, connotation, implicature, emphasis or register, formality, attitude of speaker, region, etc. These are cases of *partial synonymy*: cases in which two lexemes have *one* meaning variant in common, or words with the same descriptive meaning but different social or expressive meanings. Many of the expressions listed as synonymous in ordinary or specialized dictionaries are what may be called *near-synonyms*, expressions that are more or less similar, but not identical, in meaning. Near-synonymy is not to be confused with various kinds of what is called *partial synonymy*, which meets the criterion of identity of meaning, but which, for various reasons, fails to meet the conditions of what is generally referred to as an absolute synonymy. Typical examples of near-synonyms in English are 'mist' and 'fog', 'stream' and 'brook', and 'dive' and 'plunge'. So, taking into account that usually what we have are partial synonyms, that is, different words with the same meaning but belonging to different registers, styles, regions or even historical periods, it is quite easy to account for synonymy in DMPSs. It is enough to associate each synonym with a different membrane – understanding membrane here as register, style, sociolect, region, historical period, etc. – as shown in Figure 10. Note that ⊙, that is, the situation in which the communication between the two membranes is closed, could account for the nonexistence of total synonymy: we have the same meaning, but in different contexts (membranes).

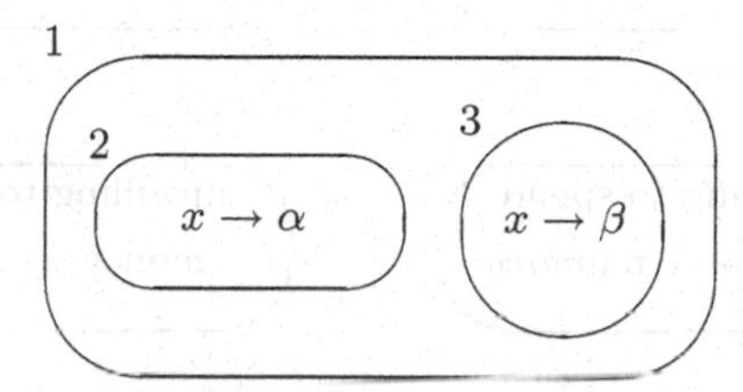

Fig. 10. Synonymy.

We have noted that examples of partial synonyms can be words with the same descriptive meaning but different social or expressive meaning. If we take into account expressions which differ in the nature of their expressive meaning, the most obvious difference is between those which imply approval or disapproval and those which are neutral with respect to expressivity. Taking this into account, we can speak of synonyms that have identical meaning but differ in their connotations, and, in this case, we will associate each element in a set of synonyms with a different membrane in order to indicate this difference in the expressivity. Consider, for example, the following four words: *thrifty, mean, stingy, economical.* If we look at a standard English dictionary we will find the following meanings for them:

- *thrifty*: if you say that someone is thrifty you are praising him for saving money, not buying unnecessary things, and not wasting things;
- *mean*: if you describe someone as mean, you are being critical of him because he are unwilling to spend money or to use very much of a particular thing;

- *stingy*: if you describe someone as stingy you are criticizing him for being unwilling to spend money (informal);
- *economical*: someone who is economical spends money sensibly and does not want to waste it on things that are unnecessary.

From the above definitions, the descriptive meaning of the four words is the same, so they can be said to be synonyms. However, it follows also from the definitions that the four words are not interchangeable in all contexts because they express approval, disapproval, or neutrality. So, the meaning is the same, but they are appropriate in different contexts, and this is perfectly accounted for by using our DMPS definition of synonymy. We associate different words (linguemes) to the same meaning (semantemes) in different membranes, accounting in this way for the identity of meaning (the same semanteme) and the difference of connotation (different membrane) as shown in Figure 11:

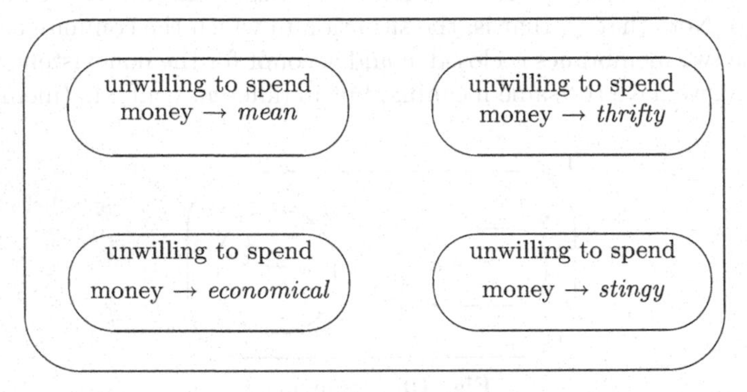

Fig. 11. Example of synonymy: *thrifty, stingy...*

If we associate membranes with different dialects, we can say that *'car'*, *'automobile'*, and *'wheels'* are synonyms, since these lexemes all denote the same kind of motor vehicle. However, *'automobile'* is restricted to American English, whereas *'wheels'*, apart from being variably labelled slang, informal, or spoken, is also grammatically different from *'car'* and *'automobile'*: as a plural-only noun, it cannot be used with the indefinite article *a* or the numeral *one*. In the DMPS approach what we have is similar to Figure 11.

We could go on with examples of synonymy: *'statesman'* versus *'politician'*, *'stink'* versus *'stench'* versus *'fragrance'* versus *'smell'*; *'crafty'* versus *'cunning'* versus *'skilful'* versus *'clever'*; and so on. But, at the end, we always have different words with the same meaning but associated with different contexts, where context can be understood as styles, dialects, collocational positions, etc. As can be seen from the above examples, in order to explain synonymy we need to postulate different membranes – standing for registers, regions, styles, or whatever can condition the choice of one synonym or another – that contain different linguemes with which the same semanteme is associated.

Meaning Relations: Hyponymy, Taxonomy, and Meronymy

By *hyponymy* we mean a sense relation between one lexeme (the superordinate term) and two or more terms (the hyponyms) which denote more specific instances of what the superordinate term denotes. The sense of a hyponym includes the sense of its superordinate term. According to [17], hyponymy can be defined as follows: an expression A is a hyponym of an expression B if, and only if, the meaning of B is part of the meaning of A, and A is a subordinate of B. In addition to the meaning of B, the meaning of A must contain more specifications, rendering the meaning of A, the hyponym, more specific than the meaning of B. If A is a hyponym of B, B is called a *hyperonym* of A. In order to provide a DMPS approach to hyponymy, we can use the *nesting* relationship between membranes. We recall here that two membranes, M_1 and M_2, are nested when $M_1 \subset M_2$. It is easy to see that this relation and its properties can account for hyponymy, that it is an inclusion relationship. We can take every membrane to be a word, so that by including one membrane into another we are accounting for the relation between words that are in a hyponymy relationship. For example, let us assume that membranes 1, 2, and 3 in Figure 12 represent respectively the words "animal," "dog," and "terrier"; by establishing a nesting relation between them ($M_3 \subset M_2 \subset M_1$), we are easily accounting for the hyponymy relation as shown in Figure 12.

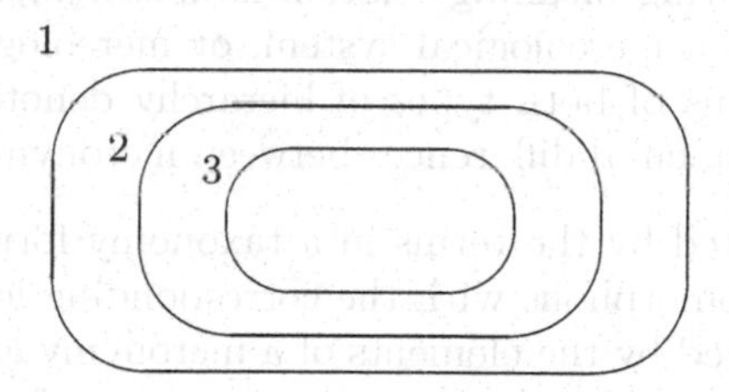

Fig. 12. Hyponymy.

Examples of hyponymy tend to be biological or botanical. Consider the hyponymic relationship between *animal, dog,* and *terrier, dalmatian, or poodle.* We say that *animal* is the superordinate term that has as hyponym *dog,* that in turn has the hyponyms *dalmatian, poodle,* and *terrier.* This nesting relationship is represented as shown in Figure 13.

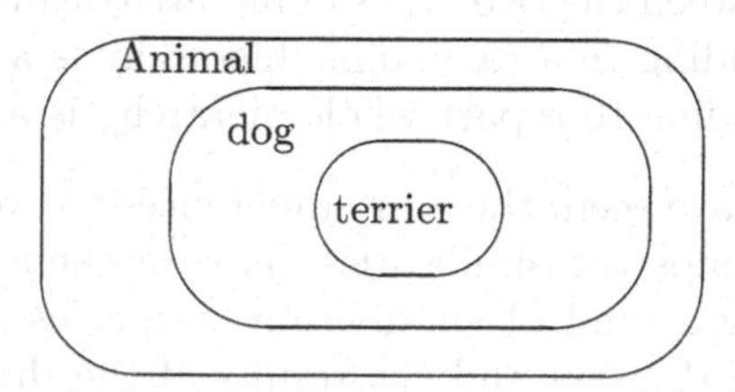

Fig. 13. Example of hyponymy: *Animal.*

Most lexical items form groups with other lexemes. Semantic theories of different orientations, in particular structuralism, have tried to capture this phenomenon by the notion of *lexical field*. We distinguish two types of lexical fields, taxonomies and meronymies. Terms for animals, plants, food, or artefacts such as furniture, vehicles, clothes, musical instruments, etc. form lexical fields of considerable size. Their underlying structure is a hierarchy with two or more levels: a topmost hyperonym like *vehicle*, a level of general terms such as *car*, *bicycle*, *boat*, *aeroplane*, and further levels of more specific types of cars, bicycles, boats, aeroplanes, etc. Such systems represent a special type of hierarchies, called *taxonomies*: subordinates in taxonomies (called taxonyms) are not just arbitrary subordinates but hyponyms that denote subtypes. According to [8], taxonomy may be regarded as a subspecies of hyponymy: the taxonyms of a lexical item are a subset of its hyponyms. Taking into account that a taxonomy may be regarded as a subspecies of hyponymy, we use again the *nesting* relationship between membranes in order to provide a DMPS approach of taxonomy. Again, we can take every membrane to be a word, so that by including one membrane into another we are accounting for the hierarchy with two or more levels that make up a taxonomy. The second major type of branching lexical hierarchy is the part-whole type. Many objects in the world are perceived as a whole consisting of different parts. Correspondingly, our concepts for complex objects contain these parts as elements. The technical term for the constituting meaning relation is *meronymy*; a system based on meronymies is called a mereological system, or mereology. According to [8], even though the terms of both types of hierarchy denote classes of entities, there are some fundamental differences between meronymies and taxonomies:

- The classes denoted by the terms in a taxonomy form a hierarchy which is more or less isomorphous with the corresponding lexical hierarchy.
- The classes denoted by the elements of a meronymy are not hierarchically related; that is to say, the hierarchical structuring of a meronymy does not originate in a hierarchy of classes. It is rather the way the individual parts of each whole are related which generates the hierarchical structuring that forms the basis of a meronymy.
- A meronymy thus has closer links with concrete reality than a taxonomy. The classes denoted by the terms of a meronymy are formed by associating together analogous parts (e.g., Arthur's nose, Tom's nose) of isomorphous wholes (Arthur's body, Tom's body).
- The difference between the two types of hierarchy can be expressed by saying that corresponding to a taxonomic hierarchy is a hierarchy of classes, whereas corresponding to a part-whole hierarchy is a class of hierarchies.

Although, as we have seen, there are differences between meronymies and taxonomies, it is perhaps the similarities between them which are the more striking. Both involve a kind of subdivision, a species of inclusion between the entity undergoing division and the results of the division, and a type of exclusion between the results of the division. Any taxonomy can be thought of in part-whole terms (although the converse is not true): a class can be seen as a whole whose parts are its subclasses. Corresponding to each of the common nouns constituting a typical taxonomy, there exists a proper noun labelling the class as an individual. The sibling relation among membranes may account for meronymies in lexical semantics as shown in Figure 14.

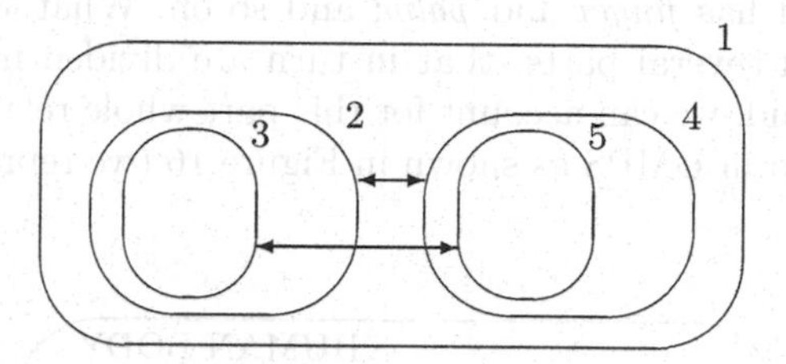

Fig. 14. Meronymy.

Examples of taxonomies are very easy to find. If, for example, we take musical instruments, we can establish a hierarchy with a topmost hyperonym *musical instrument*, a level of general terms, such as *strings*, *woodwind*, *brass*, *percussion*, and further levels of more specific kinds of strings, woodwind, brass, and percussion instruments. The result is a nesting relationship as shown in Figure 15.

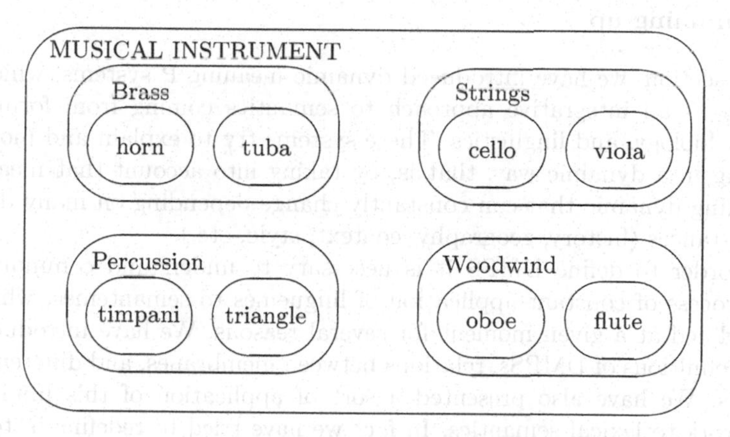

Fig. 15. Example of taxonomy: *Musical instrument.*

Another example of taxonomy can be the following: think about the things (other than food) which go on the table at mealtime that can be labelled with a hyperonym such as *tableware* and be divided into a level of general terms such as *cutlery*, *crockery*, *glassware*, *table linen*, and further levels of more specific kinds of *cutlery*, such as *knife* and *spoon*, specific kinds of *crockery*, etc. The result is again the nesting relationship, easy to represent in DMPS terms. Notice that due to the relationship between hyponymy and taxonomy the examples provided in this section are valid also as examples of hyponymy.

The division of the human body into parts serves as a prototype for all part-whole hierarchies. The human body is divided into *head, neck, trunk, legs*, and *arms*; the *head* is divided into *ears* and *face*; the *face* is made up of *eyes, mouth*; the *trunk* is made up of *belly* and *breast*; the *arm* has *hand* and *forearm*; the hand has *finger* and *palm*; and so on. What we have here is a whole made up of several parts, that in turn are divided into several parts, etc. As we have said, we can account for this part-whole relationship by using the sibling relation in DMPS as shown in Figure 16 (we represent just part of the example).

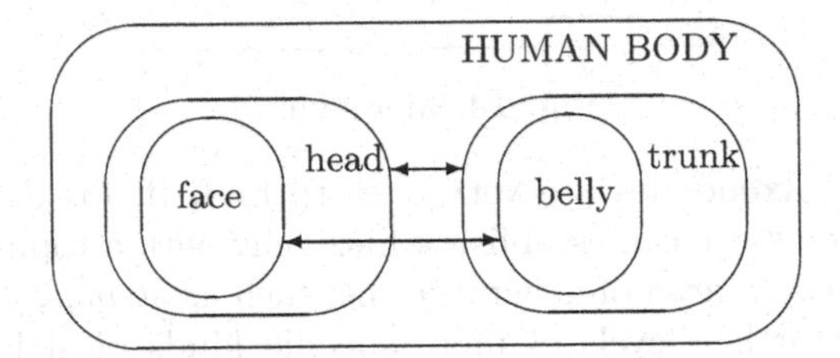

Fig. 16. Example of meronymy: *Human body.*

5.4 Summing up

In this section, we have introduced dynamic meaning P systems, which may be seen as an integrative approach to semantics coming from formal languages, biology, and linguistics. These systems try to explain and model the meaning in a dynamic way, that is, by taking into account that meaning is something dynamic that can constantly change depending on many different circumstances (history, geography, context, style, etc.).

In order to define DMPS it is necessary to understand communication as a process of constant application of linguemes to semantemes, which can be modified at a given moment for several reasons. We have introduced the basic definitions of DMPSs, relations between membranes, and different types of rules. We have also presented a sort of application of this bio-inspired framework to lexical semantics. In fact we have tried to redefine in terms of a DMPS six meaning relations: *polysemy, homonymy, synonymy, hyponymy, taxonomy*, and *meronymy*.

In all cases, the relation between membranes in a linguistic P system can account for lexical semantic relations (hyponymy, taxonomy, and meronymy), whereas different/the same linguemes associated with different/the same semantemes in different/the same membrane account for semantic relations that lead to ambiguity (polysemy, homonymy, and synonymy).

5.5 Possible Extensions

In this section, we have seen how easy it is to account for semantic relations in terms of a DMPS. Now, if we go back to our initial goals of defining a dynamic meaning framework, we should suggest how to account for evolution of meaning in DMPSs.

It is often said that there is less resistance to change in semantics than in other areas of grammar, so meaning changes relatively quickly and easily. According to [19], there seem to be different aspects of language in general, and meaning in particular, which allow semantic change to occur. Two of these aspects are:

1. *Polysemy*: words are typically polysemous; they have various meanings or cover a whole range of shades of meaning. This flexibility is necessary since words are used in a wide variety of contexts by many different speakers, who may vary in the meaning they wish to convey. Words can lose or gain meanings relatively easily due to this elasticity, and they do not have to lose an earlier meaning to gain a new one.
2. *Arbitrariness of the linguistic sign*: the sign is bipartite, made up of a signifier and a signified. These two components are arbitrarily linked. Arbitrariness allows us to regard the signifier and the signified as essentially independent; either may therefore change with time.

Semantic changes have been classified as follows:

1. *Extension* (or generalization, or broadening) increases the number of contexts in which a word can be used, reducing the amount of information conveyed about each one.
2. *Restriction* (or specialization, or narrowing) of meaning also involves an increase in information conveyed, since a restricted form is applicable to fewer situations, but tells more about each one.

The above classification refers to the evolution of meaning (signifier), but note that we have said that according to the arbitrariness of the linguistic sign the signifier and signified are independent and may also change independently.

Let us consider the above classification of semantic change and the possibility of affecting the signifier, of a word without affecting the signified and vice versa. If with this simple idea of semantic change in mind we look at DMPS, we will quickly individuate a mechanism that allows us to provide a DMPS's explanation of the semantic change: *variable domains*. We have postulated as one of the features of DMPS the possibility of having variable domains, that is, semantic and linguistic domains can evolve. The processes for changing domains in a DMPS may explain semantic change in the following way:

1. The first process we have postulated is the possibility of adding new semantemes to the domain of a membrane. Note that if we add new semantemes, they should be related, via a mapping h_n, to existing linguemes. If it is the case that such linguemes, with which we associate the new

semantemes, already exist in another membrane and are already related to different semantemes, what we obtain is an extension of meaning. So, by postulating the addition of semantemes to a membrane domain we account for the gain in the meaning of a word. This gain in the meaning may lead to polysemy. Hence, addition of semantemes describes one of the two types of semantic changes referred to above, *extension*, that is, the increase in the number of contexts in which a word can be used.
2. Semantemes can also be deleted from a domain. Note that using this process we can easily account for loss in the meaning of a word. If a lingueme is related to different semantemes in different membranes, and we delete one of these semantemes from one membrane, we are losing one meaning of this word. So, deletion of semantemes accounts for *restriction* of meaning, that is, the restriction of contexts where a lingueme can be used.

Therefore, by using the process of addition and deletion of semantemes to/from a membrane domain, we can explain extension and restriction of meaning. But we have said that signifier and signified can change independently. So, we have to account for the evolution of signifier, and we can do that by using the following two processes for changing membrane domains:

1. New linguemes can be added to a membrane domain. The addition of new linguemes can account for the introduction of a new signifier that may then be related to new or already existing meanings (semantemes). In this case h_n associates a new or already existing semanteme to the new lingueme introduced.
2. Some linguemes can be deleted from the domain. This process accounts for the elimination of a signifier whose meaning should be absorbed by another (new or existing) lingueme. Note that in order to describe this fact we just need that h_n associates the semanteme that was associated with the lost lingueme with another lingueme.

Up to now, we have accounted separately for the change (evolution) of meaning and for the change (evolution) of signifiers. We can account also for the total deletion of a word by using subscript i: all elements marked with a subscript i disappear because there is no transposition rule for such an element in h_n. Thus, every word marked will this subscript will stand for total loss of a word.

In conclusion, we can account for semantic change by using DMPSs. Actually, DMPS can account separately for semantic change – understood in terms of addition/deletion of semantemes from a membrane domain – and for 'signifier' change – understood as addition/deletion of linguemes.

6 Challenges for the Future

In the present work, we have presented a general formalism for dealing with linguistics by means of P systems. Precedents of this study, which involves computer science, general linguistics and speech acts theory, are almost nonexistent, and our effort has been directed to getting the clearest and widest possible formalization. As a consequence, many aspects could not be developed here.

The application of membrane computing to linguistics was carried out through linguistic P systems (LPSs), theoretical devices specially defined to deal with the hard features of instability, sensibility to the context, and mutual influence which characterize languages.

Starting from LPSs, we have suggested (Section 3) several intuitive applications: semantics, language evolution, sociolinguistics, dialogue, and anaphora resolution. These are just some of the branches of linguistics capable of being approached by means of P systems. Among them, only dialogue and semantics have been subjects of our interest in these pages, and, even those, from a small perspective, due mainly to space restrictions. Therefore, language evolution, sociolinguistics, and anaphora resolution remain to be studied from LPS points of view. Their formalization is important in order to achieve a complete characterization of linguistics in this framework, which can provide the tools for giving a general linguistic theory.

Other important fields of linguistics, such as syntax, have not been considered since, as an internal and structural part of linguistics, it does not have the same requirements as the fields we have dealt with here. Linguistic P systems do not seem to be the best way to analyze syntax, other approximations, such as P automata, being more useful for explaining how the sentences combine to make a complex construction.

On the other hand, another possibility is to approach linguistics with regular P systems. With special rules for membrane evolution, the changes in societies, the spread of linguistic units and ideas, and the interaction between agents in the world can be modeled. In this sense, the possibility of predicting the future evolution of events in real life is a challenging goal which fits the intrinsic algorithmic nature of P systems. The framework is very general, and linguistics is only one of the fields that can be studied in terms of membrane computing (maybe also using fuzzy mathematics, an interesting direction of research which remains to be explored).

References

1. L. Amgoud, N. Maudet, S. Parsons: Modeling Dialogues Using Argumentation. In *Proceedings of the 4th Conference on Multi-Agent Systems*, Boston, 2000.
2. J.L. Austin: *How to Do Things With Words.* New York, Oxford University Press, 1962.

3. G. Bel Enguix: Preliminaries about Some Possible Applications of P Systems in Linguistics. In *Membrane Computing. International Workshop WMC-CdeA 2002, Revised Papers* (Gh. Păun, G. Rozenberg, A. Salomaa, C. Zandron, eds.), LNCS 2597, Springer, Berlin, 2003, 74–89.
4. H.C. Bunt: *DIT–Dynamic Interpretation in Text and Dialogue.* ITK Research Report no 15, Tilburg University, The Netherlands, 1990.
5. L. Carlson: *Dialogue Games.* Reidel, Dordretch, 1983.
6. H.H. Clark: *Using Language.* Cambridge Univ. Press, 1996.
7. W. Croft: *Explaining Language Change. An Evolutionary Approach.* Longman, Singapore, 2000.
8. D.A. Cruse: *Lexical Semantics.* Cambridge Univ. Press, 1986.
9. J.A. Fodor, J.J. Katz: *The Structure of Language: Readings in the Philosophy of Language.* Prentice-Hall, New Jersey, 1964.
10. H.P. Grice: Logic and Conversation. In *Syntax and Semantics 3: Speech Acts* (P. Cole, J. Morgan, eds.), Academic Press, New York, 1975.
11. J. Groenendijk, M. Stokhof: Dynamic Predicate Logic. *Linguistics and Philosophy,* 14 (1991), 39–100.
12. M.D. Jiménez-López: Formal Languages for Conversation Analysis. In *Estudios Hispánicos y Románicos* (A. García Español, ed.), Rovira i Virgili Univ., Tarragona, 2002.
13. M.D. Jiménez-López: *Grammar Systems: A Formal-Language-Theoretic Framework for Linguistics and Cultural Evolution.* PhD Dissertation, Rovira i Virgili Univ., Tarragona, 2000.
14. J.C. Kwotko, S.D. Isard, G.M. Doherty: Conversational Games Within Dialogue. *Technical Report HCRC/RP-31*, HCRC Publications Univ. of Edinburgh, 1993.
15. Sh. Lappin: *The Handbook of Contemporary Semantic Theory.* Blackwell, Oxford, 1996.
16. S.C. Levinson: *Pragmatics.* Cambridge Univ. Press, 1983.
17. S. Löbner: *Understanding Semantics.* Arnold, London, 2002.
18. J. Lyons: *Linguistic Semantics. An Introduction.* Cambridge Univ. Press, 1995.
19. A.M.S. McMahon: *Understanding Language Change.* Cambridge Univ. Press, 1996.
20. R. Mitkov: *The Oxford Handbook of Computational Linguistics.* Oxford Univ. Press, 2003.
21. C. Morris: Foundations of the Theory of Signs. In *International Encyclopaedia of Unified Science* (R. Carnap et al., eds.), 2:1, The Univ. of Chicago Press, 1938.
22. Gh. Păun: Computing with Membranes. *Journal of Computer and System Sciences,* 61 (2000), 108–143.
23. Gh. Păun: *Membrane Computing. An Introduction.* Springer, Berlin, 2002.
24. C.A. Reed, D.P. Long: Collaboration, Cooperation and Dialogue Classification. In *Working Notes of the IJCAI97 Workshop on Collaboration, Cooperation and Conflict in Dialogue Systems* (K. Jokinen, ed.), Nagoya, 1997, 73–78.
25. H. Sacks, E.A. Schegloff, G. Jefferson: A Simplest Systematics for the Organization of Turn-Taking for Conversation. *Language,* 50, 4 (1974), 696–735.
26. J. Searle: *Speech Acts: An Essay in the Philosophy of Language.* Cambridge Univ. Press, 1069.
27. D.N. Walton, E.C.W. Krabbe: *Commitment in Dialogue: Basic Concepts of Interpersonal Reasoning.* State Univ. of New York Press, New York, 1995.
28. The P Systems Web Page: `http://psystems.disco.unimib.it`.

Chapter 14
Parsing with P Automata

Radu Gramatovici[1], Gemma Bel Enguix[2]

[1] Faculty of Mathematics and Computer Science
University of Bucharest
Academiei 14, RO 010014, Bucharest, Romania
radu@funinf.cs.unibuc.ro

[2] Research Group on Mathematical Linguistics
Rovira i Virgili University
Pl. Imperial Tárraco 1, 43005 Tarragona, Spain
gbe@astor.urv.es

Summary. P automata are membrane systems that work as accepting devices. In this chapter, two parsing methods using P automata are presented. The first method uses P automata with active membranes for parsing natural language sentences into dependency trees. The second method uses a variant of P automata with evolution and communication rules for parsing Marcus contextual languages.

1 Introduction

P automata are a special type of membrane system that work as accepting devices. Different classes of P automata have been introduced and studied recently, starting with [8, 9], where one defines P automata using communication rules and final configurations. A model of P automata with states was considered in [26] and extended in [12]. In [12], the results about other previously defined types of P automata, such as P automata with communication rules and final configurations, are improved. P automata with priorities were introduced in [5].

Another model, inspired by natural processes taking place in cells, the P automaton with membrane channels, was studied in [13, 32]. Paper [13] also extends the action of P automata on infinite words proceeding in a way similar to ω-Turing machines. A variant of P automaton using catalysts was considered in [14], where results on the descriptional complexity of P systems were obtained. Results on the computational complexity were also presented in [7]. In [3], a parallel computation model with a flexible structure that evolves in time according to a set of rules was introduced in the form of a P automaton.

For a recent complete survey on P automata, the reader is referred to [6]. For details on different features of P systems, the monograph [34] can be consulted.

In this chapter, we investigate the possible application of P automata to parsing.

There are multiple ways in which the analysis of natural languages can benefit from membrane computing. In general, language analysis does not mean only syntactic analysis, but also discourse analysis, pragmatic resolution, and semantic interpretation. Many of these tasks require an exponential time complexity that can hardly be satisfied by current silicon computers.

The highly parallel framework of membrane computing can very well accommodate these complex tasks of natural language processing. The parallel computation capacity of P automata is in this chapter the most exploited feature of membrane systems. Other characteristics of P automata that can be found useful for designing language systems are (i) the flexibility in writing the rules, (ii) the computational interplay between the membranes, seen as independent processors, (iii) the string objects, evolving and traveling through membranes, and (iv) the communication capacity and the cooperative behavior of membranes.

The chapter is split into two parts. In the first part, we study P automata with active membranes and their application to parsing natural language sentences into dependency trees. In the second part, we study P automata with evolution and communication rules and their application to Marcus contextual languages parsing.

2 Parsing with P Automata with Active Membranes

The first half of this chapter is devoted to P automata with active membranes and it is based on [3].

2.1 P Automata with Active Membranes

In P automata of the type introduced in [3], the accepting computation starts with one membrane, which contains the string that has to be recognized and possibly some other information. The computation develops according to the input string and, during the evolution, membranes can be created or dissolved.

Formally, a *P automaton* with active membranes (from now on we will simply say *P automaton*) is a construct $M = (\Sigma, O, T, \mu, M_1, \ldots, M_n, R)$, such that:

- Σ is the input alphabet; we also consider a copy of Σ denoted by $\overline{\Sigma}$.
- O is the alphabet of feature objects.
- T is the alphabet of agent objects.

- μ is a hierarchical structure of n membranes. For the rest of this section, we suppose that μ describes a linear structure, from the membrane labeled 1 which is always an elementary membrane to the membrane labeled n, with $n \geq 1$, such that the membrane i is included in the membrane $i+1$ for any $1 \leq i \leq n-1$.
- $M_i \subseteq O \cup T \cup \Sigma^+$ is the content of membrane i, for all $1 \leq i \leq n$. If i is not an elementary membrane (equivalent in our case with $i > 1$), then $M_i \subseteq O \cup T \cup \Sigma$.
- R is the set of rules applicable to all membranes. These rules can be of the following types[3]:

1. $aw \rightarrow \overline{a}; w \mid_{\neg(\overline{\Sigma} \cup T)}$ with $a \in \Sigma$, $w \in \Sigma^*$. The first symbol a is extracted from the input string (and copied to $\overline{a}$), provided that no other symbol from $\overline{\Sigma} \cup T$ is present in the membrane.
2. $x; b \rightarrow c$ with $x \in \overline{\Sigma} \cup T$, $b, c \in O$. A symbol $b \in O$ is rewritten as $c \in O$ in the presence of a symbol x from $\overline{\Sigma} \cup T$. This last symbol is consumed.
3. $x; b \rightarrow \delta$ with $x \in \overline{\Sigma} \cup T$, $b \in O$, where δ indicates the dissolution of the membrane. A symbol $b \in O$ is deleted in the presence of a symbol x from $\overline{\Sigma} \cup T$. This last symbol is also deleted and the current membrane is dissolved. The (other) content of the current membrane (particularly the input string if it is so the case) will be found after dissolution in the membrane immediately above.
4. $[x; b]_i \rightarrow [c_{i+j}[c_{i+j-1} \cdots [c_0]_i \cdots]_{i+j-1}]_{i+j}$ with $x \in \overline{\Sigma} \cup T$, $b, c_0, c_1, \ldots, c_j \in O$. $j \geq 1$ new membranes are created around the membrane i. In every new membrane $i+k$ a symbol c_{i+k} is written $(1 \leq k \leq j)$, while in the membrane i the symbol x is deleted and the symbol b is rewritten in c_j.
5. $w \rightarrow y; w \mid_{\neg(\overline{\Sigma} \cup T)}$ with $y \in T$, $w \in \Sigma^+$. A symbol y from T is generated in the membrane containing the input string if no other symbol from $\overline{\Sigma} \cup T$ is present in that membrane.
6. $[x; b]_i \rightarrow [c_{i+j}[c_{i+j-1} \cdots [y; c_0]_i \cdots]_{i+j-1}]_{i+j}$ with $x \in \overline{\Sigma} \cup T$, $b, c_0, c_1, \ldots, c_j \in O$, $y \in T$. $j \geq 1$ new membranes are created around the membrane i. In every new membrane $i+k$ a symbol c_{i+k} is written $(1 \leq k \leq j)$, while in the membrane i, x is rewritten as y and b is rewritten as c_0.
7. $x; b \rightarrow y_{out|\neg T}; c$ with $x \in \overline{\Sigma} \cup T$, $y \in T$, $b, c \in O$. A symbol $b \in O$ is rewritten in $c \in O$ in the presence of a $x \in \overline{\Sigma} \cup T$. This last symbol is rewritten to y and sent to the immediately next membrane if no other symbol from T is present in that membrane.
8. $x; b \rightarrow y\delta$ with $x \in \overline{\Sigma} \cup T$, $y \in T$, $b \in O$, where δ indicates the dissolution of the membrane. A symbol $b \in O$ is deleted in the presence of a symbol x from $\overline{\Sigma} \cup T$. The symbol x is rewritten as $y \in T$ and the current membrane is dissolved. The (other) content of the current membrane will be found after dissolution in the membrane immediately above.

---

[3] Since we have both symbol objects and string objects, we will use ';' to separate objects that occur on the same side of a rule.

The initial configuration has one membrane containing a string over Σ and a symbol from O. We consider the language $L(A)$ recognized by a P automaton A as the set of strings over Σ that lead the system to a configuration where the input string is consumed and all membranes are dissolved.

We denote by $P(i)$ the family of languages recognized by P automata using (only) rules of the types from 1 to i, for $i \in \{3, 4, 8\}$, and by $P'(7)$ the class of languages recognized by P automata using rules of the types 1 to 7 (though the rules of type 3 cannot be applied with a symbol x from T).

Example 1. We consider a P automaton M defined as above where $\Sigma = \{a, b, c\}$, $O = \{Z, A, B\}$, $T = \{t\}$, and the set R contains the rules:

$xw \rightarrow \overline{x}; w \mid_{\neg\overline{\Sigma}}$ with $x \in \Sigma$, $w \in \Sigma^+$,
$\overline{a}; Z \rightarrow A$,
$[\overline{a}; A]_1 \rightarrow [A[A]_1]_2$,
$\overline{b}; A \rightarrow B$,
$x; B \rightarrow t_{out|\neg t}; B$ with $x \in \{\overline{b}, t\}$,
$\overline{c}; B \rightarrow \delta$.

Then, $L(A) = \{a^n b^n c^n \mid n \geq 1\}$ (the triple agreement).

We denote by RE, CS, CF, and REG the families of recursively enumerable, context-sensitive, context-free, and regular languages, respectively.

The following results are from [3] and they establish the power of P automata of the previous type for different combinations of rules.

Theorem 1. $P(8) = RE$, $P'(7) = CS$, $P(4) = CF$, $P(3) = REG$.

Proofs of these equalities can be found in [3].

The rules of the P automata are inspired by the transition function of the go-through automata introduced in [17].

2.2 Bubble Trees

In [22], a hybrid tree structure was proposed for representing some linguistic constructions such as coordination or nominal or verbal nuclei. The tree structure called bubble tree was introduced by Gladkij in [15].

Formally, a *bubble tree* is a 4-tuple $(X, \mathcal{B}, \phi, \triangleleft)$, where X is the set of basic nodes, $\mathcal{B}$ is the set of bubbles, ϕ is a map from $\mathcal{B}$ to the non-empty subsets of X (which describes the content of the bubbles), and $\triangleleft$ is a relation on $\mathcal{B}$ satisfying the following properties:

P1. $\triangleleft$ is a tree relation.
P2. Any one element subset of X is the content of one and only one terminal node (a node without dependents).
P3. If $\alpha, \beta \in \mathcal{B}$, then $\phi(\alpha) \cap \phi(\beta) = \emptyset$ or $\phi(\alpha) \subseteq \phi(\beta)$ or $\phi(\beta) \subseteq \phi(\alpha)$.

P4. If $\phi(\alpha) \subset \phi(\beta)$, then $\alpha \prec \beta$, where $\prec$ is the transitive closure of $\lhd$. If $\phi(\alpha) \subset \phi(\beta)$, then $\alpha \preceq \beta$ or $\beta \preceq \alpha$, where $\preceq$ is the reflexive and transitive closure of $\lhd$.

The relation $\lhd$ is called *dependency-embedding relation.* Two subrelations of $\lhd$ are considered, the *dependency relation* $\lhd_1$ defined by $\alpha \lhd_1 \beta$ if $\alpha \lhd \beta$ and $\phi(\alpha) \cap \phi(\beta) = \emptyset$ and the *embedding relation* $\lhd_2$ defined by $\alpha \lhd_2 \beta$ if $\alpha \lhd \beta$ and $\alpha \subseteq \beta$. If $\lhd_1 = \emptyset$, then the bubble tree is called a *constituency tree.* If $\lhd_2 = \emptyset$, then the bubble tree is called a *dependency tree.*

Bubble trees have been applied by Kahane to linguistic contructions as coordination or for improving the definition of projectivity of dependency trees.

In the next section, we will define P systems with the structure of membranes described by bubble trees. We will use bubble trees to represent the dynamics of the membrane systems generating dependency trees. Embedding relations within a bubble tree will represent temporary computational relations between syntactic items, while dependency relations will characterize syntactic dependencies.

2.3 Parsing Dependency Trees

In this section, we consider *P translators* (associated with P automata as above) that map natural language phrases into syntactic structures like dependency trees. During the evolution of a P translator, the internal structure of the membrane system takes the form of a bubble tree.

A *P translator* is a modified P automaton whose membrane structure μ is characterized by a dependency-embedding relation, exactly as in the structure of a bubble tree. The rules of types 2, 3, and 8 that rewrite feature objects and/or dissolve membranes are replaced by rules of types 2′, 3′, and 8′ given below, in which instead of dissolving the current membrane, we attach it by a dependency link to the immediately higher membrane or to other membrane in the system.

2′. $(x; b; w)_i \to (c; w)_i \lhd_1 (b)_j$ with $x \in \overline{\Sigma} \cup T$, $b, c \in O$, $w \in \Sigma^*$.

3′. $((x; b; w)_i \neg z)_{i+1} \to (y; w)_{i+1} \lhd_1 (c)_i$ with $x \in \overline{\Sigma} \cup T$, $y, z \in T$, $b, c \in O$, $w \in \Sigma^*$.

8′. $((x; b; w)_i \neg y)_{i+1} \to (w)_{i+1} \lhd_1 (c)_i$ with $x \in \overline{\Sigma} \cup T$, $y \in T$, $b, c \in O$, $w \in \Sigma^*$.

The (other) contents of the current membrane (particularly the input string if it is so the case) will be found after dissolution in the immediately higher membrane. The structure of the membrane system is not any more a list as in the case of a P automaton, but a bubble tree. Also, the accepting configurations are not any more the empty ones as in P automata, but the membrane systems that consume the input string by producing a tree dependency structure.

In the following discussion, we present an example of a sentence analyzed into a dependency tree. Consider the following sentence from Dutch:

Wim Jan Marie zag leren zwemmen.

and the following P translator $M = (\Sigma, O, T, \mu, M_1, \ldots, M_n, R)$ with:

$$\Sigma = \{\texttt{Wim, Jan, Marie, zag, leren, zwemmen,.}\},$$
$$O = \left\{X, \begin{bmatrix} N \\ def \end{bmatrix}, \begin{bmatrix} V \\ V(S,C) \end{bmatrix}\right\},$$
$$T = \{\overline{\texttt{Jan}}, \overline{\texttt{Marie}}\},$$

and the set R of rules described by

1. $aw \to \overline{a}; w \mid_{\neg\overline{b}}$ with $a, b \in \Sigma$, $w \in \Sigma^+$,
2. $(\overline{a}; X)_1 \to \left(\left(\begin{bmatrix} N \\ def \end{bmatrix}\right)_1 X\right)_2$ with $a \in \{\texttt{Wim}, \texttt{Jan}\}$,
3. $\overline{\texttt{Marie}}; X \to \begin{bmatrix} N \\ def \end{bmatrix}$,
4. $\overline{a}; \begin{bmatrix} N \\ def \end{bmatrix} \to \overline{a}_{out}; \begin{bmatrix} N \\ def \end{bmatrix}$ with $a \in \{\texttt{Jan}, \texttt{Marie}\}$,
5. $\left(\overline{b}; \begin{bmatrix} N \\ def \end{bmatrix}\right)_i \to \left(\begin{bmatrix} V \\ V(S,C) \end{bmatrix}\right)_i \triangleleft_1 \left(\begin{bmatrix} N \\ def \end{bmatrix}\right)_j$ with $b \in \{\texttt{zag}, \texttt{leren}, \texttt{zwemmen}\}$,
6. $\left(\left(\overline{b}; \begin{bmatrix} V \\ V(S,C) \end{bmatrix}\right)_i\right)_j \to \left(\begin{bmatrix} V \\ V(S,C) \end{bmatrix}\right)_j \triangleleft (\overline{b})_i$ with $b \in \{\texttt{leren}, \texttt{zwemmen}\}$,
7. $\overline{.}; \begin{bmatrix} V \\ V(S,C) \end{bmatrix} \to \begin{bmatrix} V \\ V(S,C) \end{bmatrix}$

The analysis of the sentence Wim Jan Marie zag leren zwemmen. is described by the following sequence of pictures. Each transformation corresponds to an evolution step. Note that in some evolution steps several rules are applied to various membranes.

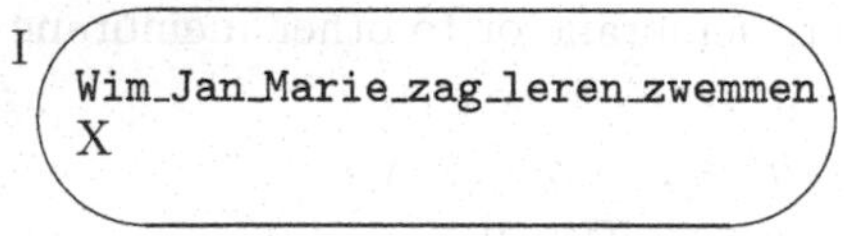

I
Wim_Jan_Marie_zag_leren_zwemmen.
X

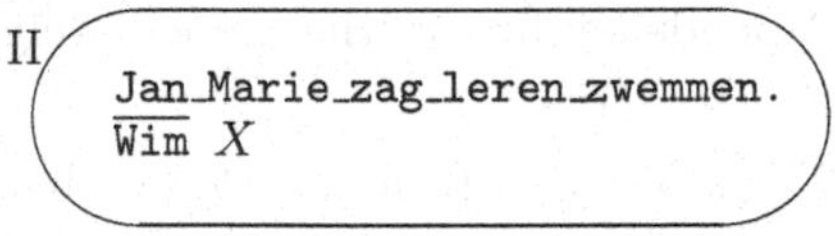

II
Jan_Marie_zag_leren_zwemmen.
$\overline{\texttt{Wim}}$ X

Configuration I represents an initial configuration of the membrane systems. In configuration II, the first item, Wim, is extracted from the input string. In configuration III, rule 2 is applied and a new embedded membrane of type N is created. The input string lies in this new membrane. In configuration IV, the second item, Jan, is extracted from the input string.

III

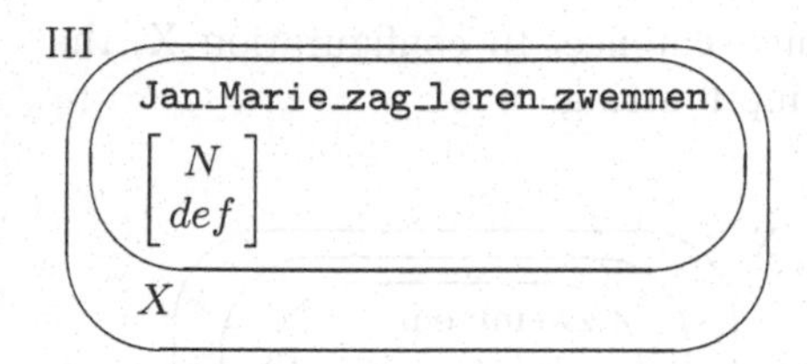

IV

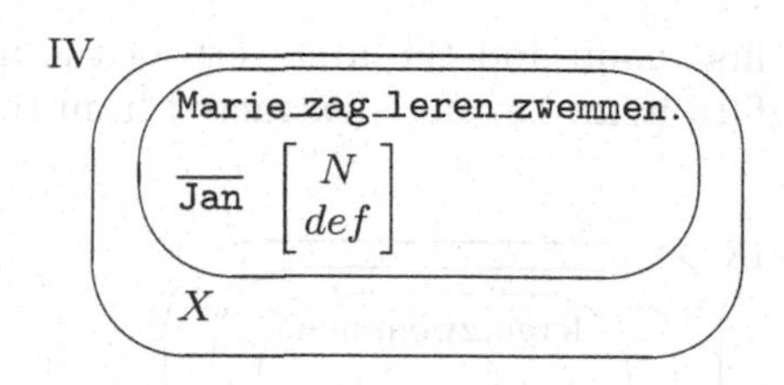

In configuration V, the symbol $\overline{\texttt{Jan}}$ is sent to the parent membrane of type X. The presence of $\overline{\texttt{Jan}}$ in this membrane produces the application of rule 2 and the creation of a new membrane, which will hold the type N. The new membrane is created immediately under the parent membrane, as shown in configuration VI. In parallel, the third item, `Marie`, is extracted from the input string.

V

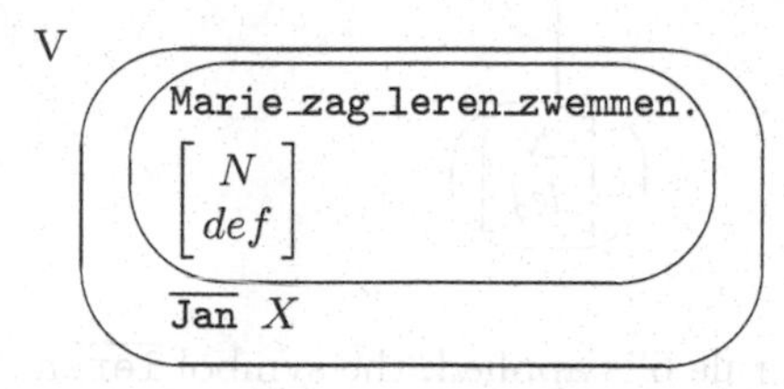

VI

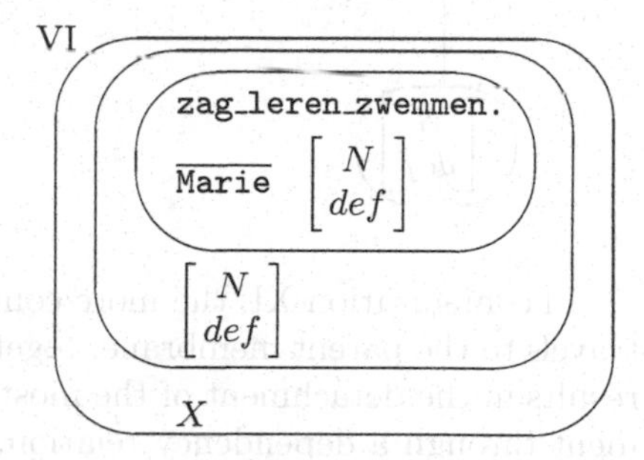

In configurations VII and VIII, the symbol $\overline{\texttt{Marie}}$ travels from the most deeply embedded membrane to the outer membrane. Also, in parallel, in configuration VIII, `zag`, the fourth item, is extracted from the input string.

VII

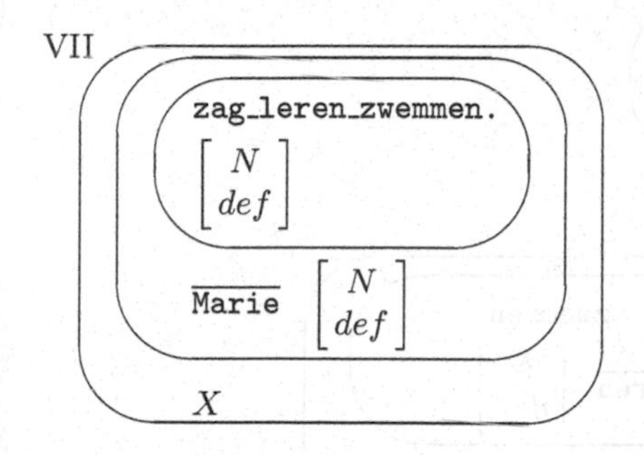

VIII

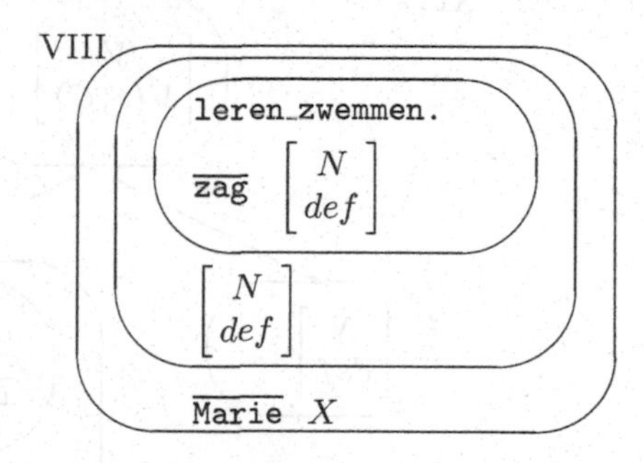

In configuration IX, the symbol $\overline{\texttt{Marie}}$ changes the type of the outer membrane from X to N. In parallel, another rule is applied to the most deeply embedded membrane: the symbol $\overline{\texttt{zag}}$ results in the creation of a new membrane of type V to which the original membrane of type N is attached through a dependency relation. This is the first dependency relation that appears in our bubble tree and it characterizes the syntactic dependency between the first noun and the first verb of the input sentence. In configuration X, the fifth item, `leren`, is extracted from the input string.

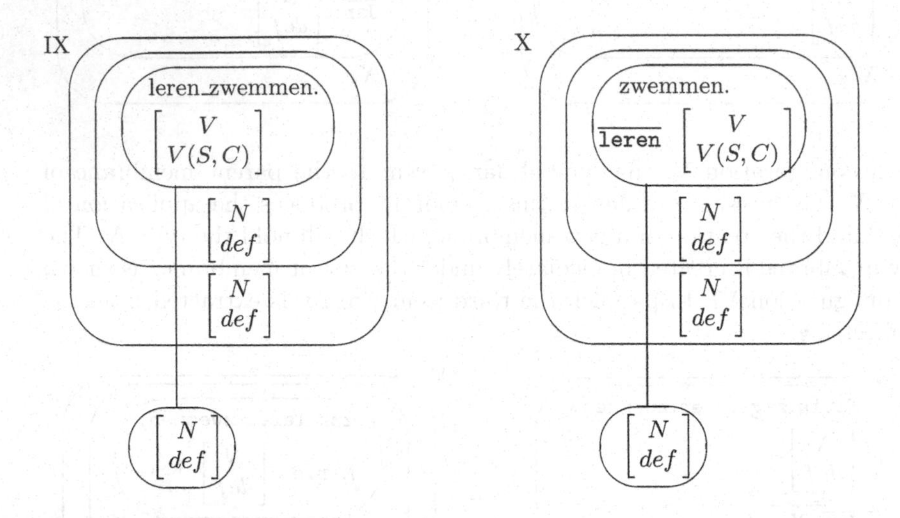

In configuration XI, the more complex rule 6 is applied: the symbol $\overline{\texttt{leren}}$ travels to the parent membrane, together with the rest of the input string and results in the detachment of the most embedded membrane and its reattachement through a dependency relation. This dependency relation characterizes the syntactic dependency between the first and the second verbs of the input sentence.

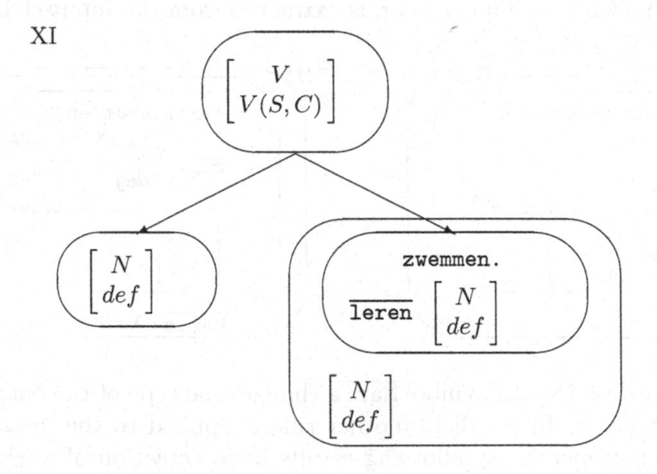

In configuration XII, the symbol $\overline{\texttt{leren}}$ present in the most deeply embedded membrane results in the creation of a new membrane of type V to which the original membrane of type N is attached through a dependency relation. This dependency relation characterizes the syntactic dependency between the second noun and the second verb of the input sentence.

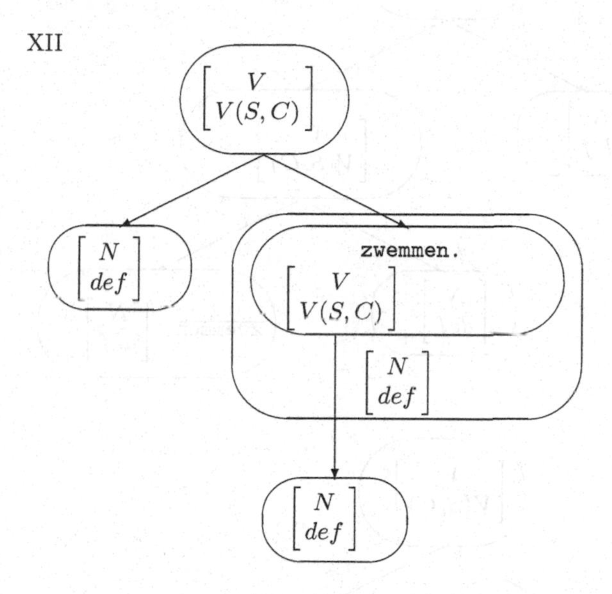

In configuration XIII, the sixth item, `zwemmen`, is extracted.

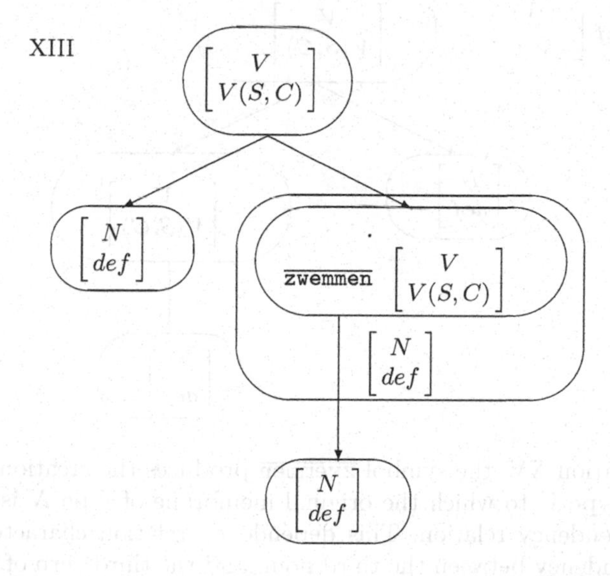

In configuration XIV, rule 6 is again applied: the symbol $\overline{\texttt{zwemmen}}$ travels to the parent membrane resulting in the reattachement of the innermost membrane through a dependency relation. The dependency relation characterizes the dependency between the second and the third verb of the input sentence.

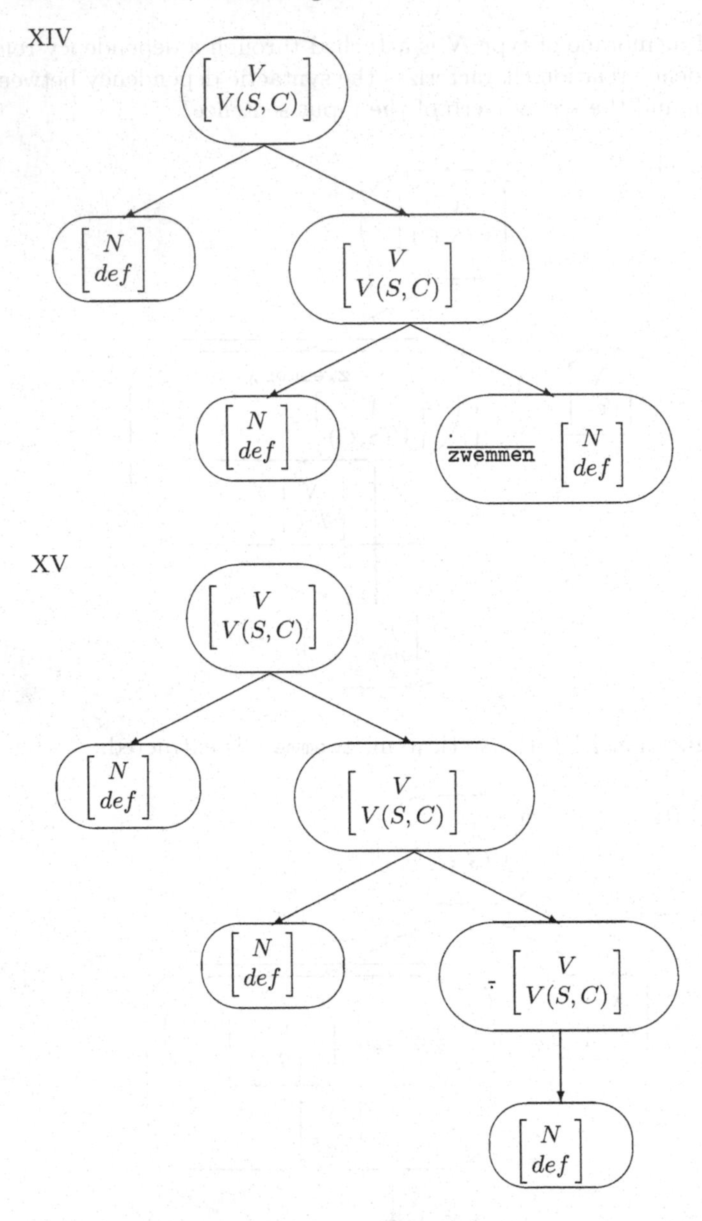

In configuration XV, the symbol $\overline{\texttt{zwemmen}}$ produces the creation of a new membrane of type V to which the original membrane of type N is attached through a dependency relation. This dependency relation characterizes the syntactic dependency between the third noun and the third verb of the input sentence.

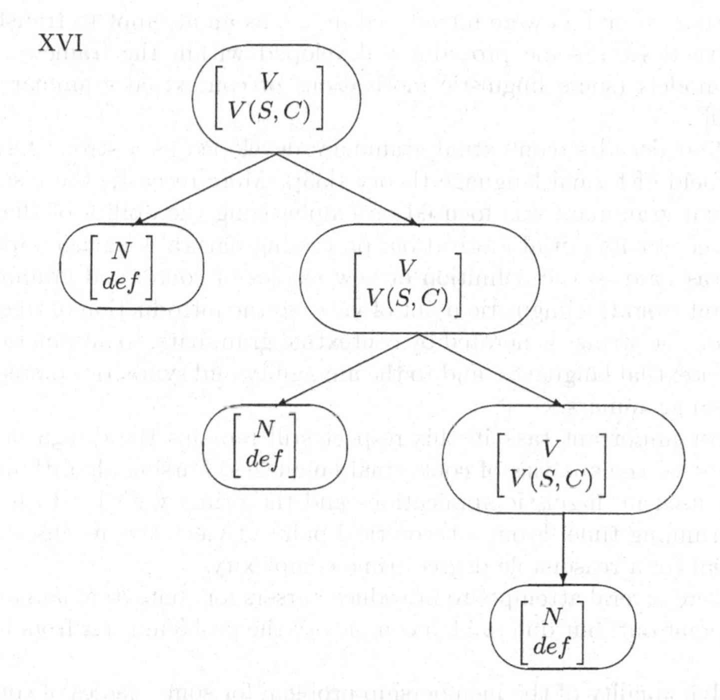

The final membrane structure, depicted as configuration XVI, represents the dependency tree corresponding to the input sentence.

The previous example illustrates some benefits of using P automata with active membranes for parsing dependency trees:

- The parallel application of the rules. Even for this deterministic analysis, by the parallel application of the rules the time complexity of the parsing is reduced from $\mathcal{O}(n^2)$ complexity, which is the obtained by a sequential automaton, to $\mathcal{O}(n)$.
- The tree representation of the intermediate generation phases. Using membrane systems with a bubble tree structure, the result of the parsing is illustrated not only as a final dependency tree, but as a chain of trees in which the constituency relations are also represented, with their dynamics that create the dependency structure.

The analysis performed in the above example is an original parsing of the classical Dutch cross-dependency linguistic construction.

3 Parsing Contextual Grammars

In the second half of this chapter, we address the problem of parsing languages generated by Marcus contextual grammars using P automata. For an introduction to Marcus contextual grammars, the reader is referred to [33].

Contextual grammars were introduced in [28] as an attempt to transform in generative devices some procedures developed within the framework of analytical models (some linguistic motivations of contextual grammars are given in [29]).

In the first decades, contextual grammars developed as a strong formalism in the field of formal language theory ([33]). More recently, the research on contextual grammars was focused on emphasizing the ability of the formalism to recover its initial motivation: processing natural language. Special attention was paid to the definition of new classes of contextual grammars, more relevant from the linguistic point of view, to the introduction of tree-like structures on the strings generated by contextual grammars, to automata recognizing contextual languages, and to the ambiguity and syntactic complexity of contextual grammars.

The most important task in this respect still remains the design of efficient parsers for the analysis of contextual languages. Parsing algorithms are extensively used in linguistic applications and the primary goal is to have a very good running time. From a theoretical point of view, this means having a polynomial (of a reasonable degree) time complexity.

There were several attempts to introduce parsers for contextual languages, as we will point out, but due to high complexity the problem is far from being solved yet.

The polynomiality of the membership problem for some classes of contextual grammars was addressed by Ilie in [24] and [25]. In [25], it was proved that the class of external contextual grammars with context-free selectors is parsable in polynomial time, while in [24] the same result was proved for a variety of internal contextual grammars with regular selectors. In the latter case, each derivation in the grammar has to verify a condition of locality with respect to the place where the new context is inserted.

The results presented in [24, 25] are proved using the construction of a nondeterministic Turing machine working in a space logarithmic in the length of the input word. Further, the polynomiality of the parsers for the two classes of contextual grammars follows from a well known complexity theorem due to Savitch that transforms the logarithmic space in polynomial time. The degree of the polynomial representing the time complexity of the parsers is not computed in either of the two papers.

Nevertheless, the two classes of contextual grammars approached in [24, 25] are rather weak and potentially inappropriate to describe all linguistic constructions. For external contextual grammars with context-free selectors, it is known that they are not able to generate all regular languages. For local internal contextual grammars with regular selectors, such a result is still unknown.

A condition similar to the local derivation is used in [16] (as "derivation that preserves the selectors") in order to define a parser for a class of internal contextual grammars with context-free selectors. The parser is proved to be quadratic in the deterministic case, but no complexity results are given for the nondeterministic case. No techniques are presented to deal with the exponential complexity of the nondeterminism. Again it is unknown how powerful this class of grammars is, or how relevant from a linguistic point of view.

Other papers in which recognizers for different classes of contextual grammars are presented are [18, 27, 31]. In [31], Mraz et al. use restarting automata for the recognition of languages generated by a type of total contextual grammar with regular selectors. The construction is extended in [18] for a class of contextual grammars generating dependency trees. In [27], go-through automata are used to recognize languages generated by shuffled contextual grammars and internal contextual grammars with finite and regular selectors. In none of these cases, are solutions provided to deal with the nondeterminism.

An interesting approach to contextual language parsing is that developed by Harbusch in a series of papers ([19, 20, 21]). However, the parser for internal contextual grammars with context-free selectors reported by Harbusch is not completely defined and was proved neither correct nor polynomial.

This section presents an approach to contextual language parsing that essentially uses the idea introduced by Harbusch in the aforementioned papers, but solves the complexity problem due to the strong parallelism exhibited by P automata.

3.1 Internal Contextual Grammars with Context-Free Choice

An *internal contextual grammar with choice in the modular presentation* is a construct $G = (V, A, (L_1, c_1), \ldots, (L_n, c_n))$, where V is an alphabet, A is a finite language over V (the set of *axioms*), while (L_i, c_i), with $L_i \subseteq V^*$ (the *selection language*) and $c_i = (u_i, v_i) \in V^* \times V^*$ (the *context*), for any $1 \leq i \leq n$ are *contextual rules*. The derivation relation in a contextual grammar is defined as:

$$x \Rightarrow y \text{ iff there is a contextual production } (L_i, (u_i, v_i)) \text{ such that } x = x_1 x_2 x_3 \text{ with } x_2 \in L_i \text{ and } y = x_1 u_i x_2 v_i x_3.$$

If $\stackrel{*}{\Rightarrow}$ is the reflexive and transitive closure of $\Rightarrow$, then $L(G) = \{x \in V^* \mid \exists w \in A, w \stackrel{*}{\Rightarrow} x\}$ denotes the language generated by G.

We say that G has F-choice, for a given family of languages F, if and only if all its selection languages S_i, $1 \leq i \leq n$, belong to F. For the rest of this section, we assume that all the selection languages are context-free; thus we will work with internal contextual grammars with context-free choice.

Any sentence generated by a contextual grammar is obtained by adjoining a sequence of contexts starting with an axiom. But the sequence of contexts, also called *control sequence*, alone does not necessarily characterize the derivation of the sentence because of the different positions in which the contexts can be inserted.

A more accurate characterization of the generation of w is given by the following sequence, which is called the *description* of the derivation:

$$\overline{\alpha}|x_{11}\overline{w_1}x_{21}\overline{t_1}x_{31}|x_{12}\overline{w_2}x_{22}\overline{t_2}x_{32}|\ldots|x_{1k}\overline{w_k}x_{2k}\overline{t_k}x_{3k} = w,$$

where $\alpha \in A$ is an axiom, (w_i, t_i) are contexts, for any $1 \leq i \leq k$, $\alpha = x_{11}x_{21}x_{31}$, and $x_{1i}\overline{w_i}x_{2i}\overline{t_i}x_{3i} = x_{1(i+1)}\overline{w_{i+1}}x_{2(i+1)}\overline{t_{i+1}}x_{3(i+1)}$, for any $1 \leq i \leq k$.

The selectors that were used at each step can be extracted from the description of the derivation. The derivation structure of the selectors is not expressed in the description of the contextual derivation, but normally the structure of the selectors is not relevant for the contextual mechanism.

Example 2. We consider a simple example of an internal contextual grammar with choice, but with a rich set of derivations associated with each generated sentence. Let $G = (V, A, (L, c))$ be an internal contextual grammar with choice such that $V = \{a\}$, $A = \{a, aa\}$, $L = \{a^n \mid n \geq 0\}$, and $c = (a, a)$. Then, $L(G) = \{a^n \mid n \geq 1\}$.

It can be proved that the number of derivations of the sentence a^n is:

$$\frac{n!}{2^{[n/2]}},$$

where $n! = 1 \cdot 2 \cdot \ldots \cdot n$ and $[x]$ is the integer part of x (the largest integer not larger than x).

3.2 Defining the Parser

Let $G = (V, A, (L_1, c_1), \ldots, (L_n, c_n))$ be an internal contextual grammar with context-free choice. For any contextual rule (L_i, c_i) with $G'_i = (N'_i, V, S'_i, P'_i)$, a context-free grammar generating L_i and $c_i = (u_i, v_i)$, we define a context-free grammar $G_i = (N_i, V, S_i, P_i)$ such that:

$$N_i = N'_i \cup \{S_i, X_i\}, \text{ where } \{S_i, X_i\} \cap N'_i = \emptyset,$$
$$P_i = P'_i \cup \{S_i \to X_i u_i S'_i v_i X_i\} \cup \{X_i \to \epsilon\} \cup \{X_i \to aX_i \mid a \in V\}.$$

Also, for the axioms contained in A, we define a context-free grammar $G_0 = (N_0, V, S_0, \{S_0 \to \alpha \mid \alpha \in A\})$, with $N_0 = \{S_0\}$.

A *dotted string* over an alphabet W is a string $u.v$ such that $uv \in W^*$. We define a marked copy of the alphabet V as $\overline{V} = \{\overline{a} \mid a \in V\}$.

A *P contextual parser* of the contextual grammar G is a construct $M = (O, \mu, M_0, M_1 \ldots, M_n, R)$, such that:

- $O = V \cup \overline{V} \cup \bigcup_{i=1}^{n} N_i \cup \{., |, <\}$ is the alphabet of objects.
- μ is the structure of $n+1$ membranes, as described by Figure 1, i.e., a skin membrane labeled by 0, which contains n membranes labeled from 1 to n;
- M_i is the content of membrane i, for all $0 \leq i \leq n$. M_i is a multiset of *string* objects over the alphabet O. Although the objects that populate the membranes have the same form, we identify three kinds of objects:
  - "description" (DES) objects. These are strings over the alphabet $V \cup \overline{V} \cup \{|\}$, which encode descriptions of (partial) derivations of sentences over V.

- "context-free production" (CFP) objects. These are strings of the form $X|\alpha$, where $X \in \bigcup_{0=1}^{n} N_i$ is a nonterminal symbol and α is a string over the alphabet $V \cup \bigcup_{0=1}^{n} N_i$. These strings encode context-free productions from the grammars G_i, $0 \le i \le n$.
- "mixed" (MIX) objects. MIX objects can be simple or double. A simple MIX object has the form $\alpha_1|\alpha_2|\ldots|\alpha_k$, where α_1 is a dotted string over $V \cup \overline{V}$, α_2 is a dotted string over $V \cup \{S\} \cup \bigcup_{i=1}^{n} N_i$, and α_i, $3 \le i \le k$, are strings over $V \cup \overline{V}$. A double MIX object has the form $r < r'$, where r and r' are simple MIX objects.

Membrane 0 contains DES objects, CFP objects induced by the grammar G_0, and simple MIX objects. Each membrane i, with $1 \le i \le n$, contains CFP objects induced by the grammar G_i and MIX objects (both simple and double). We suppose that all objects are in a sufficient number of copies so that all possible evolution rules that use these objects can be applied at any moment.

- R is the set of rules applicable to all membranes. R is defined by the following types of rules[4]:

1. $s \to s_{in}$, where s is a DES object. This rule sends a DES object in all the inner membranes i, for $1 \le i \le n$, and may occur only in the membrane 0.
2. $s;p \to r$, where $s = \alpha_1|\alpha_2|\ldots|\alpha_k$ is a DES object, $p = S_i|\beta$ is a CFP object, and $r = .\alpha_1|S_i|.\beta||\alpha_2|\ldots|\alpha_k$ is a simple MIX object. This rule combines a DES object with a CFP object into a simple MIX object and may occur in the membrane i, for any $i = \overline{0,n}$.
3. $r \to r'$, where $r = \alpha.\overline{a}\beta|\alpha_2|\ldots|\alpha_k\gamma$ and $r' = \alpha\overline{a}.\beta|\alpha_2|\ldots|\alpha_k\gamma$, are both MIX objects of the same type, either simple ($\gamma = \epsilon$) or double ($\gamma =< t$), and $a \in V$. This rule passes the dot over a marked symbol in a MIX object and may occur in any membrane where such objects reside.
4. $r \to r'$, where $r = \alpha_1.a\beta_1|S_i|\alpha_2.a\beta_2|\alpha_3|\ldots|\alpha_k$, and $r' = \alpha_1\overline{a}.\beta_1|S_i|\alpha_2a.\beta_2|\alpha_3 \overline{a}|\ldots|\alpha_k$ are simple MIX objects and $a \in V$. This rule passes the dot over an unmarked symbol in a MIX object corresponding to the contextual rule, marks the symbol as read, and registers the symbol in the description of the derivation as a marked symbol. The rule may occur in membrane i, for any $0 \le i \le n$.
5. $r \to r'$, where $r = \alpha_1.a\beta_1|X|\alpha_2.a\beta_2|\alpha_3 < t$, and $r' = \alpha_1a.\beta_1|X|\alpha_2a.\beta_2| \alpha_3a < t$ are double MIX objects, $X \notin \{S_i \mid 1 \le 0 \le n\}$, and $a \in V$. This rule passes the dot over an unmarked symbol in a MIX object corresponding to the context-free rule and registers the symbol in the description of the derivation as an unmarked symbol. The rule may occur in membrane i, for any $1 \le i \le n$.
6. $r;p \to r'$, where $r = \alpha_1|X|\alpha_2.Y\beta_2|\alpha_3|\ldots|\alpha_k\gamma$ is either a simple ($\gamma = \epsilon$) or a double ($\gamma =< t$) MIX object, $p = Y|\beta$ is a CFP object, $r' = \alpha_1|Y|.\beta| < \alpha_1|X|\alpha_2.Y\beta_2|\alpha_3|\ldots|\alpha_k$ is a double MIX object, and $X, Y \in \bigcup_{i=0}^{n} N_i$. This

[4] As above, we will use ';' to separate objects that occur on the same side of a rule.

rule combines a MIX object with a CFP object in order to create a new double MIX object. The rule may occur in membrane i, for any $1 \leq i \leq n$.

7. $r; r' \rightarrow r''$, where $r = \alpha_1|X|\alpha_2.Y\beta_2|\alpha_3|\ldots|\alpha_k\gamma$ is either a simple ($\gamma = \epsilon$) or a double ($\gamma =< t$) MIX object, $r' = \alpha'_1|Y|\alpha'_2.|\alpha'_3 < \alpha_1|X|\alpha_2.Y\beta_2|\alpha_3|\ldots|\alpha_k$ is a double MIX object, $r'' = \alpha'_1|X|\alpha_2Y.\beta_2|\alpha_3\alpha'_3|\ldots|\alpha_k\gamma$ is a MIX object of the same type with r, and $X, Y \in \bigcup_{i=0}^{n} N_i$. This rule combines two MIX objects to create a new MIX object in which the dot was moved over a nonterminal symbol. The rule may occur in membrane i, for any $1 \leq i \leq n$.
8. $r \rightarrow s_{out}$, where $r = \alpha_1.|S_i|\alpha_2.|\alpha_3|\ldots|\alpha_k$ is a parsed simple MIX object and $s = \alpha_1|\alpha_3|\ldots|\alpha_k$ is a DES object. This rule extracts a DES object from a parsed simple MIX object and sends it in the parent membrane (or to the environment, in the case of the skin membrane). The rule may occur in membrane i, for any $1 \leq i \leq n$.

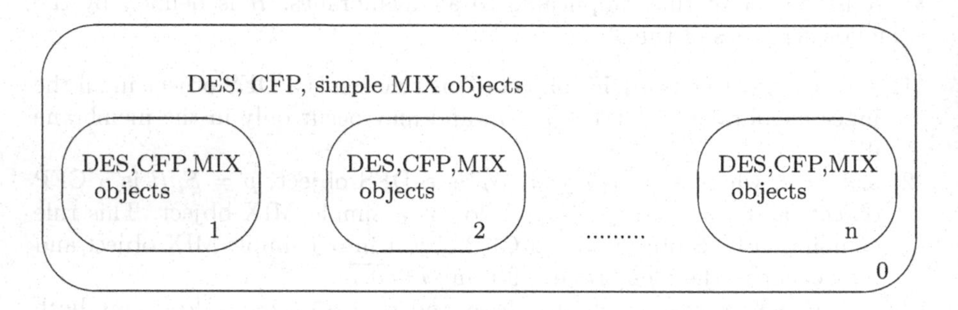

Fig. 1. The structure of a P contextual parser.

We describe now the functioning of the P contextual parser. In the preliminary configuration, M contains only CFP objects; more precisely, each membrane i contains all context-free productions from the grammar G_i in a sufficient number of copies.

If the membrane system has to recognize the sentence w, then the skin membrane 0 is populated with a sufficient number of copies of w (DES objects).

The skin membrane (membrane 0) performs two main activities:

- It tries to discover if a current input sentence is an axiom. For this purpose, the skin membrane:
  - combines a DES object with a CFP object representing an axiom (rule 2);
  - ignores the symbols already marked as processed – or overlined (rule 3);
  - tries to match the remaining input symbols (the symbols of the input sentence that were not marked as processed) with the axiom (rule 4);

- sends to the environment a DES object characterizing a successful derivation of the sentence w by the contextual grammar G whenever the match between the remaining input sentence and the axiom succeeds (rule 8).

- It sends copies of DES objects representing partial descriptions of derivations of w to all the inner membranes i, for $1 \leq i \leq n$.

Each inner membrane i, for $1 \leq i \leq n$, tries to discover in the remaining input string the context and a potential selector, corresponding to the contextual rule i. Basically, the set of rules that act in an inner membrane implements an Earley parser for context-free languages (see [10] for details). Therefore, rules 3, 4, and 5 implement the *scanner*, rule 6 implements the *predictor*, and rule 7 implements the *completer.* More specifically, an inner membrane:

- combines a DES object with a CFP object representing a contextual rule (rule 2);
- ignores the symbols already marked as processed – or overlined (rule 3);
- tries to match some of the remaining input symbols (the symbols of the input sentence that were not marked as processed) with the context represented by the contextual rule (rule 4);
- tries to match the unmarked symbols with the terminal symbols of the context-free productions that generate the selection language represented by the contextual rule (rule 5);
- applies the predictor to guess the next context-free production that contributes to the generation of the selector (rule 6);
- applies the completer to a context-free production for the generation of the selector (rule 7);
- sends to the skin membrane a DES object characterizing the partial description of a derivation of the sentence w by the contextual grammar G whenever the parsing of the remaining input sentence identifies a possible application of the corresponding contextual rule (rule 8).

3.3 An Example

We consider the internal contextual grammar $G = (V, A, (L, c))$ defined in Example 2, with $V = \{a\}$, $A = \{a, aa\}$, $L = \{a^n \mid n \geq 0\}$, and $c = (a, a)$. We construct:

- The context-free grammar $G'_1 = (N'_1, V, S'_1, P'_1)$, with $N'_1 = \{S'_1\}$, and $P'_1 = \{S'_1 \to \epsilon,\ S'_1 \to aS'_1\}$.
- The context-free grammar $G_1 = (N_1, V, S_1, P_1)$, with $N_1 = \{S_1, S'_1\}$, and $P_1 = \{S_1 \to S'_1aS'_1aS'_1, S'_1 \to \epsilon,\ S'_1 \to aS'_1\}$.
- The context-free grammar $G_0 = (N_0, V, S_0, P_0)$, with $N_0 = \{S_0\}$, and $P_0 = \{S_0 \to a,\ S_0 \to aa\}$.

In the next set of figures, we partially illustrate the parsing of the sentence *aaa*. The figures present the content of the membrane system at some moments, but, except for the first configuration, we list only the content of the skin membrane. All the objects are listed only once even if they may be present in multiple copies.

Figure 2 presents the preliminary configuration of the system, before the introduction of the input sentence.

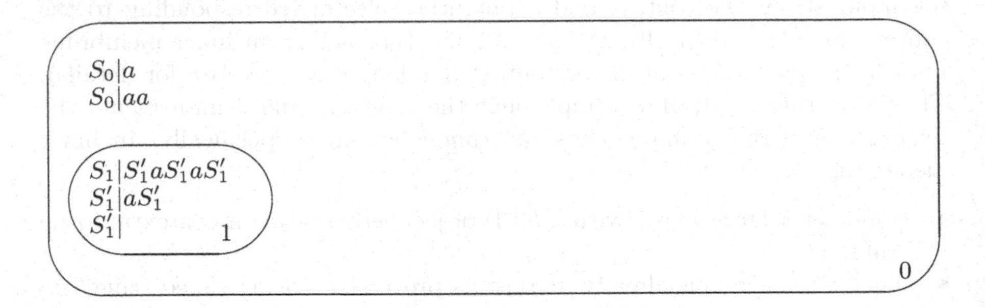

Fig. 2. Preliminary configuration of the P contextual parser.

Figure 3 presents the configuration after the first cycle of computation was performed. DES objects representing the input sentence were introduced in the skin membrane and were combined with CFP objects. On the other hand, copies the DES objects were sent to the inner membrane and as a result of the computation in the inner membrane some DES objects were expelled back in the skin membrane.

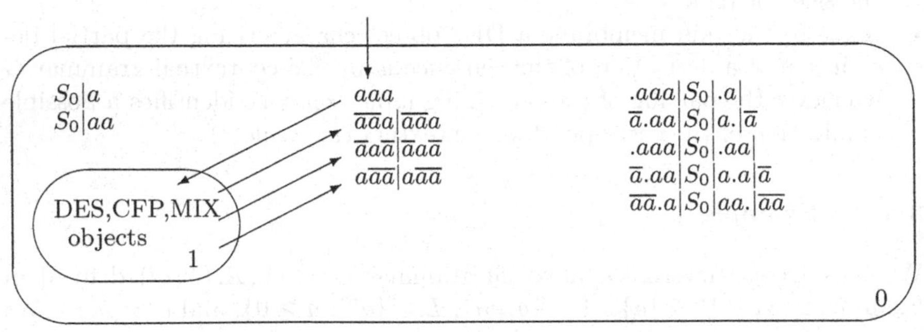

Fig. 3. Intermediate configuration of the P contextual parser.

Figure 4 presents the configuration after the second cycle of computation was performed. This configuration can be interpreted as a “final” configuration, since copies of the DES objects representing the results of the parsing have been already sent out of the skin membrane. However, there is no final configuration associated with a P contextual parser, since the computation continues until no rule can be applied, but this final moment (if such a moment ever occurs) is of no importance for the computation.

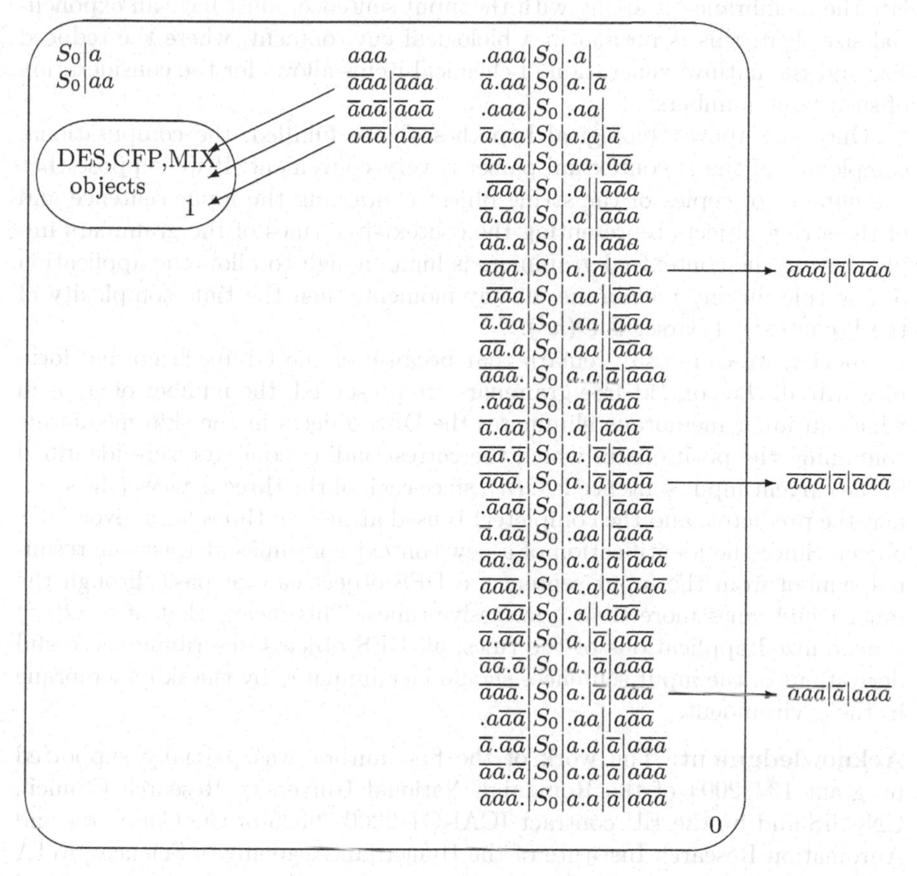

Fig. 4. "Final" configuration of the P contextual parser.

The three DES objects that are sent to the environment in Figure 4 represent the descriptions of all the possible derivations of the sentence *aaa* by the internal contextual grammar *G*.

3.4 Complexity Remarks

We finish this section by making several remarks on the complexity of the P contextual parser that we have presented.

Since we are not really interested in the structure of the selectors, i.e., of the context-free sentences that select the contexts, we may suppose that all the context-free grammars G'_i are in the Greibach normal form, i.e., they contain only productions of the form $A \to a\alpha$, with $a \in V$, $\alpha \in (N'_i)^*$, and possible a production $S_i' \rightarrow \epsilon$ provided that S_i' does not occur on the right hand side of any production of G_i.

It is clear that in the general case the multisets of string objects that populate the membranes, starting with the input sentence, must have an exponential size. But, this is normal in a biological environment, where the reduced size and the natural generation of chemical items allows for the consideration of such large numbers[5].

Once the above "biological hypotheses" are fulfilled, the computational complexity of the P contextual parser is very convenient. If we suppose that the number of copies of the string object containing the input sentence and of the string objects representing the context-free rules of the grammars implementing the contextual grammars is high enough to allow the application of any rule, in any membrane, at any moment, then the time complexity of the P contextual parser is $O(n^2)$.

Indeed, we can easily remark that because of the Greibach normal form in which all the context-free grammars are presented, the number of steps in which an inner membrane eliminates the DES objects in the skin membrane containing the positions in which the corresponding contexts were identified in the current input sentence is $O(n)$, since each of the three devices (the scanner, the predictor, and the completer) is used at most n times for a given DES object. Since the identification of a new context consumes at least one terminal symbol from the input sentence, a DES object cannot pass through the inner membranes more than n succesive times. This means that after $O(n^2)$ synchronized applications of the rules, all DES objects describing successful derivations of the input sentences should be eliminated by the skin membrane in the environment.

Acknowledgment. The work of the first author was partially supported by grant 134/2004 of the Romanian National University Research Council, CNCSIS and by the EU contract ICAI-CT-2000-70025 of the Computer and Automation Research Institute of the Hungarian Academy of Sciences, MTA SZTAKI.

References

1. A. Alhazov: Minimizing Evolution-Communication P Systems and EC P Automata. In [4], 23–31.
2. G. Bel-Enguix: Preliminaries About Some Possible Applications of P Systems in Linguistics. In [35], 74–89.
3. G. Bel-Enguix, R. Gramatovici: Parsing with Active P Automata. In [30], 31–42.

[5] In formal terms, we can consider a precomputation phase, where we can generate copies of the objects we need, by operations such as string replication and membrane division; such a phase can take linear time for generating exponentially many objects.

4. M. Cavaliere, C. Martín-Vide, Gh. Păun, eds.: *Brainstorming Week in Membrane Computing. Tarragona 2003*, Technical Report 26/03, Universitat Rovira i Virgili, Tarragona, 2003.
5. L. Cienciala, L. Ciencialova: P Automata with Priorities. In *Pre-Proceedings of the 4th Workshop on Membrane Computing*, WMC2003, Tarragona, Spain, July 2003 (A. Alhazov, C. Martín-Vide, Gh. Păun, eds.), Technical Report 28/03, URV Tarragona, 2003, 161–168.
6. E. Csuhaj-Varjú: P Automata – Models, Results, and Research Topics. In *Pre-Proceedings of the 5th Workshop on Membrane Computing*, WMC2004, Milano, Italy, 2004 (G. Mauri, Gh. Păun, C. Zandron, eds.), 1–11.
7. E. Csuhaj-Varjú, O.H. Ibarra, G. Vaszil: On the Computational Complexity of P Automata. In *Pre-Proceedings of 10th International Meeting on DNA Computing* (C. Ferretti, G. Mauri, C. Zandron, eds.), Univ. Milano-Bicocca, 2004, 97–106.
8. E. Csuhaj-Varjú, G. Vaszil: P Automata or Purely Communicating Accepting P Systems. In [35], 219–233.
9. E. Csuhaj-Varjú, G. Vaszil: New Results and Research Directions Concerning P Automata, Accepting P Systems with Communication Only. In [4], 171–179.
10. J. Earley: An Efficient Context-Free Parsing Algorithm. *Comm. of the ACM*, 13, 2 (1970), 94–102.
11. R. Freund, M. Oswald: A Short Note on Analyzing P Systems with Antiport Rules. *Bulletin of the EATCS*, 78 (October 2002), 231–236.
12. R. Freund, C. Martín-Vide, A. Obtulowicz, Gh. Păun: On Three Classes of Automata-Like P Systems. In *Proceedings of Developments in Language Theory, 7th International Conference, DLT 2003, Szeged, Hungary, July 7-11, 2003* (Z. Esik, Z. Fülöp, eds.), LNCS 2710, Springer, Berlin, 2003, 292–303.
13. R. Freund, M. Oswald, L. Staiger: ω-P Automata with Communication Rules. In [30], 203–217.
14. R. Freund, L. Kari, M. Oswald, P. Sosik: Computationally Universal P Systems without Priorities: Two Catalysts Are Sufficient. *Theoretical Computer Scienc*, 330, 2 (2005), 251–266.
15. A.V. Gladkij: On Describing the Syntactic Structure of a Sentence (in Russian). *Computational Linguistics*, 7, Budapest, 1968.
16. R. Gramatovici: An Eficient Parser for a Class of Contextual Languages. *Fundamenta Informaticae*, 33, 3 (1998), 211–238.
17. R. Gramatovici: On the Recognizing Power of Non-Expansive Iterated Go-Through Automata. *Annals of the University of Bucharest, Matematics-Informatics Series*, Vol. LII, 1(2003), 45–54.
18. R. Gramatovici, C. Martín-Vide: 1-Contextual Grammars with Sorted Dependencies. In *Proceedings of the 3rd International AMAST Workshop on Algebraic Methods in Language Processing*, AMiLP-3, Verona, Italy, 2003, 99–109.
19. K. Harbusch: A Polynomial Parser for Contextual Grammars with Linear, Regular and context-free Selectors. In *Proceedings of MOL6 – Sixth Meeting on the Mathematics of Language*, Orlando, Florida/USA, 1999, 323–335.
20. K. Harbusch: Parsing Contextual Grammars with Linear, Regular and Context-Free Selectors. In *Grammars and Automata for String Processing* (C. Martín-Vide, V. Mitrana, eds.), Taylor & Francis, London, New-York, 2003, 45–54.
21. K. Harbusch: An Efficient Online Parser for Contextual Grammars with at Most Context-Free Selectors. In *Computational Linguistics and Intelligent Text*

Processing, 4th International Conference, CICLing 2003 (A.F. Gelbukh, ed.), LNCS 2588, Springer, Berlin, 2003, 168–179.
22. S. Kahane: Bubble Trees and Syntactic Representations. In *Proceedings of Mathematics of Language (MOL5) Meeting* (T. Becker, H.-U. Krieger, eds.), DFKI, Saarbrucken, 1997, 70–76.
23. S. Kahane: A Fully Lexicalized Grammar for French Based on Meaning-Text Theory. *Computational Linguistics, CICLing 2001*, Mexico, Springer, Berlin, 2001, 18–31.
24. L. Ilie: On Computational Complexity of Contextual Languages. *Theoretical Computer Science*, 183 (1997), 33–44.
25. L. Ilie: On the Computational Complexity of Marcus External Contextual Languages. *Fundamenta Informaticae*, 30 (1997) 161–167.
26. M. Madhu, K. Krithivasan: On a Class of P Automata. *Int. J. Comput. Math.*, 80(9) (2003), 1111–1120.
27. F. Manea: Contextual Grammars and Go Through Automata. In *Proceedings of the ESSLLI 2003 Student Session* (B. ten Cate, ed.), Vienna, Austria, 2003, 159–168.
28. S. Marcus: Contextual Grammars. *Rev. Roum. Math. Pures Appl.*, 14 (1969), 1525–1534.
29. S. Marcus: Contextual Grammars and Natural Languages. Chapter 5 in vol. 2 of *The Handbook of Formal Languages*, G. Rozenberg, A. Salomaa, eds., Springer, Berlin, 1997, 215–235.
30. C. Martín-Vide, G. Mauri, Gh. Păun, G. Rozenberg, A. Salomaa, eds.: *Membrane Computing. International Workshop, WMC 2003, Tarragona, Spain, July 2003, Revised Papers.* LNCS 2933, Springer, Berlin, 2004.
31. F. Mraz, M. Platek, M. Prochazcha: Restarting Automata, Deleting and Marcus Grammars. In *Recent Topics in Mathematical and Computational Linguistics* (C. Martín-Vide, Gh. Păun, eds.), Romanian Academy Publishing House, Bucharest, 2000, 218–233.
32. M. Oswald, R. Freund: P Automata with Membrane Channels. In *Proceedings of the 8th International Symposium on Artificial Life and Robotics*, Beppu, Japan, 2003, 275–278.
33. Gh. Păun: *Marcus Contextual Grammars.* Kluwer, Dordrecht, Boston, London, 1997.
34. Gh. Păun: *Membrane Computing. An Introduction.* Springer, Berlin, 2002.
35. Gh. Păun, G. Rozenberg, A. Salomaa, C. Zandron, eds.: *Membrane Computing. International Workshop, WMC-CdeA 2002, Curtea de Argeş, Romania, August 2002, Revised Papers.* LNCS 2597, Springer, Berlin, 2003.

Chapter 15
Available Membrane Computing Software

Miguel Angel Gutiérrez-Naranjo, Mario J. Pérez-Jiménez, Agustín Riscos-Núñez

Research Group on Natural Computing
Department of Computer Science and Artificial Intelligence
University of Sevilla
Avda. Reina Mercedes s/n, 41012 Sevilla, Spain
{magutier,marper,ariscosn}@us.es

Summary. The simulation of a P system with current computers is a quite complex task. P systems are intrinsically nondeterministic computational devices and therefore their computation trees are difficult to store and handle with computers with one processor (or a bounded number of processors). Nevertheless, there exists a first generation of simulators which can be successfully used for pedagogical purposes and as assistant tools for researchers. This chapter summarizes some of these simulators, presenting the state of the art of the available software for simulating (different variants of) cell-like membrane systems.

1 Introduction

In the few years since membrane computing was initiated [39] as a new branch of natural computing, a large number of variants have been considered, concerning both the syntax and the semantics of the model.

In many of these variants, P systems are seen as devices of a *generative* nature, that is, from a given initial configuration several distinct computations may be developed (in a nondeterministic manner) and produce different outputs.

There are other approaches where P systems perform *computing tasks*. For example, if a certain number, n, is encoded somehow in the initial configuration and we consider the cardinality of the output multiset as the result of a successful computation, then we can interpret that to mean that the system *computes* a partial function from natural numbers onto sets of natural numbers.

Finally, membrane systems can also be used to deal with decision problems. In this case, special objects *yes* and *no* are included in the working alphabet, and thus the system is able to produce a boolean output (accepting or rejecting the input) in a *confluent* manner.

In all these approaches, we get the *output* of the computation from a final configuration, looking at the contents of the output membrane or, in the case of the *external output*, considering the objects that have been sent out of the system during the computation.

Unfortunately, for a machine-oriented model of computation as a P system is, it is usually a complex task to predict or to guess how a P system will behave when we are designing a cellular solution to a problem. Moreover, as there do not exist, up to now, implementations in laboratories (neither *in vitro* or *in vivo* nor in any electronical medium), it seems natural to look for software tools that can be used as assistants that are able to simulate computations of P systems.

This is the initial motivation for programming simulators. It is clear that such software tools are very useful when trying to understand how a cellular system works (both for pedagogical purposes and as an assistant tool for researchers). Another important point is that the formal verifications of the cellular solutions designed in this framework are specially hard, and having a simulator at hand allows us to quickly and easily get information about the evolution of P systems that can be used as starting point for a formal verification, maybe suggesting invariants that can be useful for the proofs. Finally, several of the existing P systems simulators were essentially used in the bio-applications of membrane computing (examples can be seen also in the first few chapters of this book).

The chapter is organized as follows. In the next section, some general considerations about the processes of the design and development of simulators of P systems are given. Section 3 is devoted to the simulators that work with transition P systems and run on sequential machines. Section 4 deals with parallel and distributed simulators (also simulating transition P systems) and Section 5 presents simulators for P systems with active membranes, including a session of one of them. The chapter ends with a section devoted to other software and some conclusions.

2 Preliminaries

The simulation of P systems with current computers is quite a complex task, but there have been several attempts in this direction in the last few years. We shall try to summarize some of them, presenting the state of the art of the available software for simulating (different variants of) cell-like membrane systems.

Generically speaking, the design and development processes for a P system simulator can be structured as follows:

2.1 Formal Definition of the Model

First of all, one has to choose which variant of membrane systems is going to be simulated, stating precisely the syntax and semantics of the model to avoid ambiguous interpretations. From a technical point of view, these models can be classified into two categories: the models of P systems where the number of membranes is bounded by the number of membranes in the initial configuration (i.e., this number does not change during the computation or decrease with the dissolution of membranes) and the models where the number of membranes can increase during the computation, via membrane creation or division.

The basic variant introduced in [39] is known as *transition P systems*. The rules in this model are of the form $u \to v$, where u is a string over the alphabet V and $v = v'$ or $v = v'\delta$, where v' is a string over

$$(V \times \{here, out\} \cup \{V \times \{in_j \,|\, 1 \le j \le n\})$$

and δ is a special symbol not in V. Besides, priority relations are considered among rules. These rules are applied in a maximally parallel way, that is, all objects which can evolve in one step must evolve (keeping in mind the priority restrictions).

This basic variant can be modified in many ways, for example, by restricting the model to non-cooperative rules or not allowing priority relations among rules, considering strings instead of multisets, or even substituting the classical tree-like membrane structure with tissue-like arrangements.

A specially relevant variant, namely, *P systems with active membranes* [41], is obtained by including rules for membrane division. Let us recall that the rules in this model are of the form

(a) $[x \to y]_h^\alpha$, for $h \in H$, $\alpha \in \{+,-,0\}$, $x \in V$, $y \in V^*$ (*Object evolution rule*). This is an internal rule, associated with a membrane labeled h and depending on the polarity α of that membrane, but not directly involving the membrane.
(b) $x[\ ]_h^{\alpha_1} \to [y]_h^{\alpha_2}$, for $h \in H, \alpha_1, \alpha_2 \in \{+,-,0\}, x, y \in V$ (*Send-in communication rule*). An object from the region immediately outside a membrane labeled h is introduced in this membrane, is possibly transformed into another object, and, simultaneously, the polarity of the membrane can be changed.
(c) $[x]_h^{\alpha_1} \to [\ ]_h^{\alpha_2} y$, for $h \in H$, $\alpha_1, \alpha_2 \in \{+,-,0\}$, $x, y \in V$ (*Send-out communication rule*). An object is sent out from a membrane labeled h to the region immediately outside, is possibly transformed into another object, and, simultaneously, the polarity of the membrane can be changed.
(d) $[x]_h^\alpha \to y$, for $h \in H$, $\alpha \in \{+,-,0\}$, $x, y \in V$ (*Dissolution rule*). A membrane labeled h is dissolved in reaction with an object. The skin is never dissolved.
(e) $[x]_h^{\alpha_1} \to [y]_h^{\alpha_2} [z]_h^{\alpha_3}$, for $h \in H, \alpha_1, \alpha_2, \alpha_3 \in \{+,-,0\}, x, y, z \in V$ (*Division rule*). An elementary membrane can be divided into two membranes with the same label but possibly different polarities. The skin cannot divide.

Note that this variant of P systems uses 2-division but no cooperation or priorities. The rules are applied according to the following principles (informal semantics of P systems with active membranes):

- The rules are used as usual in the framework of membrane computing; that is, in a maximally parallel way. In one step, each object in a membrane can be used *only by one* rule (nondeterministically chosen in case there are several possibilities), but any object which can evolve by a rule of any type should evolve.
- If a membrane is dissolved, its content (multiset and interior membranes) becomes part of the immediately external membrane (more precisely, of the closest predecessor which is not dissolved).
- All elements which are not specified in any of the operations to apply remain unchanged.
- A division rule can be applied to a membrane and, at the same time, some evolution rules can be applied to some objects inside that membrane. In this case, we can suppose that "first" the evolution rules are used, changing the objects, and "after that" the division takes place, introducing copies of the results of the evolutions in the two newly generated membranes (keeping in mind that all these processes take place in the same step of computation).
- The rules associated with label i are used for all membranes with this label. At one step, different rules can be applied to different membranes with the same label, but one membrane can be the subject of *at most* one rule of types (b) to (e).

These two models (transition and with active membranes) are widely considered in the existing simulators.

2.2 The Choice of a Programming Language

Each programming language has its own advantages and disadvantages and, up to now, there is no objective criterion to decide which is the most suitable one for simulating the evolution of a membrane system. Indeed, a large number of different languages such as Haskell, Prolog, Java, C, LISP, Visual C++, CLIPS or Scheme have been chosen by authors in the literature. The language chosen has to be able to carry out the evolution of the P system and to interact with the user in a friendly way.

It is also possible to design the interface separately from the engine that performs the evolution, using two different programming languages that are able to communicate with each other. For example, declarative languages can be appropriated for programming the inference engine, because an evolution step of a P system is nearer to a production system based upon rules than to a list of instructions to be executed in a sequential way.

2.3 A Good Way to Represent the Knowledge

The choice of a suitable data structure is a key problem in all fields of Computer Science (in particular, when dealing with the simulation of P systems). This decision is of course related to the programming language used, as specific techniques related to it have to be applied. A good representation allows a quick transition between configurations, and therefore speeds up the simulation.

There are also some designs that use two different knowledge representations, one for communicating with the user, whose goal is to implement an easy way to input the data describing the P system and to present the output in a natural way, so that the simulator can provide a better understanding of the evolution of a P system even to users who are not familiar with the programming language, and another for handling configurations and rules in order to perform the evolution steps (an efficient internal representation of P systems).

If two different representations are used, it becomes very useful to have at our disposal a parser (able to analyze syntactically the input introduced by the user) and a compiler (that translates the analyzed input into the internal grammar). Note that in some cases the internal representation is the same grammar used to input the data, so no compiler is needed.

2.4 Design of an Inference Engine to Carry out the Computation

There exists a basic difficulty intrinsic to the simulation of a P system in a conventional computer: the main power of P systems, concerning the execution of computations, is their massive parallelism. Furthermore, there are two levels of parallelism: all objects inside a membrane can be transformed simultaneously, and this process occurs in all membranes at the same time. Therefore, in one time unit (cellular step), many atomic transformations can be carried out. However, sequential conventional computers have only one processor. This means that, regardless of the programming language and the design chosen for the simulator, only one atomic transformation can be performed in each time unit (processor step).

The second feature which makes hard the design of a simulator is the intrinsic nondeterminism of P systems. If there is a large number of branches in the computation tree, the storage of the information can exceed the capacity of the computer and therefore, from a practical point of view, the simulation in this case is not feasible.

Keeping in mind these two difficulties, that is, since current computers are not able to deal with all the information related to the maximal parallelism and the nondeterminism of (relatively large) P systems, different authors have imposed several restrictions on their simulators. These constraints can be to bound the number of membranes or the cardinality of the multisets, to develop the computation tree until a prefixed depth, or to follow only one branch in the computation tree.

Usually, in the first generation of simulators, the codes are balanced between efficiency and explicitness in the following sense: the purpose of designing a simulator is to get information about the evolution of the system that is simulated and, therefore, we are interested in a software able to describe the intermediate steps and configurations. In many cases, authors have preferred to write the first versions of their simulators in code where clarity is enforced over efficiency, leaving the latter for further versions.

In spite of these limitations, the success of the first generation of simulators of P systems is beyond doubt. They are useful tools for teachers and researchers. On the one hand, one of the main utilities of this software is its use for a better understanding of membrane computing, so it is a pedagogical tool of first choice. On the other hand, it has proved to be a useful assistant tool for the design and verification of complex P systems which solve problems, relieving researchers of calculations by hand.

3 Simulators of Transition P Systems

The first simulators appeared in 2000, less than two years after Păun's foundational paper [39] was presented. All of them were focused on the basic model of transition P systems, and they pointed out one feature that has been followed by newer simulators: the balance between understandability and efficiency.

3.1 Maliţa's Simulator (2000)

In the *Workshop on Multiset Processing* which was held in Curtea de Argeş, Romania, in 2000, Mihaela Maliţa presented one of the first simulators for membrane systems [30]. It is a program written in LPA-Prolog for simulating transition P systems.

A configuration is represented as a list of labeled nested lists where objects are represented together with their multiplicities. There are also flags `x` or `y`, to distinguish between objects that can and cannot be processed.

The rules are represented by expressions explicitly mentioning four fields: the membrane (region) where the rule can be applied, the ordinal of the rule in its membrane, the initial multiset, and a multiset of products with target indicators ($here, in(j)$, or out) or, eventually, with the flag *dissolve*.

This simulator applies a restricted parallelism in the following sense: in each step, *for each* membrane, the simulator selects *only one* rule, and then this rule is applied as many times as possible.

In this simulator Maliţa pointed out one of the general ideas of the first generation of simulators: the transparency of the code in order to follow the features of membrane computing paradigm. She did not try to make programming shortcuts or tricks that might have given an optimal program. Her intention was to write a program so transparent that anyone who knows Prolog could understand how a P system works, and any person having some familiarity with membrane systems could read and understand the Prolog code of the simulator.

The simulator behaves as follows. It receives as input the configuration of a system together with a set of rules, and a parameter specifying the desired number of evolution steps. The output of the simulator shows the configurations of one branch of the computation tree until reaching the desired number of evolutions.

3.2 Suzuki and Tanaka's Simulator (2000)

In the same year, Yasuhiro Suzuki and Hiroshi Tanaka presented in [50] a program written in LISP for simulating transition P systems without membrane division and, therefore, with the number of membranes and complexity of membrane structure limited by the initial configuration, since membranes can only be dissolved.

They consider a class of P systems, which they call Artificial Cell Systems (ACSs), consisting of a membrane structure, multisets of symbols placed in its regions, and a set of rewriting rules acting in all the regions.

As we pointed out above, different authors have imposed some constraints to the design of their simulators. The specific feature from this one is to bound the size of each multiset.

This simulator has been successfully used to simulate realistic situations, such as the Brusselator model (the model of a chemical oscillation related to the Belousov-Zabotinski reaction), and in modeling and analyzing ecological systems (see [51] for more examples of applications).

3.3 Natural Computing Group from Madrid (2002)

In several papers (see [5, 6, 9, 11, 10]) some members of the Natural Computing Group of the Technical University of Madrid [54] proposed frameworks and data structures suitable for P systems, but in an abstract rather than practical context.

In [7], based on previous theoretical formalizations, they present a simulator for transition P systems written in Haskell. They consider two *layers* in a P system: on the one hand, there is a static structure, composed of the membranes and objects of the system; on the other hand, there is a dynamic structure, which refers to the set of rules of the system.

They present several specific modules (*Abstract Data Types*) to transfer to the software the concepts of multiset, rule, region, membrane, etc.

The Haskell interpreter chosen has been Hugs98 for Microsoft Windows[1]. The source code can be downloaded from the P system Web page [55].

[1] The interpreter for several operating systems can be downloaded from `http://cvs.haskell.org/Hugs/pages/downloading.htm`.

The simulator behaves as follows. It receives as input a file encoding a system (configuration and rules in each region) and produces another file encoding a system (configuration and rules in each region), obtained by the application of *one* step of the computation (via a maximal multiset of rules *randomly* selected).

3.4 Balbontín et al. Simulator (2002)

Two years after the *Workshop on Multiset Processing*, also in Curtea de Argeş, D. Balbontín-Noval, M.J. Pérez-Jiménez, and F. Sancho-Caparrini presented during the *Workshop on Membrane Computing 2002* a simulator [12] for transition P systems written in MzScheme. A library of procedures was developed for working in two stages:

(1) first a parser analyzes the input and checks if it is syntactically correct, and if so, a compiler rewrites the input introduced by the user into an internal grammar;
(2) then, the simulation is carried out up to a prefixed level (number of evolution steps) in all branches of the computation tree.

The simulator behaves as follows. It receives as input the initial configuration of a system including the set of rules and a parameter specifying the desired number of evolution steps, and it outputs the computation tree of the P system, step by step, until reaching the desired number of evolutions.

The inference engine that actually implements the evolution steps follows the formalization from [47]. That is, first of all it checks which are the applicable rules, according to the priority relations; then it calculates the applicability vectors for each membrane (that is, the multisets of applicable rules satisfying the maximal parallelism condition); and finally it combines such vectors (one vector for each region) to get the applicability matrices for the system. The simulator uses this procedure to follow *all* the possible nondeterministic choices of the computation. The expansion of the computation tree is made in a progressive way, level by level (*breadth expansion*), to a prefixed depth.

3.5 Ardelean and Cavaliere's Simulator (2003)

A very interesting tool for modeling biological processes was presented in [4]. It can be thought as a transition P system simulator because the number of membranes does not change during the computation. More precisely, the software deals with a special variant of P systems: the allowed rules are both rewriting and symport/antiport rules. This variant of P systems has been proposed in [18] and its motivations are rooted in the idea to separate the evolutive mechanism of the cell from the communicative mechanism.

The authors try to bridge the mathematical model and biological reality, indicating how one can use the P system framework to model very important processes that happen in cells.

The simulator takes as input the rules of a system, its membrane structure (which can be any graph, not only a tree) and the multisets of objects associated with the regions. The software assigns to each rule two kinds of probabilities: *probability of being available* and *probability of winning a conflict.* The simulation takes place in the following way: at each step, the simulator decides which are the available rules in that step, and this decision is taken using the above mentioned probability. Then, the available rules are applied in the maximally parallel mode by using the *weak priority* approach. This can be seen as a competition of the rules for each single occurrence of the objects.

Several biological processes have been simulated illustrating the usefulness of this software (see [19] and Chapter 4 of the present book).

3.6 Nepomuceno's Simulator (2004)

In [37], we can find a software application, *SimCM*, written in Java. This tool is a friendly application which allows us to follow the evolution of a transition P system in a visual way. Essentially, we handle transition P systems by means of three basic operations: **Create** an initial membrane system (the simulator includes a debug mode in order to avoid user errors); **Load** and **Save** previously defined membrane systems; and **Carry out** a simulation of the P system evolution. This simulation can be made in three different ways: showing the computation tree to a given maximal depth, level by level, or guided.

The main screen is divided into four basic panels:

- **Computation tree**: this panel shows the tree of configurations after the simulation is finished or during its development.
- **Current cell**: initially, this panel contains a sketch in tree form of the membrane system to be studied (the program represents the membrane structures of P systems as trees). Once the simulation is finished or when it is in development, this panel will represent the state of the membrane system according to the configuration chosen by the user in the computation tree panel. In order to select a configuration, it suffices to simply click on the chosen node in the computation tree panel.
- **Rules**: in this panel the rules associated with each membrane are listed.
- **Applicable rules**: this panel shows the applicability multiset associated with the configuration selected by the user in the computation tree.

The simulator can be downloaded from the Web page of the Research Group on Natural Computing at the University of Seville [57].

4 Parallel and Distributed Simulators

As we already said in Section 2, one of the main difficulties in the simulation of P systems in current computers is that the computational power of these devices lies in their intrinsic massive parallelism. Several authors have implemented the first versions of simulators based on parallel and distributed architectures, which is close to the membrane computing paradigm.

4.1 Ciobanu and Wenyuan's Simulator (2003)

G. Ciobanu and G. Wenyuan presented in [24] a parallel implementation of transition P systems[2]. The implementation was designed for a cluster of computers. It is written in C++ and makes use of *Message Passing Interface* (MPI) as its communication mechanism. MPI is a standard library developed for writing portable message passing applications, and it is implemented both on shared memory and on distributed memory parallel computers.

The program was implemented and tested on a Linux cluster at the National University of Singapore. The cluster consisted of 64 dual processor nodes. The implementation is object-oriented and involves three components:

- class *Membrane*, which describes the attributes and behavior of a membrane,
- class *Rule*, which stores information about a particular rule, and
- *Main method*, which acts as central controller.

The rules are implemented as threads. At the initialization phase, one thread is created for each rule. Rule applications are performed in terms of rounds. To synchronize each thread (rule) within the system, two barriers implemented as mutexes[3] are associated with the thread. At the beginning of each round, the barrier that the rule thread is waiting for is released by the primary controlling thread. After the rule application is done, the thread waits for the second barrier, and the primary thread locks the first barrier.

Since each rule is modeled as a separate thread, it should have the ability to decide its own applicability in a particular round. Generally speaking, a rule can run when no other rule with higher priority is running, and the resources required are available. When more than one rule can be applied in the same conditions, the simulator randomly picks one among the candidates.

With respect to synchronization and communication, the main communication for each membrane is done by sending and receiving messages to and from its parent and children at the end of every round. With respect to termination, when the system is no longer active there is no rule applicable in any membrane. When this happens, the designated output membrane prints out the result and the whole system halts.

In order to detect if the P system halts, each membrane must inform the other membranes about its inactivity. It can do so by sending messages to other membranes (these membranes can be *normal* or *inactive*) and by using a termination detection algorithm (see [8]).

[2] A preliminary version of this paper can be found in [23].

[3] A mutex object is a synchronization object whose state is set to signaled when it is not owned by any thread, and non-signaled when it is owned.

4.2 Syropoulos et al. Simulator (2003)

Syropoulos, Mamatas, Allilomes, and Sotiriades presented in [52] a purely distributive simulation of P systems. It is implemented using Java's *Remote Methods Invocation* to connect a number of computers that interchange data. As the authors pointed out, the idea of designing a distributed simulator for a network of computers, instead of doing so for a cluster architecture, avoids the problem of limited hardware compatibility. The class of P systems that the simulator can accept is a subset of the $NOP_2(coo, tar)$ family of systems, which have the computational power of Turing machines. This variant restricts the number of membranes to two, allows cooperation, and the symbol *tar* indicates that the communication rules use target indicators of the type in_j.

Initially, a copy of the simulator is installed on a number of different computers. Randomly, we choose a computer and assign to it the role of the external compartment, while the others play the role of the internal compartments. Upon starting, a *Membrane* object is ready to participate in the network on each computer. Threads are an essential aspect of the implementation. In particular, each membrane class runs in its own thread, which, in turn, operates on a different machine. When the system starts, the computer that plays the role of the external compartment reads the specification of a P system from an external file and stores the data.

An artificial parameter is introduced in order to prevent the system from going into an infinite loop. When the simulator has successfully parsed the P system's specification, the main computer decides whether there are enough resources or not. If the available resources match the requirements set by the description of the P system, then the simulator starts the computation. Otherwise, it aborts the execution. In order to be able to make this decision, the simulator has been designed in such a way that all membrane objects send multicast UDP packets to a well known multicast address. Each packet contains the IP address of each sender, and multicast packets are received by all objects participating in the network. Thus, each computer knows which computers are *alive* at any time. In this way, the main computer has all the necessary information to decide whether there are sufficient resources to start the computation. A universal clock is owned by the object that has the role of the external compartment. This object signals each clock tick by the time the previous macrostep is completed (i.e., when, for a given macrostep, all remote objects have finished their computation).

The source code of the system, a `jar` file, which can be used to install the simulator, as well as the documentation of the simulator, can be downloaded from [53].

5 Simulators of P Systems with Active Membranes

The third group of software tools are devoted to simulating P systems with active membranes. Polynomial time solutions to **NP**-complete problems via P systems can be reached by trading time with space. This is done by producing (via membrane division) an exponential number of membranes that can work in parallel.

The simulation of these P systems has to deal with the potential growth of the membrane structure and adapt dynamically the topology of the configurations depending on whether some membranes are added or deleted. Due to the obvious limitations of computational resources, the P systems which can be simulated are of a small size.

5.1 Ciobanu and Paraschiv's Simulator (2002)

In [21] G. Ciobanu and D. Paraschiv presented a software application which provides a graphical simulation for two variants of P systems: for the initial version of catalytic hierarchical cell systems and for P systems with active membranes (see [40, 41]). Its main functions are:

- interactive definition of a membrane system,
- visualization of a defined membrane system,
- a graphical representation of the computation and final result, and
- save and (re)load of a defined membrane system.

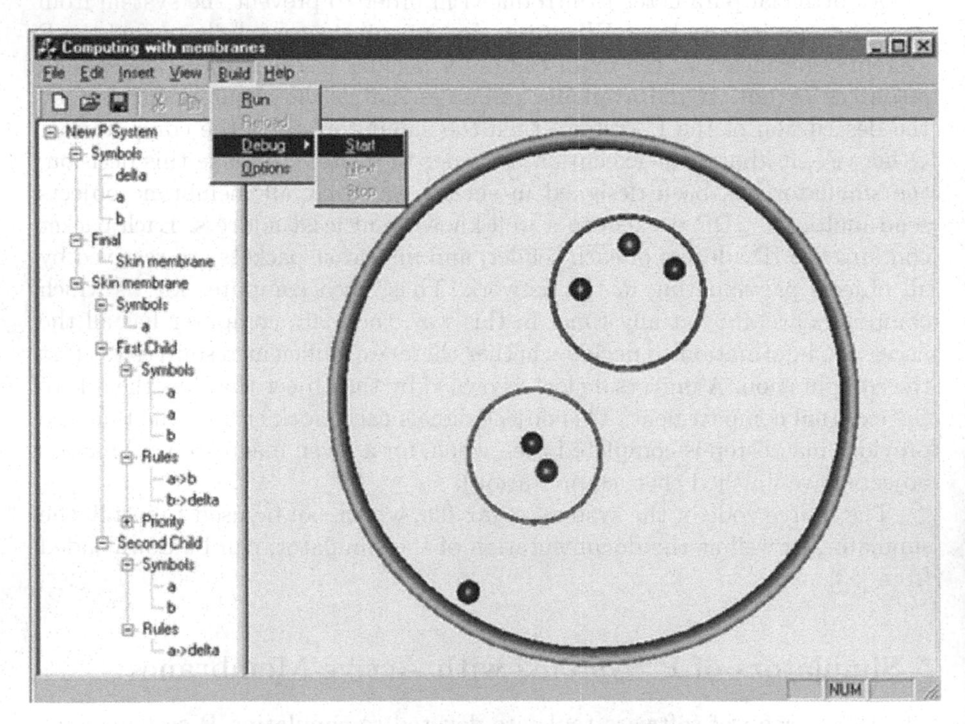

Fig. 1. Main screen of Ciobanu and Paraschiv's simulator.

The application was implemented in Microsoft Visual C++ using MFC classes. For a scalable graphical representation, Microsoft DirectX technology was used. One of the main features of this technology is that the size of each component of the graphical representation is adjusted according to the number of membranes of the system.

The system is presented to the user with a graphical interface where the main screen is divided into two windows: The left window gives a tree representation of the membrane system including objects and membranes. The right window provides a graphical representation of the membrane system given by Venn-like diagrams. A menu allows the specification of a membrane system for adding new objects, membranes, rules, and priorities. By means of the functions *Start*, *Next*, and *Stop*, the user can observe the system evolution step by step.

The following two simulators have been developed as assistant tools for the design and formal verification of cellular solutions to **NP**-complete problems via recognizer P systems [45, 49]. In this case, as we work with confluent P systems, it suffices to follow one branch of the computation tree.

5.2 Pérez and Romero's Simulator (2004)

In this case ([45]) the simulator, written in CLIPS, deals with P systems with active membranes. The design is based on representing P systems through the *production systems* programming paradigm. Generally speaking, a production system can be structured into three components:

- *Working Memory*: A set of "facts" consisting of positive literals defining what is known to be true about the world.
- *Rules*: An unordered set of user-defined "if–then" rules of the form:

$$\textbf{if } P_1 \wedge ... \wedge P_m \textbf{ then } Action_1, ..., Action_n$$

  where the P_is are facts that determine the conditions when the rule is applicable. Each *Action* adds or deletes a fact from the Working Memory.
- *Inference Engine*: Procedure for inferring changes (additions and deletions) to Working Memory.

Configurations are represented as a set of unordered facts using the following *template*:

```
(deftemplate membrane (slot number)
   (slot father)      (multislot children)
   (slot evolved)     (slot label)
   (slot polarity-0) (slot polarity-1)
   (slot multiset-0) (slot multiset-1))
```

The slots *number*, *father* and *children* are used to represent the membrane structure. The slots *label*, *polarity-0*, *polarity-1*, *multiset-0*, and *multiset-1* represent respectively the label, polarity of the membrane (current and next) and multiset (current and next) associated with each membrane.

The simulator transforms the rules of the P system into CLIPS rules as follows:

(a) $[x \rightarrow y]_h^\alpha$

```
(defrule evolution
  ?membrane <- (membrane (label h) (polarity-0 α )
                         (multiset-0 $?b0 , x , $?f0)
                         (multiset-1 $?b1 , x , $?f1))
 =>
  (modify ?membrane (multiset-0 $?b0 $?f0)
                    (multiset-1 $?b1 , y , $?f1)))
```

(b) $x[\ ]_h^{\alpha_1} \rightarrow [y]_h^{\alpha_2}$

```
(defrule send-in
  ?child <- (membrane (father ?f) (evolved 0)
                      (label h) (polarity-0 α1 )
                      (multiset-1 $?content))
  ?father <- (membrane (number ?f) (multiset-0 $?b0 , x , $?f0)
                       (multiset-1 $?b1 , x , $?f1))
 =>
  (modify ?child (evolved 1) (polarity-1 α2)
                 (multiset-1 $?content , y ,))
  (modify ?father (multiset-0 $?b0 $?f0) (multiset-1 $?b1 $?f1)))
```

(c) $[x]_h^{\alpha_1} \rightarrow [\ ]_h^{\alpha_2} y$

```
(defrule send-out
  ?child <- (membrane (father ?f) (evolved 0)
                      (label h) (polarity-0 α1 )
                      (multiset-0 $?b0 , x , $?f0)
                      (multiset-1 $?b1 , x , $?f1))
  ?father <- (membrane (number ?f) (multiset-1 $?content))
 =>
  (modify ?child (evolved 1) (polarity-1 α2)
                 (multiset-0 $?b0 $?f0) (multiset-1 $?b1 $?f1))
  (modify ?father (multiset-1 $?content , y ,)))
```

(d) $[x]_h^\alpha \rightarrow y$

```
(defrule dissolve
  ?child <- (membrane (number ?n) (evolved 0)
                      (father ?f) (children $?ch)
                      (label h) (polarity-0 α1 )
                      (multiset-0 $?b0 , x , $?f0)
```

```
                               (multiset-1 $?b1 , x , $?f1))
      ?father <- (membrane (number ?f) (children $?ch0 ?n $?ch1)
                               (multiset-1 $?content))
    =>
      (retract ?child)
      (assert (restructure (father ?f) (children $?ch)))
      (modify ?father (children $?ch0 $?ch $?ch1)
                               (multiset-1 $?content $?b1 , y , $?f1 )))
```

(e) $[x]_h^{\alpha_1} \rightarrow [y]_h^{\alpha_2}[z]_h^{\alpha_3}$

```
    (defrule division
      ?child <- (membrane (number ?n) (evolved 0)
                               (father ?f) (label h) (polarity-0 α1 )
                               (multiset-0 $?b0 , x , $?f0)
                               (multiset-1 $?b1 , x , $?f1))
      ?father <- (membrane (number ?f) (children $?ch0 ?n $?ch1)
    =>
      (retract ?child)
      (assert (membrane (number ?*number*) (evolved 1) (father ?f)
                             (label h) (polarity-1 α2)
                             (multiset-0 $?b0 $?f0)
                             (multiset-1 $?b1 , y , $?f1)))
              (membrane (number (+ ?*number* 1)) (evolved 1)
                             (father ?f) (label h) (polarity-1 α2)
                             (multiset-0 $?b0 $?f0)
                             (multiset-1 $?b1 , z , $?f1)))
      (modify ?father (children $?ch0 ?*number* (+ ?*number* 1 ) $?ch1))
      (bind ?*number* (+ ?*number* 2)))
```

In order to carry out one step of the computation, the simulator performs first an initialization step where the rules are translated into CLIPS rules and the application of the rules is then simulated.

The simulator behaves as follows. It receives as input the initial configuration of a system and a set of rules, and it simulates only one branch of the computation tree, and several options are provided to choose the degree of verbosity of the output: show all the configurations of the evolution, or show only a concrete step, or run and show only the final answer (external output) of the system. Besides, the user can also decide if the rules applied are displayed for each step or not.

This simulator has been proven useful for designing and debugging families of P systems solving strongly **NP**-complete problems like BINPACKING and the Common Algorithmic Decision Problem (CADP). Currently, variants of this simulator which provide symport-antiport rules, catalysts, and a Java interface are being developed.

5.3 Cordón-Franco et al. Simulator (2004)

In [27] and [26], a new simulator written in Prolog was presented. It is pretty different from Maliţa's simulator ([30]) in implementation, and it works with P systems with active membranes [41] instead of with transition P systems.

This simulator has been successfully used as assistant in the design of P systems with active membranes to solve **NP**-complete problems, for instance, SAT, VALIDITY, Subset Sum, Knapsack, and Partition problems (see [25, 26, 27, 28, 46, 49]).

Similarly to other programs, the simulator stores and handles the information related to the P system and tries to show the process to the user in a friendly way. One of the main features of this simulator is that both tasks (computation and relation with the user) are made in the same language. For that, one exploits the ability of Prolog to define ad hoc symbols in order to imitate natural language.

In order to give a formal representation in Prolog of the basic structures of P systems with active membranes using 2-division, the following representation is considered. A given membrane structure is expressed by means of a labeled tree, where:

1. $< >$ is the *position* to denote the root of the tree and it is associated with the skin;
2. if $< i_1, \ldots, i_n >$ is the position of a membrane h, then $< i, i_1, \ldots, i_n >$ denotes the *position* of the ith membrane placed inside membrane h.

Let us remember that to give a configuration of a P system with active membranes consists of making explicit the membrane structure and the contents of all membranes.

In this model each configuration is represented as a set of one-literal clauses, each of them representing a membrane. Hence, in this representation each clause shows label, position, polarity, multiset of objects, and current step of the computation, as well as the P system the membrane belongs to. In this way, the set of clauses gives information about the contents of the membranes and the membrane structure (by means of the position of each one).

More precisely, to denote that in the step t of its evolution the P system P has a membrane at position $[pos]$ with label h, polarity α, and m as multiset, we write

$$P \texttt{ :: } h \texttt{ ec } \alpha \texttt{ at } [pos] \texttt{ with } m \texttt{ at\_time } t$$

Note that we use the user-friendly representation of a Prolog literal instead of the functional representation.

By means of some new function symbols, the rules are also represented as literals, in the following way:

(a) $[x \to y]_h^\alpha$

$$P \texttt{ rule } x \texttt{ evolves\_to } [y] \texttt{ in } h \texttt{ ec } \alpha$$

(b) $x[\ ]_h^{\alpha_1} \rightarrow [y]_h^{\alpha_2}$

P `rule` x `out_of` h `ec` α_1 `sends_in` y `of` h `ec` α_2

(c) $[x]_h^{\alpha_1} \rightarrow [\ ]_h^{\alpha_2} y$

P `rule` x `inside_of` h `ec` α_1 `sends_out` y `of` h `ec` α_2

(d) $[x]_h^{\alpha} \rightarrow y$

P `rule` x `inside_of` h `ec` α `dissolves_and_sends_out` y

(e) $[x]_h^{\alpha_1} \rightarrow [y]_h^{\alpha_2}[z]_h^{\alpha_3}$

P `rule` x `inside_of` h `ec` α_1 `divides_into` y `inside_of` h `ec` α_2 `and` z `inside_of` h `ec` α_3

The simulator behaves as follows. The input of the program is the initial configuration of the system (which is represented as a set of literals with predicate symbol `::`, all of them `at_time 0`) and a set of rules. The Prolog algorithm to carry out the evolution of a P system works in a natural way, as explained below. It is worth mentioning that only one branch of the computation tree is simulated, and therefore the result of the simulation is faithful only in the cases of *confluent* membrane systems (that is, systems that on the same input produce the same output).

- **Step 1: Initialization**. At the beginning of each computation step, all the membranes are set to *applicable* and their objects are split into two multisets: one **usable** multiset, containing all the objects of the initial membrane, and one **used** multiset which is empty.
- **Step 2: Transition**. If there exists an applicable membrane satisfying the condition of a rule, then the rule is applied in the following way:
  - **(a) step:** At this stage, only rules of type (a) are checked. The object which triggers the rule is removed from **usable** and the resulting multiset by the application of the rule is added to **used** to prevent the use of the same object by two different rules at the same step. After that, the membrane remains *applicable*, and new evolution rules can be applied. This stage ends when no more rules of type (a) can be applied.
  - **Non-(a) step**: At this stage, only one rule of the other types (not (a)) can be applied, and Prolog selects one from the existing possibilities (remember that this simulation works only with *confluent* P systems). The action depends on the kind of rule to apply:
    - **Send out rule**: The element which triggers the rule is removed from the **usable** multiset and the new one is added to the **used** multiset of the parent membrane. The membrane changes to *not applicable* mode. If the element is sent out of the skin, it is marked with the property **outside**.
    - **Send in rule**: This is the converse of the previous action. The element which triggers the rule is removed from **usable** in the parent membrane and the new one is added to the **used** multiset. The membrane changes to *not applicable* mode.

  - **Dissolution rule**: The element which triggers the rule is removed from **usable** and the new element obtained, together with the rest of the elements of the membrane, is added to the **used** multiset of the parent membrane. The dissolved membrane is removed, and the membranes inside it become children of the parent, and their positions are arranged to be correct.
  - **Division rule**: The element which triggers the rule is removed from **usable** and the division creates two new membranes in *not applicable* mode. One of them keeps the original position and the second one gets a position which has not been occupied by any membrane.
- **End**: When no more rules can be applied to membranes in *applicable* mode, a new configuration (with `at_time` incremented by 1) is stored. At this time no membrane has *applicable* or *not applicable* state. These modes have validity only during the evolution. At this point, the P system is ready for a new evolution step.

- **Step 3: End of computation**. If there are no rules to be applied, then the evolution finishes (the P system *halts*).

As we said before, the information provided by the simulator can be helpful in the processes of designing cellular solutions to some problems. Furthermore, this simulator has been also used in [29] as a tool to study the descriptive complexity of P systems. The complexity of a computation of a P system can be described for example by a table showing the number of times that the rules of the system are applied at each step. Such tables are known as *Sevilla carpets*, and were presented in [22].

The data needed to graphically represent the Sevilla carpet associated with a P system can be extracted from the output produced when simulating the system. To illustrate this, we present in Fig. 2 the carpet associated with a P system solving the Partition problem (the parameters of the instance are $N = 5, w_1 = 5, w_2 = 4, w_3 = 1, w_4 = 8$, and $w_5 = 6$). For further details we refer the readers to [29] or [49].

The user can choose from several possible outputs. We illustrate them by showing a session as an example.

In order to launch the simulator, we open a Prolog interpreter and load the main file `simulator.pl`. Suppose that we now want to solve an instance of the Subset Sum problem. The simulator includes a tool that is able, given an instance of the problem, to automatically generate the files containing the set of rules and the initial configuration of a recognizer P system with active membranes which solves the instance.

```
?- generate(subs).
% subs_file.pl compiled 0.00 sec, 6,876 bytes

Welcome to the program to generate the used files
for a P system to solve the SUBSET SUM problem
```

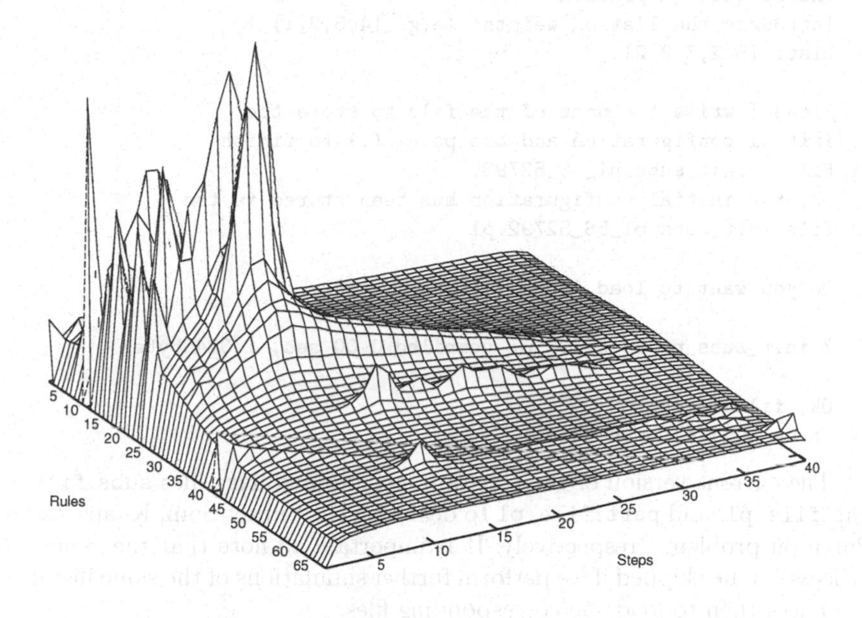

Fig. 2. Sevilla Carpet of a solution of the Partition problem through P systems with active membranes.

```
The Subset Sum problem is the following one: Given a finite
set A of N elements, a weight function w defined over it, and
a constant K determine whether or not there exists a subset of
A such that its weight is exactly k.
Please, introduce the name of the P system:
and one point (.) to end (e.g.: p1.)
Name: p1.

Next, introduce the parameters:
and one point (.) to end (e.g.: 5.)
Value of N = 5.
Value of K = 8.

The set of rules has been successfully generated
and stored in the file rules_subs_p1_5_8.pl
Do you want to load it now? (y./n.): y.

% rules_subs_p1_5_8.pl compiled 0.01 sec, 13,500 bytes

Ok, file loaded.
Next, we are going to build the initial configuration.
```

```
We need the specific INPUT for a concrete instance of
the SUBSET SUM problem
Introduce the list of weigths (e.g. [4,5,2,1].)
List: [5,2,7,9,2].

Please, write the name of the file to store the
initial configuration and one point (.) to finish
File : init_subs_p1_58_52792.
Ok, the initial configuration has been stored in the
file init_subs_p1_58_52792.pl

Do you want to load it now? (y./n.): y.

% init_subs_p1_58_52792.pl compiled 0.00 sec, 1,024 bytes

Ok, file loaded.
 Have a nice computation!
```

The current version of the program includes auxiliary files `subs_file.pl`, `knp_file.pl`, and `part_file.pl` to deal with the Subset Sum, Knapsack, and Partition problems, respectively. It is important to note that the generation process can be skipped if we perform further simulations of the same instances; it suffices then to load the corresponding files.

The user can select different types of outputs for the simulation. One can ask for a configuration in a concrete step, or to let the simulator run internally, getting only information about the number of cellular steps of the computation and the output of the P system.

```
?- go(p1).
The P system p1 stops at step 32 and returns NO
```

One can also ask the simulator to show the configurations step by step until a given point in the computation. Given a configuration of a P system `p1` at time `t`, the Prolog instruction that simulates one computation step is `evolve(p1,t)`.

```
?- evolve(p1,0).

p1 :: s ec 0 at [] with [z1-1] at_time 1
p1 :: e ec -1 at [1] with [a_-8, q-1, x1-5, x2-2, x3-7, x4-9, x5-2]
    at_time 1
p1 :: e ec 1 at [2] with [a_-8, e0-1, x1-5, x2-2, x3-7, x4-9, x5-2]
    at_time 1

Used rules in the step 0:
  * The rule 1 has been used only once
  * The rule 51 has been used only once
```

Note that the output displayed includes not only the next configuration but also information related to the rules used. Besides, the simulator informs us if any objects have been sent out to the environment.

```
?- evolve(p1,31).

p1 :: s ec 0 at [] with [# -508] at_time 32
p1 :: e ec -1 at [1] with [a-8, x1-5, x2-2, x3-7, x4-9, x5-2]
                                                      at_time 32
p1 :: e ec -1 at [2] with [a-3, x1-2, x2-7, x3-9, x4-2] at_time 32
p1 :: e ec -1 at [3] with [a-6, x1-7, x2-9, x3-2] at_time 32
p1 :: e ec -1 at [4] with [a-1, x1-7, x2-9, x3-2] at_time 32
:
p1 :: e ec 1 at [64] with [a0_-25, a_-8, e5-1] at_time 32

Used rules in the step 31:
  * The rule 83 has been used only once
The P-system has sent out d1 at step 29
The P-system has sent out no at step 31

?- evolve(p1,32).
No more evolution!
The P system p1 has already reached a halting configuration
at step 32
```

Currently, a graphical interface of this simulator is being developed using the Prolog/XPCE object-oriented library.

6 Other Software

We can also find in the literature other approaches that do not exactly fit in the previous sections.

For instance, although it is not exactly a simulator, we would like to note the work that Nicolau Jr., Solana, Fulga, and Nicolau published in *Fundamenta Informaticae* in 2002.

In [38], D.V. Nicolau Jr. et al. presented an ANSI C library developed to facilitate the implementation and simulation of P systems. Using the library proposed in this paper a user can specify an initial configuration (membrane structure and its contents) and perform actions on the objects or on the membranes. In fact, with this library membranes can be altered by dissolve, divide, and create actions. This library represents an intermediate step toward a practical implementation of P systems *in silico*.

The authors describe membrane structures as trees. A membrane is represented by a node in the tree, and the contents (symbols or strings) of that membrane are associated with the node by means of some auxiliary data structure such as an array or a list. The immediate "children" of this membrane are also included in the node contents. This is done by using a recursive *Abstract Data Type*. From a theoretical point of view, there is no limit on the number of membranes in a P system, and this represents a problem for simulating *in silico* P systems which allow division or creation of membranes. To avoid memory space violations, this software fixes an upper limit on the number of children in each membrane.

The information in the data structure also include the *name* of the membrane (its label), the number of children, and the name of its parent membrane.

Rules are not implemented explicitly as data structures. Instead, the user is supposed to write a *function* for each membrane reflecting the *program* of that membrane. In this way, the user is given complete flexibility over the way in which the rules are defined and applied in each membrane, including priority relations, and so on.

In September 2003, Alexandros Georgiou from University of Sheffield presented a simulator called SubLP-Studio. It is a software simulator for the Sub LP-Systems model, a variant of L systems and P systems. It optionally interfaces with CPFG, thus producing plant graphics using the turtle interpreter. It is available from the P systems Web page [55].

The *Group for Models of Natural Computing* [56] in Verona has developed a P systems simulator[4] based on the implementation of the metabolic algorithm introduced in [14]. The algorithm is inspired by the Law of Mass Action. This law states that the driving force of a chemical reaction is directly proportional to the active masses of all the reactants.

They propose regarding a rule $r : A_1A_2 \rightarrow B_1B_2$ as a *chemical reaction*; then the left objects A and B play the role of *reactants* while those of the right are *products*. Following this chemical interpretation, they propose regarding rules as descriptors of the changes in concentration of the reactants into products.

The simulator which implements these ideas is written in Java and the input is provided to the simulator as an XML file.

We would also like to note the implementation of catalytic P systems presented by Binder et al. in [15] and Alhazov's simulator for maximally parallel multiset-rewriting systems with promoters/inhibitors [1]. The latter was used as an engine of the communicative P systems simulator by Vladimir Rogozhin to check the theorems in [2, 32].

Although it is beyond the scope of the present chapter, we consider Petreska and Teuscher's implementation [48] interesting. Instead of developing software, they have presented a hardware-based parallel implementation that allows us to run a certain class of P systems in a highly efficient manner. The source code of the implementation and more information are available from [58].

[4] The simulator is described in [13].

It is also worth mentioning the fact that Holger Hoos from the University of British Columbia teaches a course on *Algorithms for Bioinformatics*[5] and one of the assignments for his students is to implement a P system simulator for a restricted version of transition P systems.

7 Conclusions

In this chapter we have briefly presented some programs from what we consider to be the first generation of P systems simulators. In a few years, more than a dozen software simulators have been presented. As we pointed out above, the common purpose of all of them is better understanding of the computational process of P systems for pedagogical purposes, as assistants for researchers, and for use (mainly) in biological applications. One of the most extended features is the balance between efficiency and explicitness of the code.

We are at the beginning of a new generation of simulators, whose properties have been already pointed out by some of the simulators mentioned above.

For example, it is necessary that the simulator have a friendly and intuitive graphical interface. This is not a trivial task, because problems such as division or dissolution of membranes need dynamical solutions in order to update the graphical representation of all membranes *simultaneously*.

Another important point to address is the way in which the P system is provided to the simulator. Future simulators will need parsers to check the information provided by the user and store it appropriately. Likewise, the use of tools is necessary to handle information of P systems when the number of rules, membranes, or objects in a configuration is large.

It is also desirable that the simulator be able to interact with the user by providing detailed information about the computation, for example, about the number of rules used in each step and intermediate configurations or objects sent to the environment (if any) in order to make statistical studies of the computations (see, e.g., [19], [29], [31] or [51]). Indeed, biologically inspired variants of membrane systems are not interested in looking for halting configurations, but in the evolution process itself.

Then, the simulators have to be tested when approaching new problems both with computational interest (such as solving new **NP**-complete problems) and related to applications in biology. The development of more complex simulators will also require the use of tools for their verification.

The next generation of simulators may be oriented to solve (at least partially) the problems of storage of information and massive parallelism by using parallel language programming or by using multiprocessor computers. In this framework, the emergent generation of simulators based on parallel or distributed architectures could lead to an *efficient* simulation of P systems *in silico.*

[5] http://www.cs.ubc.ca/labs/beta/Courses/CPSC545-03/

In some sense, the current P systems simulators represent a first step toward an implementation of such cellular models in electronic media. However, we can note an important limitation: the problem of finding an *efficient* implementation of P systems with active membranes (i.e., a software able to simulate computations with a polynomial number of cellular steps in polynomial processor time) is as hard as proving that **P=NP**.

Acknowledgement

The support for this research through project TIC2002-04220-C03-01 of the Ministerio de Ciencia y Tecnología of Spain, cofinanced by FEDER funds, is gratefully acknowledged.

References

1. A. Alhazov: Maximally Parallel Multiset-Rewriting Systems: Browsing the Configurations. *Proc. Third Brainstorming Week on Membrane Computing*, Sevilla, 2005, RGNC Report 01/2005, 1-10.
2. A. Alhazov, M. Margenstern, V. Rogozhin, Yu. Rogozhin, S. Verlan: Communicative P systems with Minimal Cooperation. In [35].
3. A. Alhazov, C. Martín-Vide, Gh. Păun, eds.: *Pre-Proceedings of the Workshop on Membrane Computing*, Tarragona, Spain, 2003, Report RGML 28/03.
4. I.I. Ardelean, M. Cavaliere: Modelling Biological Processes by Using a Probabilistic P System Software. *Natural Computing*, 2, 2 (2003), 173–197.
5. F. Arroyo, A.V. Baranda, J. Castellanos, C. Luengo, L.F. de Mingo: A Recursive Algorithm for Describing Evolution in Transition P Systems. In [33], 19–30.
6. F. Arroyo, A.V. Baranda, J. Castellanos, C. Luengo, L.F. de Mingo: Structures and Bio-Language to Simulate Transition P Systems on Digital Computers. In [17], 1–16.
7. F. Arroyo, C. Luengo, A.V. Baranda, L.F. de Mingo: A Software Simulation of Transition P Systems in Haskell. In [43], 19–32.
8. H. Attiya, J. Welch: *Distributed Computing: Fundamentals, Simulations and Advanced Topics*. McGraw-Hill, 2000.
9. A.V. Baranda, J. Castellanos, F. Arroyo, R. Gonzalo: Data Structures for Implementing P Systems in Silico. In [16], 21–34.
10. A.V. Baranda, J. Castellanos, F. Arroyo, R. Gonzalo: Towards an Electronic Implementation of Membrane Computing: A Formal Description of Nondeterministic Evolution in Transition P Systems. In *Proceedings of DNA-Based Computers, Tampa, Florida, 2002* (N. Jonoska, N.C. Seeman, eds.), LNCS 2340, Springer, Berlin, 2002, 350–359.
11. A.V. Baranda, J. Castellanos, R. Gonzalo, F. Arroyo, L.F. de Mingo: Data Structures for Implementing Transition P Systems in Silico. *Romanian Journal of Information Science and Technology*, 4, 1-2, (2001), 21–32.
12. D. Balbontín-Noval, M.J. Pérez-Jiménez, F. Sancho-Caparrini: A MzScheme Implementation of Transition P Systems. In [43], 58–73.
13. L. Bianco: A P System Simulator: Introduction to Psim. Unpublished manuscript, 2004.

14. L. Bianco, F. Fontana, G. Franco, V. Manca: P Systems for Biological Dynamics. In this volume.
15. A. Binder, R. Freund, G. Lojka, M. Oswald: Implementation of Catalytic P Systems. *Proceedings of CIAA 2004, Ninth International Conference on Implementation and Application of Automata*, Kingston, Canada, 2004, 24–33.
16. C.S. Calude, M.J. Dinneen, Gh. Păun, eds.: *Pre-Proceedings of Workshop on Multiset Processing*, Curtea de Argeş, Romania, CDMTCS TR 140, Univ. of Auckland, 2000.
17. C.S. Calude, Gh. Păun, G. Rozenberg, A. Salomaa, eds.: *Multiset Processing. Mathematical, Computer Science and Molecular Computing Points of View*, LNCS 2235, Springer, Berlin, 2001.
18. M. Cavaliere: Evolution-Communication P Systems. In [43], 134–145.
19. M. Cavaliere, I.I. Ardelean: Modelling Respiration in Bacteria and Respiration/Photosynthesis Interaction in Cyanobacteria by Using a P System Simulator. In this volume.
20. M. Cavaliere, C. Martín-Vide, Gh. Păun, eds.: *Proceedings of the Brainstorming Week on Membrane Computing*, Tarragona, Spain, 2003, Report RGML 26/03.
21. G. Ciobanu, D. Paraschiv: P System Software Simulator. *Fundamenta Informaticae*, 49, 1-3 (2002), 61–66.
22. G. Ciobanu, Gh. Păun, Gh. Ştefănescu: Sevilla Carpets Associated with P Systems. In [20], 135–140.
23. G. Ciobanu, G. Wenyuan: A Parallel Implementation of Transition P Systems. In [3], 169–184.
24. G. Ciobanu, G. Wenyuan: P Systems Running on a Cluster of Computers. In [34], 123–139.
25. A. Cordón-Franco, M.A. Gutiérrez-Naranjo. M.J. Pérez-Jiménez, A. Riscos-Núñez, F. Sancho-Caparrini: Implementing in Prolog an Effective Cellular Solution to the Knapsack Problem. In [34], 140–152.
26. A. Cordón-Franco, M.A. Gutiérrez-Naranjo, M.J. Pérez-Jiménez, A. Riscos-Núñez, F. Sancho-Caparrini: Cellular Solutions of Some Numerical NP-Complete Problems: A Prolog Implementation. In *Molecular Computational Models: Unconventional Approaches* (M. Gheorghe, ed.), Idea Group, Inc., 2005, 115–149.
27. A. Cordón-Franco, M.A. Gutiérrez-Naranjo, M.J. Pérez-Jiménez, F. Sancho-Caparrini: A Prolog Simulator for Deterministic P Systems with Active Membranes. *New Generation Computing*, 22, 4 (2004), 349–364.
28. M.A. Gutiérrez-Naranjo, M.J. Pérez-Jiménez, A. Riscos-Núñez: A Fast P System for Finding Balanced 2-Partition. *Soft Computing*, 9 (2005).
29. M.A. Gutiérrez-Naranjo, M.J. Pérez-Jiménez, A. Riscos-Núñez: On Descriptive Complexity of P Systems. In [36], 245–255.
30. M. Maliţa: Membrane Computing in Prolog. In [16], 159–175.
31. V. Manca: On the Dynamics of P Systems. In [36], 29–43.
32. M. Margenstern, V. Rogozhin, Yu. Rogozhin, S. Verlan: About P Systems with Minimal Symport/Antiport Rules and Four Membranes. In [36], 283–294.
33. C. Martín-Vide, Gh. Păun, eds.: *Pre-Proceedings of Workshop on Membrane Computing*, Curtea de Argeş, Romania, August 2001. Technical Report GRLMC 17/01, Rovira i Virgili University, Tarragona, Spain, 2001.
34. C. Martín-Vide, Gh. Păun, G. Rozenberg, A. Salomaa, eds.: *Membrane Computing. International Workshop WMC2003, Tarragona, Spain, 2003. Revised Papers.* LNCS 2933, Springer, Berlin, 2004.

35. G. Mauri, Gh. Păun, M.J. Pérez-Jiménez, G. Rozenberg, A. Salomaa, eds.: *Membrane Computing. International Workshop WMC5, Milano, Italy, 2004. Revised Papers.* LNCS 3365, Springer, Berlin, 2005.
36. G. Mauri, Gh. Păun, C. Zandron, eds.: *Pre-Proceedings of the Workshop on Membrane Computing WMC5*, Universitá di Milano-Bicocca, Italy, 2004.
37. I.A. Nepomuceno-Chamorro: A Java Simulator for Basic Transition P Systems. In [42], 309–315.
38. D.V. Nicolau Jr., G. Solana, F. Fulga, D.V. Nicolau: A C Library for Simulating P Systems. *Fundamenta Informaticae*, 49, 1-3 (2002), 241–248.
39. Gh. Păun: Computing with Membranes. Turku Centre for Computer Science, TUCS Technical Report, Nr.208, 1998.
40. Gh. Păun: Computing with Membranes. *Journal of Computer and System Sciences*, 61, 1 (2000), 108–143.
41. Gh. Păun: P Systems with Active Membranes: Attacking NP-Complete Problems. *Journal of Automata, Languages and Combinatorics*, 6, 1 (2001), 75–90.
42. Gh. Păun, A. Riscos-Núñez, A. Romero-Jiménez, F. Sancho-Caparrini, eds.: *Proceedings of the Second Brainstorming Week on Membrane Computing*, Sevilla, Spain, Report RGNC 01/04, 2004.
43. Gh. Păun, G. Rozenberg, A. Salomaa, C. Zandron, eds.: *Membrane Computing, International Workshop WMC-CdeA 2002, Curtea de Argeş, Romania. Revised Papers.* LNCS 2597, Springer, Berlin, 2003.
44. Gh. Păun: *Membrane Computing. An Introduction.* Springer, Berlin, 2002.
45. M.J. Pérez-Jiménez, F. Romero-Campero: A CLIPS Simulator for Recognizer P Systems with Active Membranes. In [42], 387–413.
46. M.J. Pérez-Jiménez, A. Romero-Jiménez, F. Sancho-Caparrini: Solving VALIDITY Problem by Active Membranes with Input. In [20], 279–290.
47. M.J. Pérez-Jiménez, F. Sancho-Caparrini: A Formalization of Transition P Systems. *Fundamenta Informaticae*, 49, 1-3 (2002), 261–272.
48. B. Petreska, C. Teuscher: A Reconfigurable Hardware Membrane System. In [34], 269–285.
49. A. Riscos-Núñez: *Cellular Programming: Efficient Resolution of Numerical NP-Complete Problems.* Ph.D. Thesis, University of Seville, 2004.
50. Y. Suzuki, H. Tanaka: On a LISP Implementation of a Class of P Systems. *Romanian Journal of Information Science and Technology*, 3, 2 (2000), 173–186.
51. Y. Suzuki, Y. Fujiwara, H. Tanaka, J. Takabayashi: Artificial Life Applications of a Class of P Systems: Abstract Rewriting Systems on Multisets. In [17], 299–346.
52. A. Syropoulos, E.G. Mamatas, P.C. Allilomes, K.T. Sotiriades: A Distributed Simulation of Transition P Systems. In [34], 357–368.
53. `http://research.araneous.com`
54. `http://www.lpsi.eui.upm.es/nncg/`
55. `http://psystems.disco.unimib.it/`
56. `http://www.di.univr.it`
57. `http://www.gcn.us.es`
58. `http://www.teuscher.ch/psystems`

Selective Bibliography of Membrane Computing

As we have mentioned also in the Preface, a complete bibliography of membrane computing can be found at the Web address `http://psystems.disco.unimib.it`.

In what follows, we list only the books, collective volumes, PhD theses, and special issues of various journals devoted to membrane computing; several of them are available for downloading in the above mentioned web page.

1. R. Alberich, J. Casasnovas, M. Llabrés, J. Miró-Juliá, J. Rocha, F. Rosselló, eds.: *Proceedings of the Brainstorming Workshop on Uncertainty in Membrane Computing.* Palma de Mallorca, Spain, 2004.
2. A. Alhazov, C. Martín-Vide, Gh. Păun, eds.: *Pre-Proceedings of Workshop on Membrane Computing, WMC 2003.* Tarragona, Spain, July 2003, Technical Report 28/03, Rovira i Virgili University, Tarragona, 2003.
3. F. Arroyo Montoro: *Estructuras y biolenguaje para simular computacion con membranas.* PhD Thesis, Polytechnical Univ. of Madrid, Spain, 2004.
4. D. Besozzi: *Computational and Modelling Power of P Systems.* PhD Thesis, Univ. degli Studi di Milano, Italy, 2004.
5. C. Calude, M.J. Dinneen, Gh. Păun, eds.: *Pre-Proceedings of Workshop on Multiset Processing.* Curtea de Argeş, Romania, August 2000, TR 140, CDMTCS, Univ. Auckland, New Zealand, 2000.
6. C. Calude, Gh. Păun: *Computing with Cells and Atoms. An Introduction to Quantum, DNA and Membrane Computing.* Francis and Taylor, London, 2000.
7. C.S. Calude, Gh. Păun, G. Rozenberg, A. Salomaa, eds.: *Multiset Processing. Mathematical, Computer Science, Molecular Computing Points of View. Lecture Notes in Computer Science* 2235, Springer, Berlin, 2001.
8. M. Cavaliere, C. Martín-Vide, Gh. Păun, eds.: *Proceedings of the Brainstorming Week on Membrane Computing, Tarragona, February 2003.* Technical Report 26/03, Rovira i Virgili University, Tarragona, 2003.
9. P. Frisco: *Theory of Molecular Computing. Splicing and Membrane Systems.* PhD Thesis, Leiden University, The Netherlands, 2004.

10. M.A. Gutiérrez-Naranjo, Gh. Păun, M.J. Pérez-Jiménez, eds.: *Cellular Computing. Complexity Aspects.* Fenix Editora, Sevilla, 2005.
11. M.A. Gutiérrez-Naranjo, F.J. Romero-Campero, A. Riscos-Núñez, D. Sburlan, eds.: *Proceedings of the Third Brainstorming Week on Membrane Computing, Sevilla, February 2005.* Technical Report 01/2005 of Research Group on Natural Computing, Sevilla University, Spain, 2005.
12. N. Jonoska, Gh. Păun, eds.: *New Generation Computing*, 22, 4 (2004).
13. S.N. Krishna: *Languages of P Systems. Computability and Complexity.* PhD Thesis, Madras University, India, 2001.
14. L. Lakshmanan: *On the Crossroads of P Systems and Contextual Grammars: Variants, Computability, Complexity and Efficiency.* PhD Thesis, Indian Institute of Technology, Madras, India, 2003.
15. C. Martín-Vide, G. Mauri, Gh. Păun, G. Rozenberg, A. Salomaa, eds.: *Membrane Computing, International Workshop, WMC 2003, Tarragona, Spain, July 2003, Selected Papers.* *Lecture Notes in Computer Science* 2933, Springer, Berlin, 2004.
16. C. Martín-Vide, Gh. Păun, eds.: *Pre-Proceedings of Workshop on Membrane Computing.* Curtea de Argeş, Romania, August 2001, TR 16/01, Univ. Rovira i Virgili, Tarragona, Spania, 2001.
17. C. Martín-Vide, Gh. Păun, eds.: *Natural Computing*, 2, 3 (2003).
18. G. Mauri, Gh. Păun, M.J. Pérez-Jiménez, G. Rozenberg, A. Salomaa, eds.: *Membrane Computing, International Workshop, WMC5, Milano, Italy, June 2004, Selected Papers.* *Lecture Notes in Computer Science* 3365, Springer, Berlin, 2005.
19. G. Mauri, Gh. Păun, C. Zandron, eds.: *Pre-Proceedings of Fifth Workshop on Membrane Computing, WMC5.* Milano, Italy, 2004.
20. M. Mutyam: *Studies of P Systems as a Model of Cellular Computing.* PhD Thesis, Indian Institute of Technology Madras, India, 2003.
21. M. Oswald: *P Automata.* PhD Thesis, Technical Univ. Vienna, Austria, 2003.
22. A. Păun: *Unconventional Models of Computation: DNA and Membrane Computing.* PhD Thesis, Univ. of London, Ontario, Canada, 2003.
23. Gh. Păun, ed.: *Romanian Journal of Information Science and Technology*, 4, 1-2 (2001).
24. Gh. Păun: *Membrane Computing. An Introduction.* Springer, Berlin, 2002.
25. Gh. Păun, ed.: *Fundamenta Informaticae*, 49, 1-3 (2002).
26. Gh. Păun, M.J. Pérez-Jiménez, eds.: *Journal of Universal Computer Science – J.UCS*, 10, 5 (2004).
27. Gh. Păun, M.J. Pérez-Jiménez, eds.: *Soft Computing*, 9, 7 (2005).
28. Gh. Păun, A. Riscos-Núñez, A. Romero-Jiménez, F. Sancho-Caparrini, eds.: *Proceedings of the Second Brainstorming Week on Membrane Computing, Sevilla, February 2004.* Technical Report 01/04 of Research Group on Natural Computing, Sevilla University, Spain, 2004.
29. Gh. Păun, G. Rozenberg, A. Salomaa, C. Zandron, eds.: *Membrane Computing. International Workshop WMC-CdeA 2002, Curtea de Argeş, Ro-*

mania, August 2002, Revised Papers. Lecture Notes in Computer Science 2597, Springer, Berlin, 2002.

30. Gh. Păun, C. Zandron, eds.: *Pre-Proceedings of Workshop on Membrane Computing, Curtea de Argeş, Romania, August 2002.* MolCoNet Publication No 1, 2002.
31. M.J. Pérez-Jiménez, A. Romero-Jiménez, F. Sancho-Caparrini: *Teoria de la complejidad en modelos de computacion celular con membranas.* Kronos Editorial, Sevilla, 2002.
32. M.J. Pérez-Jiménez, F. Sancho-Caparrini: *Computacion celular con membranas: Un modelo no convencional.* Kronos Editorial, Sevilla, 2002.
33. A. Riscos-Núñez: *Cellular Programming. Efficient Resolution of* **NP**-*Complete Numerical Problems.* PhD Thesis, Univ. of Sevilla, Spain, 2004.
34. A. Romero-Jiménez: *Complexity and Universality in Cellular Computing Models.* PhD Thesis, Univ. of Sevilla, Spain, 2003.
35. F. Sancho-Caparrini: *Verification of Programs in Unconventional Computing Models.* PhD Thesis, Univ. of Sevilla, Spain, 2002.
36. S. Verlan: *Head Systems and Applications to Bio-Informatics.* PhD Thesis, LITA, Univ. Metz, France, 2004.
37. C. Zandron: *Computing with Membranes. P Systems.* PhD Thesis, Univ. of Milano-Bicocca, Italy, 2002.

Natural Computing Series

W.M. Spears: **Evolutionary Algorithms. The Role of Mutation and Recombination.** XIV, 222 pages, 55 figs., 23 tables. 2000

H.-G. Beyer: **The Theory of Evolution Strategies.** XIX, 380 pages, 52 figs., 9 tables. 2001

L. Kallel, B. Naudts, A. Rogers (Eds.): **Theoretical Aspects of Evolutionary Computing.** X, 497 pages. 2001

G. Păun: **Membrane Computing. An Introduction.** XI, 429 pages, 37 figs., 5 tables. 2002

A.A. Freitas: **Data Mining and Knowledge Discovery with Evolutionary Algorithms.** XIV, 264 pages, 74 figs., 10 tables. 2002

H.-P. Schwefel, I. Wegener, K. Weinert (Eds.): **Advances in Computational Intelligence. Theory and Practice.** VIII, 325 pages. 2003

A. Ghosh, S. Tsutsui (Eds.): **Advances in Evolutionary Computing. Theory and Applications.** XVI, 1006 pages. 2003

L.F. Landweber, E. Winfree (Eds.): **Evolution as Computation.** DIMACS Workshop, Princeton, January 1999. XV, 332 pages. 2002

M. Hirvensalo: **Quantum Computing.** 2nd ed., XI, 214 pages. 2004 (first edition published in the series)

A.E. Eiben, J.E. Smith: **Introduction to Evolutionary Computing.** XV, 299 pages. 2003

A. Ehrenfeucht, T. Harju, I. Petre, D.M. Prescott, G. Rozenberg: **Computation in Living Cells. Gene Assembly in Ciliates.** XIV, 202 pages. 2004

L. Sekanina: **Evolvable Components. From Theory to Hardware Implementations.** XVI, 194 pages. 2004

G. Ciobanu, G. Rozenberg (Eds.): **Modelling in Molecular Biology.** X, 310 pages. 2004

R.W. Morrison: **Designing Evolutionary Algorithms for Dynamic Environments.** XII, 148 pages, 78 figs. 2004

R. Paton†, H. Bolouri, M. Holcombe, J.H. Parish, R. Tateson (Eds.): **Computation in Cells and Tissues. Perspectives and Tools of Thought.** XIV, 358 pages, 134 figs. 2004

M. Amos: **Theoretical and Experimental DNA Computation.** XIV, 170 pages, 78 figs. 2005

G. Ciobanu, G. Păun, M.J. Pérez-Jiménez (Eds.): **Applications of Membrane Computing.** X, 441 pages, 99 figs., 24 tables. 2005